AUTOMATIC TRANSMISSION/TRANSAXLE DIAGNOSIS AND REPAIR

by Jacques Gordon

PUBLISHED BY **HAYNES NORTH AMERICA**, Inc.

Manufactured in USA
© 1998 Haynes North America, Inc.
ISBN 0-8019-8944-2
Library of Congress Catalog Card No. 98-71359
1234567890 9876543210

Haynes Publishing Group
Sparkford Nr Yeovil
Somerset BA22 7JJ England

Haynes North America, Inc
861 Lawrence Drive
Newbury Park
California 91320 USA

ABCDE
FGHIJ
KLM

Contents

Contents

SAFETY NOTICE

Proper service and repair procedures are vital to the safe, reliable operation of all motor vehicles, as well as the personal safety of those performing repairs. This manual outlines procedures for servicing and repairing vehicles using safe, effective methods. The procedures contain many NOTES, CAUTIONS and WARNINGS which should be followed, along with standard procedures to eliminate the possibility of personal injury or improper service which could damage the vehicle or compromise its safety.

It is important to note that repair procedures and techniques, tools and parts for servicing motor vehicles, as well as the skill and experience of the individual performing the work vary widely. It is not possible to anticipate all of the conceivable ways or conditions under which vehicles may be serviced, or to provide cautions as to all possible hazards that may result. Standard and accepted safety precautions and equipment should be used when handling toxic or flammable fluids, and safety goggles or other protection should be used during cutting, grinding, chiseling, prying, or any other process that can cause material removal or projectiles.

Some procedures require the use of tools specially designed for a specific purpose. Before substituting another tool or procedure, you must be completely satisfied that neither your personal safety, nor the performance of the vehicle will be endangered.

Although information in this manual is based on industry sources and is complete as possible at the time of publication, the possibility exists that some car manufacturers made later changes which could not be included here. While striving for total accuracy, the authors or publishers cannot assume responsibility for any errors, changes or omissions that may occur in the compilation of this data.

PART NUMBERS

Part numbers listed in this reference are not recommendations by Haynes North America, Inc. for any product brand name. They are references that can be used with interchange manuals and aftermarket supplier catalogs to locate each brand supplier's discrete part number.

SPECIAL TOOLS

Special tools are recommended by the vehicle manufacturer to perform their specific job. Use has been kept to a minimum, but where absolutely necessary, they are referred to in the text by the part number of the tool manufacturer. These tools can be purchased, under the appropriate part number, from your local dealer or regional distributor, or an equivalent tool can be purchased locally from a tool supplier or parts outlet. Before substituting any tool for the one recommended, read the SAFETY NOTICE at the top of this page.

ACKNOWLEDGMENTS

The publisher expresses appreciation to the following companies for their generous assistance:

- B&M Racing and Performance - Chatsworth, CA
- Felpro Inc. - Skokie, IL
- Life Automotive Products - Memphis, TN
- Transman Transmission Specialist - Essington, PA

A special thanks to the fine companies who support the production of all our books: Hand tools, supplied by Craftsman, were used during all phases of teardown and photography. Many of the fine specialty tools used in procedures were provided courtesy of Lisle Corporation. Lincoln Automotive Products provided the industrial shop equipment including jacks, engine stands and shop presses. A Rotary lift, the largest automobile lift manufacturer in the world offering the biggest variety of surface and inground lifts available, was also used.

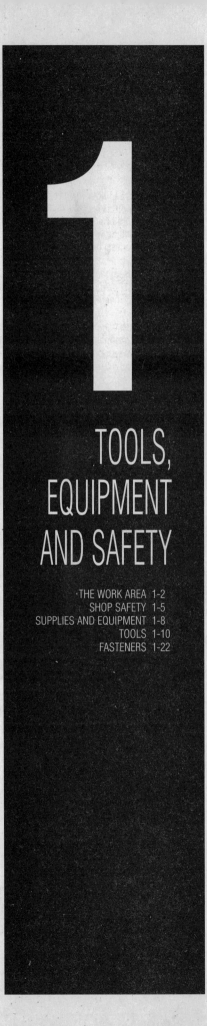

1

TOOLS, EQUIPMENT AND SAFETY

THE WORK AREA

Floor Space and Working Height

The average one car garage will give you more than enough workspace. A floor plan of 16 ft. X 12 ft. is more than sufficient for shelving, workbenches, tool shelves or boxes and parts storage areas. 12 X 16 works out to 192 square feet. You may think that this sounds like a lot of room, but when you start building shelves, and constructing work benches almost most half of that can be eaten up!

Also, you may wonder why a lot of floor space is needed. There are several reasons, not the least of which is the safety factor. You'll be working around a large, heavy, metal object—your vehicle. You don't want to be tripping, falling, crashing into things or hurting yourself, all because your vehicle takes up most of your work space. Accidents can happen! You can easily trip over a jack handle or work stand leg or drop a heavy part or tool. You'll need room to take evasive action!

Most garages have concrete floors. Your creeper rolls best on a smooth surface. If your garage floor has cracks with raised sections or blocks with deep grooves, you may have a problem. The wheels can hang up on these cracks or grooves causing you to get stuck under the vehicle.

As for working height, overhead clearance can be a problem if you have a tall vehicle such as an RV or sport utility vehicle, especially when removing a transmission. On some vehicles the engine and transmission are lifted out as a unit. This may require as much as ten feet of overhead clearance.

The Work Station

WORKBENCHES

▶ See Figure 1

Work benches can be either store-bought or homemade. The store-bought workbenches can be steel or precut wood kits. Either are fine and are available at most building supply stores or through tool catalogs.

Homemade benches have the advantage of being made-to-fit your workshop. A freestanding workbench is best, as opposed to one attached to an outside wall. The freestanding bench can take more abuse since it doesn't transfer the shock or vibration to wall supports.

A good free-standing workbench should be constructed using 4 X 4 pressure treated wood as legs, 2 X 6 planking as header boards and ¾ inch plywood sheathing as a deck. Diagonal supports can be 2 X 4 studs and it's always helpful to construct a full size ¾ inch plywood shelf under

Fig. 1 Homemade workbenches

87933512

the bench. Not only can you use the shelf for storage but also it gives great rigidity to the whole bench structure. Assembling the bench with screws rather than nails takes longer but adds strength and gives you the ability to take the whole thing apart if you ever want to move it.

If you're planning to put a transmission on it, keep two more things in mind: a transmission is heavy and full of fluid. Purpose-built transmission benches are stout enough to hold several hundred pounds while remaining rock-steady. Also the working surface is a metal tray, with short side walls and a drain at one end. There's no way to fully drain the fluid from a transmission, so the work space must be able to contain the mess and ease clean-up.

SHELVES

▶ **See Figures 2, 3 and 4**

You can't have enough shelf space. Adequate shelf space means that you don't have to stack anything on the floor, where it would be in the way.

Shelves aren't tough. You can make you own or buy modular or prefab units. The best modular units are those made of interlocking shelves and uprights. They are light weight and easy to assemble, and their load bearing capacity is more than sufficient.

Probably the cheapest and best shelves are one that you make yourself from one inch shelving with 2X4 or 4X4 uprights. You can make them as long, high and wide as you need. For at least the uprights, use pressure treated wood. Its resistance to rot is more than worth the additional cost.

LIGHTING

▶ **See Figures 5 and 6**

The importance of adequate lighting can't be over emphasized. Good lighting is not only a convenience but also a safety feature. If you can see what you're working on you're less likely to make mistakes, have a wrench slip or trip over an obstacle. On most vehicles, everything is about the same color and usually dirty. During disassembly, a lot of frustration can be avoided if you can see all the bolts, some of which may be hidden or obscured.

For overhead lighting, at least 2 twin tube 36 inch fluorescent shop lights should be in place. Most garages are wired with standard light bulbs

Fig. 2 Typical home made wood shelves, crammed with stuff. These shelves are made from spare pressure treated decking

Fig. 3 Modular plastic shelves are inexpensive, weather proof and easy to assemble

attached to the wall or ceiling studs at intervals. Four or five of these lights combined with the overhead fluorescent lighting should suffice. However, no matter where the lights are, your body is going to block some of it so a droplight or clip-on type work light is a great idea. These lights can be mounted on the engine stand, or even the engine itself.

VENTILATION

At one time or another, you'll be working with chemicals that may require adequate ventilation. Now, just about all garages have a big car-sized door and all sheds or workshops have a door. In bad weather the door will have to be closed so at least one window that opens is a necessity. An exhaust fan or regular ventilation fan is a great help, especially in hot weather.

If your garage is attached to the house, keep in mind that any fumes or odors generated there can eventually make it into the house. It's a good idea to keep all fluids in containers with lids as opposed to open pans. Be mindful of the chemicals used during the job and make sure there is proper ventilation, especially in cold weather when the house is closed up.

HEATERS

If you live in an area where the winters are cold, as do many of us, it's nice to have some sort of heat where we work. If your workshop or garage is attached to the house, you'll probably be okay. If your garage or shop is detached, then a space heater of some sort—electric, propane or kerosene—will be necessary. NEVER run a space heater in the presence of flammable vapors! When running a non-electric space heater, always allow for some means of venting the carbon monoxide!

ELECTRICAL REQUIREMENTS

Obviously, your workshop should be wired according to all local codes. As to what type of service you need, that depends on your electrical load. If you have a lot of power equipment and maybe a refrigerator, TV, stereo or whatever, not only do you have a great shop, but your amperage requirements may exceed your wiring's capacity. If you are at all in doubt, consult your local electrical contractor.

Fig. 4 These shelves were made from recycled kitchen cabinets

Fig. 5 At least two of this type of twin tube fluorescent light is essential

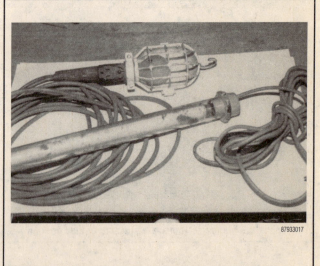

Fig. 6 Two types of droplights—incandescent and fluorescent

SHOP SAFETY

▶ **See Figures 7, 8, 9 and 10**

It is virtually impossible to anticipate all of the hazards involved with automotive maintenance and service but care and common sense will prevent most accidents.

The rules of safety for mechanics range from "don't smoke around gasoline" to "use the proper tool for the job." The trick to avoiding injuries is to develop safe work habits and take every possible precaution.

Do's

- Do keep a fire extinguisher and first aid kit handy.
- Do wear safety glasses or goggles when cutting, drilling, grinding or prying, even if you have 20–20 vision. If you wear glasses for the sake of vision, wear safety goggles over your regular glasses.
- Do shield your eyes whenever you work around the battery. Batteries contain sulfuric acid. In case of contact with the eyes or skin, flush the area with water or a mixture of water and baking soda, then seek immediate medical attention.

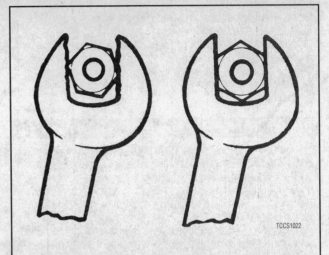

Fig. 9 Using the correct size wrench will help prevent the possibility of rounding-off a nut

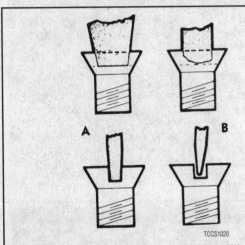

Fig. 7 Screwdrivers should be kept in good condition to prevent injury or damage that could result if the blade slips from the screw

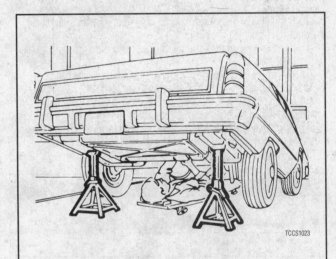

Fig. 10 NEVER work under a vehicle unless it is supported using safety stands (jackstands)

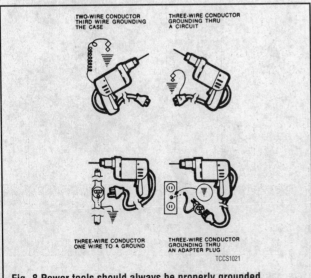

Fig. 8 Power tools should always be properly grounded

- Do use safety stands (jackstands) for any under vehicle service. Jacks are for raising vehicles; jackstands are for making sure the vehicle stays raised until you want it to come down. Whenever the vehicle is raised, block the wheels remaining on the ground and set the parking brake.
- Do use adequate ventilation when working with any chemicals or hazardous materials. Like carbon monoxide, the asbestos dust resulting from some brake lining wear can be hazardous in sufficient quantities.
- Do disconnect the negative battery cable when working on the electrical system. The secondary ignition system contains EXTREMELY HIGH VOLTAGE. In some cases it can even exceed 50,000 volts.
- Do follow manufacturer's directions whenever working with potentially hazardous materials. Most chemicals and fluids are poisonous if taken internally.
- Do properly maintain your tools. Loose hammerheads, mushroomed punches and chisels, frayed or poorly grounded electrical cords, excessively worn screwdrivers, spread wrenches (open end), cracked sockets, slipping ratchets, or faulty droplight sockets can cause accidents.
- Likewise, keep your tools clean; a greasy wrench can slip off a bolt head, ruining the bolt and often harming your knuckles in the process.

• Do use the proper size and type of tool for the job at hand. Do select a wrench or socket that fits the nut or bolt. The wrench or socket should sit straight, not cocked.

• Do, when possible, pull on a wrench handle rather than push on it, and adjust your stance to prevent a fall.

• Do be sure that adjustable wrenches are tightly closed on the nut or bolt and pulled so that the force is on the side of the fixed jaw.

• Do strike squarely with a hammer; avoid glancing blows.

• Do set the parking brake and block the drive wheels if the work requires a running engine.

Don'ts

• Don't run the engine in a garage or anywhere else without proper ventilation—EVER! Carbon monoxide is poisonous; it takes a long time to leave the human body and you can build up a deadly supply of it in your system by simply breathing in a little every day. You may not realize you are slowly poisoning yourself. Always use power vents, windows, fans and/or open the garage door.

• Don't work around moving parts while wearing loose clothing. Short sleeves are much safer than long, loose sleeves. Hard-toed shoes with neoprene soles protect your toes and give a better grip on slippery surfaces. Jewelry such as watches, fancy belt buckles, beads or body adornment of any kind is not safe working around a vehicle. Long hair should be tied back under a hat or cap.

• Don't use pockets for toolboxes. A fall or bump can drive a screwdriver deep into your body. Even a rag hanging from your back pocket can wrap around a spinning shaft or fan.

• Don't smoke when working around gasoline, cleaning solvent or other flammable material.

• Don't smoke when working around the battery. When the battery is being charged, it gives off explosive hydrogen gas.

• Don't use gasoline to wash your hands; there are excellent soaps available. Gasoline contains dangerous additives which can enter the body through a cut or through your pores. Gasoline also removes all the natural oils from the skin so that bone dry hands will suck up oil and grease.

• Don't service the air conditioning system unless you are equipped with the necessary tools and training. When liquid or compressed gas refrigerant is released to atmospheric pressure it will absorb heat from whatever it contacts. This will chill or freeze anything it touches. Although refrigerant is normally non-toxic, R-12 becomes a deadly poisonous gas in the presence of an open flame. One good whiff of the vapors from burning refrigerant can be fatal.

• Don't use screwdrivers for anything other than driving screws! A screwdriver used as a prying tool can snap when you least expect it, causing injuries. At the very least, you'll ruin a good screwdriver.

• Don't use a bumper or emergency jack (that little ratchet, scissors, or pantograph jack supplied with the vehicle) for anything other than changing a flat! These jacks are only intended for emergency use out on the road; they are NOT designed as a maintenance tool. If you are serious about maintaining your vehicle yourself, invest in a hydraulic floor jack of at least a 1½ ton capacity, and at least two sturdy jackstands.

SAFETY EQUIPMENT

▶ **See Figure 11**

Fire Extinguishers

▶ **See Figure 12**

There are many types of safety equipment. The most important of these is the fire extinguisher. You'll be well off with two 5 lbs. extinguishers rated for oil, chemical and wood (Class B & C).

First Aid

Next you'll need a good first aid kit. Any good kit that can be purchased from the local drug store will be fine. It's a good idea, in addition, to have something easily accessible in the event of a minor injury, such a hydrogen peroxide or other antiseptic that can be poured onto or applied to a wound immediately. Remember, your hands will be dirty. Just as you wouldn't want dirt entering a brake system that has been opened, you certainly don't want bacteria entering a blood stream that has just been opened!

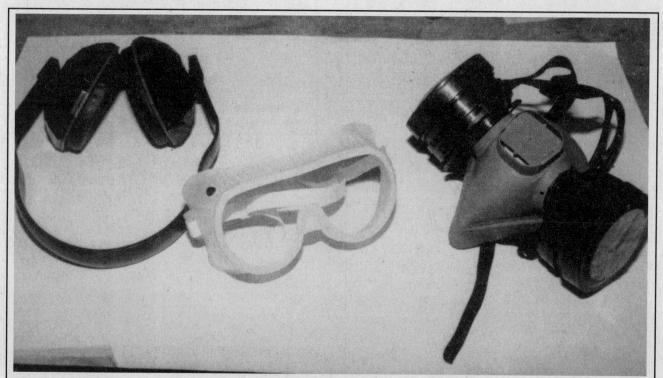

Fig. 11 Three essential pieces of safety equipment. Left to right: ear protectors, safety goggles and respirator

87932514

Fig. 12 A good, all-purpose fire extinguisher

Work Gloves

▶ **See Figure 13**

Unless you think scars on your hands are cool, enjoy pain and like wearing bandages, get a good pair of work gloves. Canvass or leather are the best. And yes, we realize that there are some jobs involving small parts that can't be done while wearing work gloves. These jobs are not the ones usually associated with hand injuries.

A good pair of rubber gloves such as those usually associated with dish washing is also a great idea. There are some liquids such as solvents and penetrants that don't belong on your skin. Avoid burns and rashes. Wear these gloves.

And lastly, an option. If you're tired of being greasy and dirty all the time, go to the drug store and buy a box of disposable latex gloves like medical professionals wear. You can handle greasy parts, perform small tasks, wash parts, etc. all without getting dirty! These gloves take a surprising amount of abuse without tearing and aren't expensive. Note however, that it has been reported that some people are allergic to the latex or the powder used inside the gloves.

Work Boots

A good, comfortable pair of steel-toed work boots is a sensible idea, primarily because heavy parts always get dropped sooner or later. A brake rotor or drum can do significant damage to a sneaker-clad foot. Good work boots also provide better support,—you're going to be on your feet a lot—are oil-resistant, and they keep your feet warm and dry. To keep the boots protected, get a spray can of silicone-based water repellent and spray the boots when new, and periodically thereafter.

Eye Protection

Don't begin any job without a good pair of work goggles or impact resistant glasses. When doing any kind of work, it's all too easy to avoid eye injury through this simple precaution. And don't just buy eye protection and leave it on the shelf. Wear it all the time! Things have a habit of breaking, chipping, splashing, spraying, splintering and flying around. And, for some reason, your eye is always in the way!

If you wear vision correcting glasses as a matter of routine, get a pair made with polycarbonate lenses. These lenses are impact resistant and are available at any optometrist.

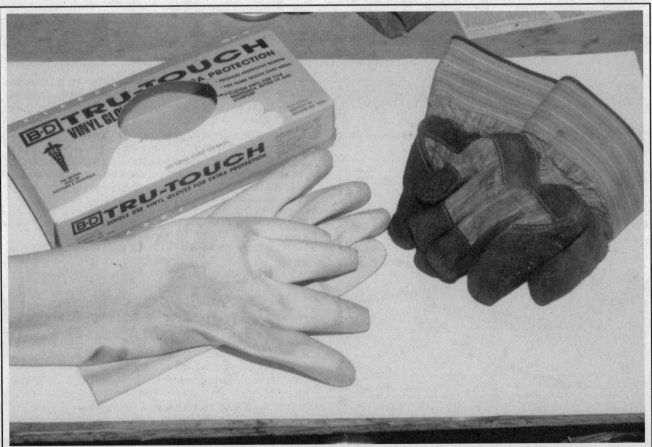

Fig. 13 Three different types of work gloves. The box contains latex gloves

Ear Protection

Often overlooked is hearing protection. Power equipment is noisy, and loud noises damage your ears every time. The effects are cumulative. The simplest and cheapest form of ear protection is a pair of noise-reducing ear plugs. More substantial protection is a good pair of noise reducing earmuffs. They protect to some extent from even the loudest sounds.

Work Clothes

Everyone has "work clothes". Usually this consists of old jeans and a shirt that has seen better days. That's fine. In addition, a denim work apron is a nice accessory. It's rugged and you don't feel bad wiping your hands or tools on it. That's what it's for.

If you're so inclined, overalls are a superb work garment. They're rugged and are equipped with numerous pockets, loops and places to put stuff. When bending or reaching, you won't have to worry about your shirt pulling out. Also, they cover your shirt like a work apron.

SUPPLIES AND EQUIPMENT

Fluid Disposal

Used fluids such as engine oil, transmission fluid, antifreeze and brake fluid are hazardous wastes and must be disposed of properly. Before draining any fluids, consult with your local authorities; in many areas waste oil, etc. is being accepted as a part of recycling programs. A number of service stations and auto parts stores are also accepting waste fluids for recycling.

Be sure of the recycling center's policies before draining any fluids, as many will not accept different fluids that have been mixed together.

Chemicals

There is a whole range of chemicals that you'll need. The most common types are, lubricants, penetrants and sealers. Keep these handy, on some convenient shelf.

When a particular chemical is not being used, keep it capped, upright and in a safe place. These substances may be flammable or irritants or caustic and should always be stored properly, used properly and handled with care. Always read and follow all label directions and wear hand and eye protection!

LUBRICANTS & PENETRANTS

▶ See Figure 14

In this category, a well-prepared automotive shop should have:
• Clean engine oil. Whatever you use regularly in your engine will be fine.

Fig. 14 A variety of penetrants and lubricants is a staple of any DIYer's garage

• Lithium grease.
• Chassis lube
• Assembly lube
• Silicone grease
• Silicone spray
• Penetrating oil

Clean engine oil is used to coat most bolts, screws and nuts prior to installation. This is always a good practice since the less friction there is on a fastener, the less chance there will be of breakage and crossthreading. Also, an oiled bolt will give a truer torque value and be less likely to rust or seize. An obvious exception would be wheel lugs. These are not oiled.

Lithium grease, chassis lube, silicone grease or a synthetic brake caliper grease can all be used pretty much interchangeably. All can be used for coating rust-prone fasteners and for facilitating the assembly of parts that are a tight fit. Silicone and synthetic greases are the most versatile and should always be used on the sliding areas of slider-type calipers or on the pins with pin-type calipers. It's also a good lubricant for the mounting pads on drum-type brake backing plates. The main advantages of silicone grease are that it's slipperier than most similar lubricants and it has a higher melting point. You don't want a grease melting and possibly contaminating the friction surfaces of your brakes.

➡**Silicone dielectric grease is a non-conductor that is often used to coat the terminals of wiring connectors before fastening them. It may sound odd to coat metal portions of a terminal with something that won't conduct electricity, but here is it how it works. When the connector is fastened the metal-to-metal contact between the terminals will displace the grease (allowing the circuit to be completed). The grease that is displaced will then coat the non-contacted surface and the cavity around the terminals, SEALING them from atmospheric moisture that could cause corrosion.**

Silicone spray is a good lubricant for hard-to-reach places and parts that shouldn't be lubricated with grease.

Penetrating oil may turn out to be one of your best friends during disassembly. The most familiar penetrating oils are Liquid Wrench® and WD-40®. These products have hundreds of uses. For your purposes, they are vital!

Before disassembling any part, check the fasteners. If any appear rusted, soak them thoroughly with the penetrant and let them stand while you do something else. This simple act can save you hours of tedious work trying to extract a broken bolt or stud.

Engine assembly lube. There are several types of this product available. Essentially it is a heavy-bodied lubricant used for coating moving parts prior to assembly. For engine work, the idea is that is stays in place until the engine starts for the first time and dissolves in the engine oil as oil pressure is achieved. This way, expensive parts receive needed protection until everything is working. For non-engine work, it comes in handy for assembling tight-fitting parts.

SEALANTS

▶ **See Figure 15**

Sealants are an indispensable part of almost all automotive work. The purpose of sealants is to establish a leak-proof bond between or around assembled parts. Most sealers are used in conjunction with gaskets, but some are used instead of conventional gasket material in newer engines.

The most common sealers are the non-hardening types such as Permatex®No.2 or its equivalents. These sealers are applied to the mating surfaces of each part to be joined, then a gasket is put in place and the parts are assembled.

One very helpful type of non-hardening sealer is the "high tack" type. This type is a very sticky material that holds the gasket in place while the parts are being assembled. This stuff is really a good idea when you don't have enough hands or fingers to keep everything where it should be.

The stand-alone sealers are the Room Temperature Vulcanizing (RTV) silicone gasket makers. On many newer vehicles, this material is used instead of a gasket. In those instances, a gasket may not be available or, because of the shape of the mating surfaces, a gasket shouldn't be used. This stuff, when used in conjunction with a conventional gasket, produces the surest bonds.

It does have its limitations though. When using this material, you will have a time limit. It starts to set-up within 15 minutes or so, so you have to assemble the parts without delay. In addition, when squeezing the material out of the tube, don't drop any into the transmission. The stuff will form and set and travel around the oil gallery, possibly plugging up a passage.

✳✳ CAUTION

Caution: Sealing chemicals will clog transmission valve body passages. Use sealers sparingly and make sure they do not squish into the inside of the transmission when installing the pan.

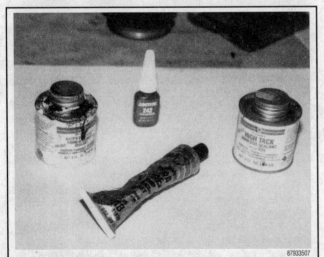

Fig. 15 Sealants are essential. These four types are all that you'll need

CLEANERS

▶ **See Figures 16, 17 and 18**

You'll have two types of cleaners to deal with: parts cleaners and hand cleaners. The parts cleaners are for the vehicle; the hand cleaners are for you. There are many good, non-flammable, biodegradable parts cleaners on the market. These cleaning agents are safe for you, the parts and the envi-

Fig. 16 Three types of cleaners. Some are caustic; some are not. Always read and follow label instructions

Fig. 17 This is one type of hand cleaner that not only works well but smells pretty good too

Fig. 18 The best thing to clean up all types of spills is "kitty litter"

ronment. Therefore, there is no reason to use flammable, caustic or toxic substances to clean your parts or tools.

As far as hand cleaners go, the waterless types are the best. They have always been efficient at cleaning, but left behind a pretty smelly odor. Recently though, just about all of them have eliminated the odor and added stuff that actually smells good. Make sure that you pick one that contains lanolin or some other moisture-replenishing additive. Cleaners not only remove grease and oil but also skin oil.

SHOP TOWELS

▶ **See Figure 19**

One of the most important elements in doing shop work is a good supply of shop towels. Paper towels just don't cut it! Most auto parts stores sell packs of shop towels, usually 50-100 in a pack. They are relatively cheap and can be washed over and over. Always keep them handy.

One of the best shop towels known to science, is the old-fashioned cloth diaper. They're highly absorbent and rugged, but, in these days of disposable diapers, are hard to find.

Fig. 19 A pack of shop towels

TOOLS

▶ **See Figures 20, 21, 22, 23 and 24**

Every do-it-yourselfer loves to accumulate tools. So gathering the tools necessary for engine work can be real fun!

When buying tools, the saying "You get what you pay for" is absolutely true! Don't go cheap! Any hand tool that you buy should be drop forged and/or chrome vanadium. These two qualities tell you that the tool is strong enough for the job. With any tool, power or not, go with a name that you've heard of before, or, that is recommended buy your local professional retailer. Let's go over a list of tools that you'll need.

Most of the world uses the metric system. So, if you have an imported vehicle, you can be pretty certain that it was built with metric fasteners and put together using metric measured clearances and adjustments. In the United States, most people still use the English system. However, if your U.S. made vehicle was built after 1980, most if not all of the fasteners and measurements are metric.

So, accumulate your tools accordingly. Any good DIYer should have a good set of both U.S. and metric measure tools. Don't be confused by terminology. Most advertising refers to "SAE and metric", or "standard and metric". Both are misnomers. The Society of Automotive Engineers (SAE) did not invent the English system of measurement; the English did. The SAE likes metrics just fine. Both English (U.S.) and metric measurements are SAE approved. Also, the current "standard" measurement in automotive building IS metric. So, if it's not metric, it's U.S. measurement.

Hands Tools

SOCKET SETS

▶ **See Figures 25, 26, 27, 28 and 29**

Socket sets are the most basic, necessary hand tools for automotive work. For our purposes, socket sets basically come in three drive sizes: ¼ inch, ⅜ inch and ½ inch. Drive size refers to the size of the drive lug on the ratchet, breaker bar or speed handle.

You'll need a good ½ inch set since this size drive lug assure that you won't break a ratchet or socket. Also, torque wrenches with a torque scale high enough are usually ½ inch drive. The socket set that you'll need should range in sizes from ⁷⁄₁₆ inch through 1 inch for older American models, or 6mm through 19mm on imports and late-model American vehicles.

A ⅜ set is very handy to have since it allows you to get into tight places that the larger drive ratchets can't. Also, this size set gives you a range of smaller sockets that are still strong enough for heavy duty work.

¼ inch drive sets aren't usually necessary or applicable for brake work, but they're good to have for other, light work around the vehicle or house. Besides, they're tools . . . you NEED them!

As for the sockets themselves, they come in standard and deep lengths as well as standard and thin walled, in either 6 or 12 point.

Standard length sockets are good for just about all jobs, however, some stud-head bolts, hard-to-reach bolts, nuts on long studs, etc., require the deep sockets.

Thin-walled sockets are not too common and aren't usually needed in most work. They are exactly what you think, sockets made with thinner wall to fit into tighter places. They don't have the wall strength of a standard socket, of course, but their usefulness in a tight spot can make them worth it.

Six and 12 points. This refers to how many sides are in the socket itself. Each has advantages. The 6 point socket is stronger and less prone to slipping which would strip a bolt head or nut. 12 point sockets are more common, usually less expensive and can operate better in tight places where the ratchet handle can't swing far.

Fig. 20 The most important tool you need to do the job is the proper information, so always have a Chilton manual handy

Fig. 21 The well-stocked garage pegboard. Pegboards can store most tools and other equipment for ease of access. Besides, they're cool looking

Fig. 22 You can arrange the pegboard any way you like, but it's best to hang the most used tools closest to you

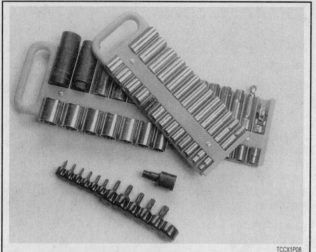

Fig. 23 Socket holders, especially the magnetic type, are handy items to keep tools in order

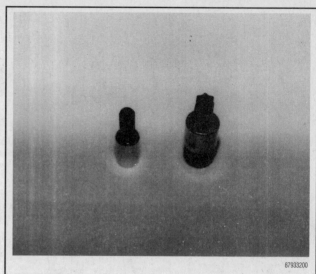

Fig. 26 Left, a hex drive socket; right, a Torx® drive socket

Fig. 24 A good set of handy storage cabinets for fasteners and small parts makes any job easier

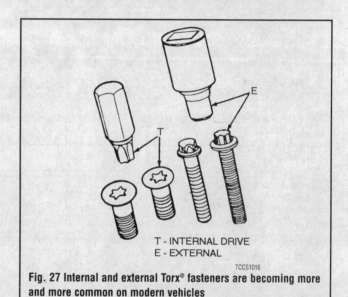

T - INTERNAL DRIVE
E - EXTERNAL

Fig. 27 Internal and external Torx® fasteners are becoming more and more common on modern vehicles

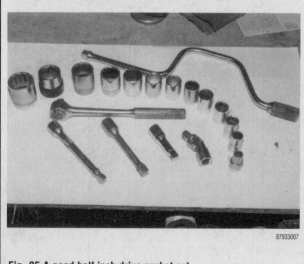

Fig. 25 A good half inch drive socket set

Fig. 28 Two types of drive adapters and a swivel (U-joint) adapter

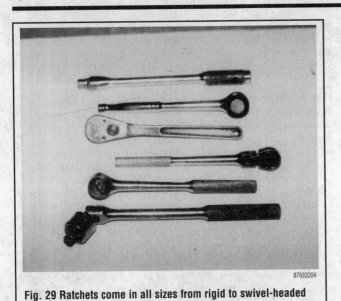

Fig. 29 Ratchets come in all sizes from rigid to swivel-headed

Many manufacturers use recessed hex-head fasteners to retain caliper pins. These fasteners require a socket with a hex shaped stud or a large sturdy hex key. To help prevent torn knuckles, we would recommend that you stick to the sockets and leave the hex keys for lighter applications. Hex stud sockets are available individually or in sets just like conventional sockets. Any complete tool set should include hex stud sockets.

More and more, manufacturers are using Torx® head fasteners, which were once known as tamper resistant fasteners (because many people did not have tools with the necessary odd driver shape). They are still used on parts of the vehicle where the manufacturer would prefer only knowledge-able technicians or advanced Do-It-Yourselfers (DIYers) be working. One automotive example would be some headlight adjustment screws, though it is possible to find these fasteners just about anywhere.

There are currently three different types of Torx® fasteners; internal, external and a new tamper resistant. The internal fasteners require a star-shaped driver. The external fasteners require a star-shaped socket. And, the new tamper resistant fasteners use a star-shaped driver with a small hole drilled through the center. The most common are the internal Torx® fasteners, but you might find any of them on your vehicle.

Torque Wrenches

▶ **See Figure 30**

In most applications, a torque wrench can be used to assure proper installation of a fastener. Torque wrenches come in various designs and most automotive supply stores will carry a variety to suit your needs. A torque wrench should be used any time we supply a specific torque value for a fastener. A torque wrench can also be used if you are following the general guidelines in the accompanying charts. Keep in mind that because there is no worldwide standardization of fasteners, the charts are a general guideline and should be used with caution. Again, the general rule of "if you are using the right tool for the job, you should not have to strain to tighten a fastener" applies here.

BEAM TYPE

▶ **See Figure 31**

The beam type torque wrench is one of the most popular types. It consists of a pointer attached to the head that runs the length of the flexible beam (shaft) to a scale located near the handle. As the wrench is pulled, the beam bends and the pointer indicates the torque using the scale.

CLICK (BREAKAWAY) TYPE

▶ **See Figure 32**

Another popular design of torque wrench is the click type. To use the click type wrench you pre-adjust it to a torque setting. Once the torque is reached, the wrench has a reflex signaling feature that causes a momentary breakaway of the torque wrench body, sending an impulse to the operator's hand.

PIVOT HEAD TYPE

▶ **See Figure 33**

Some torque wrenches (usually of the click type) may be equipped with a pivot head that can allow it to be used in areas of limited access. BUT, it must be used properly. To hold a pivot head wrench, grasp the handle lightly, and as you pull on the handle, it should be floated on the pivot

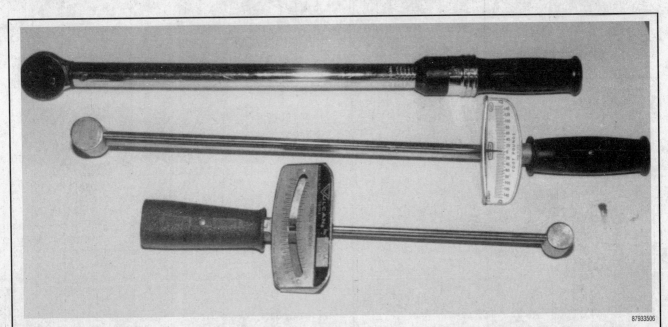

Fig. 30 Three types of torque wrenches. Top to bottom: a ½ inch drive clicker type, a ½ inch drive beam type and a ⅜ inch drive beam type that reads in inch lbs.

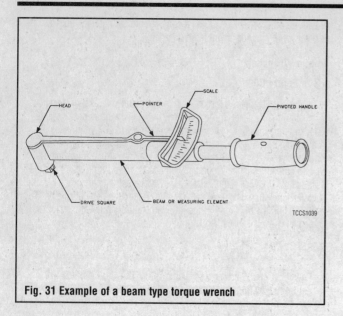

Fig. 31 Example of a beam type torque wrench

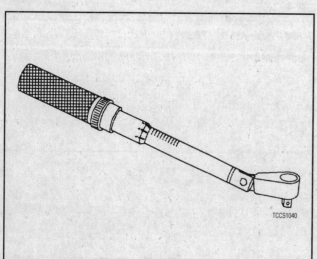

Fig. 32 A click type or breakaway torque wrench—note this one has a pivoting head

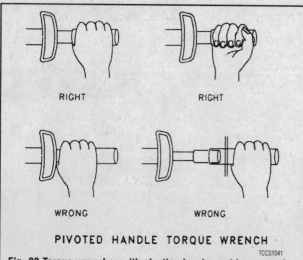

PIVOTED HANDLE TORQUE WRENCH

Fig. 33 Torque wrenches with pivoting heads must be grasped and used properly to prevent an incorrect reading

point. If the handle comes in contact with the yoke extension during the process of pulling, there is a very good chance the torque readings will be inaccurate because this could alter the wrench loading point. The design of the handle is usually such as to make it inconvenient to deliberately misuse the wrench.

→ It should be mentioned that the use of any U-joint, wobble or extension would have an effect on the torque readings, no matter what type of wrench you are using. For the most accurate readings, install the socket directly on the wrench driver. If necessary, straight extensions (which hold a socket directly under the wrench driver) will have the least effect on the torque reading. Avoid any extension that alters the length of the wrench from the handle to the head/driving point (such as a crow's foot). U-joint or wobble extensions can greatly affect the readings; avoid their use at all times.

RIGID CASE (DIRECT READING)

▶ See Figure 34

A rigid case or direct reading torque wrench is equipped with a dial indicator to show torque values. One advantage of these wrenches is that they can be held at any position on the wrench without affecting accuracy. These wrenches are often preferred because they tend to be compact, easy to read and have a great degree of accuracy.

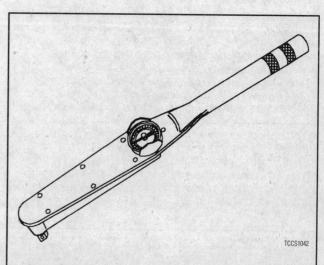

Fig. 34 The rigid case (direct reading) torque wrench uses a dial indicator to show torque

Torque Angle Meters

▶ See Figure 35

Because the frictional characteristics of each fastener or threaded hole will vary, clamp loads which are based strictly on torque will vary as well. In most applications, this variance is not significant enough to cause worry. But, in certain applications, a manufacturer's engineers may determine that more precise clamp loads are necessary (such is the case with many aluminum cylinder heads). In these cases, a torque angle method of installation would be specified. When installing fasteners that are torque angle tightened, a predetermined seating torque and standard torque wrench are usually used first to remove any compliance from the joint. The fastener is then tightened the specified additional portion of a turn measured in degrees. A torque angle gauge (mechanical protractor) is used for these applications.

Breaker Bars

Breaker bars are long handles with a drive lug. Their main purpose is to provide extra turning force when breaking loose tight bolts or nuts. They

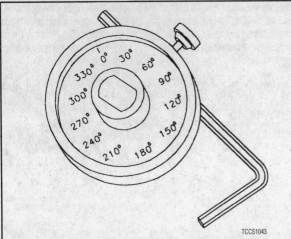

Fig. 35 Some assembly procedures (mostly on engines not brakes) require the use of a torque angle meter (mechanical protractor)

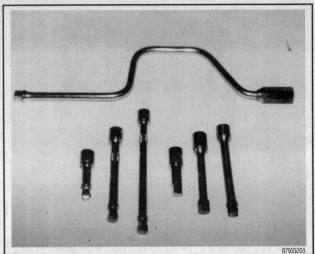

Fig. 36 A speed driver and extensions. The 3 on the left are called "wobbles" since they allow some lateral movement

come in all drive sizes and lengths. Always wear gloves when using a breaker bar.

Speed Handles
▶ See Figure 36

Speed handles are tools with a drive lug and angled turning handle that allow you to quickly remove or install a bolt or nut. They don't, however have much torque ability. You might consider one when installing a number of similar fasteners such as brake backing plate bolts or nuts.

WRENCHES

▶ See Figures 37, 38, 39 and 40

Basically, there are 3 kinds of fixed wrenches: open end, box end, and combination.

Open end wrenches have 2-jawed openings at each end of the wrench. These wrenches are able to fit onto just about any nut or bolt. They are extremely versatile but have one major drawback. They can slip on a worn or rounded bolt head or nut, causing bleeding knuckles and a useless fastener.

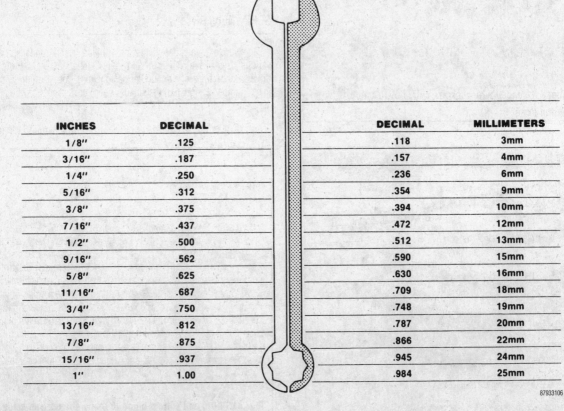

INCHES	DECIMAL		DECIMAL	MILLIMETERS
1/8"	.125		.118	3mm
3/16"	.187		.157	4mm
1/4"	.250		.236	6mm
5/16"	.312		.354	9mm
3/8"	.375		.394	10mm
7/16"	.437		.472	12mm
1/2"	.500		.512	13mm
9/16"	.562		.590	15mm
5/8"	.625		.630	16mm
11/16"	.687		.709	18mm
3/4"	.750		.748	19mm
13/16"	.812		.787	20mm
7/8"	.875		.866	22mm
15/16"	.937		.945	24mm
1"	1.00		.984	25mm

Fig. 37 Comparison of U.S. measure and metric wrench sizes

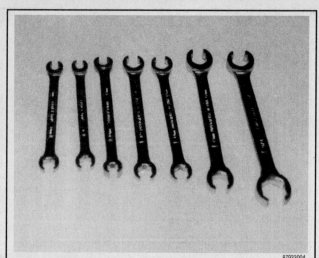

87933004

Fig. 38 Flarenut wrenches are critical for brake lines or tubing, to make sure the fittings do not become rounded

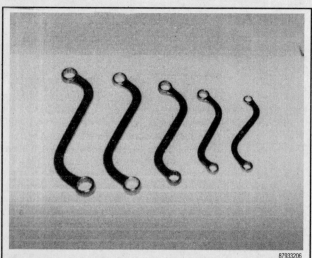

87933206

Fig. 39 These S-shaped wrenches are called obstruction wrenches

87933005

Fig. 40 Several types and sizes of adjustable wrenches

Box-end wrenches have a 360° circular jaw at each end of the wrench. They come in both 6 and 12 point versions just like sockets and each type has the same advantages and disadvantages as sockets.

Combination wrenches have the best of both. They have a 2-jawed open end and a box end. These wrenches are probably the most versatile.

As for sizes, you'll need a range of ¼ inch through 1 inch. As for numbers, you'll need 2 of each size, since, in many instances, one wrench holds the nut while the other turns the bolt. On most fasteners, the nut and bolt are the same size.

One extremely valuable type of wrench is the adjustable wrench. An adjustable wrench has a fixed upper jaw and a moveable lower jaw. The lower jaw is moved by turning a threaded drum. The advantage of an adjustable wrench is its ability to be adjusted to just about any size fastener. The main drawback of an adjustable wrench is the lower jaw's tendency to move slightly under heavy pressure. This can cause the wrench to slip. Adjustable wrenches come in a large range of sizes, measured by the wrench length.

PLIERS

▶ **See Figures 41 and 42**

At least 2 pair of standard pliers is an absolute necessity. Pliers are simply mechanical fingers. They are, more than anything, an extension of your hand

In addition to standard pliers there are the slip-joint, multi-position pliers such as ChannelLock® pliers and locking pliers, such as Vise Grips®.

Slip joint pliers are extremely valuable in grasping oddly sized parts and fasteners. Just make sure that you don't use them instead of a wrench too often since they can easily round off a bolt head or nut.

Locking pliers are usually used for gripping bolt or stud that can't be removed conventionally. You can get locking pliers in square jawed, needle-nosed and pipe-jawed. Pipe jawed have slightly curved jaws for gripping more than just pipes. Locking pliers can rank right up behind duct tape as the handy-man's best friend.

SCREWDRIVERS

Screwdrivers are either standard or Phillips. Standard blades come in various sizes and thickness for all types of slotted fasteners. Phillips screwdrivers come in sizes with number designations from 1 on up, with the lower number designating the smaller size. Screwdrivers can be purchased separately or in sets.

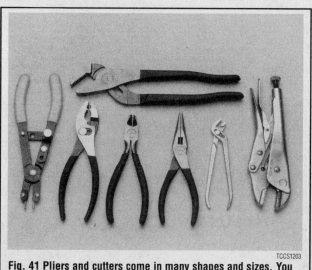

TCCS1203

Fig. 41 Pliers and cutters come in many shapes and sizes. You should have an assortment on hand

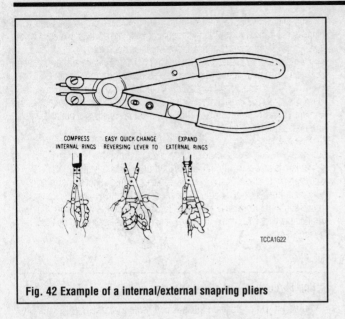

Fig. 42 Example of a internal/external snapring pliers

HAMMERS

▶ **See Figure 43**

You always need a hammer—for just about any kind of work. For most metal work, you need a ball-peen hammer for using drivers and other like tools, a plastic hammer for hitting things safely, and a soft-faced dead-blow hammer for hitting things safely and hard.

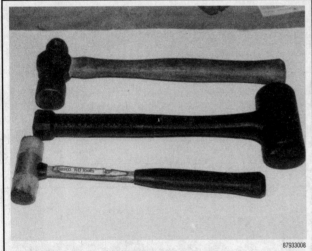

Fig. 43 Three types of hammers. Top to bottom: ball peen, rubber dead-blow, and plastic

OTHER COMMON TOOLS

▶ **See Figures 44 thru 55**

There are a lot of other tools that every workshop should have for automotive work. They include:

- Chisels
- Punches
- Files
- Hacksaw
- Bench Vise
- Tap and Die Set
- Flashlight

- Magnetic Bolt Retriever
- Gasket scraper
- Putty Knife
- Screw/Bolt Extractors
- Prybar

A large bench vise, of at least 4 inch capacity, is essential. A vise is needed to hold anything being worked on.

A tap and die set might be something you've never needed, but you will eventually. It's a good rule, when everything is apart, to clean-up all threads, on bolts, screws and threaded holes. Also, you'll likely run across a situation in which stripped threads will be encountered. The tap and die set will handle that for you.

Gasket scrapers are just what you'd think, tools made for scraping old gasket material off of parts. Old gasket material can be removed with a putty knife or single edge razor blade. However, putty knives may not be sharp enough for some really stuck gaskets and razor blades have a knack of breaking.

For scraping gaskets from aluminum, a hard plastic scraper is highly recommended. They are becoming more commonly available since many manufacturers require factory technicians to have them. Metal scrapers almost always damage aluminum gasket surfaces and the lighter the fluid to be sealed, the more critical the surface preparation.

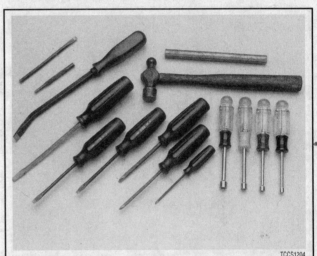

Fig. 44 Various drivers, chisels and prybars are great tools to have in your box

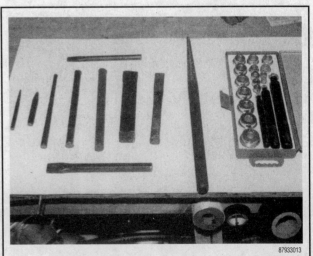

Fig. 45 Punches, chisels and drivers can be purchased separately or in sets

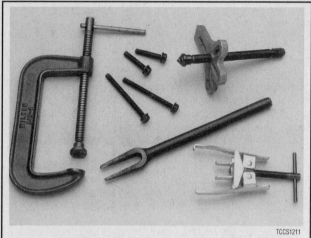

Fig. 46 An assortment of pullers, clamps and separator tools are also needed for many larger repairs (especially engine and suspension work)

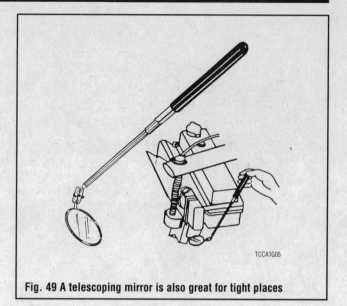

Fig. 49 A telescoping mirror is also great for tight places

Fig. 47 A good quality, heavy-duty bench vise, like this 5½ in. type, with reversible jaws, is ideal for shop work

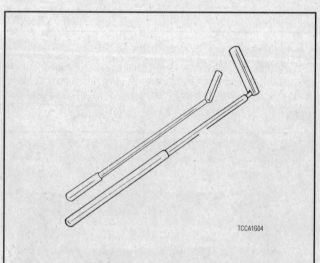

Fig. 50 A magnetic pick-up tool pays for itself the first time you need it

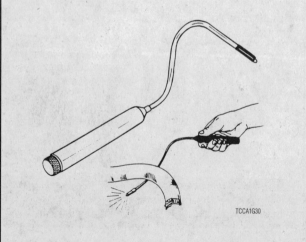

Fig. 48 A flexible flashlight can become invaluable in tight places

Fig. 51 Two good tap and die sets; U.S. measure (left) and metric

87933517

Fig. 52 A set of drill bits and a set of screw extractors

87933003

Fig. 53 A really handy tool is the nut splitter. When a frozen nut simply won't budge, use one of these

TCCX1P10

Fig. 54 Code scanners are available for many of the major vehicle manufacturers

TCCS4P06

Fig. 55 Most hand-held scanners, like this AutoXray®, read all codes on some vehicles

Putty knives really do have a use in automotive work. Just because you remove all the bolts from a component sealed with a gasket doesn't mean it's going to come off. Most of the time, the gasket and sealer will hold it tightly. Lightly driving a putty knife at various points between the two parts will break the seal without damage to the parts.

A small—8–10 inches long—prybar is extremely useful for removing stuck parts such as cylinder heads, timing cases, intake manifolds, etc. NEVER, NEVER, use a screwdriver as a prybar! Screwdrivers are not meant for prying. Screwdrivers, used for prying, can break, sending the broken shaft flying!

Screw/bolt extractors are used for removing broken bolts or studs that have broke off flush with the surface of the part.

On vehicles with fuel injection and/or electronically controlled transmissions, many problems can be diagnosed only with code scanners. These scanners plug into the vehicle's diagnostic connector and read trouble codes that are stored in the on-board computer's memory. Since these scanners are relatively expensive, we don't recommend that you purchase a scanner unless you intend to use it on a regular basis. If you intend to do extensive work on a computer controlled vehicle, a scanner may be worthwhile since it can be used to read codes associated with all operating systems. See the Chilton Total Car Care manual for your particular vehicle when using a scanner.

MICROMETERS & CALIPERS

Outside Micrometers

Outside micrometers are used to check the diameters of such components as shafts and clutch plates. The most common type of micrometer reads in 1/1000 of an inch. Micrometers that use a vernier scale can estimate to 1/10 of an inch. Micrometers and calipers are devices used to make extremely precise measurements. The success of any rebuild is dependent, to a great extent on the ability to check the size and fit of components as specified by the manufacturer. These measurements are made in thousandths and ten-thousandths of an inch.

A micrometer is an instrument made up of a precisely machined spindle which is rotated in a fixed nut, opening and closing the distance between the end of the spindle and a fixed anvil.

To make a measurement, you back off the spindle until you can place the piece to be measured between the spindle and anvil. You then rotate the spindle until the part is contacted by both the spindle and anvil. The measurement is then found by reading the gradations in the handle of the micrometer.

Unless the tool has a digital read-out, reading a micrometer requires some education. The spindle is threaded. Most micrometers use a thread pitch of 40 threads per inch. One complete revolution of the spindle move the spindle toward or away from the anvil 0.025 in. (¼₀ in.).

The fixed part of the handle (called, the sleeve) is marked with 40 gradations per inch of handle length, so each line is 0.025 in. apart. Okay so far?

Every 4th line is marked with a number. The first long line marked 1 represents 0.100 in., the second is 0.200 in., and so on.

The part of the handle that turns is called the thimble. The beveled end of the thimble is marked with gradations, each of which corresponds to 0.001 in. and, usually, every 5th line is numbered.

Turn the thimble until the 0 lines up with the 0 on the sleeve. Now, rotate the thimble one complete revolution and look at the sleeve. You'll see that one complete thimble revolution moved the thimble 0.025 in. down the sleeve.

To read the micrometer, multiply the number of gradations exposed on the sleeve by 0.025 and add that to the number of thousandths indicated by the thimble line that is lined up with the horizontal line on the sleeve. So, if you've measured a part and there are 6 vertical gradations exposed on the sleeve and the 7th gradation on the thimble is lined up with the horizontal line on the sleeve, the thickness of the part is 0.157 in. (6 x 0.025 = 0.150 . Add to that 0.007 representing the 7 lines on the thimble and you get 0.157). See?

If you didn't understand that, try the instructions that come with the micrometer or ask someone that knows, to show you how to work it.

Inside Micrometers

Inside micrometers are used to measure the distance between two parallel surfaces. Inside micrometers are graduated the same way as outside micrometers and are read the same way as well.

Remember that an inside micrometer must be absolutely perpendicular to the work being measured. When you measure with an inside micrometer, rock it gently from side to side and tip it back and forth slightly so that you span the widest part of the bore. Just to be on the safe side, take several readings. It takes a certain amount of experience to work any micrometer with confidence.

Metric Micrometers

▶ **See Figures 56 and 57**

Metric micrometers are read in the same way as inch micrometers, except that the measurements are in millimeters. Each line on the main scale equals 1 mm. Each fifth line is stamped 5, 10, 15, and so on. Each line on the thimble scale equals 0.01 mm. It will take a little practice, but if you can read an inch micrometer, you can read a metric micrometer.

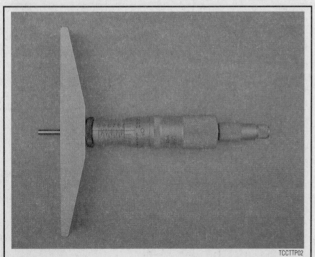

Fig. 56 Depth gauges, like this micrometer, can be used to measure the amount of pad or shoe remaining above a rivet

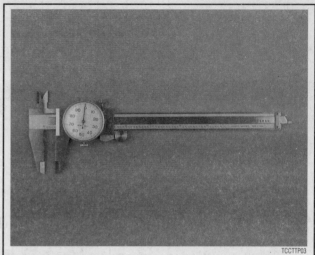

Fig. 57 Outside calipers are fast and easy ways to measure pads or rotors

Inside and Outside Calipers

Inside and outside calipers are useful devices to have if you need to measure something quickly and precise measurement is not necessary. Simply take the reading and then hold the calipers on an accurate steel rule.

DIAL INDICATORS

A dial indicator is a gauge that utilizes a dial face and a needle to register measurements. There is a movable contact arm on the dial indicator. When the arms moves, the needle rotates on the dial. Dial indicators are calibrated to show readings in thousandths of an inch and typically, are used to measure end-play and runout on camshafts, crankshafts, gears, and so on.

Dial indicators are quite easy to use, although they are relatively expensive. A variety of mounting devices are available so that the indicator can be used in a number of situations. Make certain that the contact arm is always parallel to the movement of the work being measured.

TELESCOPING GAUGES

A telescope gauge is used to measure the inside of bores. It can take the place of an inside micrometer for some of these jobs. Simply insert the gauge in the hole to be measured and lock the plungers after they have contacted the walls. Remove the tool and measure across the plungers with an outside micrometer.

Special Tools

Normally, the use of special factory tools is avoided for repair procedures, since these are not readily available for the do-it-yourself mechanic. When it is possible to perform the job with more commonly available tools, it will be pointed out, but occasionally, a special tool was designed to perform a specific function and should be used. Before substituting another tool, you should be convinced that neither your safety nor the performance of the vehicle would be compromised.

There are upwards of fifty different automatic transmission cases on the road today in cars and light duty trucks. There are about five or six different versions of each case, making about 300 different automatic transmissions. Special tools are required to assemble each one of them. These tools include holding fixtures, alignment and measuring tools, gauges, wrenches, test fixtures, special press tools and quite a few more. This does not mean you can't do the job. Most of these tools are available through parts stores, dealer parts departments or tool catalogs. However, for some of the import cars, the special tools are available only to the authorized dealer. Many special tools, like a transmission jack, are common to most transmissions and can often be rented from an equipment rental service.

You should make sure the tools are available before removing the transmission from the car. When necessary, the service procedures in this book will specify and describe any special tools required.

Electric Power Tools

▶ **See Figures 58 and 59**

Power tools are most often associated with woodworking. However, there are a few which are very helpful in automotive work.

The most common and most useful power tool is the bench grinder. You'll need a grinder with a grinding stone on one side and a wire brush wheel on the other. The brush wheel is indispensable for cleaning parts and the stone can be used to remove rough surfaces and for reshaping, where necessary.

Almost as useful as the bench grinder is the drill. Drills can come in very handy when a stripped or broken fastener is encountered.

Power ratchets and impact wrenches can come in very handy. Power ratchets can save a lot of time and muscle when removing and installing long bolts or nuts on long studs, especially where there is little room to swing a manual ratchet. Electric impact wrenches can be invaluable in a lot of automotive work, especially wheel lugs and axle shaft nuts. They don't have much use on brakes, though.

Air Tools and Compressors

▶ **See Figures 60 and 61**

Air-powered tools are not necessary. They are, however, useful for speeding up many jobs and for general clean-up of parts. If you don't have air tools and you want them, be prepared for an initial outlay of a lot of money.

The first thing you need is a compressor. Compressors are available in electrically driven and gas engine driven models. As long as you have electricity, you don't need a gas engine driven type.

The common shop-type air compressor is a pump mounted on a tank. The pump compresses air and forces it into the tank where it is stored until you need it. The compressor automatically turns the pump on when the air pressure in the tank falls below a certain preset level.

There are all kinds of air powered tools, including ratchets, impact wrenches, saws, drills, sprayers, nailers, scrapers, riveters, grinders and sanders. In general, air powered tools are much cheaper than their electric counterparts.

87933006

Fig. 58 Three types of common power tools. Left to right: a hand-held grinder, drill and impact wrench

87933062

Fig. 60 This compressor operates off ordinary house current and provides all the air pressure you'll need

87933060

Fig. 59 The bench grinder can be used to clean just about every part removed from the vehicle

87933022

Fig. 61 An air storage tank

When deciding what size compressor unit you need, you'll be driven by two factors: the Pounds per Square Inch (PSI) capacity of the unit and the deliver rate in Cubic Feet per Minute (CFM). For example, most air powered ratchets require 90 psi at 4 to 5 cfm to operate at peak efficiency. Grinders and saws may require up to 7 cfm at 90 psi. So, before buying the compressor unit, decide what types of tools you'll want so that you don't short-change yourself on the compressor purchase.

If you decide that a compressor and air tools isn't for you, you can have the benefit of air pressure rather cheaply. Purchase an air storage tank, available in sizes up to 20 gallons at most retail stores that sell auto products. These storage tanks can safely store air pressure up to 125 psi and come with a high pressure nozzle for cleaning things and an air chuck for filling tires. The tank can be filled using the common tire-type air compressor.

Jacks and Jackstands

▶ **See Figures 62 and 63**

Jacks and safety stands (jackstands) will be needed for just about anything that you'll do on the lower end of a vehicle.

Your vehicle was supplied with a jack for emergency road repairs. This jack is fine for changing a flat tire or other short-term procedures not requiring you to go beneath the vehicle. For any real work, you must use a floor jack.

Never place the jack under the radiator, engine or transmission components. Severe and expensive damage will result when the jack is raised. Additionally, never jack under the floorpan or bodywork; the metal will deform.

Check your owner's manual or a Chilton Total Car Care for proper jacking and support locations on your vehicle. Many vehicles have crossmembers at the front and rear of the sub-frames that are suitable for jacking, but be careful not to mistake a thin metal skid plate or plastic trim piece as a crossmember.

There are usually reinforced pinch welds along the sides of the vehicle (just in front of the rear wheel and just behind the front wheel) that are used with the vehicle's emergency jack and can be used to raise the vehicle or that can be used with a pair of jackstands. In this case, a block of wood with a cut down the middle in order to cradle the pinch weld will help prevent stress or damage to the metal.

If you have a truck or an older framed vehicle, jackstands can be used almost anywhere along the frame for support. Always use a pair of stands directly across from each other (no closer to the front or rear of the vehicle than the other stand) to help keep the vehicle properly balanced.

☀☀ WARNING

Always position a block of wood or small rubber pad on top of the jack or jackstand to protect the lifting point's finish when lifting or supporting the vehicle.

Whenever you plan to work under the vehicle, you must support it on jackstands or ramps. Never use cinder blocks or stacks of wood to support the vehicle, even if you're only going to be under it for a few minutes. Never crawl under the vehicle when it is supported only by the tire-changing jack or other floor jack.

☀☀ CAUTION

Refer to the jacking precautions and the safety information earlier in this chapter before attempting to raise or support the

Fig. 62 Floor jacks come in all sizes and capacities. Top is a large 2¼ ton models; underneath is a compact 2 ton model

Fig. 63 Jackstands are necessary for holding your vehicle up off the ground. Top are 6 ton models; bottom are 4 ton models

vehicle. Failure to follow proper jacking procedures could result in severe injury or death.

Small hydraulic, screw, or scissors jacks are satisfactory for raising the vehicle. Drive-on trestles or ramps are also a handy and safe way to both raise and support the vehicle. Be careful though, some ramps may be too steep to drive your vehicle onto without scraping the front bottom panels. Never support the vehicle on any suspension member (unless specifically instructed to do so by a repair manual) or by an underbody panel.

FASTENERS

▶ **See Figures 64 and 65**

Although there are a great variety of fasteners found in the modern vehicle, the most commonly used retainer is the threaded fastener (nuts, bolts, screws, studs, etc). Most threaded retainers may be reused, provided that they are not damaged in use or during the repair. Some retainers (such as stretch bolts or torque prevailing nuts) are designed to deform when tightened or in use and should not be reinstalled.

Whenever possible, we will note any special retainers which should be replaced during a procedure. But you should always inspect the condition of a retainer when it is removed and replace any that show signs of damage. Check all threads for rust or corrosion that can increase the torque necessary to achieve the desired clamp load for which that fastener was originally selected. Additionally, be sure that the driver surface of the fastener has not been compromised by rounding or other damage. In some

Fig. 64 Here are a few of the most common screw/bolt driver styles

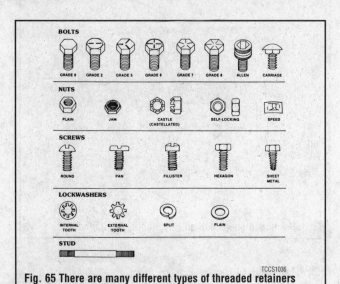

Fig. 65 There are many different types of threaded retainers found on vehicles

cases a driver surface may become only partially rounded, allowing the driver to catch in only one direction. In many of these occurrences, a fastener may be installed and tightened, but the driver would not be able to grip and loosen the fastener again. (This could lead to frustration down the line should that component ever need to be disassembled again).

If you must replace a fastener, whether due to design or damage, you must ALWAYS be sure to use the proper replacement. In all cases, a retainer of the same design, material and strength should be used. Markings on the heads of most bolts will help determine the proper strength of the fastener. The same material, thread and pitch must be selected to assure proper installation and safe operation of the vehicle afterwards.

Thread gauges are available to help measure a bolt or stud's thread. Most automotive and hardware stores keep gauges available to help you select the proper size. In a pinch, you can use another nut or bolt for a thread gauge. If the bolt you are replacing is not too badly damaged, you can select a match by finding another bolt that will thread in its place. If you find a nut that threads properly onto the damaged bolt, then use that nut to help select the replacement bolt. If however, the bolt you are replacing is so badly damaged (broken or drilled out) that its threads cannot be used as a gauge, you might start by looking for another bolt (from the same assembly or a similar location on your vehicle) which will thread into the damaged

bolt's mounting. If so, the other bolt can be used to select a nut; the nut can then be used to select the replacement bolt.

In all cases, be absolutely sure you have selected the proper replacement. Don't be shy, you can always ask the store clerk for help.

❊❊ WARNING

Be aware that when you find a bolt with damaged threads, you may also find the nut or drilled hole it was threaded into has also been damaged. If this is the case, you may have to drill and tap the hole, replace the nut or otherwise repair the threads. NEVER try to force a replacement bolt to fit into the damaged threads.

Bolts and Screws

▸ **See Figure 66**

Technically speaking, bolts are hexagon head or cap screws. For the purposes of this book, however, cap screws will be called bolts because that is the common terminology for them. Both bolts and screws are turned into drilled or threaded holes to fasten two parts together. Frequently, bolts require a nut on the other end, but this is not always the case. Screws seldom, if ever, require a nut on the other end.

Screws are supplied with slotted or Phillips heads. For obvious reasons, screws are not generally used where a great deal of torque is required. Most of the screws you will encounter will be used to retain components, such as brake hose connection clamps, where strength is not a factor.

Threaded retainers (such as bolts and screws) come in various sizes, designated as 8-32, 10-32, or ¼-32. The first number indicates the minor diameter, and the second number indicates the number of threads per inch (or distance between threads in mm).

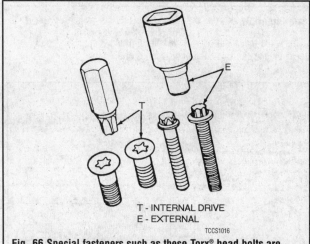

T - INTERNAL DRIVE
E - EXTERNAL

Fig. 66 Special fasteners such as these Torx® head bolts are used by manufacturers to discourage people from working on vehicles without the proper tools

Nuts

Nuts have only one use: they simply hold the other end of the bolt or stud and, thereby, hold the two parts together. There are a variety of nuts used on vehicles, but a standard hexagon head (six-sided) nut is the most common.

Castellated and slotted nuts are designed for use with a cotter pin and are usually used when it is extremely important that the nuts do not work loose (in wheel bearings, for example). Other nuts are self-locking nuts that have a slot cut in the side.

When the nut is tightened, the separated sections pull together and lock the nut onto the bolt. Interference nuts have a collar of soft metal or fiber.

The bolt cuts threads in the soft material that then jams in the threads and prevents the nut and bolt from working loose.

A jam nut is a second hexagon nut that is used to hold the first nut in place. They are usually found where some type of adjustment is needed; parking brake cables, for instance.

Pawlnuts are single thread nuts that provide some locking action when they have been turned down on the nut.

Speed nuts are simply rectangular bits of sheet metal that are pushed down over a bolt, screw, or stud to provide locking action.

Studs

Studs are simply pieces of threaded rod. They are similar to bolts and screws in their thread configuration, but they have no heads. One end is turned into a threaded hole and the other end is generally secured by same type of nut. Unless the nut is self-locking, a lockwasher or jam nut is generally used underneath it.

Lockwashers

Lockwashers are a form of washer. They may be either split or toothed, and they are always installed between a nut or screw head and the actual part being held. The split washer is crushed flat and locks the nut in place by spring tension. The toothed washer provides many edges to improve the locking effect and is usually used on smaller bolts and screws.

Screw and Bolt Terminology

Bolts and screws are identified by type, major diameter, minor diameter, pitch or threads per inch, class, length, thread length, and the size of the wrench required.

MAJOR DIAMETER

▶ **See Figure 67**

This is the widest diameter of the bolt as measured from the top of the threads on one side to the top of the threads on the other side.

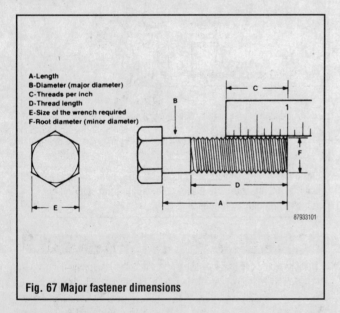

A-Length
B-Diameter (major diameter)
C-Threads per inch
D-Thread length
E-Size of the wrench required
F-Root diameter (minor diameter)

87933101

Fig. 67 Major fastener dimensions

MINOR DIAMETER

This is the diameter obtained by measuring from the bottom of the threads on one side of the bolt to the bottom of the threads on the other side. In other words, it is the diameter of the bolt if it does not have any threads.

PITCH OR THREADS PER INCH

▶ **See Figure 68**

Thread pitch is the distance between the top of one thread to the top of the next. It is simply the distance between one thread and the next. There are two types of threads in general use today. Unified National Coarse thread, and Unified National Fine. These are usually known simply as either fine or coarse thread.

Anyone who has been working on vehicles for any length of time can tell the difference between the two simply by looking at the screw, bolt, or nut. The only truly accurate way to determine thread pitch is to use a thread pitch gauge. There are some general rules to remember, however.

Coarse thread screws and bolts are used frequently when they are being threaded into aluminum or cast iron because the finer threads tend to strip more easily in these materials. Also, as a bolt or screw's diameter increases, thread pitch becomes greater.

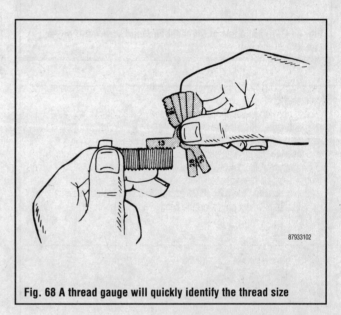

87933102

Fig. 68 A thread gauge will quickly identify the thread size

THREAD CLASS

Thread class is a measure of the operating clearance between the internal nut threads and the external threads of the bolt. There are three classes of fit, 1, 2, or 3. In addition, there are letter designations to designate either internal (class A) or external (class B) threads.

Class 1 threads are a relatively loose fit and are used when ease of assembly and disassembly are of paramount importance.

Class 2 bolts are most commonly encountered in automotive applications and give an accurate, but not an overly tight, fit.

Class 3 threads are used when utmost accuracy is needed. You might find a class 3 bolt and nut combination on an airplane, but you won't encounter them very often on a vehicle.

LENGTH & THREAD LENGTH

Screw length is the length of the bolt or screw from the bottom of the head to the bottom of the bolt or screw. Thread length is exactly that, the length of the threads.

TYPES OR GRADES OF BOLTS & SCREWS

▶ **See Figure 69**

The tensile strength of bolts and screws varies widely. Standards for these fasteners have been established by the Society of Automotive Engi-

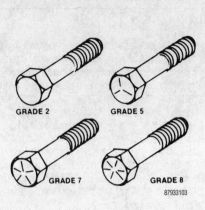

Fig. 69 Markings on U.S. measure bolts indicate the relative strength of the bolt

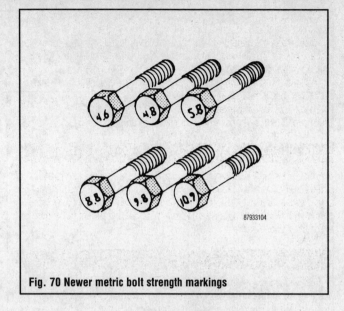

Fig. 70 Newer metric bolt strength markings

neers (SAE). Distinctive markings on the head of the bolt will identify its tensile strength.

These outward radiating lines are normally called points. A bolt with no points on the head is a grade 1 or a grade 2 bolt. This type of bolt is suitable for applications in which only a low-strength bolt is necessary.

On the other hand, a grade 5 bolt is found in a number of automotive applications and has double the tensile strength of a grade 2 bolt. A grade 5 bolt will have three embossed lines or points on the head.

Grade 8 bolts are the best and are frequently called aircraft grade bolts. Grade 8 bolts have six points on the head.

METRIC BOLTS

▶ **See Figure 70, 71, 72 and 73**

While metric bolts may seem to be the same as their U.S. measure counterparts, they definitely are not. The pitch on a metric bolt is different from that of an U.S. measure bolt. It is entirely possible to start a metric bolt into a hole with U.S. measure threads and run it down a few turns. Then it is going to bind. Recognizing the problem at this point is not going to do much good. It is also possible to run a metric nut down on an U.S. measure bolt and find that it is too loose to provide sufficient strength.

Metric bolts are marked in a manner different from that of U.S. measure bolts. Most metric bolts have a number stamped on the head. This metric grade marking won't be an even number, but something like 4.6 or 10.9. The number indicates the relative strength of the bolt. The higher the number, the greater the strength of the bolt. Some metric bolts are also marked with a single-digit number to indicate the bolt strength. Metric bolt sizes are also identified in a manner different from that of U.S. measure fasteners.

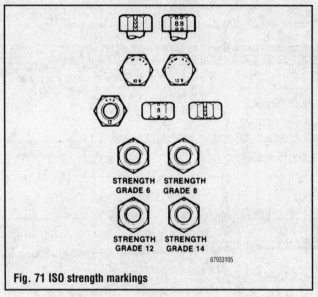

Fig. 71 ISO strength markings

If, for example, a metric bolt were designated 14 x 2, that would mean that the major diameter is 14 mm (.56 in.), and that the thread pitch is 2 mm (.08 in.). More important, metric bolts are not classified by number of threads per inch, but by the distance between the threads, and the distance between threads does not quite correspond to number of threads per inch. For example, 2 mm between threads is about 12.7 threads per inch.

	Mark	Class		Mark	Class
Hexagon head bolt	Bolt head No. 4 — 5 — 6 — 7 — 8 — 9 — 10 — 11 —	4T 5T 6T 7T 8T 9T 10T 11T	Stud bolt	No mark	4T
	No mark	4T			
Hexagon flange bolt w/ washer hexagon bolt	No mark	4T		Grooved	6T
Hexagon head bolt	Two protruding lines	5T			
Hexagon flange bolt w/ washer hexagon bolt	Two protruding lines	6T	Welded bolt		4T
Hexagon head bolt	Three protruding lines	7T			
Hexagon head bolt	Four protruding lines	8T			

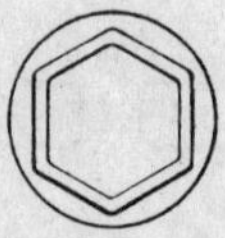

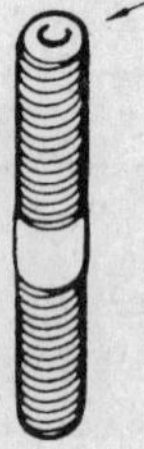

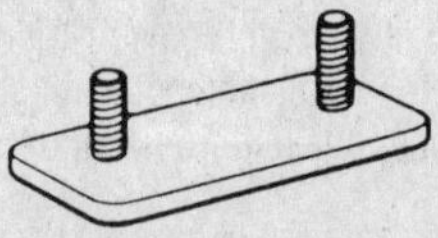

TCCS1240

Fig. 72 Metric bolt strength indicator marks

| Class | Diameter mm | Pitch mm | Specified torque | | | | | |
| | | | Hexagon head bolt | | | Hexagon flange bolt | | |
			N·m	kgf·cm	ft·lbf	N·m	kgf·cm	ft·lbf
4T	6	1	5	55	48 in.·lbf	6	60	52 in.·lbf
	8	1.25	12.5	130	9	14	145	10
	10	1.25	26	260	19	29	290	21
	12	1.25	47	480	35	53	540	39
	14	1.5	74	760	55	84	850	61
	16	1.5	115	1,150	83	—	—	—
5T	6	1	6.5	65	56 in.·lbf	7.5	75	65 in.·lbf
	8	1.25	15.5	160	12	17.5	175	13
	10	1.25	32	330	24	36	360	26
	12	1.25	59	600	43	65	670	48
	14	1.5	91	930	67	100	1,050	76
	16	1.5	140	1,400	101	—	—	—
6T	6	1	8	80	69 in.·lbf	9	90	78 in.·lbf
	8	1.25	19	195	14	21	210	15
	10	1.25	39	400	29	44	440	32
	12	1.25	71	730	53	80	810	59
	14	1.5	110	1,100	80	125	1,250	90
	16	1.5	170	1,750	127	—	—	—
7T	6	1	10.5	110	8	12	120	9
	8	1.25	25	260	19	28	290	21
	10	1.25	52	530	38	58	590	43
	12	1.25	95	970	70	105	1,050	76
	14	1.5	145	1,500	108	165	1,700	123
	16	1.5	230	2,300	166	—	—	—
8T	8	1.25	29	300	22	33	330	24
	10	1.25	61	620	45	68	690	50
	12	1.25	110	1,100	80	120	1,250	90
9T	8	1.25	34	340	25	37	380	27
	10	1.25	70	710	51	78	790	57
	12	1.25	125	1,300	94	140	1,450	105
10T	8	1.25	38	390	28	42	430	31
	10	1.25	78	800	58	88	890	64
	12	1.25	140	1,450	105	155	1,600	116
11T	8	1.25	42	430	31	47	480	35
	10	1.25	87	890	64	97	990	72
	12	1.25	155	1,600	116	175	1,800	130

TCCS1241

Fig. 73 Determining the strength of metric fasteners

ENGLISH TO METRIC CONVERSION: LENGTH

To convert inches (ins.) to millimeters (mm): multiply number of inches by 25.4

To convert millimeters (mm) to inches (ins.): multiply number of millimeters by .04

Inches		Decimals	Milli-meters	Inches to millimeters inches	mm		Inches		Decimals	Milli-meters	Inches to millimeters inches	mm
	1/64	0.051625	0.3969	0.0001	0.00254			33/64	0.515625	13.0969	0.6	15.24
	1/32	0.03125	0.7937	0.0002	0.00508		17/32		0.53125	13.4937	0.7	17.78
	3/64	0.046875	1.1906	0.0003	0.00762			35/64	0.546875	13.8906	0.8	20.32
1/16		0.0625	1.5875	0.0004	0.01016		9/16		0.5625	14.2875	0.9	22.86
	5/64	0.078125	1.9844	0.0005	0.01270			37/64	0.578125	14.6844	1	25.4
	3/32	0.09375	2.3812	0.0006	0.01524		19/32		0.59375	15.0812	2	50.8
	7/64	0.109375	2.7781	0.0007	0.01778			39/64	0.609375	15.4781	3	76.2
1/8		0.125	3.1750	0.0008	0.02032		5/8		0.625	15.8750	4	101.6
	9/64	0.140625	3.5719	0.0009	0.02286			41/64	0.640625	16.2719	5	127.0
	5/32	0.15625	3.9687	0.001	0.0254		21/32		0.65625	16.6687	6	152.4
	11/64	0.171875	4.3656	0.002	0.0508			43/64	0.671875	17.0656	7	177.8
3/16		0.1875	4.7625	0.003	0.0762		11/16		0.6875	17.4625	8	203.2
	13/64	0.203125	5.1594	0.004	0.1016			45/64	0.703125	17.8594	9	228.6
	7/32	0.21875	5.5562	0.005	0.1270		23/32		0.71875	18.2562	10	254.0
	15/64	0.234375	5.9531	0.006	0.1524			47/64	0.734375	18.6531	11	279.4
1/4		0.25	6.3500	0.007	0.1778		3/4		0.75	19.0500	12	304.8
	17/64	0.265625	6.7469	0.008	0.2032			49/64	0.765625	19.4469	13	330.2
	9/32	0.28125	7.1437	0.009	0.2286		25/32		0.78125	19.8437	14	355.6
	19/64	0.296875	7.5406	0.01	0.254			51/64	0.796875	20.2406	15	381.0
5/16		0.3125	7.9375	0.02	0.508		13/16		0.8125	20.6375	16	406.4
	21/64	0.328125	8.3344	0.03	0.762			53/64	0.828125	21.0344	17	431.8
	11/32	0.34375	8.7312	0.04	1.016		27/32		0.84375	21.4312	18	457.2
	23/64	0.359375	9.1281	0.05	1.270			55/64	0.859375	21.8281	19	482.6
3/8		0.375	9.5250	0.06	1.524		7/8		0.875	22.2250	20	508.0
	25/64	0.390625	9.9219	0.07	1.778			57/64	0.890625	22.6219	21	533.4
	13/32	0.40625	10.3187	0.08	2.032		29/32		0.90625	23.0187	22	558.8
	27/64	0.421875	10.7156	0.09	2.286			59/64	0.921875	23.4156	23	584.2
7/16		0.4375	11.1125	0.1	2.54		15/16		0.9375	23.8125	24	609.6
	29/64	0.453125	11.5094	0.2	5.08			61/64	0.953125	24.2094	25	635.0
	15/32	0.46875	11.9062	0.3	7.62		31/32		0.96875	24.6062	26	660.4
	31/64	0.484375	12.3031	0.4	10.16			63/64	0.984375	25.0031	27	690.6
1/2		0.5	12.7000	0.5	12.70							

ENGLISH TO METRIC CONVERSION: TORQUE

To convert foot-pounds (ft. lbs.) to Newton-meters: multiply the number of ft. lbs. by 1.3

To convert inch-pounds (in. lbs.) to Newton-meters: multiply the number of in. lbs. by .11

in lbs	N-m	in lbs	N-m	in lbs	N-m	in lbs	N-m	in lbs	N-m
0.1	0.01	1	0.11	10	1.13	19	2.15	28	3.16
0.2	0.02	2	0.23	11	1.24	20	2.26	29	3.28
0.3	0.03	3	0.34	12	1.36	21	2.37	30	3.39
0.4	0.04	4	0.45	13	1.47	22	2.49	31	3.50
0.5	0.06	5	0.56	14	1.58	23	2.60	32	3.62
0.6	0.07	6	0.68	15	1.70	24	2.71	33	3.73
0.7	0.08	7	0.78	16	1.81	25	2.82	34	3.84
0.8	0.09	8	0.90	17	1.92	26	2.94	35	3.95
0.9	0.10	9	1.02	18	2.03	27	3.05	36	4.0

ENGLISH TO METRIC CONVERSION: MASS (WEIGHT)

Current **mass** measurement is expressed in pounds and ounces (lbs. & ozs.). The metric unit of mass (or weight) is the kilogram (kg). Even although this table does not show conversion of masses (weights) larger than 15 lbs, it is easy to calculate larger units by following the data immediately below.

To convert ounces (oz.) to grams (g): multiply th number of ozs. by 28
To convert grams (g) to ounces (oz.): multiply the number of grams by .035

To convert pounds (lbs.) to kilograms (kg): multiply the number of lbs. by .45
To convert kilograms (kg) to pounds (lbs.): multiply the number of kilograms by 2.2

lbs	kg	lbs	kg	oz	kg	oz	kg
0.1	0.04	0.9	0.41	0.1	0.003	0.9	0.024
0.2	0.09	1	0.4	0.2	0.005	1	0.03
0.3	0.14	2	0.9	0.3	0.008	2	0.06
0.4	0.18	3	1.4	0.4	0.011	3	0.08
0.5	0.23	4	1.8	0.5	0.014	4	0.11
0.6	0.27	5	2.3	0.6	0.017	5	0.14
0.7	0.32	10	4.5	0.7	0.020	10	0.28
0.8	0.36	15	6.8	0.8	0.023	15	0.42

ENGLISH TO METRIC CONVERSION: TEMPERATURE

To convert Fahrenheit (°F) to Celsius (°C): take number of °F and subtract 32; multiply result by 5; divide result by 9

To convert Celsius (°C) to Fahrenheit (°F): take number of °C and multiply by 9; divide result by 5; add 32 to total

Fahrenheit (F)	Celsius (C)			Fahrenheit (F)	Celsius (C)			Fahrenheit (F)	Celsius (C)		
°F	°C	°C	°F	°F	°C	°C	°F	°F	°C	°C	°F
−40	−40	−38	−36.4	80	26.7	18	64.4	215	101.7	80	176
−35	−37.2	−36	−32.8	85	29.4	20	68	220	104.4	85	185
−30	−34.4	−34	−29.2	90	32.2	22	71.6	225	107.2	90	194
−25	−31.7	−32	−25.6	95	35.0	24	75.2	230	110.0	95	202
−20	−28.9	−30	−22	100	37.8	26	78.8	235	112.8	100	212
−15	−26.1	−28	−18.4	105	40.6	28	82.4	240	115.6	105	221
−10	−23.3	−26	−14.8	110	43.3	30	86	245	118.3	110	230
−5	−20.6	−24	−11.2	115	46.1	32	89.6	250	121.1	115	239
0	−17.8	−22	−7.6	120	48.9	34	93.2	255	123.9	120	248
1	−17.2	−20	−4	125	51.7	36	96.8	260	126.6	125	257
2	−16.7	−18	−0.4	130	54.4	38	100.4	265	129.4	130	266
3	−16.1	−16	3.2	135	57.2	40	104	270	132.2	135	275
4	−15.6	−14	6.8	140	60.0	42	107.6	275	135.0	140	284
5	−15.0	−12	10.4	145	62.8	44	112.2	280	137.8	145	293
10	−12.2	−10	14	150	65.6	46	114.8	285	140.6	150	302
15	−9.4	−8	17.6	155	68.3	48	118.4	290	143.3	155	311
20	−6.7	−6	21.2	160	71.1	50	122	295	146.1	160	320
25	−3.9	−4	24.8	165	73.9	52	125.6	300	148.9	165	329
30	−1.1	−2	28.4	170	76.7	54	129.2	305	151.7	170	338
35	1.7	0	32	175	79.4	56	132.8	310	154.4	175	347
40	4.4	2	35.6	180	82.2	58	136.4	315	157.2	180	356
45	7.2	4	39.2	185	85.0	60	140	320	160.0	185	365
50	10.0	6	42.8	190	87.8	62	143.6	325	162.8	190	374
55	12.8	8	46.4	195	90.6	64	147.2	330	165.6	195	383
60	15.6	10	50	200	93.3	66	150.8	335	168.3	200	392
65	18.3	12	53.6	205	96.1	68	154.4	340	171.1	205	401
70	21.1	14	57.2	210	98.9	70	158	345	173.9	210	410
75	23.9	16	60.8	212	100.0	75	167	350	176.7	215	414

ENGLISH TO METRIC CONVERSION: TORQUE

Torque is now expressed as either foot-pounds (ft./lbs.) or inch-pounds (in./lbs.). The metric measurement unit for torque is the Newton-meter (Nm). This unit—the Nm—will be used for all SI metric torque references, both the present ft./lbs. and in./lbs.

ft lbs	N-m	ft lbs	N-m	ft lbs	N-m	ft lbs	N-m
0.1	0.1	33	44.7	74	100.3	115	155.9
0.2	0.3	34	46.1	75	101.7	116	157.3
0.3	0.4	35	47.4	76	103.0	117	158.6
0.4	0.5	36	48.8	77	104.4	118	160.0
0.5	0.7	37	50.7	78	105.8	119	161.3
0.6	0.8	38	51.5	79	107.1	120	162.7
0.7	1.0	39	52.9	80	108.5	121	164.0
0.8	1.1	40	54.2	81	109.8	122	165.4
0.9	1.2	41	55.6	82	111.2	123	166.8
1	1.3	42	56.9	83	112.5	124	168.1
2	2.7	43	58.3	84	113.9	125	169.5
3	4.1	44	59.7	85	115.2	126	170.8
4	5.4	45	61.0	86	116.6	127	172.2
5	6.8	46	62.4	87	118.0	128	173.5
6	8.1	47	63.7	88	119.3	129	174.9
7	9.5	48	65.1	89	120.7	130	176.2
8	10.8	49	66.4	90	122.0	131	177.6
9	12.2	50	67.8	91	123.4	132	179.0
10	13.6	51	69.2	92	124.7	133	180.3
11	14.9	52	70.5	93	126.1	134	181.7
12	16.3	53	71.9	94	127.4	135	183.0
13	17.6	54	73.2	95	128.8	136	184.4
14	18.9	55	74.6	96	130.2	137	185.7
15	20.3	56	75.9	97	131.5	138	187.1
16	21.7	57	77.3	98	132.9	139	188.5
17	23.0	58	78.6	99	134.2	140	189.8
18	24.4	59	80.0	100	135.6	141	191.2
19	25.8	60	81.4	101	136.9	142	192.5
20	27.1	61	82.7	102	138.3	143	193.9
21	28.5	62	84.1	103	139.6	144	195.2
22	29.8	63	85.4	104	141.0	145	196.6
23	31.2	64	86.8	105	142.4	146	198.0
24	32.5	65	88.1	106	143.7	147	199.3
25	33.9	66	89.5	107	145.1	148	200.7
26	35.2	67	90.8	108	146.4	149	202.0
27	36.6	68	92.2	109	147.8	150	203.4
28	38.0	69	93.6	110	149.1	151	204.7
29	39.3	70	94.9	111	150.5	152	206.1
30	40.7	71	96.3	112	151.8	153	207.4
31	42.0	72	97.6	113	153.2	154	208.8
32	43.4	73	99.0	114	154.6	155	210.2

TCCS1C03

ENGLISH TO METRIC CONVERSION: FORCE

Force is presently measured in pounds (lbs.). This type of measurement is used to measure spring pressure, specifically how many pounds it takes to compress a spring. Our present force unit (the pound) will be replaced in SI metric measurements by the Newton (N). This term will eventually see use in specifications for electric motor brush spring pressures, valve spring pressures, etc.

To convert pounds (lbs.) to Newton (N): multiply the number of lbs. by 4.45

lbs	N	lbs	N	lbs	N	oz	N
0.01	0.04	21	93.4	59	262.4	1	0.3
0.02	0.09	22	97.9	60	266.9	2	0.6
0.03	0.13	23	102.3	61	271.3	3	0.8
0.04	0.18	24	106.8	62	275.8	4	1.1
0.05	0.22	25	111.2	63	280.2	5	1.4
0.06	0.27	26	115.6	64	284.6	6	1.7
0.07	0.31	27	120.1	65	289.1	7	2.0
0.08	0.36	28	124.6	66	293.6	8	2.2
0.09	0.40	29	129.0	67	298.0	9	2.5
0.1	0.4	30	133.4	68	302.5	10	2.8
0.2	0.9	31	137.9	69	306.9	11	3.1
0.3	1.3	32	142.3	70	311.4	12	3.3
0.4	1.8	33	146.8	71	315.8	13	3.6
0.5	2.2	34	151.2	72	320.3	14	3.9
0.6	2.7	35	155.7	73	324.7	15	4.2
0.7	3.1	36	160.1	74	329.2	16	4.4
0.8	3.6	37	164.6	75	333.6	17	4.7
0.9	4.0	38	169.0	76	338.1	18	5.0
1	4.4	39	173.5	77	342.5	19	5.3
2	8.9	40	177.9	78	347.0	20	5.6
3	13.4	41	182.4	79	351.4	21	5.8
4	17.8	42	186.8	80	355.9	22	6.1
5	22.2	43	191.3	81	360.3	23	6.4
6	26.7	44	195.7	82	364.8	24	6.7
7	31.1	45	200.2	83	369.2	25	7.0
8	35.6	46	204.6	84	373.6	26	7.2
9	40.0	47	209.1	85	378.1	27	7.5
10	44.5	48	213.5	86	382.6	28	7.8
11	48.9	49	218.0	87	387.0	29	8.1
12	53.4	50	224.4	88	391.4	30	8.3
13	57.8	51	226.9	89	395.9	31	8.6
14	62.3	52	231.3	90	400.3	32	8.9
15	66.7	53	235.8	91	404.8	33	9.2
16	71.2	54	240.2	92	409.2	34	9.4
17	75.6	55	244.6	93	413.7	35	9.7
18	80.1	56	249.1	94	418.1	36	10.0
19	84.5	57	253.6	95	422.6	37	10.3
20	89.0	58	258.0	96	427.0	38	10.6

TCCS1C04

ENGLISH TO METRIC CONVERSION: LIQUID CAPACITY

Liquid or fluid capacity is presently expressed as pints, quarts or gallons, or a combination of all of these. In the metric system the liter (l) will become the basic unit. Fractions of a liter would be expressed as deciliters, centiliters, or most frequently (and commonly) as milliliters.

To convert pints (pts.) to liters (l): multiply the number of pints by .47
To convert liters (l) to pints (pts.): multiply the number of liters by 2.1
To convert quarts (qts.) to liters (l): multiply the number of quarts by .95

To convert liters (l) to quarts (qts.): multiply the number of liters by 1.06
To convert gallons (gals.) to liters (l): multiply the number of gallons by 3.8
To convert liters (l) to gallons (gals.): multiply the number of liters by .26

gals	liters	qts	liters	pts	liters
0.1	0.38	0.1	0.10	0.1	0.05
0.2	0.76	0.2	0.19	0.2	0.10
0.3	1.1	0.3	0.28	0.3	0.14
0.4	1.5	0.4	0.38	0.4	0.19
0.5	1.9	0.5	0.47	0.5	0.24
0.6	2.3	0.6	0.57	0.6	0.28
0.7	2.6	0.7	0.66	0.7	0.33
0.8	3.0	0.8	0.76	0.8	0.38
0.9	3.4	0.9	0.85	0.9	0.43
1	3.8	1	1.0	1	0.5
2	7.6	2	1.9	2	1.0
3	11.4	3	2.8	3	1.4
4	15.1	4	3.8	4	1.9
5	18.9	5	4.7	5	2.4
6	22.7	6	5.7	6	2.8
7	26.5	7	6.6	7	3.3
8	30.3	8	7.6	8	3.8
9	34.1	9	8.5	9	4.3
10	37.8	10	9.5	10	4.7
11	41.6	11	10.4	11	5.2
12	45.4	12	11.4	12	5.7
13	49.2	13	12.3	13	6.2
14	53.0	14	13.2	14	6.6
15	56.8	15	14.2	15	7.1
16	60.6	16	15.1	16	7.6
17	64.3	17	16.1	17	8.0
18	68.1	18	17.0	18	8.5
19	71.9	19	18.0	19	9.0
20	75.7	20	18.9	20	9.5
21	79.5	21	19.9	21	9.9
22	83.2	22	20.8	22	10.4
23	87.0	23	21.8	23	10.9
24	90.8	24	22.7	24	11.4
25	94.6	25	23.6	25	11.8
26	98.4	26	24.6	26	12.3
27	102.2	27	25.5	27	12.8
28	106.0	28	26.5	28	13.2
29	110.0	29	27.4	29	13.7
30	113.5	30	28.4	30	14.2

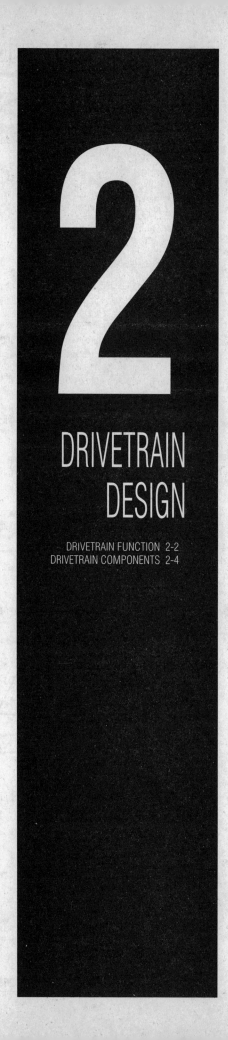

2
DRIVETRAIN
DESIGN

DRIVETRAIN FUNCTION

Drivetrain Designs

♦ See Figures 1, 2 and 3

A transmission can be defined as the system used to transmit power from an engine (motor) and send it to the output device. This may be a simple belt and pulley arrangement or a very complex system of gears, hydraulics and electronic controls. With a very few exceptions, every engine uses some kind of transmission.

An electric drill uses a gear train to convert the high rpm of the motor to a slower speed at the chuck, with the net result being an increase in torque. In an outboard boat engine, the crankshaft is vertical but the propeller must be horizontal, so a driveshaft and gears might be used to change the plane of rotation. Even in rockets and jet aircraft that use the engine's thrust directly to move the vehicle with no other output device, the engine still has a power take-off gear train for driving accessories such as pumps and generators. Mechanical clocks, turbine engines, washing machines, starter motors, almost every power source around us is equipped with a transmission of some type. It can be an interesting and educational exercise to look at the motors and engines around us and think about how they are equipped with a transmission.

Any transmission system has at least two main functions: change the rotational speed from the power source to the output device, and increase or decrease the torque delivered to the output device. A third function may be to change the directional rotation or plane on which the driveshaft turning, as in outboard boat motors. To operate efficiently, most engines spin at a relatively high rpm. With a small gear attached to the output shaft driving a larger gear, the shaft speed of the larger gear is less but the torque or twisting force is greater.

The automotive drivetrain is defined as all the components required to transmit the power from the engine to the drive wheel hubs. This includes the transmission and all its components, the differential assembly and the driveshafts. How these components are arranged depends on the location of the engine relative to the drive wheels. In today's vehicles, there are six basic designs:

- Front engine/front wheel drive
- Front engine/rear wheel drive
- Front engine/all wheel drive
- Rear engine/rear wheel drive
- Rear engine/all wheel drive
- Mid engine/rear wheel drive

Each of these designs has a purpose and to a great extent defines the vehicle's intend use. For example, front engine/rear wheel drive is most often used for heavier vehicles because the weight makes it impractical to use the same tires for driving and steering. Front engine/front wheel drive is often used for small economy cars to reduce weight and production costs. Mid-engine designs are usually high-performance sports cars, with the location of the heaviest components (engine and drivetrain) optimized for handling.

To appreciate the how the engine and transmission power an automobile, we need to review some basic fundamentals of operation. Although the engine is the source of power, the term drivetrain includes all the drive components between the engine flywheel and the drive wheels. As its name implies, the main function of the drivetrain is transmit power to move the vehicle. It is the link between the engine and drive wheels that permits the engine to start the vehicle moving and keep it moving.

The engine and transmission are designed to work together as a balanced team to best utilize the engine power. Available axle torque to drive the vehicle, is therefore a function of engine torque multiplied by the drivetrain. Since the crankshaft spins at relatively high speeds, the drivetrain must transmit engine power at reduced speed and at the same time multiply the torque made by the engine.

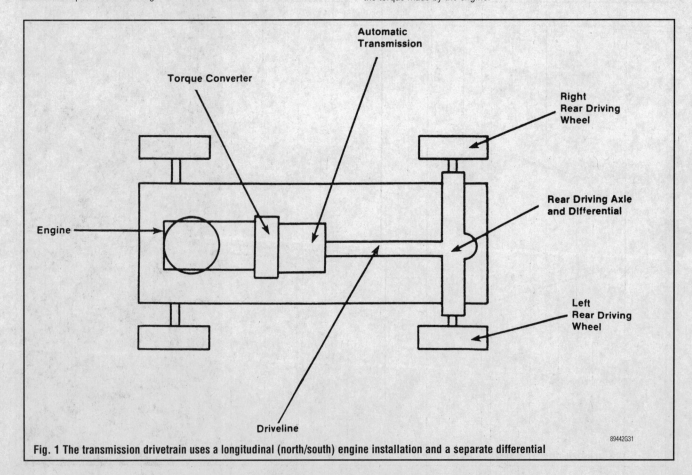

Fig. 1 The transmission drivetrain uses a longitudinal (north/south) engine installation and a separate differential

89442G31

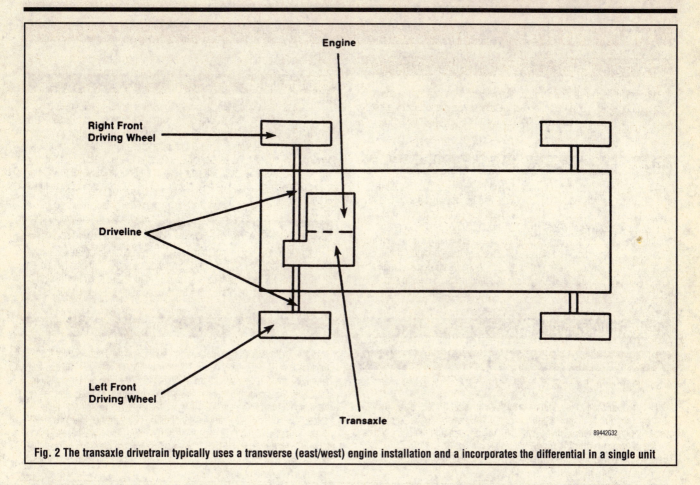

89442G32

Fig. 2 The transaxle drivetrain typically uses a transverse (east/west) engine installation and a incorporates the differential in a single unit

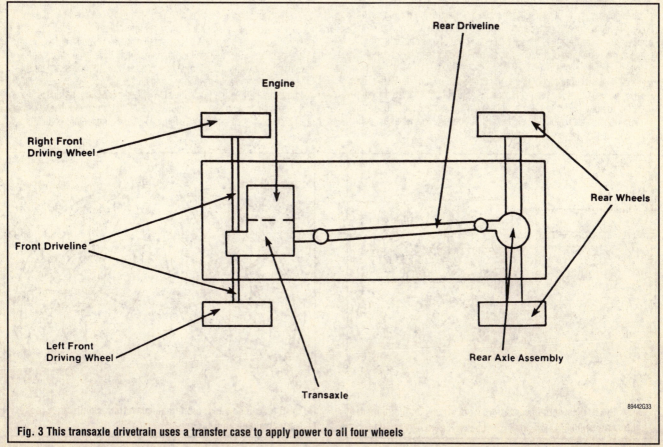

89442G33

Fig. 3 This transaxle drivetrain uses a transfer case to apply power to all four wheels

For example, at sixty miles per hour, a fifteen-inch wheel rotates at about 1340 rpm. If the crankshaft is rotating at 2200 rpm, the difference in rotating speeds is about 1.64:1. The entire drivetrain is used to accomplish this speed difference and at the same time deliver the correct power needed to maintain vehicle speed.

In a vehicle equipped with an automatic transmission, the torque converter and final drive or differential each modify speed and torque by a fixed amount. This leaves the transmission to select the proper gear range to maintain correct vehicle speed.

Transaxle

A transaxle includes the transmission and final drive differential all in one case. It is usually bolted directly to the engine and will be mounted in the car the same way as the engine, either transverse or longitudinal. It is most often used on front wheel drive vehicles, offering great savings in weight, space and cost. Front engine/front wheel drive is practical even in small vans, large sedans and wagons. In all-wheel drive designs, the necessary transfer components are often housed in the transaxle case, making it much simpler and lighter than using a separate transfer case. In some vehicles, transaxles have been used in front engine/rear wheel drive layout to optimize balance and handling. A driveshaft connects the engine flywheel at the front of the vehicle to the torque converter at the rear of the vehicle.

Transmission

The transmission is used on front engine/rear wheel or all-wheel drive layouts and requires at least one driveshaft to transmit power to a separate final drive unit (differential). On longer vehicles, the driveshaft is split and has an additional universal joint and support bearing in the center. This drivetrain layout offers great flexibility in design of the chassis and suspension and is readily adaptable to part-time and full-time all wheel drive and heavy duty applications. It is also heavy, can be expensive and takes up lots of space compared to a transaxle.

Transfer Case

Both the transmission and transaxle drivetrains can utilize an auxiliary transmission called a transfer case to provide either part-time or full-time all wheel drive. The transfer case is usually mounted to the transmission or transaxle and splits the power flow to both the front and rear wheels.

DRIVETRAIN COMPONENTS

Torque Converter

▶ **See Figures 4, 5 and 6**

The predecessor to the automotive torque converter first appeared in Germany in 1908 as a simple fluid coupling. This consists of two properly shaped components, one driven by an engine or motor and the other connected to the load, with a fluid trapped between them. The driven component puts a fluid into motion, and the moving fluid causes the other component to turn the load.

Its main advantage over a mechanical coupling is that it can be engaged or disengaged simply by changing the speed of the driving force, making fluid couplings a useful replacement for a manually operated clutch. Another major advantage is the excellent damping of torsional vibration. As noted at the beginning of this chapter, a fluid coupling can be a type of transmission all by itself.

Work on fluid coupling continued in Europe and by the late 1930's, they were being used to drive superchargers on large aircraft engines. Development led to a fluid drive torque multiplier with slip control for adjusting the amount of boost and a lock-up clutch for full boost. At this point the coupling and multiplier were two separate units.

In 1939, Chrysler Corporation introduced the first fluid coupling for the transmission. The company described the new "Fluid Drive" as similar to two halves of a metallic grapefruit from which the fruit has been removed without damage to the sectional membrane. The impeller (pump) half was attached to the engine flywheel. The turbine half was attached to a conventional single disc clutch. The clutch, in turn, drove the input shaft of the manual transmission.

The Fluid Drive didn't replace the manual clutch and shift lever with an automatic shifting system. It did, however, reduce the number of times that the driver had to shift. He could eliminate shifting altogether in the forward speeds if he chose to do so.

As you can see from its meager beginnings, the torque converter is the one component that makes the automatic transmission possible. It eliminates the driver-operated clutch, allowing fully automatic operation of the transmission once the driver has selected a forward gear. Over the history of automobiles, other self-shifting transmissions have been successfully developed, but without the torque converter, it was still necessary for the driver to take some action when starting and stopping.

In the late 1940's, the Buick division of General Motors produced the first automatic transmissions with fluid coupling/torque multiplier all in one unit. With continued refinements in the U.S., the simple fluid coupling dis-

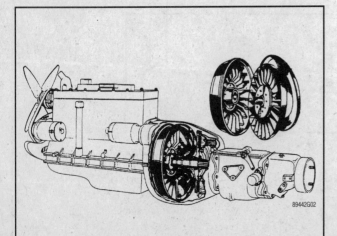

Fig. 4 The earliest torque converters were simply mated to manual transmissions

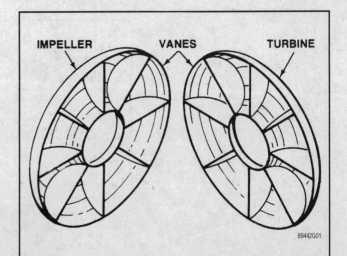

Fig. 5 As can be seen, the torque converter actually does look like a mechanical grapefruit

Fig. 6 Modern day torque converters are composed of several components which assist in multiplying torque

appeared from automatic transmissions by the mid 1960's and true torque converters became part of every automatic transmission. Since then, larger heavy duty torque converters have also been developed, making it possible to use automatic transmissions in almost any vehicle.

A detailed description of torque converter operation appears in Chapter 3.

Planetary Gear Set

▶ See Figure 7

The ability of the torque converter to multiply torque is limited. The converter tends to be more efficient when the engine is operating at relatively high speeds. Therefore, a planetary gearbox is used to carry the power output of the torque converter to the driveshaft(s).

In a manual transmission, the gears are arranged along two shafts mounted next to each other along parallel axes. Shifting gears actually moves certain components forward and back along the shafts and gear speeds must change to engage each other. In a planetary gear set , all gears are arranged around a single shaft axis. This allows the automatic transmission components to be more compact.

A simple planetary gear set consists of one gear in the center, at least three smaller gears arranged evenly around it, and one outer ring gear

encircling them all, like the planets in orbit around the sun. All gears are in constant mesh, nothing changes its position in the gear system when "shifting gears" but the power flow through the gear system changes. With this geographic arrangement, two gear reduction ratios are available, along with direct drive, reverse and neutral, all in one set of gears.

Planetary gears are strong for their size and offer other advantages over spur gear systems. However, they are more complicated to make and changing power flows through the system requires some very sophisticated control devices. This adds weight, complexity and expense, explaining why an automatic transmission is an option and not the standard on many vehicles.

A more detailed description of planetary gears appears in Chapter 3.

Output Section

▶ See Figures 8 and 9

A transmission output section is everything after the gear train, including the output shaft and any sensors or governors that may be connected to it. In a typical Ford transmission, this includes the output shaft that has the speedometer gear, parking gear, governor spline and output spline that connects to the driveshaft. It is enclosed in an extension housing with the appropriate bearings, bushings and mounts.

Fig. 8 This extension housing is being removed from a Ford AOD transmission

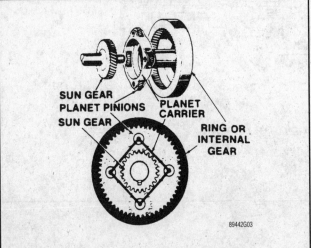

SUN GEAR
PLANET PINIONS
SUN GEAR
PLANET CARRIER
RING OR INTERNAL GEAR

Fig. 7 Planetary gears work in a similar fashion to manual transmission gears, but are composed of three parts

Speedometer Gear

Output Shaft

Fig. 9 The output shaft and speedometer gear can be easily seen once the extension housing is removed

A transaxle output section includes gears or a chain driving a shaft that transfers the power to the differential, which is in the same case. On full-time all-wheel drive vehicles, there is also a transfer section, which may include another differential, clutch and control devices.

On transmissions, the output components attached to the output shaft are easily accessible after removing the extension housing. However, the output shaft must be removed by disassembling the transmission. On transaxles, the output section is generally not serviceable without disassembling the unit.

Transfer Case

♦ **See Figures 10, 11 and 12**

A transfer case is an auxiliary transmission used to apply transmission output shaft torque to both the front and rear driveshafts. Engine torque from the transmission output shaft is fed into the transfer case. through a set of gears, chain and gears or a viscous coupling, power is transmitted to each driveshaft.

On part-time four wheel drive transfer cases, the output shaft is made of two parts. The long section is driven by the sliding splined clutch. This long section of the output shaft in turn will drive the rear driveshaft. The shorter front section is connected to the long section only when the four wheel drive sliding splined clutch is moved into engagement with the splined end of the short shaft.

When two wheel drive is desired, the short output shaft is disconnected by moving the four wheel sliding clutch out of engagement. The long shaft will then drive the rear driveshaft but will not apply torque to the front. Four wheel drive mode is accomplished by merely moving the four wheel clutch into engagement with the short shaft.

Full-time four wheel drive transfer cases permit the vehicle to constantly operate in four wheel drive on any and all road surfaces. A drive chain transmits power from the drive sprocket to the differential sprocket. A differ-

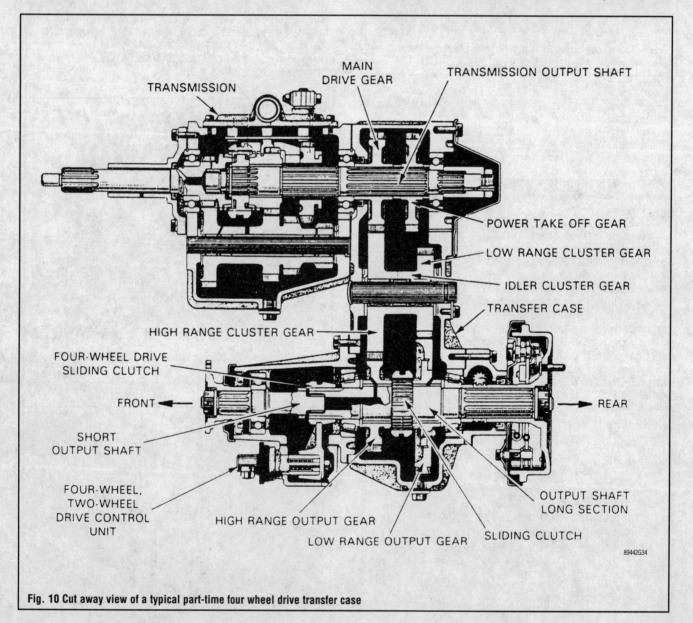

Fig. 10 Cut away view of a typical part-time four wheel drive transfer case

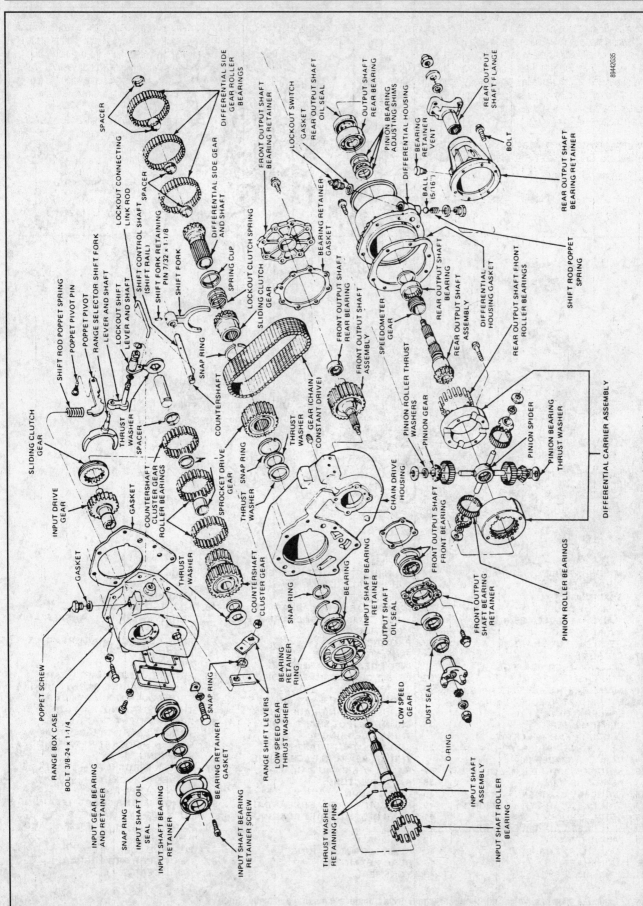

Fig. 11 Exploded view of a typical full-time four wheel drive transfer case using a mechanical differential

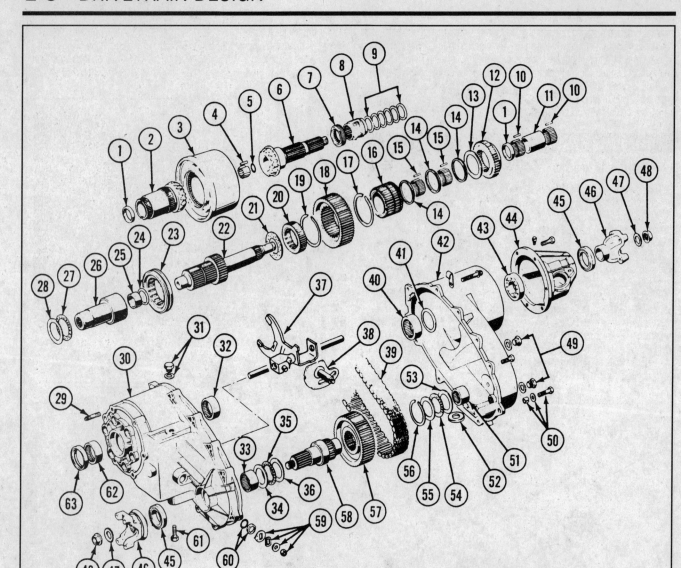

1. **MAINSHAFT BEARING SPACERS (SHORT) (2)**
2. **SIDE GEAR**
3. **VISCOUS COUPLING**
4. **MAINSHAFT PILOT BEARINGS**
5. **MAINSHAFT O-RING**
6. **REAR OUTPUT SHAFT**
7. **OIL PUMP**
8. **SPEEDOMETER DRIVE GEAR**
9. **DIFFERENTIAL SHIMS**
10. **MAINSHAFT NEEDLE BEARINGS (82)**
11. **MAINSHAFT NEEDLE BEARING SPACER (LONG) (1)**
12. **CLUTCH GEAR**
13. **CLUTCH GEAR THRUST WASHER**
14. **SPROCKET CARRIER NEEDLE BEARING SPACER (3)**
15. **SPROCKET CARRIER NEEDLE BEARINGS (120)**
16. **SPROCKET CARRIER**
17. **SPROCKET CARRIER SNAP RING**
18. **DRIVE SPROCKET**
19. **SPROCKET CARRIER SANP RING**
20. **SPLINE GEAR**
21. **MAINSHAFT THRUST WASHER**
22. **MAINSHAFT**
23. **CLUTCH SLEEVE**
24. **MAINSHAFT THRUST WASHER**
25. **MAINSHAFT BUSHING**
26. **INPUT GEAR**
27. **INPUT GEAR THRUST BEARING**
28. **INPUT GEAR THRUST BEARING RACE**
29. **MOUNTING GEAR**
30. **FRONT CASE**
31. **PLUG AND WASHER**
32. **INPUT GEAR REAR BEARING**
33. **FRONT OUTPUT SHAFT FRONT BEARING**
34. **FRONT OUTPUT SHAFT FRONT THRUST BEARING RACE (THICK)**
35. **FRONT OUTPUT SHAFT FRONT THRUST BEARING**
36. **FRONT OUTPUT SHAFT FRONT THRUST BEARING RACE (THIN)**
37. **RANGE FORK AND RAIL**
38. **RANGE SECTOR**
39. **DRIVE CHAIN**
40. **REAR OUTPUT SHAFT BEARING**
41. **REAR OUTPUT SHAFT BEARING SEAL**
42. **REAR CASE**
43. **REAR OUTPUT BEARING**
44. **REAR RETAINER**
45. **YOKE SEAL**
46. **YOKE**
47. **SEAL WASHER**
48. **YOKE NUT**
49. **FILL AND DRAIN PLUGS**
50. **ALIGNMENT DOWEL, WASHER AND BOLT**
51. **FRONT OUTPUT SHAFT REAR BEARING**
52. **MAGNET**
53. **FRONT OUTPUT SHAFT REAR THRUST BEARING RACE (THICK)**
54. **FRONT OUTPUT SHAFT REAR THRUST BEARING**
55. **FRONT OUTPUT SHAFT REAR THRUST BEARING RACE (THIN)**
56. **DRIVEN SPROCKET RETAINING SNAP RING**
57. **DRIVEN SPROCKET**
57. **DRIVEN SPROCKET**
58. **FRONT OUTPUT SHAFT**
59. **RANGE SECTOR SHAFT RETAINING LOCKNUT AND WASHERS**
60. **RANGE SECTOR SHAFT SEAL AND RETAINER**
61. **POSITIVE LOCK DETENT BOLT**
62. **INPUT GEAR FRONT BEARING**
63. **INPUT GEAR SEAL**

89442G36

Fig. 12 Exploded view of a typical full-time four wheel drive transfer case using a viscous clutch

ential unit inside the case apply driving torque to both the front and rear transfer case output shafts which in turn apply power to the driveshafts.

Two types of differentials are used in full-time four wheel drive—a Mechanical Limited Slip Differential and a Viscous Coupling. Each allows the front and rear driveshafts to rotate at different speeds while still applying power. This difference in rotation speed prevents damage to the drivetrain if the vehicle should be driven on dry land.

Differential

▶ See Figures 13 and 14

The power from the transmission or transaxle is distributed to the drive wheels through the differential. As its name implies, it allows the drive wheels to turn at different speeds to facilitate turning the vehicle. The most common designs in use today are a simple open differential, consisting of a ring gear, two pinion gears and two side gears that engage the output shafts. Open differentials are relatively trouble free, inexpensive to manufacture and work well in most applications. The main disadvantage is there can be a 100 percent torque split that allows one output shaft to be held still by the load (tire on solid ground) while all of the available power goes to the no-load side (tire in the snow). Electronic traction control that utilizes the braking system is one method devised to over come this problem, but it is not always the best solution.

Other types of differential have been developed to limit the amount of torque difference allowed between the two output shafts. Earlier designs used clutches that allowed a limited, pre-set amount of torque split (limited

slip differential). More recent developments include locking differentials that lock the output shafts together, forcing a complete 50-50 torque split. These are often used in full-time all wheel drive systems on the center and rear differentials, but the driver must engage the lock-up when desired and

Fig. 14 This transaxle differential is buried deep inside the case and can only be serviced by disassembling the transaxle completely

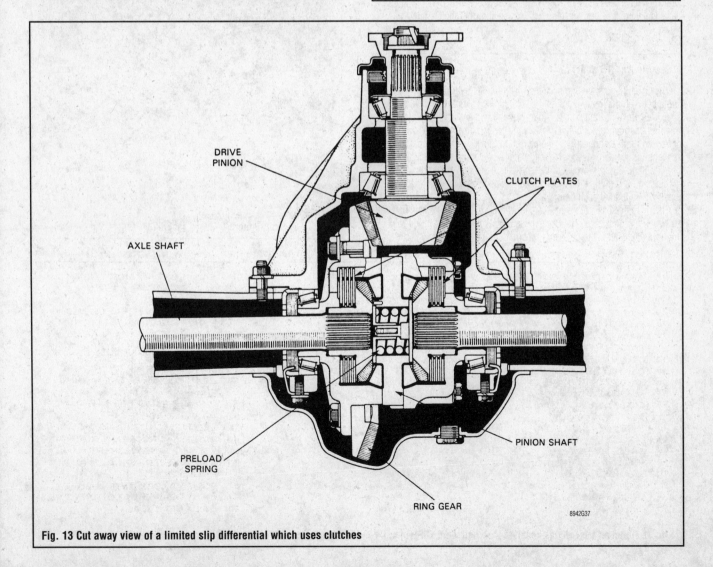

DRIVE PINION

CLUTCH PLATES

AXLE SHAFT

PRELOAD SPRING

PINION SHAFT

RING GEAR

8942G37

Fig. 13 Cut away view of a limited slip differential which uses clutches

remember to unlock the differentials on dry roads or risk damage to the driveline. Another center differential design is the viscous coupling, generally used on more sophisticated high-performance European cars. See the section on Torque Converters for more details on viscous couplings.

Other designs, such as the Torrsen® and Quaif® differentials, use a completely different type of gear shape and gear-train design. These units are capable of infinite torque split from fifty to 100 percent, sending all the power to the load side and none to the wheel in the snow. These can be difficult to manufacture, tend to be expensive and are most often used in racing or other high power applications. However they are also quite strong and able to accomplish fully automatic torque split with no locks, electronic controls or intervention from the driver. Some production vehicles are built with one of these designs as standard equipment, and they are also available as an after market item for many vehicles.

Driveshafts and Halfshafts

HALFSHAFTS

▶ **See Figure 15**

On all front wheel drive vehicles, power is transferred from the differential to the wheels using halfshafts. Rear wheel drive cars with independent rear suspension also use halfshafts as opposed to a solid axle assembly. A halfshaft is just that, an axle shaft that spans only half the total axle distance. It connects the differential to only one drive wheel, with some type of flexible joint at each end. They have been used for some time but only became common with the development of the constant velocity joint, also known as the CV-joint.

CV-JOINTS

▶ **See Figures 16 and 17**

The CV-joint gets it name from the way it transfers torque. At any the angle between the input and output shafts on each side of the joint, shaft speed remains constant. The CV-joint is the one innovation that made front wheel drive practical for almost any weight or type of vehicle. In low power or limited motion applications, the uneven torque delivery and changing suspension geometry associated with other types of halfshaft joint are less of a problem. But front wheel drive halfshafts operate at angles as high as 40 percent for turning and 20 percent for suspension movement. Universal joints would bind at such radical angles.

There are five basic types of CV-joint, two used at the outer halfshaft end and three for inner ends. The Rzeppa fixed joint and the fixed tripod joint are normally used at the outer end, where the drive deflection angles are

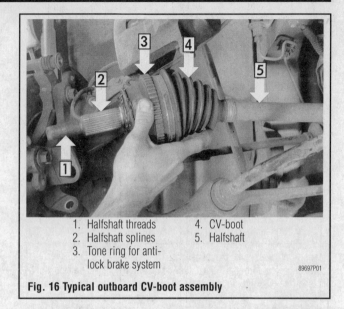

1. Halfshaft threads
2. Halfshaft splines
3. Tone ring for anti-lock brake system
4. CV-boot
5. Halfshaft

89697P01

Fig. 16 Typical outboard CV-boot assembly

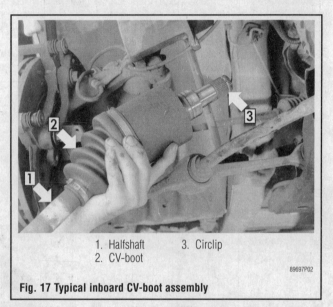

1. Halfshaft
2. CV-boot
3. Circlip

89697P02

Fig. 17 Typical inboard CV-boot assembly

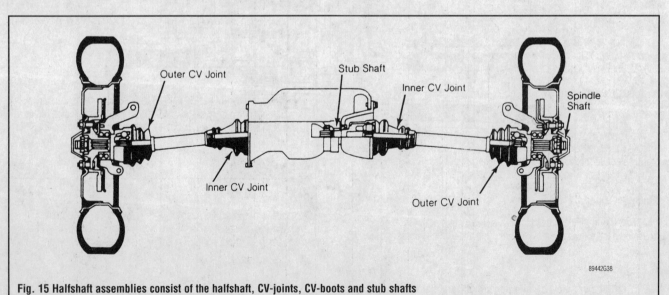

Fig. 15 Halfshaft assemblies consist of the halfshaft, CV-joints, CV-boots and stub shafts

89442G38

greater. At the inner end, a double offset joint, plunging tripod joint or cross groove plunge joint allows more axial motion. Each type is further developed for a specific application as determined by vehicle weight, performance level, etc. While most CV-joints can be disassembled for cleaning and lubrication, only the rubber boot can replaced separately. In every case, a worn or damaged CV-joint must be replaced as an assembly.

Rzeppa CV-Joint

▶ See Figure 18

This is an outer joint, also called a ball type or Birfield joint. The inner race is attached to the halfshaft. The cage is pressed into the outer race. Six balls ride in the inner race and are kept in place by slots or holes in the cage. As the shaft turns, the inner race and balls turn with it, forcing the cage and outer race to turn. There is no axial movement available, the halfshaft cannot move in and out on this joint.

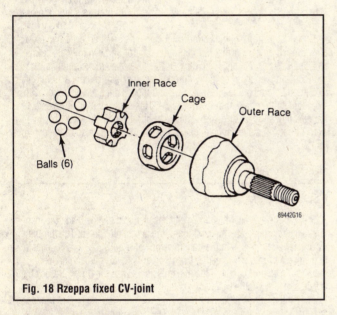

Fig. 18 Rzeppa fixed CV-joint

Fixed Tripod CV-Joint

▶ See Figure 19

This is an outer joint commonly used on smaller European cars. Three roller and ball assemblies are mounted on a hub or "spider" that is splined to the halfshaft. The housing is attached to the wheel hub or stub axle and the rollers ride in the grooves of the housing. There is not supposed to be any axial movement but the often only thing holding this joint together is the halfshaft being attached to the transaxle. Should the suspension fail and allow the wheel to move too far away from the transaxle, the spider will be pulled out of the housing.

Double Offset CV-Joint

▶ See Figure 20

This is an inner joint used on many domestic and Asian import vehicles. It is based on the Rzeppa, but modified to allow plunging or axial movement. The outer race is extended and the longer grooves in the outer race allow the halfshaft to move in and out with suspension travel.

Plunging Tripod CV-Joint

A fixed tripod joint can be made into a plunging joint by lengthening the housing so the rollers can move axially. This allows the necessary halfshaft

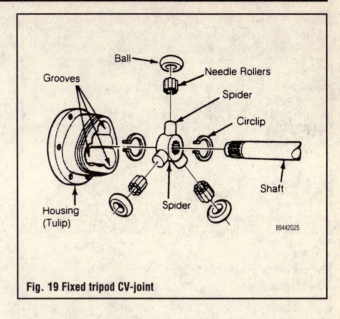

Fig. 19 Fixed tripod CV-joint

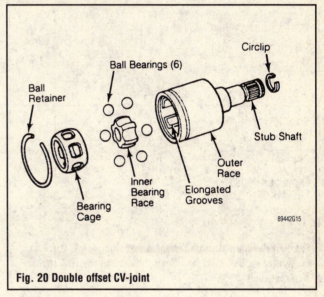

Fig. 20 Double offset CV-joint

movement for suspension travel. There is usually some kind of retainer ring to prevent the joint from pulling apart.

Cross Groove CV-Joint

▶ See Figure 21

The cross groove joint has six balls in a cage with inner and outer races like the Rzeppa joint. However the grooves in the races are angled rather than straight, and they're long enough to allow some axial movement. This joint is common on German cars.

DRIVESHAFTS

▶ See Figure 22

Vehicles with front engine/rear-wheel drive use driveshafts to transfer torque from the transmission to the rear wheels and, if equipped, from the transfer case to the front wheels. A flexible joint, usually a universal joint at

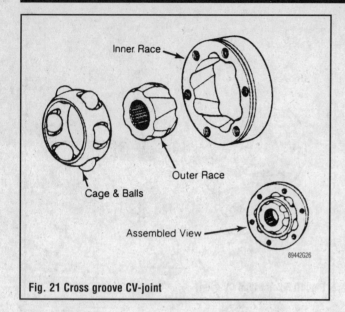

Fig. 21 Cross groove CV-joint

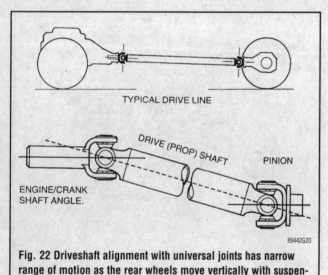

Fig. 22 Driveshaft alignment with universal joints has narrow range of motion as the rear wheels move vertically with suspension movement

each end, allows the necessary change in drive angle. With universal joints, when the deflection angle increases, torque is transferred unevenly and the output side of the joint tends to speed up and slow down with each revolution. Shafts with universal joints (or variations) must be aligned so that at rest, the deflection is between zero and three degrees. The joints can tolerate short excursions of up to eight degrees deflection, but continued or frequent operation in such extreme range causes vibration and early joint failure.

Hydraulic Controls

VALVE BODY

▶ **See Figures 23, 24 and 25**

All gear selection in every transmission is controlled through the valve body. Made from cast iron or aluminum, the body itself has many precision bores, orifices and fluid paths and includes valves, springs, seals and control levers. This one assembly has complete control over the transmission and is one of the most sophisticated components of any vehicle. The driver has several inputs to the valve body, and in newer applications with electronic controls, the engine's control unit also has some input, but ultimate control remains with the valve body.

GEAR SELECTOR LEVER

▶ **See Figure 26**

When the driver selects a gear, the lever moves the manual valve, which supplies regulated pressure to the appropriate circuits in the valve body. In Drive, gear selection becomes automatic and is regulated according to vehicle speed and load. When the selector lever is placed into a lower gear, the manual valve selects different circuits in the valve body to "lock-out" high gear. On some transmissions, the selector lever linkage may be acting directly on the band or other apply device to prevent shifting into high gear.

On transmissions with electronic controls, the selector lever also moves a lever position sensor that reports to the control unit. On fully electronically controlled transmissions, the selector lever may move only the parking lock and a position sensor with no direct mechanical connection to the valve body. Some of the latest automatic transmissions are equipped with a manual shift-up and shift-down feature. This allows the driver to select the next higher or lower gear by moving a spring-loaded lever either forward or back in a special gate, operating switches connected to the control unit.

THROTTLE VALVE

▶ **See Figure 27**

Shift points are determined by vehicle speed and engine load. The load is largely determined by the position of the accelerator pedal. The throttle valve referred to here is in the transmission valve body, but it is connected to the engine throttle by a cable. Its movement provides an engine load input to the control circuit of the transmission. When the driver presses the accelerator, the throttle valve adjusts fluid pressure in the appropriate circuit to delay the next upshift or cause a downshift, allowing the engine to reach a higher rpm and develop more power. The cable connecting these two throttle valves is often called the T.V. cable, and the pressure adjustment is often called T.V. pressure. Not all automatic transmissions are equipped with a throttle valve.

MODULATOR VALVE

▶ **See Figure 28**

The modulator is used to sense engine load and adjust shift points appropriately. It senses load directly with a diaphragm connected to intake manifold. When manifold pressure is low (high vacuum, throttle closed), the diaphragm moves the valve and shift points will happen at a lower engine speed. When the throttle is opened and manifold pressure increases, the spring pushes the modulator valve in and shift points are adjusted higher. This system also compensates automatically for high altitudes because manifold pressure is naturally reduced, requiring higher rpm shift points to compensate for the loss of engine power.

SHIFT VALVE

▶ **See Figure 29**

The shift valves control fluid pressure supplied to the actuators. They are inside the valve body and are controlled by springs and various fluid pres-

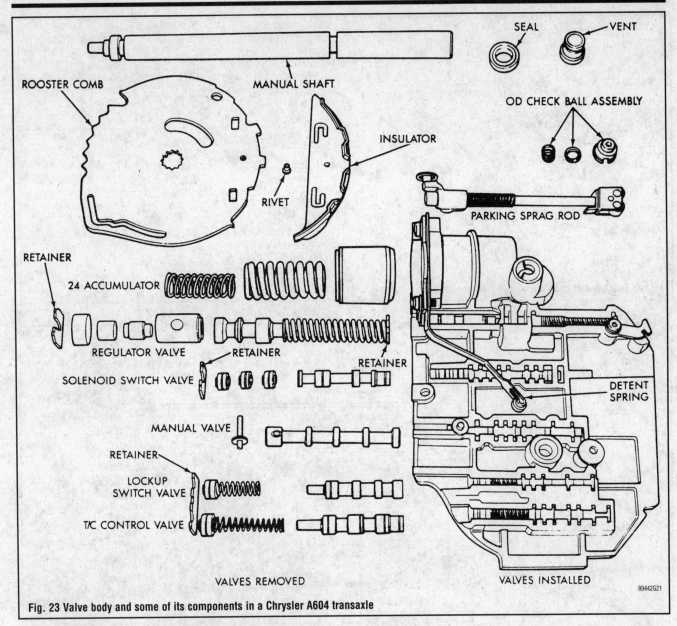

ROOSTER COMB

MANUAL SHAFT

SEAL

VENT

OD CHECK BALL ASSEMBLY

INSULATOR

RIVET

PARKING SPRAG ROD

RETAINER

24 ACCUMULATOR

REGULATOR VALVE

RETAINER

RETAINER

SOLENOID SWITCH VALVE

DETENT SPRING

MANUAL VALVE

RETAINER

LOCKUP SWITCH VALVE

T/C CONTROL VALVE

VALVES REMOVED

VALVES INSTALLED

Fig. 23 Valve body and some of its components in a Chrysler A604 transaxle

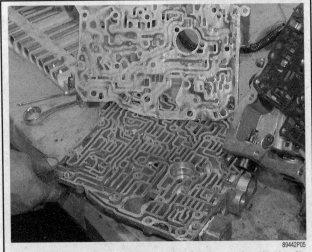

Fig. 24 The channels in the valve body act as a hydraulic circuit board and control fluid flow through the transmission

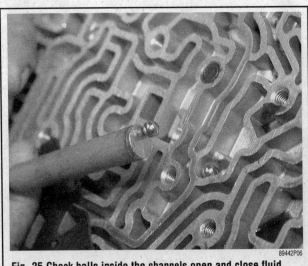

Fig. 25 Check balls inside the channels open and close fluid passages to redirect fluid

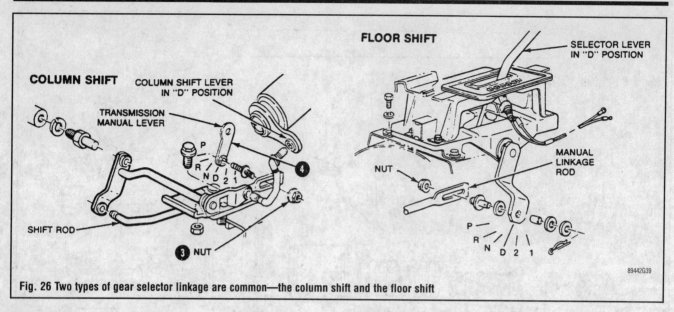

Fig. 26 Two types of gear selector linkage are common—the column shift and the floor shift

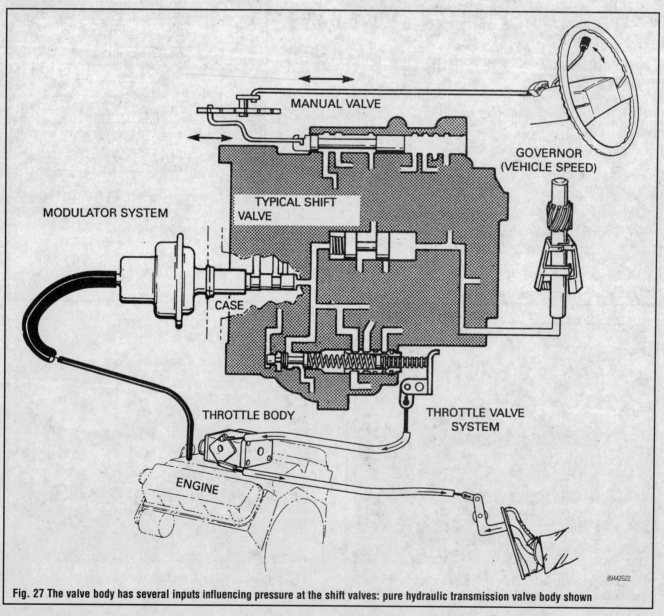

Fig. 27 The valve body has several inputs influencing pressure at the shift valves: pure hydraulic transmission valve body shown

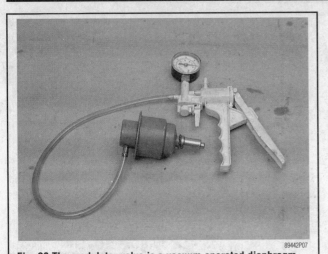

Fig. 28 The modulator valve is a vacuum operated diaphragm which senses engine load. It is shown here being tested with a hand vacuum pump

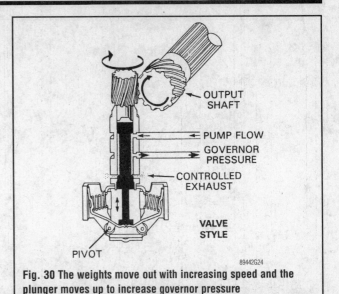

Fig. 30 The weights move out with increasing speed and the plunger moves up to increase governor pressure

sures. In fully electronically controlled transmissions, the shift valves are actually solenoid valves.

GOVERNOR

▶ **See Figures 30 and 31**

The governor senses vehicle speed. It is geared to turn with the transmission output shaft, often to the same gear as the mechanical speedometer drive. A hydraulic governor is purely mechanical, with springs and fly weights used to control the size of an orifice. All fluid pressure in the transmission is generated by the main pump. The governor essentially provides a second, speed sensitive pressure circuit in order to have pressure differentials for moving valves and actuators.

Electronic Controls

VALVE BODY CONTROLS

▶ **See Figures 32 and 33**

On almost every vehicle with electronic fuel injection, the transmission is controlled to some extent using electronic sensors and solenoid valves. Many of the early automatic-overdrive transmissions are hydraulically controlled three-speed designs plus an overdrive gear activated with a solenoid in the valve body.

Modern transmissions have been designed from the start to be fully electronically controlled and operated through solenoid valves. There is still a

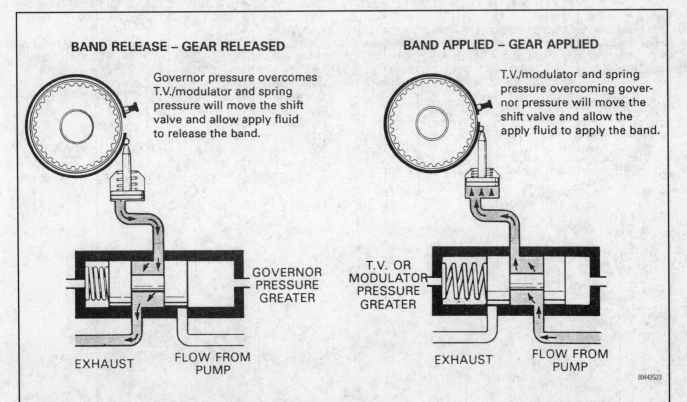

BAND RELEASE – GEAR RELEASED

Governor pressure overcomes T.V./modulator and spring pressure will move the shift valve and allow apply fluid to release the band.

GOVERNOR PRESSURE GREATER

EXHAUST   FLOW FROM PUMP

BAND APPLIED – GEAR APPLIED

T.V./modulator and spring pressure overcoming governor pressure will move the shift valve and allow the apply fluid to apply the band.

T.V. OR MODULATOR PRESSURE GREATER

EXHAUST   FLOW FROM PUMP

Fig. 29 Shift valves control fluid pressures to actuators which operate bands

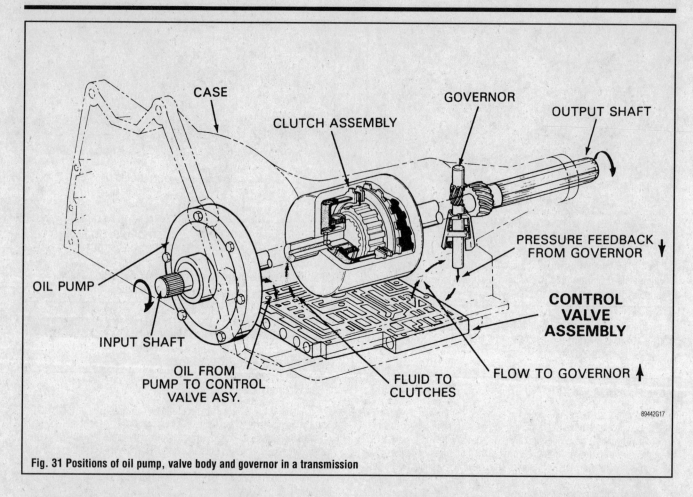

CASE

CLUTCH ASSEMBLY

GOVERNOR

OUTPUT SHAFT

PRESSURE FEEDBACK
FROM GOVERNOR ↓

**CONTROL
VALVE
ASSEMBLY**

OIL PUMP

INPUT SHAFT

OIL FROM
PUMP TO CONTROL
VALVE ASY.

FLUID TO
CLUTCHES

FLOW TO GOVERNOR ↑

89442G17

Fig. 31 Positions of oil pump, valve body and governor in a transmission

valve body, but instead of the internal governor and T.V. cable supplying speed and load information to a hydraulic control circuit, the PCM (Powertrain Control Module) interprets data from a vehicle speed sensor and a throttle position sensor, then operates solenoid valves to shift gears. In the current generations of powertrain management systems, engine and transmission control is integrated into a single computer to produce smooth, precise shifts under all speed/load conditions, something not possible with pure hydraulic controls.

SHIFT LEVER CONTROLS

▶ **See Figures 34 and 35**

All new vehicles sold with automatic transmissions in North America after 1989 are equipped with a shift lever lock. This was brought about by

several highly publicized accidents resulting from driver error. The lock requires the driver to have the brake pedal depressed before the shift lever will move out of Park.

In most vehicles, the shift lock is a simple plunger that fits into a slot in the lever only when the lever is in the Park position. A solenoid, usually activated with the brake light circuit, retracts the plunger and allows the lever to move. There is also some method of manually retracting the plunger in the event of electrical failure, such as a dead battery. Unfortunately on some vehicles, manual operation requires partial disassembly of the center console to reach the plunger.

All the newer electronically controlled transmissions also have a manual lever position sensor (MLPS) that signals the PCM the position of the lever. This sensor replaces the neutral/start switch and reverse light switch and the PCM uses the position information to know which gear the driver has selected.

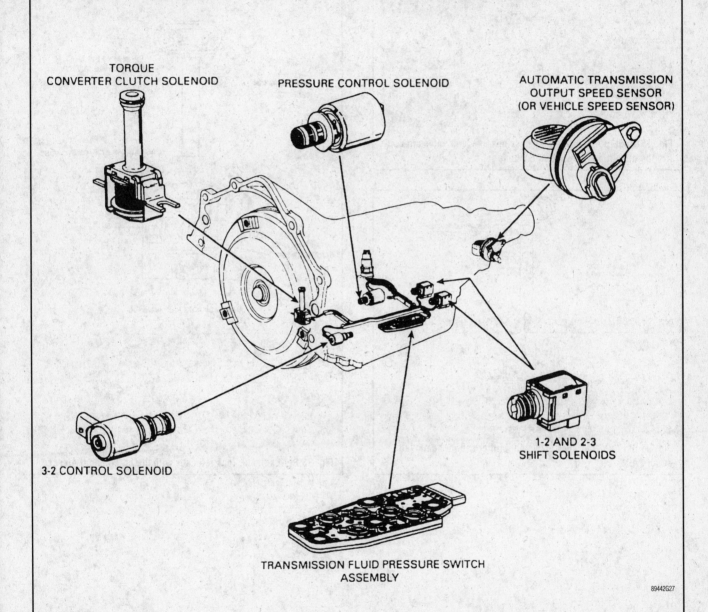

TORQUE
CONVERTER CLUTCH SOLENOID

PRESSURE CONTROL SOLENOID

AUTOMATIC TRANSMISSION
OUTPUT SPEED SENSOR
(OR VEHICLE SPEED SENSOR)

3-2 CONTROL SOLENOID

1-2 AND 2-3
SHIFT SOLENOIDS

TRANSMISSION FLUID PRESSURE SWITCH
ASSEMBLY

89442G27

Fig. 32 Control components in a General Motors 4L60E transmission

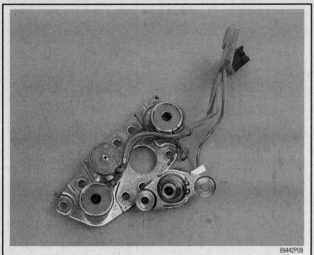

Fig. 33 Modern electronic transmissions use shift valves to control the upshift and downshift functions

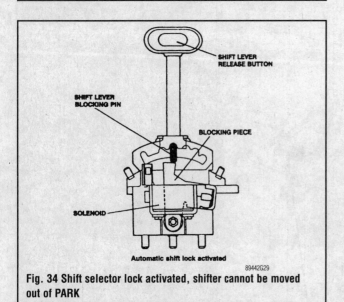

SHIFT LEVER RELEASE BUTTON

SHIFT LEVER BLOCKING PIN

BLOCKING PIECE

SOLENOID

Automatic shift lock activated

Fig. 34 Shift selector lock activated, shifter cannot be moved out of PARK

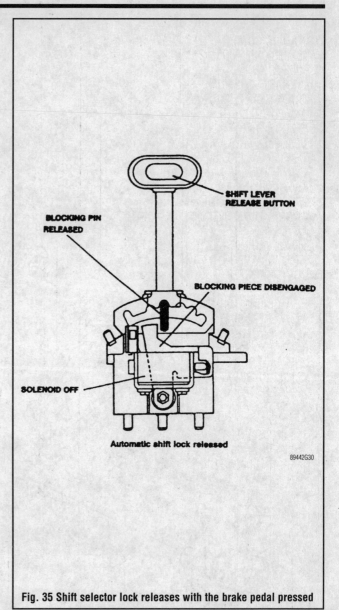

SHIFT LEVER RELEASE BUTTON

BLOCKING PIN RELEASED

BLOCKING PIECE DISENGAGED

SOLENOID OFF

Automatic shift lock released

Fig. 35 Shift selector lock releases with the brake pedal pressed

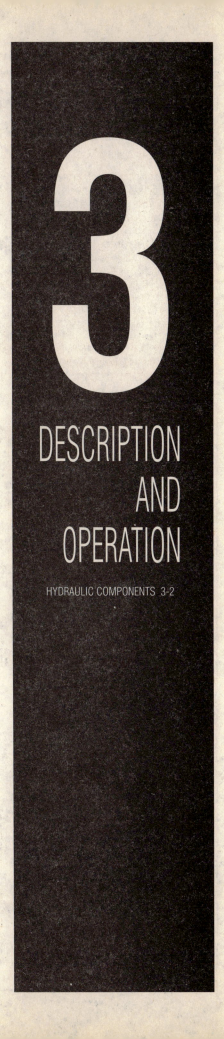

3

DESCRIPTION AND OPERATION

HYDRAULIC COMPONENTS 3-2

HYDRAULIC COMPONENTS

Transmission Fluid

♦ **See Figures 1, 2 and 3**

Transmission fluid is a highly refined petroleum product, which is why it is often referred to as oil. A red dye is added specifically to help differentiate it from other oils and fluids. Its primary functions include cooling, lubrication, hydraulic pressure application and the fluid component in a fluid coupling.

There is also a very extensive additive package, as much as 10 percent of the total volume that imparts specific qualities to the fluid. One of the additives is a friction modifier. This may sound strange because of the need for lubrication, but clutches operate submerged within this fluid. Other qualities the additive package imparts are needed to:

- improve fluid flow
- swell and lubricate seals
- stabilize fluid viscosity over a wide temperature range
- inhibit rust and corrosion of metal components
- resist fluid foaming and oxidation
- suspend impurities (detergents)

Fig. 1 High friction modifiers are available to turn regular Dexron® III/Mercon® fluid into highly friction-modified fluid

Fig. 2 Always use the correct fluid that meets the transmission/transaxle manufacturer's specifications for quality as well as type

Fig. 3 Transmission/transaxle coolers come in many shapes and sizes to fit all vehicles. These are manufactured by B & M

Consistant performance with large changes in temperature may be the most important quality of a transmission fluid . Normally, transmission fluid operates at about the same temperature as engine coolant temperature. In fact, there is a coolant-to-transmission fluid heat exchanger in most car and truck radiators. However, under heavy duty use, such as when towing or in heavy city traffic, fluid temperatures can reach 300°F. Vehicles that often see this kind of service are usually equipped with an additional transmission/transaxle cooler, usually a small fluid-to-air heat exchanger mounted somewhere in the air stream.

The fluid developed for use in the earliest automatic transmissions was designated Type A, which was modified several times as demands changed. By the mid 1960's certain Ford and JATCO transmissions required unique qualities in the fluid and Type F was introduced to be used specifically in those transmissions.

Mercon is another type of transmission fluid developed overseas that is used in almost every import. While there are some minor differences, Dexron® III/Mercon can be substituted for Mercian if the later is not available. Later developments of both Dexron and Mercon will eliminate the need for fluid change as a regular maintenance item.

Today there are several different types of transmission fluid . The difference between these fluids is the additive package, primarily the friction modifiers. While any of these will work in any transmission/transaxle (for at least the short term), it is important to use the correct fluid. Using the wrong type will compromise shift quality and eventually damage the transmission. Make sure to use only the fluid specified by the vehicle manufacturer any time fluid is changed or added. In the absence of such specific information, the following can provide a basic application outline.

Type F can be used in the following Ford applications only:
- All models 1964 and earlier
- C3, 1974–1980
- C4, 1964–1979
- C6, 1966–1976
- FMX, 1968–1981

Dexron® III®/Mercon:
- Any previous Dextrin application, including all GM and most imports
- All Ford beginning with 1985

Mopar 7176:
- All Chrysler products with a lock-up torque converter

While there are many fluids on the market, it is important to use the correct fluid that meets the transmission/transaxle manufacturer's specifications for quality as well as type. Even if the type is correct, using poor quality fluid will definitely effect the shift quality and longevity of a transmission.

➡It is important to remember that some vehicles require specific transmission fluid s. Not using these fluids will effect shift quality and may damage the transmission.

Oil Pumps

▶ **See Figures 4 and 5**

Hydraulic pressure for operating the transmission/transaxle comes from a dedicated oil pump. While the torque converter impeller is also referred to as a pump, the fluid flow and pressure it generates are used only within the torque converter, not by the rest of the transmission. Fluid does flow into and out of the torque converter, but the movement of that fluid is generated by the oil pump, not the impeller.

The oil from the pump flows immediately to a pressure regulator, which maintains a constant line pressure. All control functions of the transmission/transaxle are operated with full or partial line. There are several pressure taps on the outside of most transmission/transaxle cases for reading line pressure and the various control pressures under specific conditions, but all pressure ultimately comes from the pump.

There are two ways commonly used to drive the oil pump. In most transmissions, the pump is driven directly by the torque converter hub.

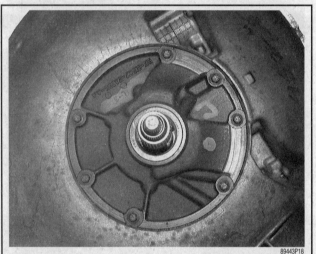

Fig. 4 Transmission fluid pumps are usually located in the bell housing behind the torque converter

Fig. 5 Pressure taps are usually located on the side of the transmission/transaxle and are used in diagnosing pump problems

The pump engages two slots or flats in the hub and turns with the impeller at engine crankshaft speed. This generally makes installing the torque converter somewhat tricky because the lugs cannot be seen when the converter is slid into the bell housing. There is however, a distinct sound and feel when the lugs align the drive engages.

The other drive method is common on transaxles. A hex or splined shaft protrudes from the front of the pump and fits into a matching recess in the torque converter impeller. It turns at crankshaft speed and is generally quite easy to assure proper engagement.

IX ROTARY LOBE PUMP

▶ **See Figure 6**

The lobe pump is sometimes referred to as a "gerotor". The prefix "IX" (the letters, not the number) refer to the internal cut lobe of the outer pump member and the external lobe of the inner member. The pump is composed of two different size rotors, one turning inside the other. The pump drive turns the inner rotor, which then drives the outer rotor by meshing its lobes. Because they are not the same size and their centers are not concentric, the lobes move closer together on one side of their rotation and separate on the other side. Where the lobes move close together, the oil is forced out through the outlet port at one end of pump chamber. As they move apart, oil is drawn into the pumping chamber.

This is a fixed displacement pump, meaning it moves the same amount of fluid for each revolution. Since the pump must supply flow and pressure even at idle speed, its supply is much more than is needed at high speed. This results in a significant amount of fluid being bypassed through the pressure control valve.

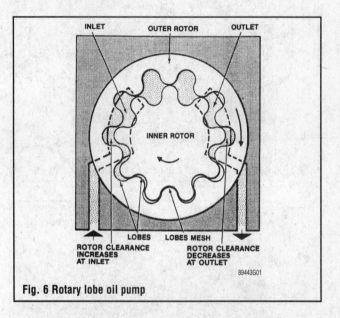

Fig. 6 Rotary lobe oil pump

ROTARY GEAR PUMP

▶ **See Figures 7, 8 and 9**

This pump is similar to the rotary lobe pump in that it uses an inner rotor turned by the pump drive to turn the larger outer rotor. The inner rotor is held in place by the pump drive, and the outer rotor is centered in a bore in the oil pump housing.

Unlike the lobe pump, these rotors are actually gears. On the "far" side of the circle where the teeth are separated, a crescent shaped vane channels the oil flow. The space at the inlet side is larger and gradually decreases, forcing the oil into a smaller space, then out the outlet port. This is also a

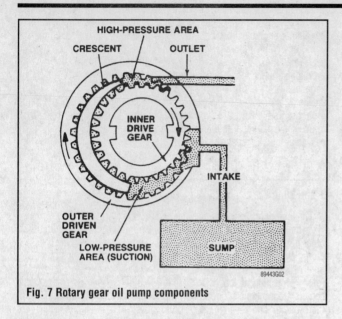

Fig. 7 Rotary gear oil pump components

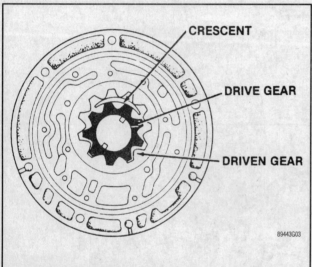

Fig. 8 Rotary gear oil pump mounted in the pump housing

Fig. 9 Once the stator support is removed, the gears are plainly visible inside the pump

fixed displacement pump, meaning its pumping capacity is far more than needed at even moderate speeds.

VANE PUMP

▶ See Figures 10, 11 and 12

The vane pump is also called a variable displacement pump. It consists of a slotted rotor with vanes that are free to move through the slots. The outer ring is called the slide and is anchored at one point on a pivot pin. As the rotor turns, oil entering the inlet port is moved around the slide towards the outlet port. The rotor is not centered in the slide, so the size of the pumping chamber decreases towards the outlet port, pressurizing the oil.

The position of the slide on the pivot determines pump output. For starting and low speed, the slide is held all the way to one side by a spring. As speed increases and flow demand is met, pressure is supplied to one side of the slide and it pivots to collapse the spring. As the slide moves toward the center of the housing bore, some of the oil is allowed to return to the inlet side of the pump. Pressure is maintained but flow is reduced. The pressure used to move the slide is controlled by the line pressure regulator valve and enters the pump at what is called the signal cavity.

Valve Body

▶ See Figures 13 and 14

The valve body contains all the control valves and circuits that supply fluid pressures to the apply devices. The body itself is a very sophisticated component, usually made of a cast metal and then precision machined and bored. The empty casting can be thought of as a hydraulic circuit board that provides the pathways for the various fluid pressures. With the valves, springs, check balls, control orifices and other components, the valve body is indeed a hydraulic computer. Programmed by the manual valve and accepting input signals from the throttle system and the governor system, the throttle body controls every function of the automatic transmission.

There are three basic input circuits to the valve body: governor pressure (road speed), throttle valve (engine load) and range selection (shift lever). The output circuits depend on the number of gears in the transmission. There is one output circuit for each apply device and the lock-up torque converter (if equipped). Since it is a mechanical device, the basic program of the valve body is determined by its physical construction. However, the program is constantly modified by the three input circuits, especially the throttle system. On the most advanced electronically controlled transmissions, several more input variables enter the equation and several more output actions are available, but the valve body as an assembly still determines the fluid pressure distribution to the apply devices.

Understanding the components that make up the valve body will go a long way towards understanding how the transmission/transaxle operates.

PRESSURE REGULATOR VALVE

▶ See Figure 15

Oil from the pump flows directly to the main pressure regulator valve, which controls basic line pressure. This is the highest continuous pressure in the whole system, and every other control operation begins with this pressure. The regulator valve may also be used to directly supply oil flow to the torque converter. It is often in the oil pump housing or may be in the valve body.

Like almost all the valves in the transmission, the pressure regulator valve is a spool valve, getting its name from the shape of the valve piston inside the bore. It operates against a spring, which ultimately determines the maximum pressure. Under some conditions, line pressure

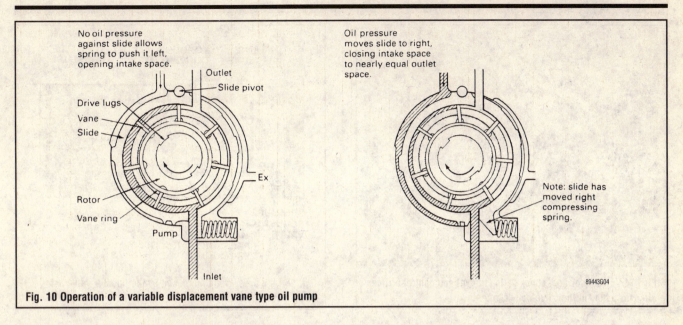

No oil pressure against slide allows spring to push it left, opening intake space.

Outlet
Slide pivot
Drive lugs
Vane
Slide
Ex
Rotor
Vane ring
Pump
Inlet

Oil pressure moves slide to right, closing intake space to nearly equal outlet space.

Note: slide has moved right compressing spring.

89443G04

Fig. 10 Operation of a variable displacement vane type oil pump

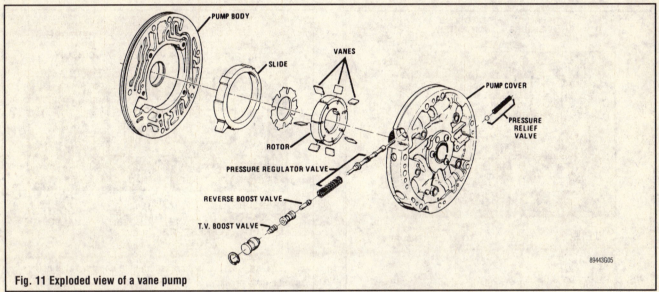

PUMP BODY
SLIDE
VANES
PUMP COVER
ROTOR
PRESSURE RELIEF VALVE
PRESSURE REGULATOR VALVE
REVERSE BOOST VALVE
T.V. BOOST VALVE

89443G05

Fig. 11 Exploded view of a vane pump

89443P20

Fig. 12 Vane pump and regulator valve installed in the side cover of a GM 440T4 transaxle

89443P21

Fig. 13 The valve body is a hydraulic circuit board controlling various fluid pressures

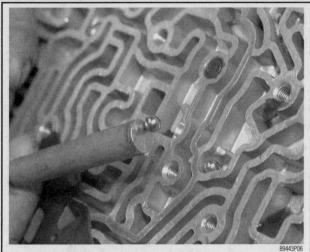

Fig. 14 Valves, springs, check balls (shown) and control orifices alter fluid flow in the valve body channels

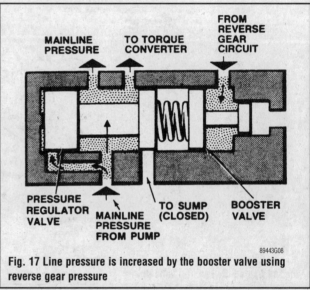

Fig. 16 Line pressure is increased by the booster valve using throttle valve pressure

Fig. 15 Schematic of pressure regulator valve

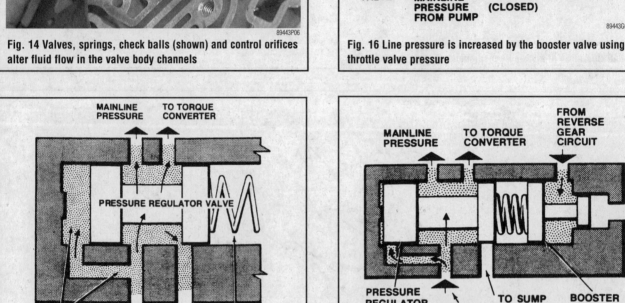

Fig. 17 Line pressure is increased by the booster valve using reverse gear pressure

may be increased by a booster valve that is part of the pressure regulator valve.

BOOSTER VALVE

▶ **See Figures 16 and 17**

For some driving conditions, it may be necessary to generate a tighter than normal hold on clutches or bands. This can be accomplished by increasing the fluid pressure at the apply device. The booster valve allows a temporary increase in main line pressure for this purpose. Typically, fluid pressure is supplied to the spring side of the regulator valve, providing an assist to the spring and increasing line pressure by a predetermined amount. Depending on the design or application, the extra pressure can be supplied by several circuits in the valve body, but the throttle valve is always one of them because it "reads" engine load.

There can be more than one level of increase in line pressure. By making the booster valve with more than one size piston for the auxiliary pressure to work against, there can be different levels of assist for the main regulator spring. Assist pressure is supplied by different circuits in the valve body, so the line pressure increase can be adjusted for different functions.

RELAY VALVE

The function of a hydraulic relay valve is much like that of an electrical relay valve: to switch a circuit OFF or ON. It may be a simple two position/two circuit valve with a spring to move it to the OFF position, or it may be quite sophisticated, controlling several circuits with the movement of one spool. The shift valves and the manual valve are both relay valves, even though the manual valve is far more complex.

Manual Valve

▶ **See Figure 18**

The manual valve is moved by the driver with the shift lever. It controls most but not all functions within the transmission. Most circuits that connect directly to line pressure are operated with the manual valve, but in newer transmissions, the function of this valve is reduced. In fully electronically controlled transmissions, there may not even be a manual valve. Moving the shift lever sends commands to the powertrain control unit, which operates electric solenoid valves to control fluid pressures in the valve body circuits.

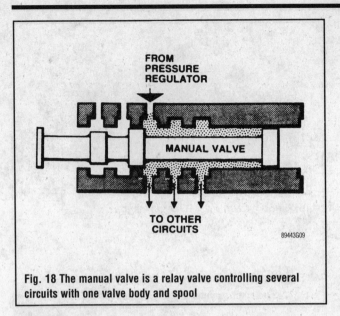

Fig. 18 The manual valve is a relay valve controlling several circuits with one valve body and spool

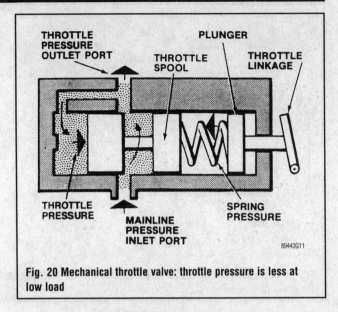

Fig. 20 Mechanical throttle valve: throttle pressure is less at low load

Shift Valve

▶ See Figure 19

The typical mechanical shift valve (as opposed to a shift solenoid in electronically controlled transmissions) is controlled by governor pressure, throttle pressure and a spring, and is used to supply fluid pressure to an apply device. As throttle pressure increases (driver pushing the pedal down), pressure holding the valve closed increases to delay the upshift. When governor pressure increases above throttle pressure, the shift valve is moved to the open position allowing the upshift. These valves are designed to move quickly, providing line pressure to the apply device instantly for a crisp upshift.

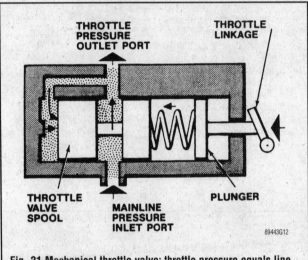

Fig. 21 Mechanical throttle valve: throttle pressure equals line pressure at full load

supplies opposes governor pressure in the shift valves to determine the shift point relative to vehicle speed and engine load. Since load is equal to the driver's demand for power, wen load is low, throttle pressure is low to allow upshifts at lower vehicle speed (governor pressure). When load is high, throttle pressure is higher, delaying upshift until governor pressure (vehicle speed) is high enough to overcome throttle pressure.

Modulator Valve

▶ See Figure 22

The modulator valve is another type of throttle valve. It supplies a shift point control pressure based on engine manifold vacuum, which is an extremely accurate and consistent way to sense engine load. It usually has its own bore in the valve body. The operation and installation of this valve is simple because there is no adjustment relative to engine throttle linkage. The vacuum signal to the diaphragm can also be controlled electronically, providing yet another layer of shift control. One disadvantage is that if the vacuum diaphragm develops a small leak, transmission fluid is drawn into the intake manifold and burned in the engine.

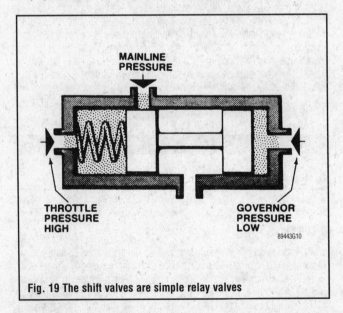

Fig. 19 The shift valves are simple relay valves

Throttle Valve

▶ See Figures 20 and 21

As its name implies, the throttle valve supplies a controlling pressure by throttling line pressure. The valve senses engine throttle position through a connection with the accelerator pedal linkage. The controlling pressure it

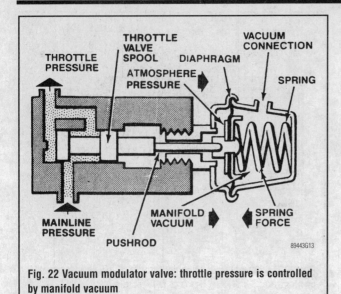

Fig. 22 Vacuum modulator valve: throttle pressure is controlled by manifold vacuum

Like the throttle valve, fluid pressure supplied by the modulator valve opposes governor pressure in the shift valves. As load increases, so does the modulator valve pressure, until governor pressure (speed) is high enough to overcome modulator pressure in the shift valve and cause an upshift.

Downshift Valve

The downshift valve always works along with the throttle valve or vacuum modulator valve to force a downshift. Also called a detent valve or kickdown, it temporarily increases throttle pressure above governor pressure on the appropriate shift valve. The upshift can be delayed by as much as ten or twenty mph, or a downshift can be forced at almost any road speed. This valve operates only when the accelerator is at wide open throttle (WOT).

GOVERNOR

▶ **See Figures 23 and 24**

The governor is geared to turn with the output shaft and is therefore directly tied to vehicle speed. In fact, many earlier automatic transmissions use the same output shaft gear to drive the mechanical speedometer. A

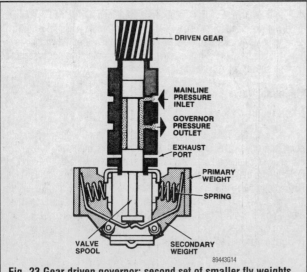

Fig. 23 Gear driven governor: second set of smaller fly weights if for fine low speed control

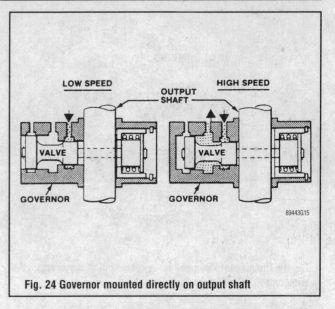

Fig. 24 Governor mounted directly on output shaft

mechanical governor consists of a set of fly weights that move out as rotational speed increases. In a transmission, each weight moves a lever which moves a spool valve in the valve body. In the gear driven governor shown here, the port that admits line pressure to the valve body is throttled by the spool. A governor may also admit full line pressure and throttle the outlet pressure port instead. In any case, the position of the spool in the body, as determined by the fly weights, controls the governor pressure.

Two different types of mechanical governor are common. Earlier designs use a separate unit mounted at right angles to the output shaft. Newer, more compact transmissions mount the governor right on the output shaft. The valve spool is connected directly to the single fly-weight on the other side of the shaft. As the shaft rotates faster, the weight moves away from the shaft to move the spool and increase governor pressure.

Governor pressure opposes throttle pressure. When throttle pressure is higher, the shift valve remains in the position dictated by its spring. When governor pressure is higher, it pushes the shift valve against the spring and throttle pressure to allow fluid into the apply device. The greater the vehicle speed, the greater the governor pressure and the more throttle pressure is required to delay the shift. Eventually governor pressure (vehicle speed) will increase beyond the highest throttle pressure available and the transmission/transaxle will shift into the next higher gear.

Apply Devices

BANDS & CLUTCHES

Bands

▶ **See Figures 25, 26, 27, 28 and 29**

To "change gears" on a planetary gear set, it is necessary to hold one element of the gear set. A band is a contracting brake wrapped around one element. More accurately, the band is wrapped around a drum that connects directly or indirectly with one or two elements of the gear set. In its simplest form, one end of the band is anchored to the transmission/transaxle housing and the other is moved by a servo. When pressurized, the servo pushes the band tighter around the drum to bring the drum to a gradual stop, causing a change in power flow through the gear.

Bands are made of flexible steel and coated on the inside surface with a friction material. Three basic designs are used for different jobs. A split band, also called a double-wrap, will hold tighter than other designs and tends to engage more smoothly, providing smoother shifts. A single-wrap band can be made wider and of heavier gauge steel. It is commonly used with more powerful servos in heavy duty applications. Single-wrap bands can also be very light and thin for very light duty, low cost applications.

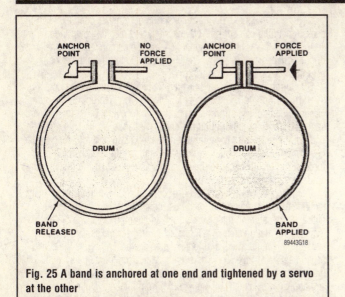

Fig. 25 A band is anchored at one end and tightened by a servo at the other

Fig. 28 This band was removed from a Ford AOD transmission. You can see the drum it holds in the background

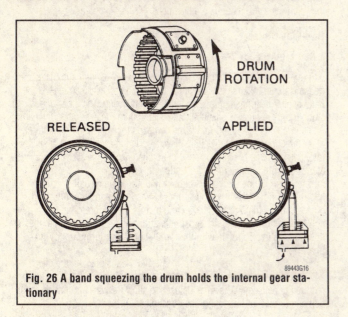

Fig. 26 A band squeezing the drum holds the internal gear stationary

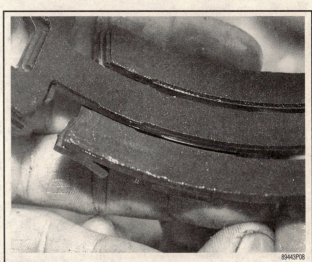

Fig. 29 The friction material on this band has taken quite a beating in normal service

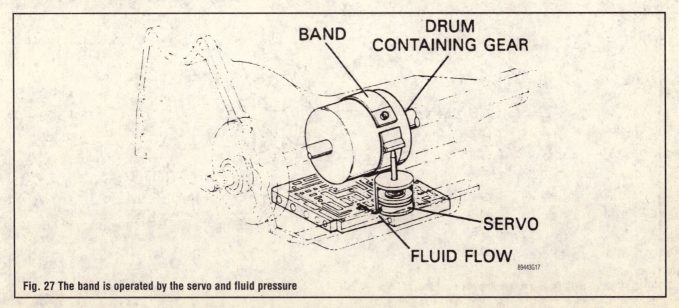

Fig. 27 The band is operated by the servo and fluid pressure

In operation, a band is applied much like a clutch in a manual transmission. If applied too quickly, the shift would be harsh and may damage other components. If allowed to slip too long during application, the heat would damage the friction lining and cause early failure. As bands age, that lining wears away and the band will slip more during engagement and not hold as tightly when engaged. So as they age, bands need to be adjusted. Previous designs provided for adjustment without any disassembly or special equipment. Newer transmissions have fully automatic band adjustment.

Clutch Pack

▶ **See Figures 30, 31 and 32**

The multi-disc clutch pack consists of a set of discs coated with friction material alternating with another set of smooth steel plates, commonly called "steels". One set is splined on the inside and the other set splined on the outside. At least one piston is used to press all the discs and plates together and at least one return spring is used to disengage the clutch pack. These components will usually be assembled into a drum and held in place with a snapring. When hydraulic pressure is applied to the piston, the plates and discs are squeezed together. When pressure is removed, the spring will push the piston away and the clutch will disengage. The movement needed is very slight but enough to prevent any drag between the plates.

Fig. 30 Multi-disc clutch uses internally and externally splined discs that squeeze together

Several interesting and innovative spring designs are used in clutch packs, including coil springs, wave springs and diaphragm springs. Wave springs look like bent washers that flatten when compressed, then take their original shape when released. They may be used to hold another component in place or as a return spring.

Diaphragm springs are also called Belleville springs. They act as both a return spring and as the fingers that the piston acts upon. The outer circumference of the spring is a flat disc with several large tangs evenly spaced around the inner circumference. The tangs work like a diaphragm, deflecting together when pushed upon and returning to flatness when released. This particular spring can multiply the clamping force of the piston because the tangs act as levers that the piston works against.

A clutch pack can offer a great deal of clamping force for a small diameter. Increasing clamping force can be accomplished by increasing the hydraulic pressure on the piston or by adding more discs and plates to increase the friction area. When one set of discs or plates is splined to something stationary, squeezing the discs together will cause the clutch pack to act as a brake. This has been used in aircraft braking systems since the 1930's. In a transmission, the stationary discs are splined to the case and applying the clutch will stop a drum from rotating. While they are more complex and expensive, clutches are far more versatile than bands because they can rotate and require no adjustment.

One-Way Clutch

▶ **See Figures 33, 34, 35 and 36**

There are two types of one-way clutch commonly used in automatic transmissions: roller clutch and sprag clutch. The roller clutch uses ether ball or roller bearings between an inner and outer ring. The inner ring is smooth but the outer is shaped to allow the roller to release when the parts rotate in one direction and wedge the two parts together when rotating the other direction. Some applications have a spring at each roller to assist the lock-up. This type of clutch is used more often on gear sets, usually to allow the drum to rotate in one direction only.

The sprag clutch works the same way but instead of ball or roller bearings, a figure-eight shaped sprag is used to wedge the inner and outer sections together. This type of clutch is used more often on torque converter stators. The stator must be stationary during the torque phase but must be free to rotate with the torque converter during coupling phase.

Each of these clutches has its own specific advantages, but the main advantage of a one-way clutch is that no piston or hydraulic equipment is required to actuate it. They may be used to allow freewheeling in one direction and to prevent rotation in one direction altogether.

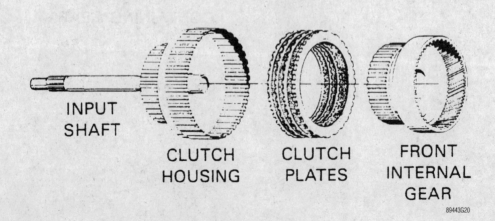

Fig. 31 Clutch pack splined to housing and front gear

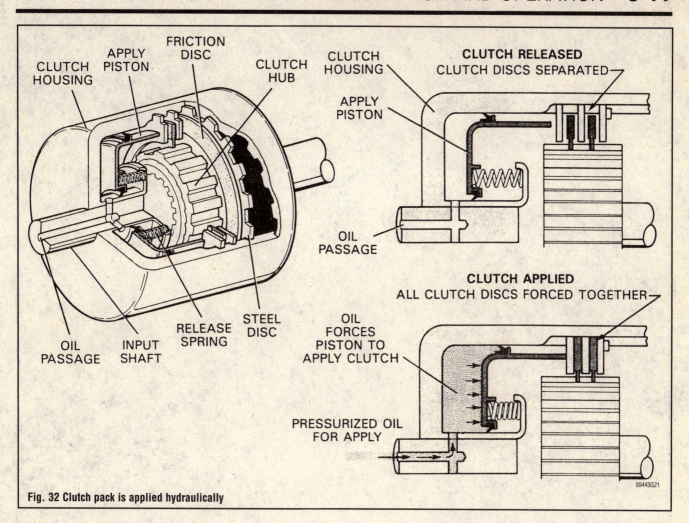

Fig. 32 Clutch pack is applied hydraulically

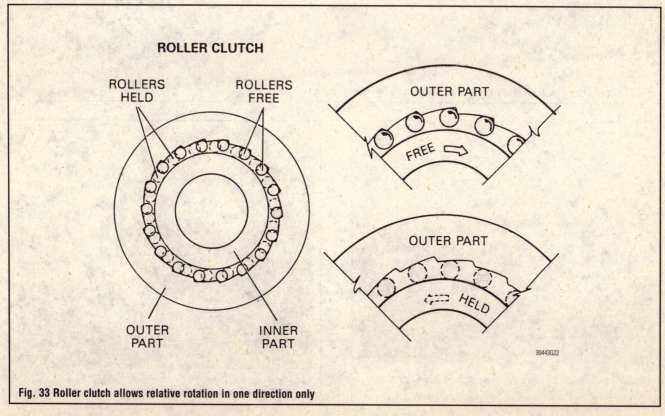

Fig. 33 Roller clutch allows relative rotation in one direction only

SPRAG CLUTCH

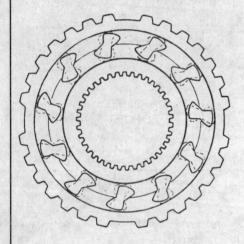

INNER AND OUTER PARTS FREE TO ROTATE

FREE

INNER AND OUTER PARTS HELD

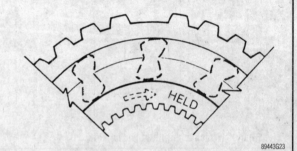

HELD

89443G23

Fig. 34 Sprag clutches use shaped sprags instead of balls or roller bearings

89443P01

Fig. 35 This sprag type one way clutch is used in a Ford AOD transmission

89443P02

Fig. 36 This one way roller clutch is used in a torque converter

SERVOS

▶ **See Figures 37, 38 and 39**

Hydraulic pressure that is applied through the shift valve pushes a piston in a cylinder to tighten the band or clutch pack. The piston and cylinder are called a servo, and while they can have many different configurations and locations within the transmission, they all function basically the same way. The piston pushes a rod, which moves the band directly or moves some other linkage to tighten the band or clutch. When hydraulic pressure is released, a strong spring in the cylinder pushes the piston to release the clutch or band.

Servos are sized according to the force needed by the band or clutch and the amount of hydraulic pressure available to hold the piston. A common servo would use a piston with an area of three square inches. With a hydraulic pressure of 50 psi, this would generate 150 pounds of force to tighten the band (area multiplied by pressure), a relatively light duty application. If hydraulic pressure is increased to 100 psi., this would generate 300 pounds of force to hold the band tight around the drum.

Two different ways for operating servos are equally common. The simplest servo uses hydraulic pressure to apply and spring pressure to release. The other type uses hydraulic pressure on both sides of the piston, either with or without a return spring. This allows for very precise control and a

Fig. 39 The two components in the background are servos, while the one in the foreground is an accumulator

much quicker and more positive release. These are important in some applications because sudden or sloppy band applications can be felt by vehicle occupants.

SERVO LINKAGE

▶ **See Figures 40, 41, 42 and 43**

Another factor in generating the clamping force of a band is the linkage connecting the servo to the band. Four different types of linkage are common. The simplest is when the operating rod of the servo piston pushes a simple strut directly against the free end of the band. This would more likely be used with a large servo. If the servo rod pushes against a lever instead of directly against the band, mechanical force is multiplied. This would also allow more design options for the servo such as placement, size and hydraulic pressures used to achieve the same clamping force. On transmissions with external band adjustment screws, these are the types of linkage used.

On a cantilever type linkage, the servo pushes a lever that moves both ends of the band at the same time. The piston moves a lever, which pushes one end of the band and also moves a second lever, which moves the other end of the band. This is a common linkage used in Chrysler Torqueflite transmissions, and the adjustment is inside the oil pan.

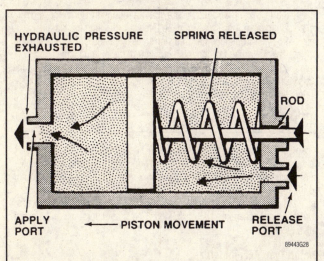

Fig. 37 Schematic of a servo with hydraulic pressure on both sides of the piston

Fig. 38 This low-reverse servo from a Ford AOD transmission/transaxle is held in by a snapring

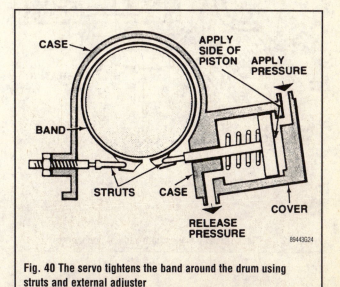

Fig. 40 The servo tightens the band around the drum using struts and external adjuster

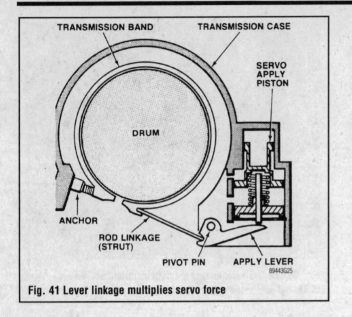

Fig. 41 Lever linkage multiplies servo force

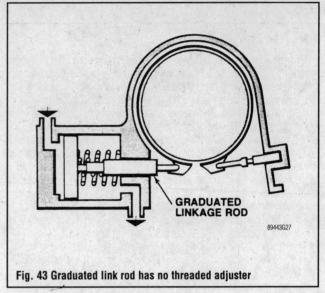

Fig. 43 Graduated link rod has no threaded adjuster

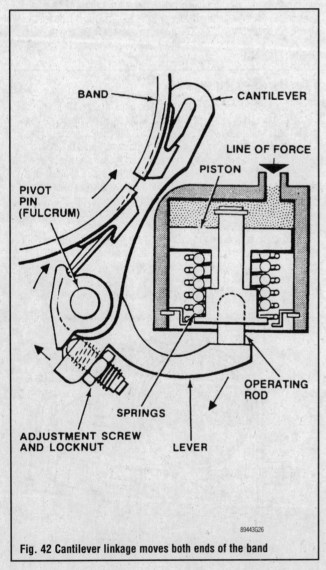

Fig. 42 Cantilever linkage moves both ends of the band

The fourth type of linkage is just a straight rod inserted into the servo called a graduated rod. These come in three specific lengths and are selected by hand when the transmission/transaxle is built. There is no adjustment other than the changing position of the piston as the band wears, much like the pistons in disc brake calipers change positions as the pads wear. The band still reaches the specific clamping force dictated by the piston size and hydraulic pressure.

ACCUMULATORS

An accumulator can be described as a shock absorber. While the shift valve must open and close sharply to provide correct shifting action, sudden engagement of the band or clutch is not desirable. To provide smooth, continuous and gradual application, the pressure to the servo is "tapped off" to another chamber so that hydraulic pressure builds in the servo at a controlled rate. There are two basic kinds of accumulators: piston type and valve type.

Piston Type Accumulator

▶ See Figures 44, 45 and 46

Like so many other components in the valve body, this accumulator uses a piston and spring. But unlike the other components, moving the piston does not activate anything, it's only there to absorb energy. When the hydraulic circuit is pressurized, the fluid must flow into the accumulator chamber and move the piston against the spring before reaching full pressure. This slows the build-up of pressure in the servo and dampens any reflected pressure waves. If hydraulic pressure is also applied to the spring side of the accumulator piston, it is also possible to control the rate of dampening. This control pressure is varied according to load and other demands and may come from the throttle valve circuit, line pressure or the accumulator regulator, a separate unit in the valve body.

Piston type accumulators can be separate chambers in the valve body or they can be part of the servo. When they are built into the servo, it means one less circuit in the valve body, but a larger servo. When separate accumulators are used, they can be mounted anywhere in the valve body and can be "tuned in the field" by changing the spring for a softer or faster engagement.

Refer to the illustration showing the low-reverse band servo. In A, accumulator regulator pressure holds the accumulator and servo piston in the release position. In B, pressure that applies the second gear clutch is also applied between the accumulator piston and servo piston, forcing them apart

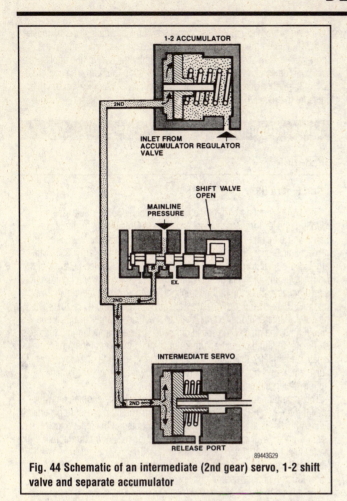

Fig. 44 Schematic of an intermediate (2nd gear) servo, 1-2 shift valve and separate accumulator

Fig. 46 This 1-2 shift piston type accumulator is used in the GM 440T4 (4T60E) transaxle

against the spring and accumulator regulator pressure. The first gear band is still held in the release position until pressure is applied to the back side of the servo piston, when it must also move the fluid out of the accumulator piston chamber. By controlling pressure built-up in both directions, this one accumulator is used to cushion engagement of two different clutches.

Valve Type Accumulator

▶ See Figures 47, 48 and 49

Typically found on Ford transmissions, a valve type accumulator is equally effective, efficient and versatile as the piston type but works in a much different way. The spool in the valve type accumulator is held against the spring by line pressure. When the shift valve opens to apply pressure to

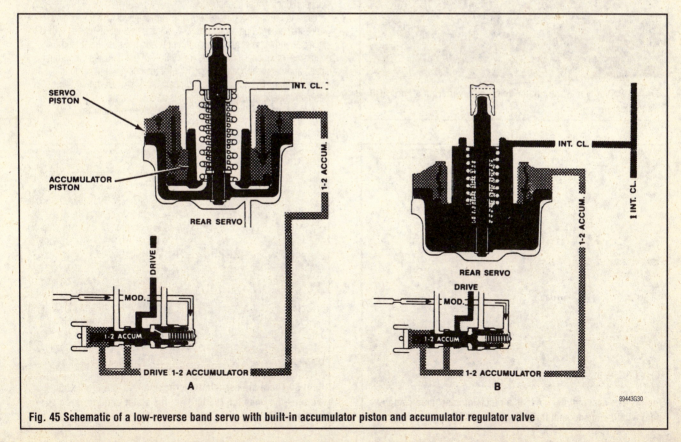

Fig. 45 Schematic of a low-reverse band servo with built-in accumulator piston and accumulator regulator valve

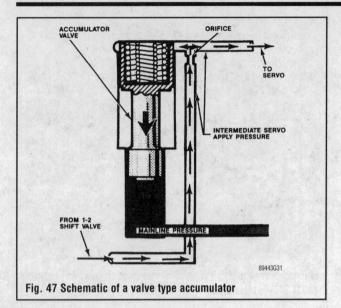

Fig. 47 Schematic of a valve type accumulator

the clutch servo, fluid must pass through an orifice, which slows its movement. The circuit then branches and fluid moves to the servo and to the spring side of the accumulator bore. Assisted by the spring, the apply pressure overcomes line pressure to move the spool in the bore. The combination of the orifice and the accumulator valve delay the application of full pressure in the servo to provide a gradual movement of the servo piston.

Torque Converter

OPERATION

▶ **See Figure 50**

There are two phases of torque converter operation, torque phase and coupling phase. Torque phase occurs when the impeller is turning faster than the turbine and the torque converter is, in effect, a gear reduction unit. In mechanical gear reduction, the smaller gear drives the larger gear. The larger gear's shaft turns more slowly but torque is multiplied because of the longer lever arm of the larger gear. In a torque converter, when the impeller is turning faster than the turbine, most of the energy is in the vortex flow of the fluid. Because the turbine vanes are curved, more energy is absorbed at the outer diameter of the turbine, making it act like a larger gear. Up to a certain limit, the greater the speed difference between impeller and turbine, the greater the fluid pressure farther out towards the end of the turbine vanes. Additionally, the stator is held in place against the one-way roller clutch and recycling most of the un-absorbed vortex flow energy back to the impeller, multiplying torque again.

The maximum torque multiplication occurs when the engine is at wide-open throttle but the turbine is not yet turning. The fluid pressure generated by the impeller is at its maximum and concentrated at the outer ends of the turbine vanes (lever arms). This will produce the maximum torque rating of the torque converter, often called stall torque. The maximum torque multiplication in most transmissions is usually designed in the range of 2:1 to 2.5:1. Much higher torque ratings are sometimes used but the trade-off is higher operating temperatures and higher vehicle speed during coupling phase, reducing fuel economy.

Once the turbine begins to turn, the vortex flow is exerting force on vanes that are moving away from it. The force against the vanes begins to decrease and there is less torque multiplication. As turbine speed increases, more energy is in the rotary flow of the fluid. As the turbine speed approaches impeller speed, the torque converter acts more like a fluid coupling. The actual coupling phase occurs when the difference between impeller and turbine speeds is at its minimum. The speed difference in a fluid coupling is never zero. The driven member (turbine)

Fig. 48 Piston type accumulator being removed from a Ford AOD transmission

Fig. 49 The accumulator valve is a precision machined component with O-rings to seal it tight against the bore

Fig. 50 There are essentially for main parts to a torque converter—the impeller, the turbine, the stator and the sprag or roller clutch

achieves some percentage of the driving member speed, called a plus speed ratio. In a torque converter, the turbine typically reaches a plus speed ratio of 90%, meaning for every ten revolutions of the impeller, the turbine makes nine. At this point the stator is not needed and must turn freely on the clutch bearings so it does not interfere with the coupling. Once the vehicle is moving, the torque converter can shift instantly between torque phase and coupling phase, depending entirely on load (driver's demand for power).

NON-LOCKING TORQUE CONVERTER

▶ **See Figures 51 thru 56**

The torque converter is located in the bell housing of the transmission. It is normally bolted to the flexplate on the crankshaft and then splined to transmission/transaxle input shaft. There are three main components of the torque converter: the impeller (sometimes called a pump), the turbine and the stator. The impeller is built into the part of the converter housing that is attached to the flexplate. As it rotates with the crankshaft, the fluid is put into motion and pushes against the turbine to turn the transmission/trans-axle input shaft. These two elements alone comprise a fluid coupling. With proper shaping of the vanes in both elements, the fluid will push against the turbine vanes with almost as much force as the engine generates. But engines generate most of their torque far above idle speed. With an engine producing 200 ft. lbs. of torque in a vehicle weighing 3000 pounds, even with the most efficient fluid coupling, acceleration would be quite leisurely at any engine operating condition other than wide open throttle at peak torque rpm.

The stator is the torque multiplier. It improves the efficiency of the fluid coupling at all speed and load conditions. It allows more of the fluid flow generated by the impeller to effectively push against the turbine vanes. To understand how this works, it is necessary to understand fluid flow within the torque converter.

Picture a hollow ring filled with fluid, mounted on a shaft through its hole. As it spins around the shaft, the fluid inside will eventually spin with it in the same direction. The fluid is said to have rotary flow. But the fluid near the inner diameter of the ring is also being forced towards the outer

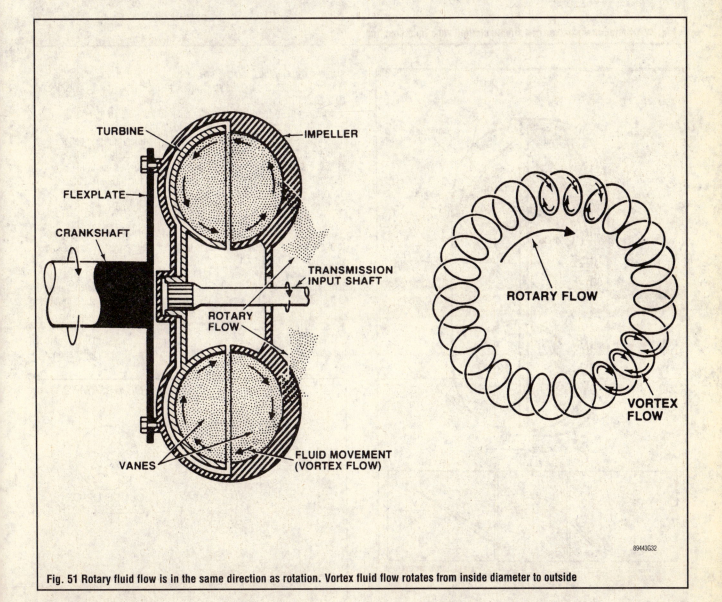

Fig. 51 Rotary fluid flow is in the same direction as rotation. Vortex fluid flow rotates from inside diameter to outside

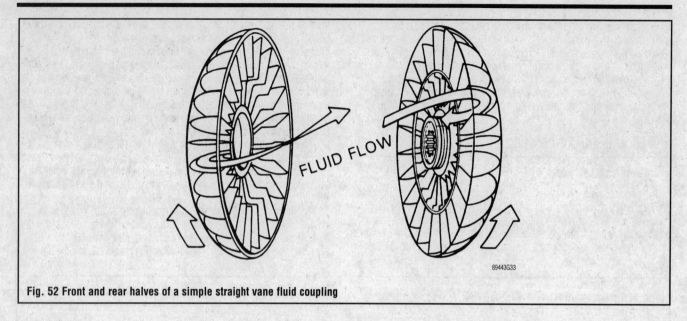

Fig. 52 Front and rear halves of a simple straight vane fluid coupling

diameter by centrifugal force. Since the walls of the ring are curved, the fluid will circulate around the walls, from the inside towards the outside and back again. This is called vortex flow.

Now cut the ring in half, perpendicular to the hole so there is a front half and rear half. If vanes are welded into each half running from inside-to-outside diameter, the vanes in the spinning impeller will generate fluid flow that pushes against the vanes in the turbine. This is a simple fluid coupling. The vortex flow moves the fluid from the outer diameter of the impeller towards the turbine's outer diameter, where most of the energy transfer takes place. By curving the vanes, the turbine is able to absorb more of the energy in the motion of the fluid.

It takes energy to set the fluid in motion. The energy that generates the rotary flow is what makes the turbine turn, but the energy that generates the vortex flow contributes little to moving the turbine. In fact, as the fluid recirculates back to the impeller, it is pushing back against the impeller vanes, increasing the load on the engine. The stator is a set of vanes placed in the vortex flow that channels the fluid returning from the turbine towards the rotary flow direction, so when it returns to the impeller, it actually assists its rotation instead of impeding it.

The stator is mounted on a one-way roller clutch on a shaft that is part of the case, so it can spin freely in one direction only. When impeller and tur-

Fig. 53 The stator redirects a portion of the vortex fluid flow to assist the impeller instead of opposing it

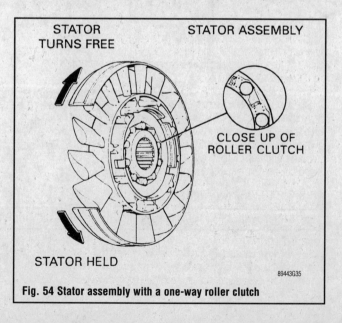

Fig. 54 Stator assembly with a one-way roller clutch

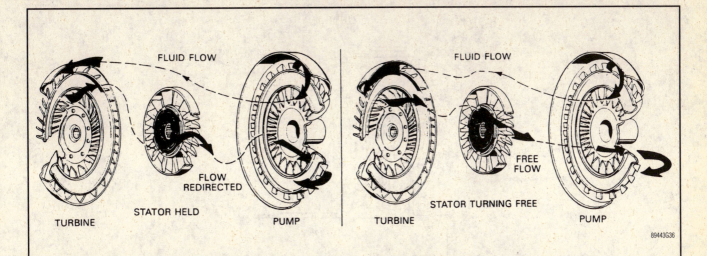

Fig. 55 Torque converter assembly showing fluid flows with stator held (torque phase) and with stator free (coupling phase)

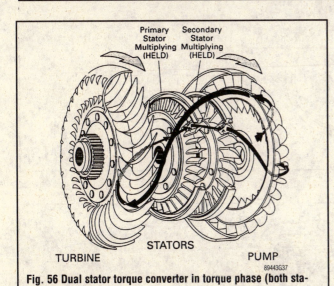

Fig. 56 Dual stator torque converter in torque phase (both stators held)

bine speed are nearly equal, such as in cruise or deceleration, the stator is free to turn on the clutch bearings or rollers. When impeller speed is greater than turbine speed, the stator is held in place against the one-way clutch by the force of the fluid against its vanes. This is when it re-directs fluid flow to multiply torque.

In some applications requiring more converter slip, dual stators are used to increase efficiency over a variety of speed and load conditions. The primary stator faces the impeller and the secondary faces the turbine. During high load conditions, both stators are locked against their one-way clutches and more fluid thrust is directed back to the impeller. As turbine speed increases and load demand is less, the secondary stator freewheels, reducing torque transfer but also decreasing slip to improve efficiency.

LOCK-UP TORQUE CONVERTER

The point of a fluid coupling is to allow slippage between the driving force and the driven unit. This allows not only gradual engagement between power and load but also dampens torsional vibrations. However every engineering solution is a compromise, and the price for this utility is a reduction in efficiency and fuel economy over the direct mechanical connection of a clutch. Even the most efficient torque converters in production today

achieve only 90% speed ratio, meaning for every ten revolutions of the impeller, the turbine makes only nine revolutions.

In the early 1950's, Packard and Studebaker, introduced torque converters equipped with a clutch that locked the impeller and turbine together in high gear. However the added complexity and expense made this a rarely ordered option and production lasted only a few years. In the 1970's when the price and availability of gasoline became important, Chrysler re-introduced lock-up torque converters, using a clutch and pressure plate operated by an electric solenoid. All modern cars and light trucks with automatic transmissions are equipped with a lock-up torque converter to improve fuel economy.

A lock-up torque converter incorporates all the same elements as non-locking converters but adds a mechanical clutch to lock the crankshaft to the transmission/transaxle input shaft. Three basic types are currently in production.

Hydraulically Applied Clutch Piston

▶ **See Figure 57**

This is the simplest type of locking torque converter, using a hydraulically actuated piston to lock the turbine to the torque converter housing. On most designs, it uses a pressure plate somewhat like that found in manual transmissions, with a spline hub and torsion damping springs. The spline fits onto the turbine so the pressure plate assembly always turns with the turbine section of the torque converter. The friction surface of the pressure plate faces the cover or converter housing that bolts to the flexplate. When the clutch is engaged, the pressure plate is held against the cover and the turbine is locked to the flexplate. To engage the clutch, hydraulic pressure is exerted against the entire rear face of the pressure plate to push it away from the turbine and hold it against the cover. To disengage the clutch, hydraulic pressure pushes the plate away from the cover and towards the turbine. In effect, the entire pressure plate moves like a piston that slides back and forth on the splines.

Chrysler uses a similar concept but a much different design. The torsion damper springs are in a cage around the outside diameter of the lock-up piston (clutch plate) and they lock the piston and turbine together directly. This replaces the spline and inner hub used on other domestic vehicles. When activated, hydraulic pressure holds the piston against the torque converter cover that is bolted to the flexplate.

Viscous Converter Clutch

▶ **See Figure 58**

This type is used in some GM transaxles. During certain operating conditions, torque converter clutch engagement can be somewhat abrupt and felt inside the vehicle. A viscous type clutch always engages more gradually

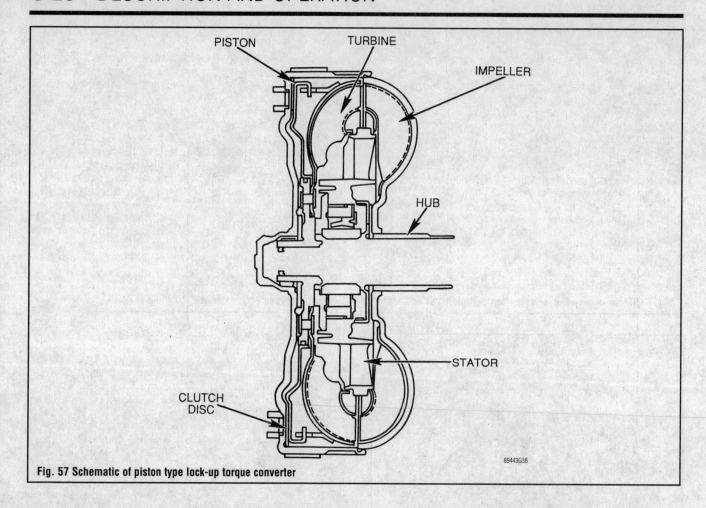

Fig. 57 Schematic of piston type lock-up torque converter

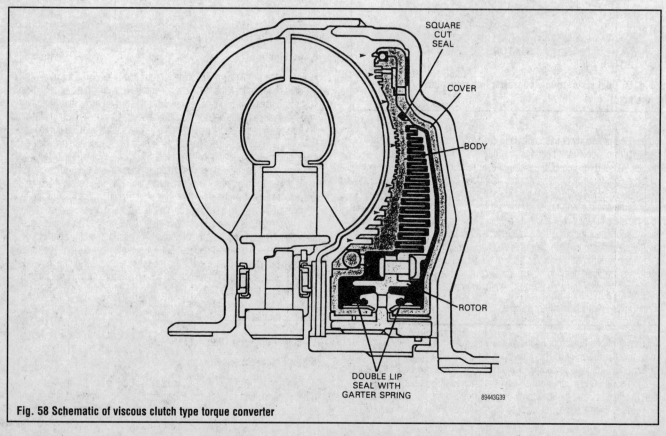

Fig. 58 Schematic of viscous clutch type torque converter

and provides a very smooth application feel during lock-up. Some slipping is still present with this design but during cruise operation, it improves fuel economy significantly. There are three basic components, the body, the rotor, and a special silicone fluid sealed between them. The rotor is splined to the turbine. When applied, transmission fluid presses against the body and the silicone fluid presses the rotor firmly against the converter cover. The silicone acts as the torsional damper spring. The smooth engagement allows this clutch to be used in all but first gear.

Since there is still some amount of slippage, the converter cover can heat up quickly under high load conditions. Fluid temperature is read directly at the impeller body and if high enough, the clutch will be disengaged by the electronic control unit and allow some of the torque converter fluid to flow directly to the cooler.

Direct Mechanical Converter

▶ See Figure 59

This lock-up technique is used in Ford AOD four-speed transmissions and Chrysler's ZF Automatic Transaxle. The converter cover has a torque damper spring set and spline hub built in. The transmission/transaxle input shaft is hollow, and a solid direct driveshaft runs inside it. The converter end of the direct driveshaft engages the hub spline in the cover and the other end engages the third and fourth gear clutch set inside the transmission. In third gear, 40% of the torque flows through the torque converter and 60% through the direct driveshaft. In fourth gear, all torque flows through the direct input shaft.

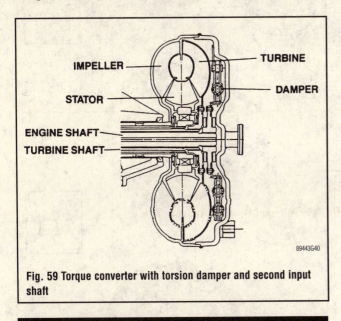

Fig. 59 Torque converter with torsion damper and second input shaft

Gear Set

GEAR RATIOS

Picture a shaft with a pulley on the end that has a radius of one foot. Wrap a cable around the pulley attached to a 100 pound weight. Attach a one foot lever to the other end of the shaft. Push the lever to move the shaft ¼ turn to lift the weight. It will require one hundred pounds of force (lever length and pulley radius are the same). If the lever is two feet long, only fifty pounds of force is required. The longer the lever, the more torque can be generated at the output shaft for any given input force. But while the shaft turns the same number of degrees to move the load, the end of the lever will move a greater linear distance. If the hand pushing the lever moves at the same speed, the weight will lift more

slowly but less force is required to move the same weight the same distance.

A gear can be described as a series of levers (each tooth is a lever) used to turn a shaft. It requires less force to make a larger gear turn against any given load. When a small gear drives a larger gear, the shaft speed of the larger gear is less but the torque is multiplied. This is gear reduction, which makes it possible to use only 120 ft. lbs. of torque to move a 3000 pound vehicle. Once the power source of the drive gear reaches its maximum speed, switching to a smaller driven gear will move the load even faster, but the load will gain speed at a reduced rate because the power source has less mechanical advantage (is pushing against a shorter lever).

When dealing with gear reduction, the difference in size produces direct speed and load differences. If the drive gear is half the diameter of the driven gear, the torque will be doubled but the speed will be cut in half. If the drive gear is twice the size of the driven gear, the torque will be half but speed doubled. When the drive gear and driven gear are the same size, the gear ratio is 1:1 or direct drive. With each change in ratio, the driven gear is expressed relative to the drive gear. A ratio of 2:1 means the driven gear is twice the size of the drive gear.

In an every application using pulleys or gears for a transmission, the ratios are selected according to the power available versus the load and desired performance. If a turbine engine, which turns at about 20,000 rpm, is used to drive a propeller that is most efficient at about 1500 rpm, the speed reduction is about 13:1, but the increase in torque at the propeller shaft is increased by almost the same ratio (there are some mechanical losses). This may make automotive transmissions seem rather modest by comparison, but the typical automotive transmission/transaxle deals with a wide range of constantly varying speeds and loads, temperature extremes and some degree of operator abuse, for years on end with little or no maintenance. In many ways they are the most sophisticated piece of engineering in any automobile.

One final point about selecting gears relative to the power and load. Gears are used to change speeds but cannot change the ultimate amount of power available. Power is the rate at which work is done. With the right gears, one horse power will move a 33,000 pound vehicle one foot in about one minute, or a 550 pound vehicle one foot in one second. Accelerating a car up to highway speeds requires a powerful engine regardless of the transmission/transaxle used.

Basic Planetary Gear Set

▶ See Figure 60

In an automotive transmission, planetary gears offer several advantages over simple spur gears.
- They are compact because they offer two different ratios and reverse in the length of a single gear on a single shaft.
- They can transmit greater torque for their size because the load is distributed over several teeth at the same time.
- They are in constant mesh so there is no damage or wear associated with changing ratios. Constant mesh also allows for smooth and instant ratio changes.
- Because they are located around a common shaft, the mechanism to change ratios is greatly simplified.

COMPONENTS

The three main components of the assembly are the internal gear, pinion gears in a planet carrier, and the sun gear. The sun gear is at the center and all other gears rotate around it. The pinion gears are mounted on short shafts in a planet carrier. Even though each pinion gear rotates on its own shaft, the carrier always rotates as a single unit. The internal gear contains the whole assembly and its name comes from the fact that its teeth are on the inside of the gear. Since all gears are in constant mesh, any one component can be the drive gear or the driven gear, depending on which is driven and which is held stationary.

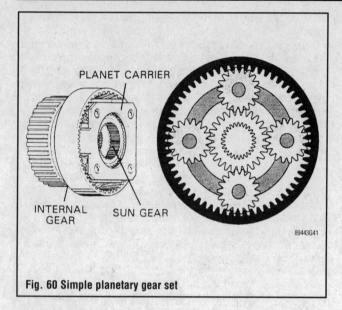

Fig. 60 Simple planetary gear set

POWER FLOWS

▶ **See Figure 61**

There are five basic power flows for all planetary gear sets.
• Direct drive: When any two gears are connected to the input shaft and none are held stationary, the entire assembly will turn as one. Output shaft speed and direction will be the same as the input shaft, gear ratio is 1:1.
• Reduction: The internal gear is held stationary and the sun gear is turned by the input shaft. The planet carrier will be driven in the same direction but at a slower speed than the input shaft.
• Overdrive: The internal gear is held stationary and the planet carrier is turned by the input shaft. The sun gear will be driven in the same direction but at a faster speed than the input shaft.
• Reverse: When the planet carrier is held stationary, turning either the sun gear or the internal gear will cause the other to turn the opposite direction. In almost all applications, the sun gear drives the pinion gears to turn the internal gear in the opposite direction, causing a speed reduction at the same time. Reverse overdrive is never used in automotive transmissions.
• Neutral: When the output shaft is held stationary by the load and no other gear is held stationary, the input shaft can turn any gear and the others will simply rotate around the load.

REDUCTION RATIOS

Two different forward reduction ratios are available with any given planetary gear set, however only can be used in any given transmission/transaxle application.

Sun Gear Drive By holding the internal gear stationary and turning the sun gear with the input shaft, the sun gear will drive the pinion gears and the planet carrier will be driven at the **greater of the two reductions available**.

Internal Gear Drive By holding the sun gear stationary and turning the internal gear with the input shaft, the internal gear will drive the pinion gears and the planet carrier will be driven at the **lesser of the two reductions available**.

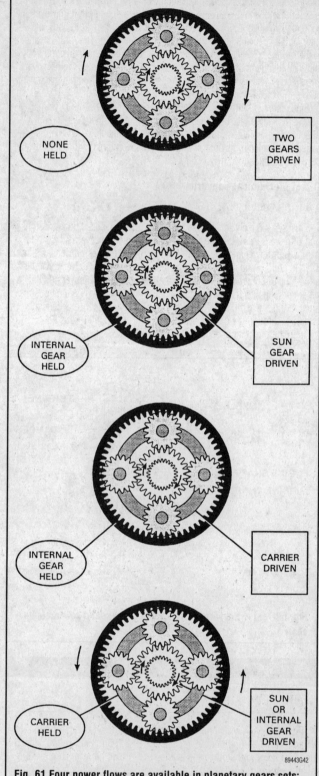

Fig. 61 Four power flows are available in planetary gears sets: direct drive, reduction (planet carrier is output), overdrive (sun gear is output) and reverse (sun or internal gear is output)

SIMPSON PLANETARY GEAR SYSTEM

▶ **See Figure 62**

Every modern automatic transmission/transaxle uses one of two variations on the basic planetary gear set. Each was developed to obtain more gear speeds in less axial space using a single input and output shaft. The Simpson planetary gear set is also called a compound gear set. Two planetary carriers and internal gears are mounted end-to-end so they share one elongated sun gear. The two planetary carriers are different ratios, providing two reductions in one gear set. The output shaft is generally connected to the internal gear of one set and the planet carrier on the other set.

With two holding bands, two disc clutches and a one-way roller clutch, this system is used in most three-speed automatic transmissions. It provides two reduction ratios, direct drive and reverse in one gear assembly. If the application calls for overdrive, a simple planetary gear set is added to the output shaft end of the compound set. While there are many different designs, the Simpson system can always be identified by its the bell-shaped housing that couples the sun gear to the direct drive drum, usually on the forward gear set.

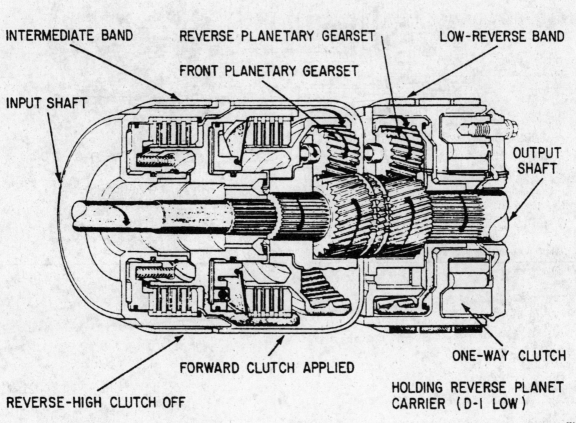

Fig. 62 Simpson type planetary gear set always has a long drum on one gear set

RAVIGNEAUX PLANETARY GEAR SET

▶ **See Figure 63**

The Ravigneaux system is also a compound gear set. It uses two sun gears, one planet carrier and one internal gear. There is only one planet carrier but there are two sets of pinion gears to mesh with the two different size sun gears. One set of pinion gears is double length so they can engage the second sun gear at one end and the internal gear at the other end. The out-put shaft is connected to either the internal gear or the planet carrier. To change power flows, there are a front and rear clutch, two holding bands and a one-way roller clutch.

The difference in the Ravigneaux system is the two independent sun gears and extra set of long pinion gears. This allows using a single internal gear, leaving more room for clutches and other components. Like the other compound system, it provides two reduction ratios, direct drive and reverse in one gear set.

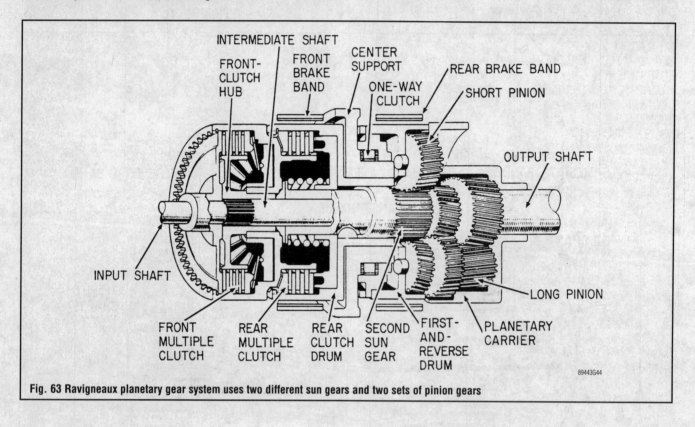

89443G44

Fig. 63 Ravigneaux planetary gear system uses two different sun gears and two sets of pinion gears

4

INSPECTION, MAINTENANCE AND ACCESSORIES

INSPECTION

Road Test

▶ **See Figures 1, 2, 3 and 4**

➥Detailed procedures and diagnostic information are provided in Section 7 of this manual. The information presented here is a general description of preliminary inspections.

If your transmission/transaxleis not performing as it should, the first thing to do is define what it's doing wrong. Manufacturers have developed a shift schedule chart for each engine/transmission/vehicle combination. This chart describes the speeds and driving conditions (load) at which the transmission/transaxleshould shift up or down. A more complete chart will include specifications for governor pressure and clutch and band application. It may also show how to interpret the information gained from a test

---

DATE _____ / _____ / _____

TRANSMISSION/TRANSAXLE CONCERN CHECK SHEET

(★ — INFORMATION REQUIRED FOR TECHNICAL ASSISTANCE)

S E R V I C E   A D V I S O R

★ VIN _____   ★ MILEAGE _____   R.O.# _____

★ MODEL YEAR _____   ★ VEHICLE MODEL _____   ★ ENGINE_____

★ TRANS. MODEL _____   ★ TRANS. SERIAL # _____

★ CUSTOMER'S CONCERN

_____

_____

_____

_____

CHECK THE ITEMS THAT DESCRIBE THE CONCERN —

WHAT:	WHEN:	OCCURS:	USUALLY NOTICED:
__ NO POWER	__ VEHICLE WARM	__ ALWAYS	__ IDLING
__ SHIFTING	__ VEHICLE COLD	__ INTERMITTENT	__ ACCELERATING
__ SLIPS	__ ALWAYS	__ SELDOM	__ COASTING
__ NOISE	__ NOT SURE	__ FIRST TIME	__ BRAKING
__ SHUDDER			AT _____ MPH

T E C H N I C I A N

PRELIMINARY CHECK PROCEDURES	NOTE FINDINGS
INSPECT	
• FLUID LEVEL & CONDITION	_____
• ENGINE PERFORMANCE — VACUUM & ECM CODES	_____
• TV CABLE AND/OR MODULATOR VACUUM	_____
• MANUAL LINKAGE ADJUSTMENT	_____
• ROAD TEST TO VERIFY CONCERN ★	

NOTE: DUPLICATE THE CONDITIONS UNDER WHICH CUSTOMER'S CONCERN WAS OBSERVED

★ PROPOSED OR COMPLETED REPAIRS

ON-CAR   BENCH

_____   _____

_____   _____

_____   _____

_____   _____

(OVER)

Fig. 1 A concern check sheet helps organize your troubleshooting

89444G25

Shift Lever Position	Start Safety	Park Sprag	CLUTCHES				
			Underdrive	Overdrive	Reverse	2/4	Low/Reverse
P — PARK	X	X					X
R — REVERSE					X		X
N — NEUTRAL	X						X
OD — OVERDRIVE							
First			X				X
Second			X			X	
Direct			X	X			
Overdrive				X		X	
D — DRIVE*							
First			X				X
Second			X			X	
Direct			X	X			
L — LOW*							
First			X				X
Second			X			X	
Direct			X	X			

*Vehicle upshift and downshift speeds are increased when in these selector positions.

89444G29

Fig. 2 Charts showing elements in use at each position of the selector lever will help to pinpoint problems in a specific gear range

TYPE OF SHIFT	APPROXIMATE SPEED
4-3 coast downshift	13 mph
3-2 coast downshift	9 mph
2-1 coast downshift	5 mph
1-2 upshift	6300 engine rpm
2-3 upshift	6300 engine rpm
4-3 kickdown shift	13-31 mph w/sufficient throttle

89444G30

Fig. 3 Upshift charts specify the proper rpm or speed at which the transmission/transaxleshould shift

TYPE OF SHIFT	APPROXIMATE SPEED
3-4 upshift	Below 15 mph
3-2 downshift	Above 74 mph @ closed throttle or 70 mph otherwise
2-1 downshift	Above 41 mph @ closed throttle or 38 mph otherwise

89444G31

Fig. 4 Downshift charts specify when a forced downshift should be allowed by the transmission

drive. With a good shift schedule chart and a test drive, it is possible to pinpoint or eliminate many components for further diagnosis.

There are two ways to take the test drive. The first is with the owner (if that's not you). It can be valuable to see how the owner operates the vehicle because it's possible that person's driving style is the real problem, or maybe caused the real problem. A second drive in a quiet area with no traffic and a helper to take notes will allow you to compare specific performance checks with the shift schedule chart.

When checking shift speeds, you are really checking throttle pressure and governor pressure at different load conditions. In addition to speed and load information, you also must pay attention to shift quality. It is possible for a transmission/transaxleto perform as specified at every point on the shift chart and still have accumulator malfunctions that would make shifting very harsh or take too long.

Overall, your road test should include driving in as many varied conditions as possible. This will provide you with all the information necessary to perform a complete diagnosis. If a specific condition is mentioned by the owner, attempt to reproduce this condition during the test drive. Knowing when the conditions occurs will go a long way in helping solve the problem.

Stall Test

▶ See Figure 5

The stall test checks items on a stall test chart without driving the car, however it does carry some risk. Many technicians believe that since this will not test shift quality, the information gained from this test is better obtained with a

Engine Liter	Transaxle Type	Stall Speed Engine rpm
2.2 EFI	A-413	2050-2250
2.2 EFI (turbocharged)	A-413	3150-3350
2.5 EFI	A-413	2250-2450
3.0 EFI	A-413	2500-2700

89444G1B

Fig. 5 As can be seen on this stall test chart, the more torque and engine makes, the higher the converter will stall

88261P48

Fig. 6 Remove the dipstick and wipe clean. Reinsert the dipstick all the way. Remove it again and check fluid level

road test, but sometimes a road test is not possible. If done properly, a stall test can provide valuable information with minimal risk of damage.

The stall test measures the maximum engine rpm attainable with the transmission/transaxle in gear, the drive wheels locked and the engine held briefly at wide open throttle. The engine speed attainable in this way should ultimately be a function of the torque converter (stall speed). The actual specification is usually different for each engine/transmission/vehicle combination. In general though, the stall speed should be the same for each gear and within specification. If it is not, this test may help pinpoint the reason.

The main risk of this test is overheating the transmission/transaxle and causing more damage. For this reason, not all vehicle manufacturers have a stall test specifications chart. If proper precautions are taken and the transmission/transaxle is cooled between steps, it can provide valuable diagnostic information. For this reason, transmission/transaxle rebuilders associations have done this research and developed such charts of their own.

Transmission fluid

DIPSTICK INSPECTION

▶ See Figures 6, 7 and 8

The condition of the transmission fluid is one of the most basic inspections you can do. It's also one of the most informative but potentially misleading. With earlier fluid types, the fluid would retain its clear red color and oily smell if the transmission/transaxle was in good condition. Dark red and burnt smelling fluid usually meant problems. However the current fluids take on those qualities almost at the first heat cycle. This is the result of the changes in friction materials used in clutches. It is still possible to gain valuable information from looking at the fluid.

When removing the transmission/transaxle dip stick, carefully lay the tip on white paper or a paper towel. The fluid will soak into the paper and any discoloration or contamination will show on the white background. The following is a short list of things you may see that indicate transmission/transaxle problems.

• Burnt fluid will appear black and may have friction material particles suspended in it. This usually indicates excess wear of the clutches.

• The fluid will soak into the paper and leave metal particles behind, further evidence of clutch or band wear.

• Fluid may be heavily varnished and no longer clear. It may be tacky or stick to the dipstick and not soak into the paper. Place a drop of new fluid next to it and compare the way they soak into the paper.

• If the fluid is milky, it is mixing with engine coolant. Water and antifreeze will swell the seals and soften the friction materials, causing leaks

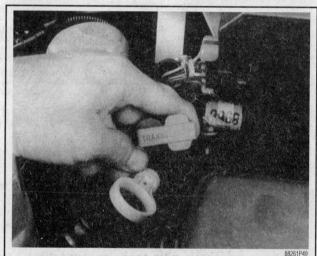

88261P49

Fig. 7 Some dipsticks are hinged and lock into place to seal the filler tube

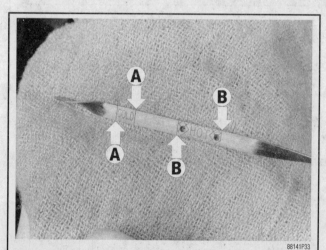

88141P33

Fig. 8 Fluid level should be between the ADD and FULL marks depending upon transmission/transaxle temperature. COLD (A) marks are low on the dipstick while HOT (B) marks are higher up

TRANSMISSION FLUID INDICATIONS

The appearance and odor of the transmission fluid can give valuable clues to the overall condition of the transmission. Always note the appearance of the fluid when you check the fluid level or change the fluid. Rub a small amount of fluid between your fingers to feel for grit and smell the fluid on the dipstick.

If the fluid appears:	It indicates:
Clear and red colored	• Normal operation
Discolored (extremely dark red or brownish) or smells burned	• Band or clutch pack failure, usually caused by an overheated transmission. Hauling very heavy loads with insufficient power or failure to change the fluid, often result in overheating. Do not confuse this appearance with newer fluids that have a darker red color and a strong odor (though not a burned odor).
Foamy or aerated (light in color and full of bubbles)	• The level is too high (gear train is churning oil) • An internal air leak (air is mixing with the fluid). Have the transmission checked professionally.
Solid residue in the fluid	• Defective bands, clutch pack or bearings. Bits of band material or metal abrasives are clinging to the dipstick. Have the transmission checked professionally.
Varnish coating on the dipstick	• The transmission fluid is overheating

8852KC21

and slipping. If further inspection confirms the presence of coolant, the transmission/transaxle and torque converter must be rebuilt. Don't forget to find the source of coolant contamination.

• Foam or bubbles in the fluid usually indicate either incorrect fluid level or there is an air leak on the suction side of the pump. If the pan and filter were ever removed, it's possible the filter was not properly re-installed. Damaged or missing gaskets or O-rings will cause the same air leak.

1. Park the vehicle on a level surface.

2. The transaxle should be at normal operating temperature when checking fluid level. To ensure the fluid is at normal operating temperature, drive the vehicle at least 20 miles.

3. With the selector lever in **P** and the parking brake applied, start the engine.

4. Open the hood and locate the transaxle fluid dipstick. Pull the dipstick from its tube, wipe it clean, and reinsert it. Make sure the dipstick is fully inserted.

5. Pull the dipstick from its tube again. Holding it horizontally, read the fluid level. The fluid should be between the marks on the dipstick. If the fluid is below the MIN mark, add fluid through the dipstick tube.

6. Replace the dipstick, and check the level again after adding any fluid.

➡**Never overfill the transmission/transaxle.**

PAN INSPECTION

▶ **See Figures 9, 10 and 11**

If examination of the fluid does not produce enough information, removing the pan and filter will show much more. Drain the transmission fluid and carefully examine the residue in the pan. Some residue is normal, but

lots of friction material or any plastic or shiny metal particles can indicate real problems. If the metal particles are magnetic, the torque converter may be damaged. If you see gasket material or RTV, the pan or transmission/transaxle has been service before and the RTV was used carelessly upon reassembly. Sometimes this will cause a clogged passage in the valve body without any other damage.

89444P02

Fig. 9 On Ford transmissions, a plastic "lollipop" may be found when changing fluid for the first time. This is plug is used on the assembly line and can be removed

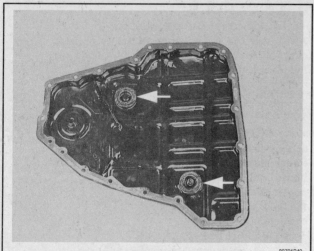

Fig. 10 Most transmission/transaxle pans contain magnets that trap metallic debris

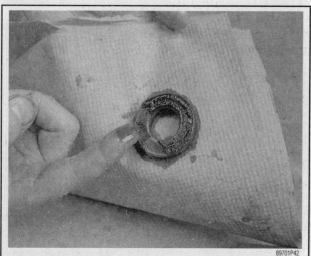

Fig. 11 Remove and clean the magnets carefully. Inspect all material to determine its source

Even if the pan seems clean and the fluid looks good, remove the filter and open it up for examination. Often the signs of wear or damage you're looking for will be concealed inside the filter and not easily seen from the outside.

Even if there is no real evidence of excessive wear or fluid damage, some transmissions use an extra fine filter that suddenly clog enough to prevent proper fluid flow through the valve body. Sometimes just carefully installing a new filter and fluid after this inspection will solve the problem.

Fluid Leaks

▶ See Figures 12 and 13

Locating the source of fluid leaks can be one of the most challenging parts of transmission/transaxle repair and maintenance. Movement of air under the car moves the fluid around, making it difficult to locate the original source of leakage. If engine oil or some other fluid is also leaking, the job is almost impossible without steam cleaning first. However difficult it seems, there are a few basic steps that are a good starting point.

• First check the transmission fluid level. If it is overfilled, it will find a way out. If caught early enough, returning to the proper fluid level can avoid permanent damage to seals and gaskets. Overfilling a transmission/transaxle is one of the most common reasons for leakage.

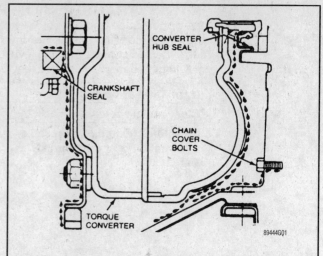

Fig. 12 Locating the source of fluid leakage at the torque converter area—automatic transmission

• Make sure the fluid leak is from the transmission. Some power steering systems use the same type of fluid. Engine oil, coolant and even fluid from hydraulic motor mounts can mask the true source of the leak.

• Drive the car to reach normal operating temperature, then park it and place a clean sheet of cardboard underneath with the engine not running. Take note if the leak appears immediately and then stops, or if it takes time for the first drip. If it drips right away, it might only leak when running, indicating cooling lines or pressure taps. If it takes time for the first drip, that means the fluid must drain down into the pan to find the leak path. Also pay attention to where on the cardboard the drips appear.

• With the clean card board underneath, run the engine. If there are no drips, hold the brake and drop it into gear. If it drips now, check for a leaking servo cover or torque converter.

A leak from the front, at or near the bell housing, could be the torque converter or the transmission/transaxle pump or the engine crankshaft seal. Either way the transmission/transaxle must be removed for repair, but it helps to know the source before starting the job. If you are able to determine it is not the pump or torque converter, you may have more time to decide just how to deal with the repair. If the leak is definitely the pump or torque converter, it will become worse quickly and should be dealt with immediately.

Linkage, Wiring and Vacuum Connections

Linkage that is not properly adjusted can cause what would seem to be much more serious problems. It is possible for simple mis-adjustments to cause anything from improper shifting to the vehicle not moving at all. The reason for the linkage being out of adjustment might be more significant than the adjustment itself. Worn or broken linkage, a broken motor mount, accident damage, improper service or many other things that have little to do with the transmission/transaxle can effect linkage adjustment. Fortunately it is easy to check and might be easy to rectify, as long as adjustment is the real answer to the problem.

Sit in the vehicle with the engine not running and move the selector lever through the full range, carefully stopping at each position. Check to see that the lever moves easily and stops in the detents that align properly with a position on the indicator. Check for the feel of each detent dropping firmly into place at each position. Make sure it moves freely but firmly and positively into PARK and that the starter operates only in PARK and NEUTRAL.

If there is a TV cable or linkage, make sure this is not damaged and is properly adjusted. Check for freedom of movement and a full return to the stop when released. Make sure the engine throttle itself opens all the way and returns to idle position. If equipped with a vacuum modulator, disconnect the vacuum line from the intake manifold and connect a handheld vacuum pump. If the vacuum cannot be held, physically check the line and the

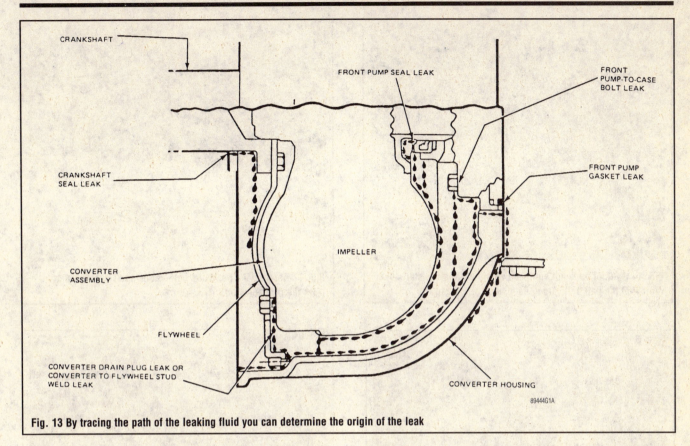

CRANKSHAFT

FRONT PUMP SEAL LEAK

FRONT PUMP-TO-CASE BOLT LEAK

CRANKSHAFT SEAL LEAK

FRONT PUMP GASKET LEAK

CONVERTER ASSEMBLY

IMPELLER

FLYWHEEL

CONVERTER DRAIN PLUG LEAK OR CONVERTER TO FLYWHEEL STUD WELD LEAK

CONVERTER HOUSING

89444G1A

Fig. 13 By tracing the path of the leaking fluid you can determine the origin of the leak

modulator. If the diaphragm is ruptured, there will be signs of transmission fluid in the vacuum line.

With the engine running, repeat all the linkage checks. Check for the appropriate reaction from the transmission/transaxle when moving in and out of gear. If possible, do this with the engine and transmission/transaxle cold, then repeat it with the engine fully warmed up. Pay attention to any differences that seem related to temperature.

If the linkage is out of adjustment or does not work properly, look for the worn bushing, broken mount, improper assembly or some other problem that caused this before deciding to adjust the linkage or cables. If you do decide to that adjustment is the correct course of action, make sure you have the proper specifications on hand before starting. Even though the same transmission/transaxle may be used on several different cars, the dif-

ferences are significant and will effect adjustment specifications, especially TV cables or linkage. On some models, the TV cable must be adjusted using a gauge in a fluid pressure test port.

Wiring problems are usually obvious if you know what to look for. As with linkage problems, first look for damage from an accident, previous repair or some other seemingly unrelated malfunction. If wiring seems improperly stretched tight, look for a broken mount or missing bracket. Many factory service bulletins deal with wiring failures resulting from abrasion or heat damage that could not be foreseen when the car was designed and built. If wiring is near steering or suspension components, the exhaust system or any other moving parts, look for signs of wear or burning. Check each connector, especially the ground connections.

MAINTENANCE

It's necessary to mention the difference between maintenance and repair. Maintenance includes routine inspections, adjustments, and replacement of parts, which show signs of normal wear. Maintenance compensates for wear or deterioration. Repair implies that something has broken or is not working. A need for repair is often caused by lack of maintenance.

For example: draining and refilling the automatic transmission fluid is maintenance recommended by the manufacturer at specific mileage intervals. Failure to do this can ruin the transmission/transaxle, requiring very expensive repairs. While no maintenance program can prevent items from breaking or wearing out, a general rule can be stated: MAINTENANCE IS CHEAPER THAN REPAIR.

Changing Fluid and Filter

▶ **See Figures 14 thru 20**

➡**Used fluids such as transmission fluid, are hazardous wastes and must be disposed of properly. Before draining any fluids, consult with your local authorities; in many areas, waste oil is being**

accepted as a part of recycling programs. A number of service stations and auto parts stores are also accepting waste fluids for recycling. Be sure of the recycling center's policies before draining any fluids, as many will not accept different fluids that have been mixed together. .

This is a general service procedure that may not apply to your specific transmission. Instructions for your specific transmission/transaxle can be found in your owners manual, or a Chilton's Total Car Care manual. Always consult your owner's manual or a Chilton's Total Car Care manual for the correct fluid capacity of your specific vehicle.

1. Raise and safely support the vehicle. Place a large drain pan under the pan. This job can be messy. Eye protection is recommended.

2. If there is an inspection cover on the bottom of the bellhousing, remove it and rotate the crankshaft with a socket and breaker bar to look for a drain plug. If you find one, remove the plug to drain the torque converter, then install the plug and torque to specification.

3. If the pan is equipped with a drain plug, remove the plug and drain the fluid.

Fig. 14 On fluid pans not equipped with a drain plug, start by removing the pan bolts along both sides of the pan

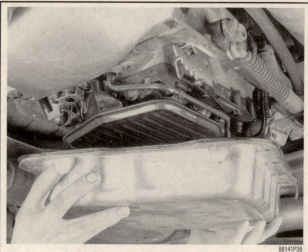

Fig. 17 Remove the pan bolts completely and lower the pan from the transmission

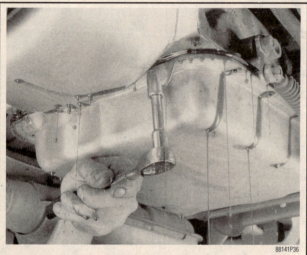

Fig. 15 Then slowly loosen the front and rear pan bolts in a one to two ratio. One turn on the front to two turns on the rear

Fig. 18 Remove and discard the gasket from the pan. Take care not to damage aluminum pans when scraping old gasket material

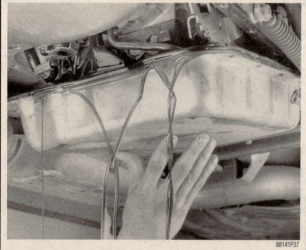

Fig. 16 After loosening the bolts, lower the rear of the pan to allow the fluid to drain completely

Fig. 19 Most pans contain some kind of magnet to trap debris in the fluid. Thoroughly clean the magnet prior to installation

Fig. 20 Remove the filter and thoroughly inspect it for debris

4. If there is no drain plug, remove the bolts from the sides of the transmission/transaxle pan.

5. Loosen the bolts at the front and rear in a one to two ratio so fluid will drain from the rear of the pan. This will be messy.

6. When all the fluid is drained, remove the remaining bolts and remove the pan. Discard the gasket. Carefully examine the residue in the pan. Some black or gray sludge is normal. Excessive sludge or metal particles indicate the need for further diagnosis and repair.

➡ **Many Ford transmissions will have a small plastic "lollipop" in the pan. This is used during the assembly proceed and remains inside the pan until the first time the fluid is changed. It can be discarded.**

7. Remove the filter.

a. On some vehicles, the filter is held in place with bolts or screws.

b. On some vehicles, the filter is held with clips. Move the clips only as needed to remove the filter without bending them permanently out of shape.

c. If no bolts or clips are visible, the filter is a press fit and can be simply pulled off.

8. Clean the pan and wipe the inside dry with a clean cloth. If necessary, scrape off any old gasket material. If material must be scraped from the transmission, be careful not to damage the aluminum sealing surfaces.

9. If there is a magnet inside the pan, clean it off and examine the metal. A small amount of metal shavings is normal, but if there is an excessive amount or any large pieces, the transmission/transaxle may need to be removed and overhauled.

10. Install the new filter. If held in place with bolts or screws, torque to 100 inch lbs. (11 Nm).

11. Carefully fit the gasket onto the pan. If it is necessary to use adhesive to hold it in place, use as little as possible and make sure none will ooze inside the pan when it's installed. Even the smallest amount of adhesive or RTV inside a transmission/transaxle can clog a fluid passage or orifice and prevent proper shifting.

12. Fit the pan into place, install the bolts and snug each one before final tightening. Torque the bolts to specification.

13. Lower the vehicle and fill the transmission/transaxle with the proper quantity and type of fluid.

➡ **Do not overfill the transmission/transaxle.**

14. Start the vehicle and allow the transmission fluid to come up to operating temperature.

15. Test drive the vehicle to ensure proper operation.

16. Recheck the fluid level and adjust as necessary.

ADJUSTMENTS

▶ **See Figures 21, 22, 23 and 24**

➡ **This procedure should be considered a guide. For more specific information on adjusting the transmission/transaxle bands or shift linkage on your vehicle, consult the Chilton's Total Car Care manual for your specific make, model and year.**

On most modern transmissions and transaxles, there are no more than three adjustments: the selector lever cable or linkage, the throttle valve (TV) cable or linkage and the neutral safety switch. There are also band adjustments on earlier transmissions, but bands and clutches are now self-adjusting. On fully electronically controlled units, there may be no adjustments at all.

Normally no adjustments are required unless the transmission/transaxle or linkage has been removed or disconnected. If something "goes out of adjustment", look for another problem before disturbing the transmission/transaxle adjustments. A failed motor mount, collision damage or even electrical problems can cause symptoms indicating the need for adjusting the transmission.

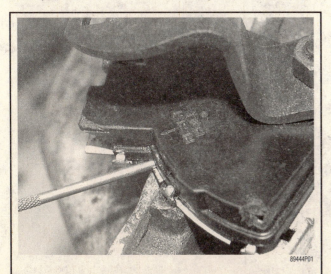

Fig. 21 Inserting an adjusting pin to align the linkage and switch

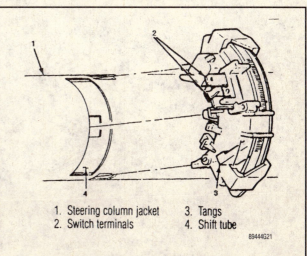

| 1. Steering column jacket | 3. Tangs |
| 2. Switch terminals | 4. Shift tube |

Fig. 22 Some vehicles have a combination back-up light switch and neutral safety switch in the steering column

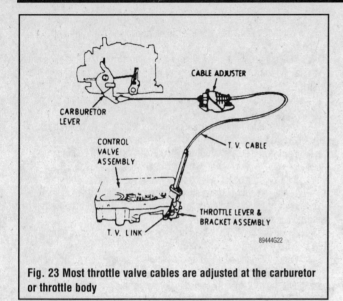

Fig. 23 Most throttle valve cables are adjusted at the carburetor or throttle body

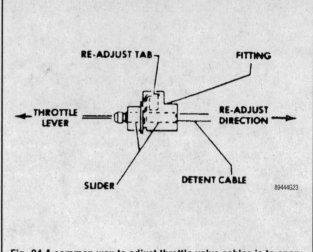

Fig. 24 A common way to adjust throttle valve cables is to open the throttle wide open and then set the cable adjustment

There is one adjustment that is fairly universal to all transmissions and transaxles: the neutral/safety switch. This prevents starter operation unless the selector is in PARK or NEUTRAL. On some electronically controlled transmissions, this switch is part of the shift lever position sensor and cannot be adjusted. On most transmissions the neutral/safety switch is mounted on the transmission/transaxle itself and often includes the switch for the back-up lights. It can be replaced if needed and there are two basic ways to adjust them.

On earlier domestic transmissions, adjustment involved simply positioning the switch so the starter operates only in PARK and NEUTRAL. Newer units have a hole in the case that must be aligned with a corresponding hole in the shift linkage or lever, greatly simplifying accurate adjustment. A drill bit or other small rod inserted in the two holes will properly align the switch and linkage, and tightening the screws in that position will secure the switch in the proper position.

TRANSMISSION/TRANSAXLE COOLER SERVICE

Cooler Function

▶ **See Figure 25**

Inside the radiator is a small cooler in a metal enclosure with lines to the transmission/transaxle through which the transmission fluid is circulated. As the fluid moves through the transmission/transaxle and torque converter, it heats up as it's exposed to hot surfaces. Fluid leaves the transmission/transaxle by cooler lines. It goes into the sealed compartment which has a series of tines, or metal fingers that are inserted all the way through the tubes so that the water that's in the radiator cools this capsule and the transmission fluid. Cooled fluid leaves the capsule through another line, returning to the transmission, starting the circle all over again.

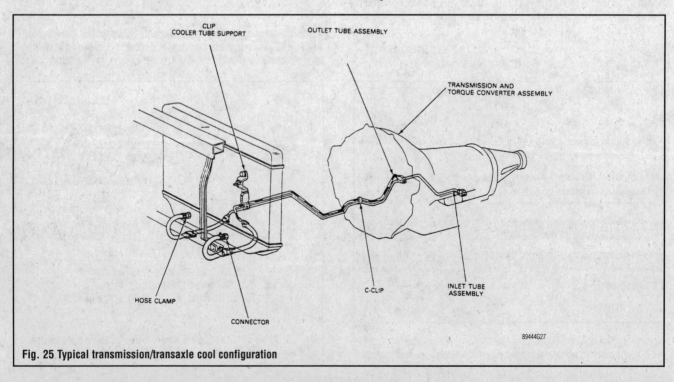

Fig. 25 Typical transmission/transaxle cool configuration

Cooler Failure

The transmission/transaxle pump circulates not only the fluid but also any contamination that may be present in the system. Part of that debris will lodge in the cooler lines and the cooler capsule itself. When a failed transmission/transaxle is removed, the cooler lines are disconnected and stay with the vehicle. This is why cooler cleaning is so often overlooked. After the transmission/transaxle has been completely cleaned and repaired and the technician is ready to put it back in, if he simply reconnects the cooler lines then contaminants that remained from the old transmission fluid circulate through the new transmission fluid and eventually can cause yet another failure.

Cooler Flushing

▶ See Figures 26 and 27

➡A complete flushing of the cooler and lines should be performed whenever the transmission/transaxle is out for servicing. A flushing should especially be performed in excessive material was found in the transmission/transaxle pan.

The cooler and cooler lines can be flushed by machine or an aerosol cleaner. The aerosol cans offer two media, depending on the manufacturer, mineral spirits or water-based solvents.

The process for both manual and aerosol flushing are similar. The inlet and outlet hoses are disconnected from the transmission. One hose is attached to the cleaning mechanism and the other to a container to receive the solvent after it passes through the cooling system. Compressed air can be blown through the system to dry it and remove any solvent that may have been left behind.

❄❄ CAUTION

Wear eye protection and take appropriate fire safety measures. Make sure there is proper ventilation.

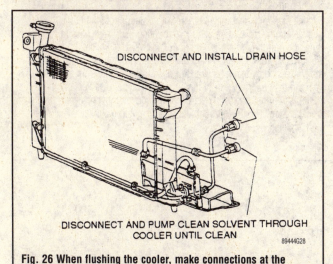

DISCONNECT AND INSTALL DRAIN HOSE

DISCONNECT AND PUMP CLEAN SOLVENT THROUGH COOLER UNTIL CLEAN

89444G28

Fig. 26 When flushing the cooler, make connections at the transmission/transaxle so both the lines and cooler are cleaned

Fig. 27 Aerosol flushing solutions are available to ease the job of removing contamination from the system

89444P07

1. Disconnect transmission/transaxle cooler lines at transaxle.
   a. If a hand-held suction gun is being used, fill the gun with mineral spirits solvent, force the solvent into transmission/transaxle cooler return line until it flows from transmission/transaxle cooler supply line. Continue flushing until solvent is clear and no sign of contamination exists.
   b. If an aerosol flushing is being performed, connect the aerosol can to the transmission/transaxle cooler supply line. Depress the button on the can to force the solvent into the transmission/transaxle cooler and out through the return line. Continue flushing until solvent is clear and no sign of contamination exists.
2. Once the solvent flow is clear, apply compressed air to the transmission/transaxle cooler return line in light applications until remaining solvent is blown from transmission/transaxle cooler and lines.
3. Pump at least one quart of transmission fluid through the system to remove any traces of solvent and contamination.
4. Replace transmission/transaxle cooler if fluid does not flow freely from transmission/transaxle cooler.

FLOW TESTING

After the cooler has been flushed, it is important to check the flow rate to make sure it is not clogged with debris that could not be removed by flushing. This must be done with the new transmission/transaxle installed. This will also ensure that the system is free of solvents and water.

1. Connect the cooler inlet line to the transmission.
2. Attach a hose to the outlet fitting or line and secure the other end to an oil-drain container.
3. Fill the transmission/transaxle to the proper level with transmission fluid.
4. Start the engine and run for about 20 seconds.
5. The minimum flow rate with the engine at 1,000 rpm should be about one quart in 20 seconds.
6. Reconnect the outlet hose and fill the transmission/transaxle to the proper fluid level.

ACCESSORIES

Auxiliary Transmission/Transaxle Cooler

▶ See Figure 28

INSTALLATION

Auxiliary transmission/transaxle coolers are designed to cool the automatic transmission fluid during sustained high speed driving, pulling heavy loads, mountain driving, or any other situation that places stress on the transmission/transaxle or torque converter. The cooler will guard against excessive overheating, but will not overcool the transmission/transaxle in winter weather.

Cooler kits can be installed in a few hours by carefully following the manufacturer's instructions. A minimum of mechanical ability and tools are necessary for a successful installation. Always read the manufacturer's instructions first to familiarize yourself with the parts and procedures.

Fig. 28 Transmission/transaxle coolers are available in many shapes and sizes to fit all types of vehicles. These are manufactured by B & M

➡**Automatic transmissions operate at temperatures between 150°and 250°F. It is suggested that the vehicle be allowed to cool for a few hours to avoid being burned by hot fluid.**

It is recommend to mounting your cooler in series with the existing transmission/transaxle cooler in the radiator tank. This method utilizes the existing cooling system for maximum efficiency and complies with all new car warranties. Using the auxiliary cooler alone should be done only if the vehicle is not equipped with an OEM cooler (as in the case of a vehicle originally having a stick shift transmission), or if the stock cooler is damaged beyond any reasonable repair.

Betore you begin the installation of an auxiliary cooler, make sure that there is adequate room on your vehicle to put your cooler in the desired position. Test fit the cooler before beginning installation.

The best location for cooler installation is in front of both the radiator and the air conditioning condenser. This location permits 100% rated cooling efficiency . Alternate locations are between the air conditioning condenser and the radiator (permits 75% rated cooling efficiency) and behind both the air conditioning condenser and the radiator (permits 60% rated cooling efficiency).

➡**Installing your cooler behind the air conditioning condenser and the radiator may require the use of an auxiliary electric fan or the use of a cooler that is one size larger than the size originally calculated for the vehicle.**

The most important part of auxiliary transmission/transaxle cooler installation is the identification of the transmission/transaxle oil lines. Oil flow direction must be determined prior to installation. The flow through the system must be correct to ensure maximum efficiency .

The most effective way to determine oil flow direction is to remove the cooler lines from the radiator and place them in a container. When disconnecting oil lines hold the adapter fitting with another wrench to avoid damage to the radiator and fitting. Place a short piece of rubber hose over or into the exposed end of the radiator cooler fitting. Put the rubber hose and the cooler line in a container of sufficient capacity. Start the engine and let it run at idle. Determine which line the oil is coming from. Stop the engine immediately.

An alternate method is to check oil line temperatures. Start the engine when cold. Place the transmission/transaxle in Drive and keep the brakes applied to heat oil. Stop the engine and feel the temperature of both oil lines. The warmer line contains the oil flowing from the transmission/transaxle to the cooler.

1. Most auxiliary coolers should be installed so that it is at least 1 inch away from the fan and 1/8 inch from the radiator. Also ensure the cooler is at least 2 inches from the hood, wheel well or firewall and 6 inches from any exhaust manifold, pipe or header pipe.

2. Decide on the best mounting position for your cooler. Remove the shipping covers from the fittings at the cooler.

3. Attach adhesive cushion pads to the cooler mounting flanges. Hold the cooler in position with the pads between the cooler and the radiator or condenser.

4. Insert the mounting rods through the radiator or condenser and through the mounting holes in the flange of the cooler. Install the locking tabs onto the mounting rods. Push the locking tab up to the cooler flange and cut off excess mounting rod. Repeat for the remaining mounting rods.

5. Identify the transmission fluid return line at the radiator.

6. Install the cooler inline with the return line. In other words, the fluid should flow from the transmission, to the radiator and then to the auxiliary cooler. After exiting the auxiliary cooler the fluid should flow back the transmission.

7. After installation, start the engine and place the transmission/transaxle in Neutral. Check the transmission fluid level and correct as necessary.

8. Fluid level must be checked with the engine running and the fluid hot. Do not over fill as this will cause foaming and overheating. Check for leaks.

9. Secure hoses so that they won't be damaged by road debris or other hazards. Retighten clamps if necessary.

Transmission Fluid Thermostat

INSTALLATION

▶ **See Figure 29**

A transmission fluid thermostatic control valve is designed to help maintain proper fluid temperatures required in vehicles outfitted with auxiliary fluid coolers. When fluid temperature is below 180° F, the valve bypasses 90% of the fluid, allowing 10% to flow through the cooler to prevent possible air pockets. Above 180° F, the valve opens fully and flows through the cooler.

When installing the thermostat, keep it away from hot sources such as exhaust manifolds that may interfere with proper operation of the thermostat. Determine the best location for mounting your valve based on temperature and fluid line routing. Both fluid lines must be spliced into so mount the valve where the lines run along side of each other. The valve can be easily attached by using tie wraps or hose clamps around the body.

The most important part of this procedure is the identification of the transmission/transaxle oil lines. Oil flow direction must be determined prior to installation. The flow through the system must be correct to ensure maximum efficiency.

Fig. 29 Transmission fluid thermostats helps maintain proper fluid temperatures required in vehicles outfitted with auxiliary fluid coolers

The most effective way to determine oil flow direction is to remove the cooler lines from the radiator and place them in a container. When disconnecting oil lines hold the adapter fitting with another wrench to avoid damage to the radiator and fitting. Place a short piece of rubber hose over or into the exposed end of the radiator cooler fitting. Put the rubber hose and the cooler line in a container of sufficient capacity. Start the engine and let it run at idle. Determine which line the oil is coming from. Stop the engine immediately.

An alternate method is to check oil line temperatures. Start the engine when cold. Place the transmission/transaxle in Drive and keep the brakes applied to heat oil. Stop the engine and feel the temperature of both oil lines. The warmer line contains the oil flowing from the transmission/transaxle to the cooler.

✳✳ CAUTION

Transmission fluid can become quite hot! Allow engine to cool for several hours before beginning installation.

1. Most thermostats come with a variety of adapters to connect the different size fluid hoses. To adapt the thermostat to your size hose, choose the appropriate size fitting and install it into the thermostat using Teflon® tape or a suitable sealer. Do not overtighten the fittings.
2. Position the valve in the desired location. Do not install the valve yet.
3. Carefully cut the transmission fluid cooler lines and splice the thermostat into place as per manufacturer's instructions.
4. The thermostatic valve has four ports. Connect the lines as per manufacturers instructions. If the lines are not properly connected, fluid will not flow to the transmission.
5. Place the car in neutral with the parking brake on and idle until the transmission fluid is warm.
6. Check for possible leaks.
7. Feel both ends of the cooler to be sure that they are warm. If not, and oil temperature is above 185° F, the oil is not flowing through the thermostat.
8. Check the oil level. Some additional fluid may be needed
9. After road testing the vehicle, recheck the system for leaks and correct operation.

Deep Transmission/Transaxle Pan

INSTALLATION

▶ See Figure 30

Deep oil pans provide several advantages over the stock factory pan. The extra capacity provides increased oil volume for added cooling capacity and ensures a large volume of oil for the transmission/transaxle oil pump, preventing oil pick-up starvation. Aluminum oil pans have the added feature of additional cooling from inner and outer fins and provide increased case rigidity. Finally, the added feature of a drain plug allows regular transmission/transaxle maintenance and oil changes without the usual mess of removing the oil pan.

Deep, aluminum oil pan kits can be installed in a few hours by carefully following directions in the box. Read all instructions first to familiarize yourself with the parts and procedures. Work slowly and do not force any parts.

➡**Automatic transmissions operate at temperatures between 150°F and 250°F. It is suggested that the vehicle be allowed to cool for a few hours to avoid burns from hot oil and parts.**

1. Raise and support the vehicle safely. Try to raise the vehicle at least 1-2 feet so you have plenty of room to work easily. Also, have a small box or pan handy to put bolts in so they won't be lost and a drain pan to catch oil.
2. Drain oil pan. Loosen and remove oil pan bolts one at a time, working toward the rear of the transmission.
3. Remove the last of the bolts slowly and the pan will tilt down to allow the last of the fluid to drain. If the pan sticks to the old gasket, pry it down slightly before removing the last two bolts to break the seal. After the last bolt is removed the pan can be lowered and set aside.
4. The oil filter will now be exposed. Two types are common:
5. Inspect your oil filter. If it has varnish on it, or the transmission/transaxle has more than 20,000 miles since its last service, the filter should be replaced.
6. On most deep pan installations, a filter tube extension will be necessary. This will allow the filter to be positioned correctly at the bottom of the fluid pan. If the filter extension is not installed, certain driving conditions could result in the transmission/transaxle pump sucking air as the filter becomes uncovered.
7. Install oil filter into case.
8. Scrape old gasket material off surface of case. Old gasket material can cause leaks.
9. Install deep pan and new gasket onto transmission. Tighten the pan bolts to specification.

➡**Do not overtighten the pan bolts as this can damage pan.**

10. Install drain plug and tighten securely.
11. Lower vehicle but try to keep rear wheels off the ground, if possible.
12. Fill the transmission/transaxle with the proper type of transmission fluid.
13. Start engine. Shift transmission/transaxle through all gear positions. If the rear wheels are off the ground, allow the transmission/transaxle to shift through all gears several times.
14. Place selector in Neutral and check fluid level.
15. Do not overfill as this will cause foaming and overheating.
16. Check for leaks.
17. Lower the vehicle.

Oil Temperature Gauge

INSTALLATION

▶ See Figure 31

A transmission/transaxle oil temperature gauge will accurately monitor transmission/transaxle oil temperature and warn you before excessive heat ruins transmission fluid or causes damage.

High heat conditions are destructive of both transmission/transaxle friction materials and transmission/transaxle components. Transmission/transaxle operation in the normal operating range of 150°F to 250°F will ensure long life and dependable service.

1. Disconnect negative battery cable to prevent accidental shorts and/or damage.
2. Install the gauge where it can easily be viewed from the driving position.
3. Locate the oil return line to the transmission.

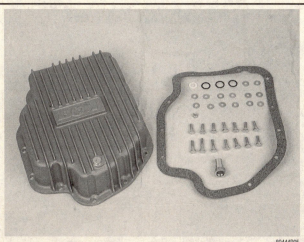

89444P05

Fig. 30 Deep oil pans provide several advantages over the stock factory pan. This is a B & M pan for a TH-400

Fig. 31 A transmission/transaxle oil temperature gauge will warn you before excessive heat ruins transmission fluid or causes serious and costly transmission/transaxle damage

4. On Chrysler and most Ford transmissions, this is the line to the rear of the transmission/transaxle case. On GM TH-200, TH400, TH700-R4 and Powerglide transmissions, this is the, upper oil line to the transmission. On GM TH-350, TH-200-4R and Ford AOD transmissions, this is the lower oil line to the transmission/transaxle case. For all other transmissions and transaxles, this will be the cooler (temperature) of the two lines leading from the transmission.

5. Connect the sending unit from the temperature gauge in-line on the transmission fluid return line. Follow the gauge manufacturer's instructions.

6. Connect the gauge electrical wires by following the manufacturer's instructions. This usually involves a power wire, a ground wire and a signal wire. Additionally a light wire may be installed for viewing the gauge at night.

➡**Ensure the sending unit is well grounded. A sending unit installed in a rubber cooler line is not properly grounded. Improper grounding will cause a false reading.**

7. Connect the negative battery cable and turn the ignition key ON. Check the gauge for proper operation.

GASKET IDENTIFICATION

The following section contains illustrations of the most common transmission and transaxle gaskets sold today. These illustrations can be helpful in determining what transmission/transaxle you have in your vehicle. They also come in helpful for determining the correct gasket at the parts store.

Not all manufacturers gaskets are listed here. If you do not find a gasket that matches your particular transmission/transaxle, consult the gasket manufacturer's catalog in the parts store for further information.

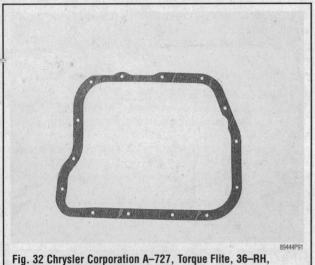

Fig. 32 Chrysler Corporation A–727, Torque Flite, 36–RH, 46–RH, 47–RH and A–518 Transmissions

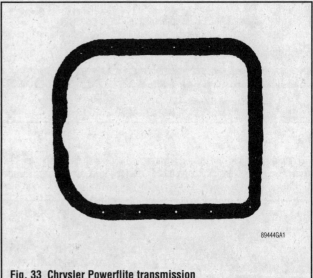

Fig. 33 Chrysler Powerflite transmission

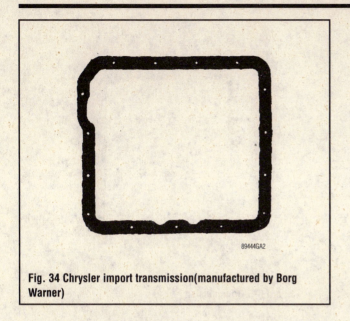

Fig. 34 Chrysler import transmission(manufactured by Borg Warner)

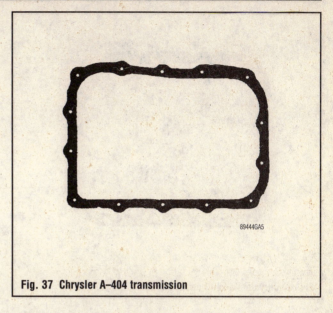

Fig. 37 Chrysler A–404 transmission

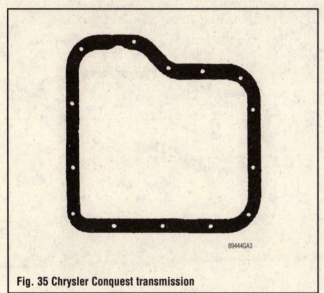

Fig. 35 Chrysler Conquest transmission

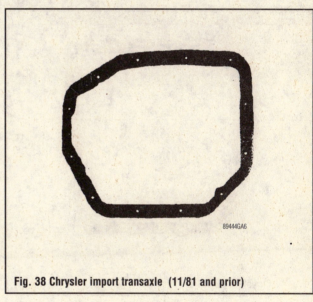

Fig. 38 Chrysler import transaxle (11/81 and prior)

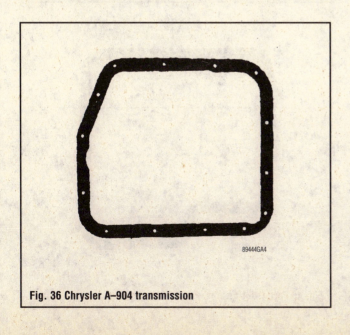

Fig. 36 Chrysler A–904 transmission

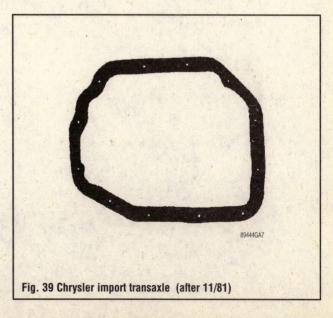

Fig. 39 Chrysler import transaxle (after 11/81)

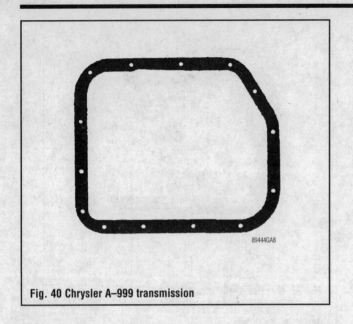

Fig. 40 Chrysler A–999 transmission

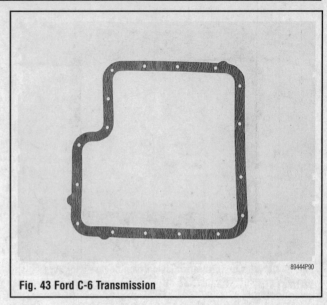

Fig. 43 Ford C-6 Transmission

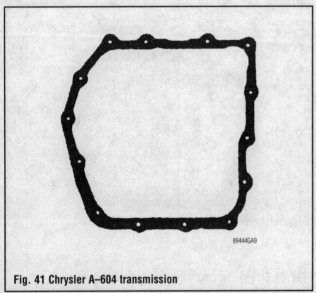

Fig. 41 Chrysler A–604 transmission

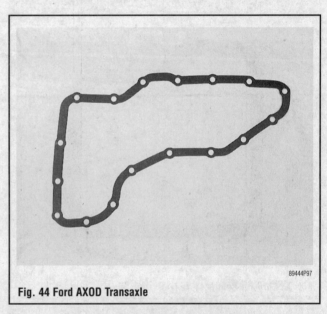

Fig. 44 Ford AXOD Transaxle

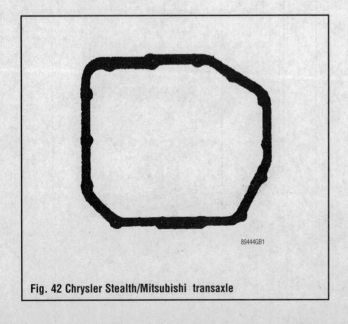

Fig. 42 Chrysler Stealth/Mitsubishi transaxle

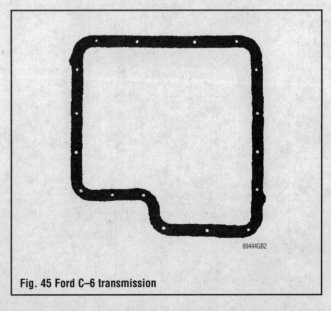

Fig. 45 Ford C–6 transmission

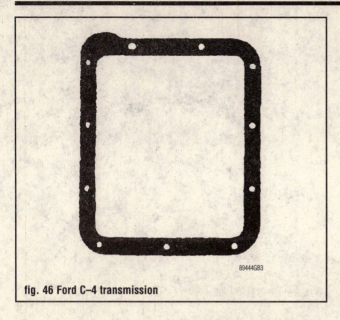

89444GB3

fig. 46 Ford C–4 transmission

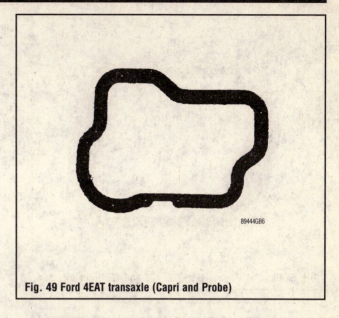

89444GB6

Fig. 49 Ford 4EAT transaxle (Capri and Probe)

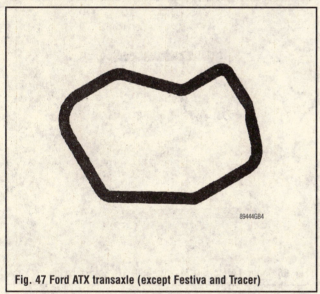

89444GB4

Fig. 47 Ford ATX transaxle (except Festiva and Tracer)

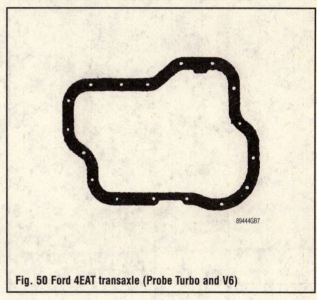

89444GB7

Fig. 50 Ford 4EAT transaxle (Probe Turbo and V6)

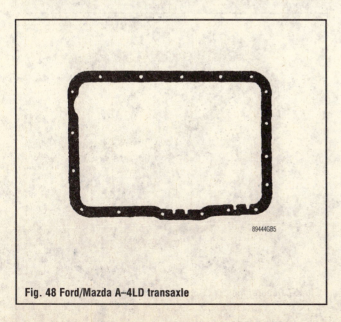

89444GB5

Fig. 48 Ford/Mazda A–4LD transaxle

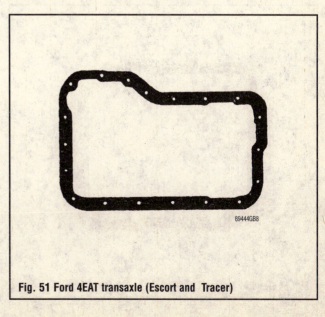

89444GB8

Fig. 51 Ford 4EAT transaxle (Escort and Tracer)

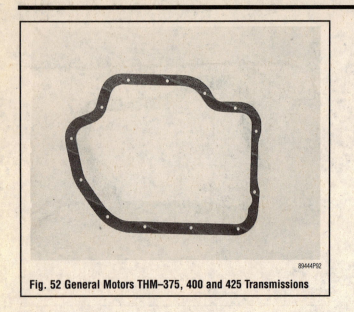

89444P92

Fig. 52 General Motors THM–375, 400 and 425 Transmissions

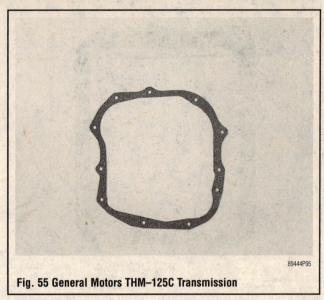

89444P95

Fig. 55 General Motors THM–125C Transmission

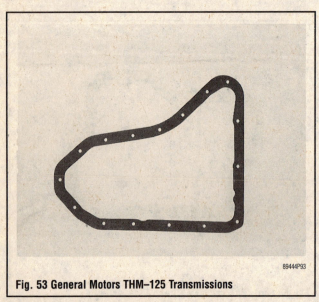

89444P93

Fig. 53 General Motors THM–125 Transmissions

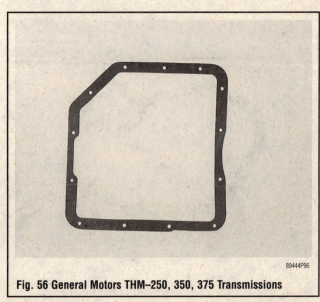

89444P96

Fig. 56 General Motors THM–250, 350, 375 Transmissions

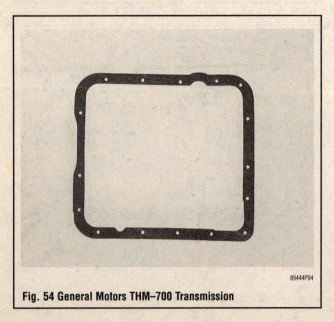

89444P94

Fig. 54 General Motors THM–700 Transmission

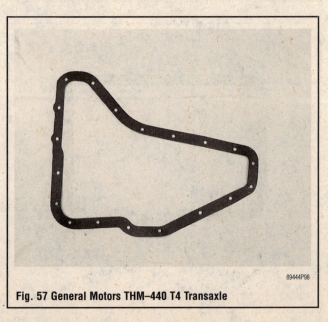

89444P98

Fig. 57 General Motors THM–440 T4 Transaxle

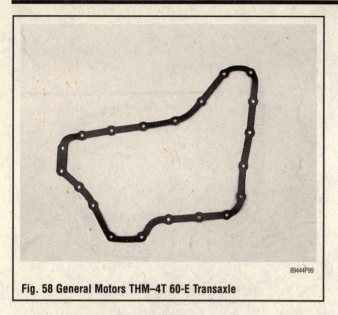

89444P99

Fig. 58 General Motors THM–4T 60-E Transaxle

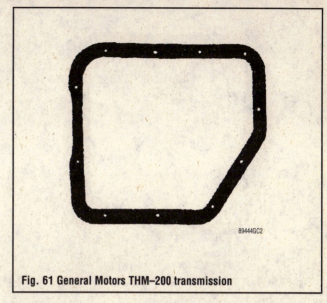

89444GC2

Fig. 61 General Motors THM–200 transmission

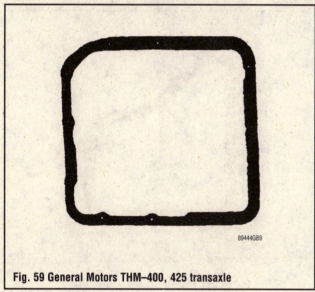

89444GB9

Fig. 59 General Motors THM–400, 425 transaxle

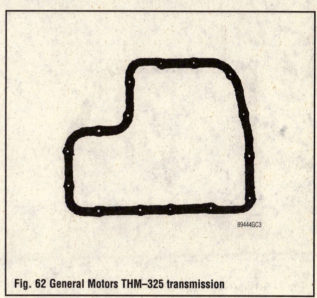

89444GC3

Fig. 62 General Motors THM–325 transmission

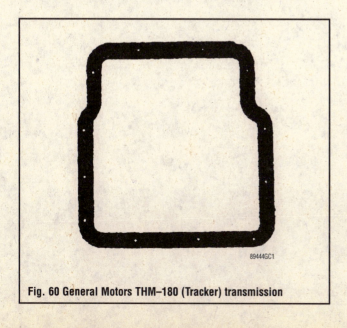

89444GC1

Fig. 60 General Motors THM–180 (Tracker) transmission

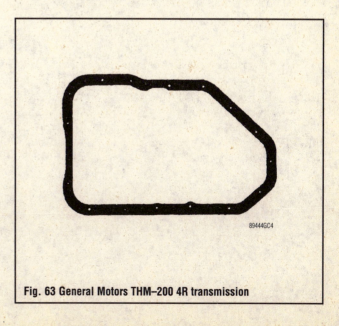

89444GC4

Fig. 63 General Motors THM–200 4R transmission

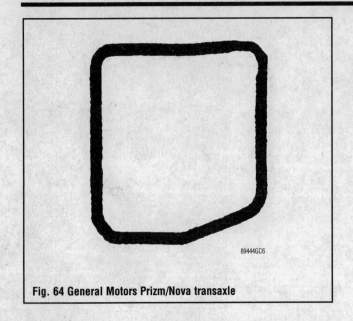

89444GC6

Fig. 64 General Motors Prizm/Nova transaxle

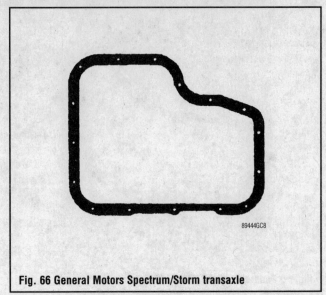

89444GC8

Fig. 66 General Motors Spectrum/Storm transaxle

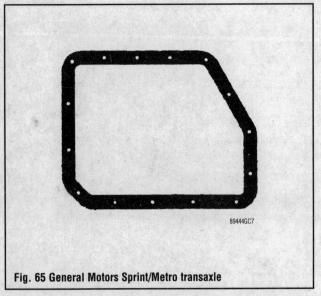

89444GC7

Fig. 65 General Motors Sprint/Metro transaxle

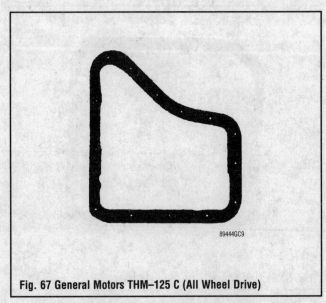

89444GC9

Fig. 67 General Motors THM–125 C (All Wheel Drive)

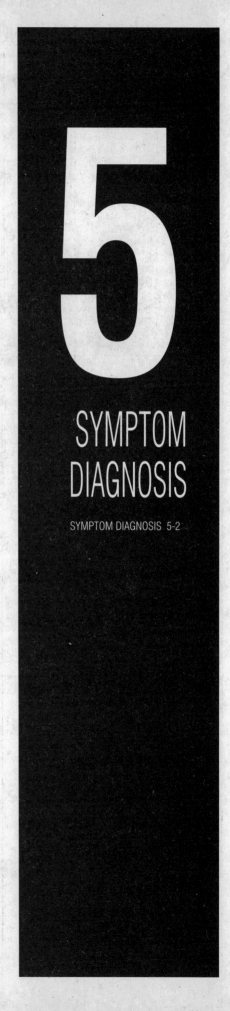

5

SYMPTOM

DIAGNOSIS

SYMPTOM DIAGNOSIS

General Information

The proper diagnosis of an automatic transmission/transaxle requires a through understanding of the transmission/transaxle and it requires a systematic, step-by-step (logical) approach. A hit-or-miss, shot-in-the-dark approach is dangerous and leads to creating more problems and wasting valuable time and money.

An automatic transmission/transaxle that has a malfunction will signal distress in one of the following ways: slipping, display of a shifting problem, noise, display of a Malfunction Indicator Light (MIL), or fluid leakage. A symptom does not necessarily mean that a major malfunction has developed. Slipping describes a noticeable delay between the time the transmission/transaxle is shifted into gear and accelerator is depressed to the time the vehicle actually accelerates; noticeable lag in the application of power from the engine to the appropriate wheels. Shifting problems occur in a number of ways, including refusal of the transmission/transaxle to shift, irregular shifting, and shifting that doesn't occur smoothly but is hard enough to jolt the vehicle. Noise from the automatic transmission/transaxle is reason for immediate investigation. The automatic transmission/transaxle should not make any noise — not a whir, clunk, whine, or thunk. Vehicle equipped with electronically controlled transmissions will incorporate the MIL to indicate a fault detected by the automatic transmission/transaxle computer. New technology requires the use of scanners, analyzers, digital and analog meters, and transmissions testers to help diagnose problems. Also, an electronically controlled transmission/transaxle relies on information from the engine control computer to determine shift point and other transmission/transaxle functions. Fluid leakage from the transmission/transaxle should be repaired as soon as possible, because a low fluid condition could cause terminal internal damage. When diagnosing transmission/transaxle trouble you should keep in mind the usual safety precautions, such as proper clothing, and work habits. Use jack stands to safely support the vehicle while accessing the underside of the vehicle. Some additional safety points to remember are: disconnect the battery when not actually required for that particular test, be careful of HOT engine, exhaust and transmission/transaxle parts, HOT transmission/transaxle fluid can cause severe burns, use only the equipment to service a transmission/transaxle , many internal parts are machined to close tolerances and are easily damaged, place parts so that will not roll off the work bench, absolute cleanliness is imperative because the smallest particle of dirt or lint could cause problems, and ensure that what ever your doing will not result in consequences that can cause personal injury or damage to parts or vehicle.

Diagnostic Fundamentals

▶ **See Figures 1 and 2**

An automatic transmission/transaxle problem is typically caused by one or more of the following conditions: hydraulic failure, mechanical failure, friction element failure, torque converter failure, electronic failure, poor engine performance or improper engine signal to the transmission, or improper linkage adjustments.

Accurate diagnosis of these conditions and related problems will result in a cost-effective repair. The success of diagnosis is the ability to gather information. It may be simpler to complete a diagnosis chart to gather the information.

Terminology

When diagnosing automatic transmission/transaxle conditions, it is important to use a common terminology to describe related operating conditions. The following are terms used to describe throttle position and shift condition.

THROTTLE POSITIONS

- Minimum throttle—described as the least amount of throttle opening required for an upshift (change to the next higher gear and increase of vehicle speed); normally close to zero throttle
- Light throttle—approximately one-fourth of the accelerator pedal travel
- Medium throttle—approximately one-half of the accelerator pedal travel
- Heavy throttle—approximately three-fourths of accelerator pedal travel
- Wide Open Throttle (WOT)—full travel of the accelerator pedal
- Full throttle detent downshift—a quick application of WOT, forcing the transmission/transaxle to downshift (change to the next lower gear) and increase the vehicles speed
- Zero-throttle coast down—a full release of the accelerator pedal while the vehicle is in motion and in drive range
- Engine braking—use of the engine's compression to slow the vehicle by manually downshifting during zero-throttle coast down

SHIFT CONDITIONS

- Bump—a sudden and forceful application of a clutch or band
- Chuggle—bucking or jerking condition that may be engine related and may be most noticeable when the torque converter clutch is engaged
- Delayed (late or extended)—condition where the shift is expected but does not occur for a period of time, where a clutch or band engagement does not occur as quickly as expected during part or wide open throttle or when manually downshifting to a lower range
- Double bump—two sudden and forceful applications of a clutch or band
- Early—this condition occurs when the shift happens before the vehicle has reached proper speed, which tends to labor the engine after an upshift
- End bump—a firmer feel at the end of a shift when compared with the feel at the start of the shift
- Firm—a noticeable quick application of a clutch or band that is considered normal with a medium to heavy throttle shift; should not be confused with harsh or rough shift conditions
- Flare (slipping)—a quick increase in engine rpm accompanied by a momentary loss of torque, generally occurs during a shift
- Garage shift—initial engagement feel of the automatic transmission/transaxle , such as neutral to reverse or neutral to forward gear
- Harsh (rough)—an application of a clutch or band that is more noticeable than a firm, considered undesirable at any throttle position
- Hunting (busyness)—repeating a quick series of upshifts and downshifts that cause a noticeable change in engine rpm, such as 4–3–4 shift pattern
- Initial feel—a distinct firmer feel at the start of a shift when compared to the feel at the end of the shift
- Late—a shift that occur when the engine is at a higher than normal rpm for the given amount of throttle
- Mushy—similar to soft, a slow and drawn out clutch application with very little shift feel
- Shudder—repeated jerking sensation, similar to chuggle but more severe and rapid in frequency, that may be most noticeable during certain ranges of vehicle speed, also used to define the condition after the torque converter clutch engages
- Slipping—a noticeable increase in engine rpm without vehicle speed increase, usually occurs during or after the initial clutch or band engagement

VIN _____   MILEAGE _____   R.O.# _____

MODEL YEAR _____   VEHICLE MODEL _____   ENGINE _____

TRANS MODEL _____   TRANS DATE CODE/OR SERIAL # _____

CUSTOMER'S CONCERN

_____

_____

_____

_____

WHEN: OCCURS: RECENT WORK:

___ VEHICLE WARM   ___ ALWAYS   COLLISION   Yes ___ No ___

___ VEHICLE COLD   ___ INTERMITTENT   TRANSMISSION   Yes ___ No ___

___ ALWAYS   ___ SELDOM   ENGINE   Yes ___ No ___

___ NOT SURE   ___ FIRST TIME   ACCESSORY
                                                    INSTALLATION Yes ___ No ___

Notes: _____

_____

PRELIMINARY QUICK CHECKS

. Fluid Level   _____ Pass _____ High _____ Low

. Fluid Condition   _____ Pass _____ Fail

. Cooler Lines Visual   _____ Pass _____ Fail

Notes: _____

_____

89445G01

Fig. 1 Diagnosis guide

AUTOMATIC TRANSMISSION	**CUSTOMER QUESTIONNAIRE**

1. How long have you had the condition?   R. O. _____

   ☐ Since car was new
   ☐ Recently (when?) _____
   ☐ Came on gradually   ☐ Suddenly

2. Describe the condition?

	P-R-N-D-2-1 SELECTOR POSITION(S)	CHECK AS APPROPRIATE WHICH GEAR?		
		HIGH	INTERMEDIATE	LOW
☐ Slow Engagement	_____	_____	_____	_____
☐ Rough Engagement	_____	_____	_____	_____
☐ Slip	_____	_____	_____	_____
☐ No Drive	_____	_____	_____	_____
☐ No Upshift	_____	_____	_____	_____
☐ No Downshift	_____	_____	_____	_____
☐ Slip During Shift	_____	_____	_____	_____
☐ Wrong Shift Speed(s)	_____	_____	_____	_____
☐ Rough Shift	_____	_____	_____	_____
☐ Mushy Shift	_____	_____	_____	_____
☐ Erratic Shift	_____	_____	_____	_____
☐ Engine "runaway or "buzzy"	_____	_____	_____	_____
☐ No Kickdown	_____	_____	_____	_____
☐ Starts in high gear in D				
☐ Starts in intermediate gear in D				
☐ Oil leak (where?) _____				

3. Which of the following cause or affect the condition?

   ☐ Transmission cold   ☐ Engine at fast (cold) idle
   ☐ After warm-up   ☐ Normal idle
   ☐ High speed   ☐ Wet road
   ☐ Cruising speed   ☐ Dry road
   ☐ Low Speed   ☐ Braking
   ☐ Accelerating   ☐ Coasting down

4. Does the engine need a tune-up?

   ☐ Yes   ☐ No   ☐ When was last tune-up? _____

5. Describe any strange noises

   ☐ Rumble   ☐ Squeak
   ☐ Knock   ☐ Grind
   ☐ Chatter   ☐ Hiss
   ☐ Snap or pop   ☐ Scrape
   ☐ Buzz   ☐ Other (describe)_____
   ☐ Whine   _____

89445G13

Fig. 2 Diagnostic checksheet used by a technician to gather information

• Soft—a slow, almost unnoticeable; clutch application with very little shift feel

• Surge—repeated engine related feeling of acceleration and deceleration that is less intense than the chuggle condition

• Tie-up—a condition where two opposing clutches are attempting to be applied at the same time, causing the engine to labor with a noticeable loss of engine rpm

Basic Inspection

Once you have determined what the transmission/transaxle is doing or not doing, the next step is to perform some simple and basic checks on the vehicle.

FLUID

The fluid level and condition are very important to proper automatic transmission/transaxle operation. The fluid can hold some immediate and important clues to the general condition of the automatic transmission/transaxle . Improper fluid level can be responsible for many malfunctions. If the automatic transmission/transaxle is even one pint low, it can produce malfunctioning symptoms.

Level

◆ **See Figure 3**

The fluid level should be checked with the automatic transmission/transaxle at normal operating temperature, approximately 180°F (85°C). Driving at least twenty minutes at highway speeds or the equivalent heavy city driving can achieve this temperature. The operating temperature can not be achieved by with the automatic transmission/transaxle in neutral or park and the engine at high idle. Because it is not always possible to check the fluid level at operating temperature, most dipsticks provide a HOT and COLD measurement markings. As part of the fluid level check, perform a visual inspection of the automatic transmission/transaxle for leaks and deteriorated hoses or other fluid connections.

Check the fluid level as follows:
1. Bring fluid to operating temperature.
2. Position the vehicle so that it is level.
3. Full apply the parking brake.
4. With the engine idling, move the selector lever through all the gear positions.
5. Leave the engine at idle (most vehicles), and place the selector lever in the position specified by the manufacturer (park or neutral).

➡**Vehicles can produce a different level when in the park position or the neutral position.**

6. Pull the dipstick from the tube and wipe the blade clean. Next, insert the dipstick into the tube and ensure that it is seated. Withdrawal the dipstick and check the level and condition (color, smell, contamination).

✳✳ CAUTION

DO NOT road test the vehicle if the fluid level is excessively high or low, further damage to the automatic transmission/transaxle could result.

7. Correct the fluid level, then road test.

A low fluid level condition can result in the automatic transmission/transaxle oil pump to cavitate (suck air) which will result in insufficient fluid pressure to apply the friction clutches, causing clutch and band slippage which will overheat and seriously damage the internals of the automatic transmission/transaxle . Low fluid levels can also cause delayed engagement and mushy shifts. It also can cause disturbing noises, such as regulator valve buzz and pump whine.

The cause of low fluid level should be determined and the problem then solved. External leakage, vacuum modulator diaphragm leakage, or

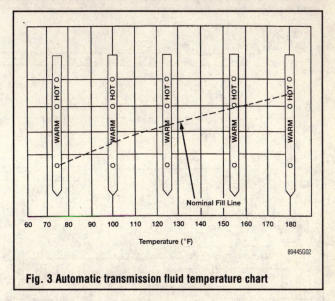

Fig. 3 Automatic transmission fluid temperature chart

improper filings are some of the causes of low fluid level. If no external leakage or vacuum modulator diaphragm leakage can be found, then possibly the fluid has not been properly filled and the fluid level should be checked and filled. If the fluid level appears to be normal, then care should be taken to fill to the appropriate hot or cold marks. At this point, the transmission/transaxle should perform as normal. The automobile should be taken for a road test and then given the okay to drive.

A high fluid level condition creates a situation where the fluid is chuned up by the rotating gear train, causing fluid aeration (foaming) and overheating of the automatic transmission/transaxle . The combination of foaming and high temperature causes premature brake-down of the fluid's lubricating and hydraulic properties, interfering with normal valve, clutch, and servo operation. Overfilling the automatic transmission/transaxle causes the fluid to push out through the automatic transmission/transaxle breather or the fill tube. To avoid overfilling the automatic transmission/transaxle , check with the vehicle's recommended fluid capacities for the proper amount of fluid needed. The overhaul fluid requirements are based on the total automatic transmission/transaxle , which includes the torque converter and oil pan. When performing routine maintenance, only the pan fluid is all that can be changed.

When replacing the fluid during the operations of overhaul or fluid changing, it is important to follow the recommendations of the car manufacturer. In most cases, the requirements are based on the condition of the transmission/transaxle and this involves the converter, oil pan and drain plugs. If the requirements are met, oil pan fluid levels will not have high fluid level conditions because most of the time this is the only fluid that can be changed. The drain plug is not always present on the torque converter. When refilling the automatic transmission/transaxle , stop at least one quart short of the fill recommendation and check the dipstick. Because this is a cold check, add fluid to the cold level or no further than the add mark, then bring the automatic transmission/transaxle up to operating temperatures and recheck the fluid level.

A common cause of overfilling the automatic transmission/transaxle is misjudgment of the temperature when checking the fluid level. When adding fluid, always allow the fluid level to stabilize over a three-minute period before rechecking the level.

Condition

◆ **See Figure 4**

With modern automatic transmission fluids, the color and smell are no longer definite when examining the fluid condition. Fluid colors range from an almost a clear fluid to a deep reddish color. Some fluids tend to discolor and darken somewhat after use. This darkening used to indicate fluid failure. The smell of the fluid is also a false indication of failure with

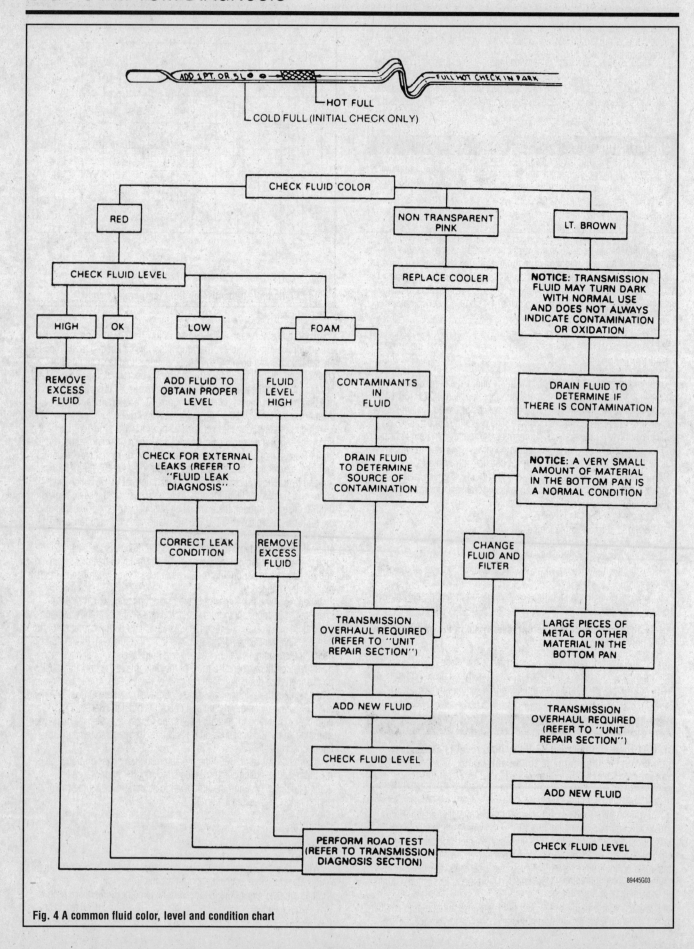

Fig. 4 A common fluid color, level and condition chart

current automatic transmission fluids. The changes in current fluids are caused by reformulating the new fluids and the addition of resins in the friction material for bonding. Due to the reduction of asbestos fibers, these new fluids have been introduced. Some failure clues can still be detected from fluid diagnosis. Depositing several drop of fluid from the dipstick onto a clean white cloth helps in diagnosing the fluid condition and, to some extent, the internal failure of the automatic transmission/transaxle . If the test reveals that the fluid is black in appearance, and possibly contaminated with particles of metal and/or friction material, sever internal damage has occurred. If the fluid is heavily varnished, loosing its fluid like properties, due to overheating, transmission/transaxle overhaul is certain. A simple test for fluidity, is to compare the soak rates of the bad fluid to the good fluid. The questionable fluid will pool-up and slowly soak into a paper towel, the good fluid will soak easily into the paper towel. If equipped with an internal trans cooler in the radiator and a milky appearance to the fluid has occurred, the engine coolant has contaminated the automatic transmission fluid. This causes the seals to swell up and the friction material to become unglued from its backing. To rectify this failure, repair the cause of the contamination, flush out the cooler lines and the nonlockup torque converter (lockup torque converters must be replaced), then completely overhaul the automatic transmission/transaxle . Fluid that appears foamy or bubbly can be the result of a high or low fluid condition. If the level appears to be within range of the dipstick, then suspect an air leak on the suction side of the fluid pump. Begin with an inspection of the fluid filter installation, noting the gasket or O-ring damage or improper installation.

If further verification of the fluid and transmission/transaxle condition is required, the automatic transmission/transaxle oil pan should be removed and the fluid/deposits examined.

There is always some normal residue in the pan, but watch for significant particles of steel, bronze or plastic, indicating damage to a bushing, thrust washer or other internal hard parts. Also, look for friction material, indicating clutch or band failure. Steel particles or friction material, not associated with clutch or band failure, indicates torque converter failure. It may be necessary to cut open the filter to inspect for residue, because the filter will retain and conceal the particles to prevent them from doing further damage.

Leakage

▶ See Figures 5 thru 10

Fluid leakage form the transmission/transaxle can be a simple repair, if corrected when first noticed. The cause of most external leaks can generally be located and repaired with the transmission/transaxle in the vehicle. Leaking fluid will cause a low fluid condition, and if compensated by overfilling the unit, opens the way for terminal damage, not to mention polluting the environment.

In order to repair the fluid leak, you must verify that the leak is transmission/transaxle related and not engine related. A general method of locating leaks is as follows: Thoroughly clean the suspect area of the leak, run the vehicle approximately 15 minutes or until normal operating temperatures are achieved, park the vehicle over a clean cardboard or paper towel, shut the vehicle **OFF** and look for spots on the paper, then make necessary repairs. A method using aerosol powder (foot powder) is an alternative way to pinpoint a fluid leak. Thoroughly clean the suspect area of the leak with a solvent, apply an aerosol type powder (foot powder works just fine) to the suspected leak area, run the vehicle approximately 15 minutes or until normal operating temperatures are achieved, shut the vehicle **OFF**, inspect the leak area and trace the leak path through the powder to find the source of the leak, then make the appropriate leak repair. A method utilizing fluorescent dye and black light to locate the origin of a leak can be adapted to locate more than just automatic transmission/transaxle leaks, like engine oil, power steering gear, transfer case and differential fluid leaks. Following the manufacturer's recommendation, add the dye to the fluid, run the vehicle approximately 15 minutes or until normal operating temperatures are achieved, then shine the black light until you find the leak area (the dye will glow with the light shining on it).

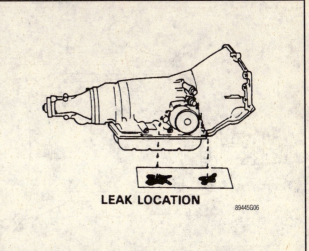

Fig. 5 Placing a clean piece of paper or cardboard under the vehicle to help locate the leak

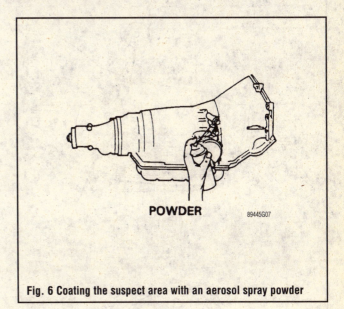

Fig. 6 Coating the suspect area with an aerosol spray powder

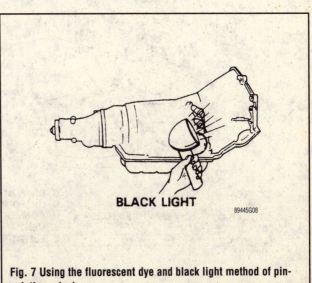

Fig. 7 Using the fluorescent dye and black light method of pinpointing a leak

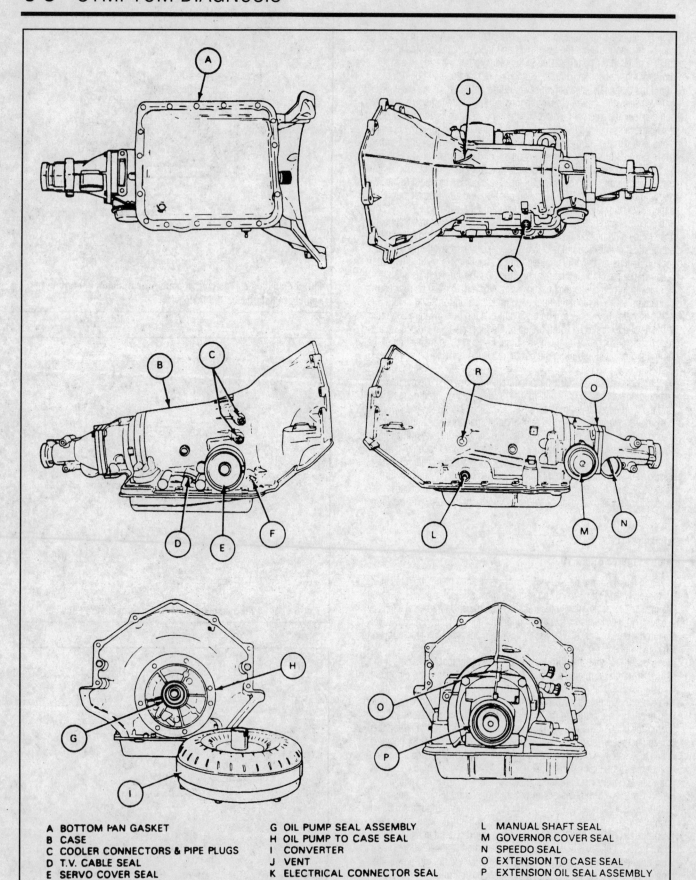

A	BOTTOM PAN GASKET	G OIL PUMP SEAL ASSEMBLY	L MANUAL SHAFT SEAL
B	CASE	H OIL PUMP TO CASE SEAL	M GOVERNOR COVER SEAL
C	COOLER CONNECTORS & PIPE PLUGS	I CONVERTER	N SPEEDO SEAL
D	T.V. CABLE SEAL	J VENT	O EXTENSION TO CASE SEAL
E	SERVO COVER SEAL	K ELECTRICAL CONNECTOR SEAL	P EXTENSION OIL SEAL ASSEMBLY
F	OIL FILL TUBE SEAL		R LINE PRESSURE TAP

89445G04

Fig. 8 Typical leak points of a transmission

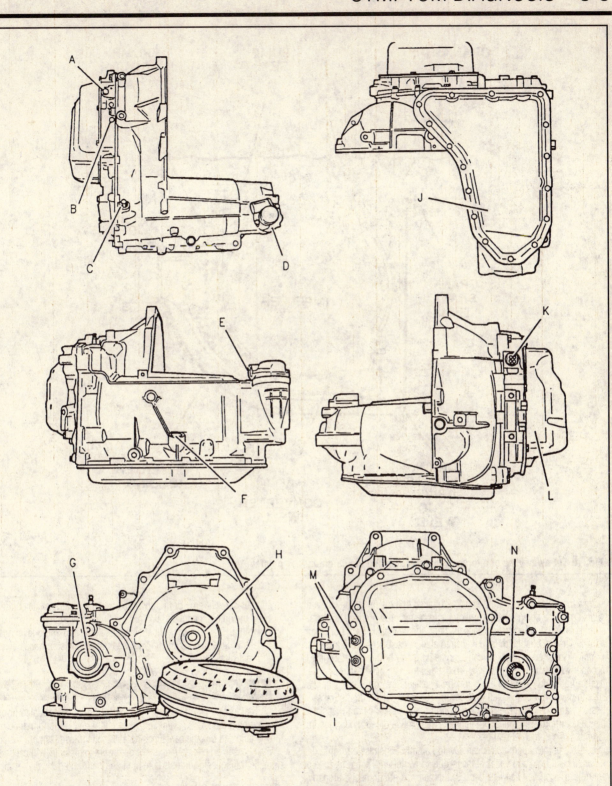

A T.V. CABLE AND/OR CASE SEAL
B FLUID VENT
C MANUAL LEVER SEAL
D GOVERNOR OR SPEED SENSOR COVER
E SPEEDOMETER GEAR ASSEMBLY
F FILLER PIPE AND/OR SEAL
G AXLE SEAL (R.H.)

H CONVERTER TO CASE SEAL
I TORQUE CONVERTER
J FLUID PAN (BOTTOM)
K ELECTRICAL CONNECTOR
L VALVE BODY COVER (SIDE PAN)
M COOLER FITTINGS
N AXLE SEAL (L.H.)

89445G05

Fig. 9 Typical leak points of transaxle

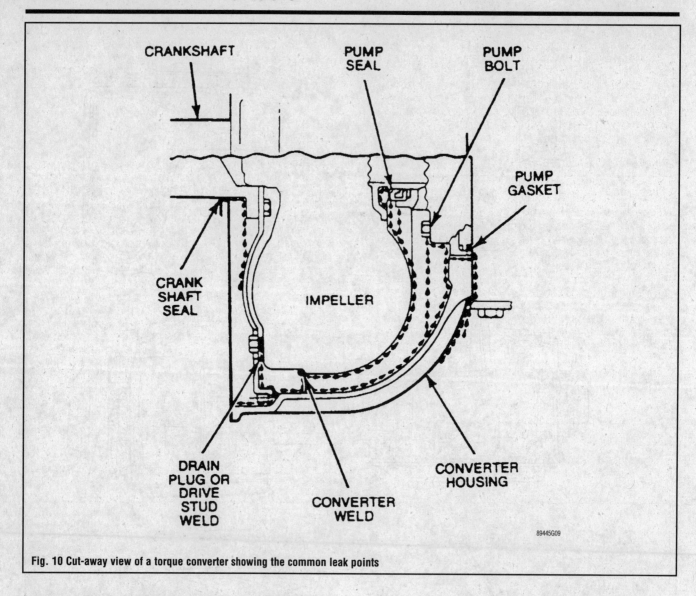

Fig. 10 Cut-away view of a torque converter showing the common leak points

Once the leak has been pinpointed and traced to the source, the cause of the leak must be determined in order for it to be repaired properly. If a gasket is replaced, but the sealing flange is bent, the new gasket will not repair the leak. The bent flange must be repaired also. Gaskets and seals are areas where leaks can occur, gaskets can become hard and brittle, then fail (leak), improperly tightened fasteners, warped flanges from an overheated transmission/transaxle , and use of the incorrect sealant, are conditions where a leak can develop. Seal can fail from several reasons, overfilling the transmission/transaxle, fluid pressure too high, plugged vent tube/drain back holes, age, improper installation, improper seal, loose or worn bearings in the unit, and a damaged seal bore.

Possible points where a leak can develop are the transmission/transaxle oil pan, a case leak, and a leak at the torque converter end, at the fill tube or vent hole. The oil pan can leak because the pan was not tightened properly, an improperly installed or wrong gasket, the oil pan or mounting surface is not flat or sealer was used when not required by the manufacturer. The case can leak from many different seal, such as the filler tube multi-lip seal being damaged or missing, the filler tube misaligned, speedometer/speed sensor seal damage, shifter seal missing or damaged, oil cooler connections loose, driveshaft/halfshaft seal worn or damaged, and the line pressure check port fitting loose. To repair a leak at the torque converter end of the transmission/transaxle , the unit will have to be removed, a converter seal lip can be cut or wear out, the converter can crack at the weld points or mounting points because of improperly tightened fasteners. If fluid is com-

ing out the vent pipe or fill tube, the transmission/transaxle could have been overfilled, a coolant leak into the fluid cooling lines is contaminating the unit, using the wrong dipstick to check the trans fluid level or the vent hole is plugged could possibly cause the transmission/transaxle to leak out the vent pipe.

Road Test

The operating condition of the vehicle should be evaluated prior to taking it on a road test. A vehicle that has poor engine performance or improperly adjusted linkage can affect the automatic transmission/transaxle . In many cases an engine that is not performing efficiently will cause the automatic transmission/transaxle to shift and perform erratically.

ENGINE PERFORMANCE

By placing the engine under load, some performance problems will become more apparent. A quick method for checking rough engine performance is as follows:
1. Start the engine and bring it to normal operating temperature.
2. Apply the parking brake.
3. Fully apply the service brakes.
4. Place the gear selector in the drive (D) position.
5. Raise the engine rpm's until stall speed is reached.

6. Listen for an engine misfire.

7. Release the accelerator pedal.

8. With the engine at idle, move the selector to the neutral (N) position.

9. Raise the engine speed to 1000–1200 rpm for a few minutes to cool the ATF.

If the engine has a weak or dead cylinder, the load test will make the symptoms easier to detect. This type of engine problem is easier to notice than other types of malfunctions.

LINKAGE & CABLE ADJUSTMENTS

▶ **See Figure 11**

Proper adjustment of the manual linkage and throttle valve cable is essential for proper automatic transmission/transaxle operation. The linkage connecting the gear selector inside the vehicle to the manual valve in the transmission/transaxle should be inspected to verify proper operation and adjustment. If the transmission/transaxle utilizes a throttle valve cable, its operation and adjustment should be checked. If equipped with a vacuum modulator, the vacuum lines should be inspected and the operation of the modulator should be verified by removing the vacuum line from the modulator, then increasing the engine speed to 1000 rpm. You should be able to feel suction at the end of the hose. Reconnect all lines before continuing.

Once you have established that the engine is performing properly and that the shift linkage and vacuum systems are properly functioning and adjusted correctly, you can proceed to the road test.

ROAD TEST

▶ **See Figures 12, 13 and 14**

To road test the transmission, you may need a scan tool to display engine rpm and feedback from the engine control and/or

transmission/transaxle control computers. Connecting pressure testers to the appropriate ports on the transmission/transaxle will indicate what is occurring inside the transmission/transaxle as it is actually happening.

The engine and transmission fluid levels must be correct. Road test the vehicle under the conditions as described, such as:

- Always
- Cold only
- Hot only
- Morning only
- Startup only

Operate the transmission/transaxle in each selector range to check for clutch/band slipping and engagement quality and any abnormal variations in the shift schedule. Performance testing should be made at minimum, medium, and heavy throttle modes. When performance testing the transmission/transaxle , observe the whether the shifts are harsh or long and drawn out. Check the engine speeds (rpm) at which the upshifts and down shifts occur, also the amount of rpm change between shifts. Look for as engine flare up during a shift, especially on medium and heavy throttle testing. What rpm and road speed does the torque converter engage and disengage.

To attain your minimum shift points, the throttle should be applied just enough to keep the vehicle moving through 20 mph (32 km/h). Engine manifold vacuum must be kept high. For best results, find a flat stretch of road or one that is slightly downgrade. The minimum shift pattern checkout requires a careful and very slight throttle manipulation for accurate test results.

The diagnostic road test requires collecting information and data. Due to the wide variety of transmission/transaxle s covered, it is recommended to refer to the manufacturer's applicable shift speed specifications. Because of the many items and readings that have to be recorded, it is best to chart the results rather than depending on memory. A comprehensive road test program for several models are outlined in this section.

Sometimes the transmission/transaxle shift feel is difficult to sense, especially at the lower shift schedules. A special tester called a Shiftalizer

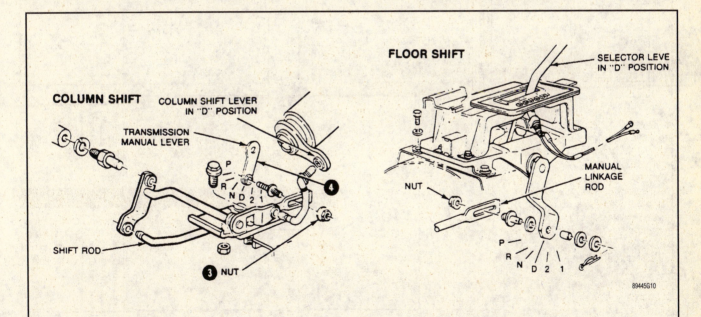

Fig. 11 Be sure the shift linkage is properly adjusted to eliminate that potential problem area

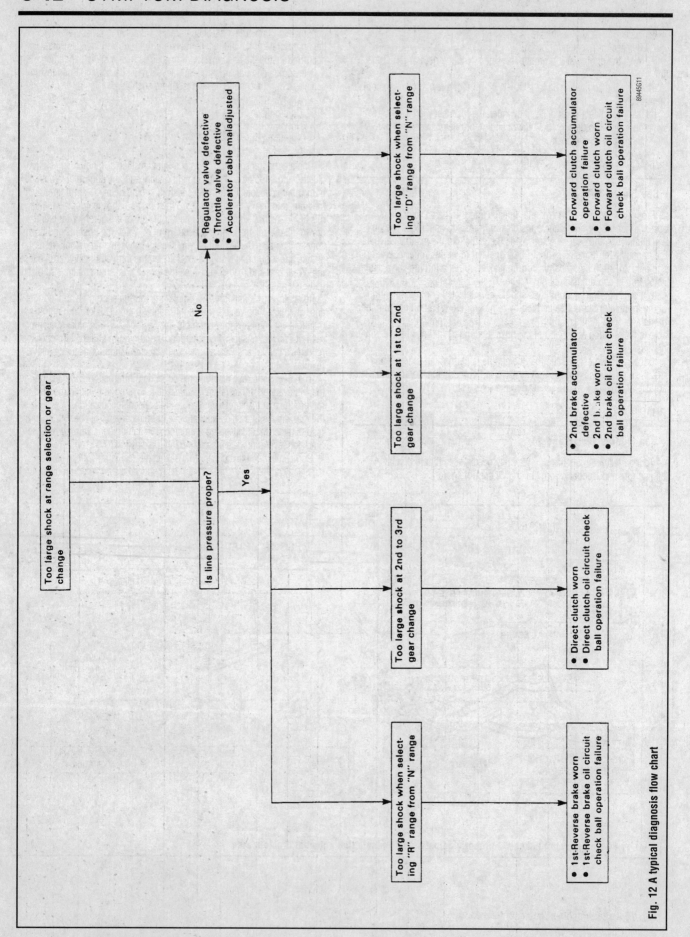

Too large shock at range selection or gear change

Is line pressure proper?

No →
- Regulator valve defective
- Throttle valve defective
- Accelerator cable maladjusted

Yes

Too large shock when selecting "R" range from "N" range →
- 1st-Reverse brake worn
- 1st-Reverse brake oil circuit check ball operation failure

Too large shock at 2nd to 3rd gear change →
- Direct clutch worn
- Direct clutch oil circuit check ball operation failure

Too large shock at 1st to 2nd gear change →
- 2nd brake accumulator defective
- 2nd brake worn
- 2nd brake oil circuit check ball operation failure

Too large shock when selecting "D" range from "N" range →
- Forward clutch accumulator operation failure
- Forward clutch worn
- Forward clutch oil circuit check ball operation failure

89445G11

Fig. 12 A typical diagnosis flow chart

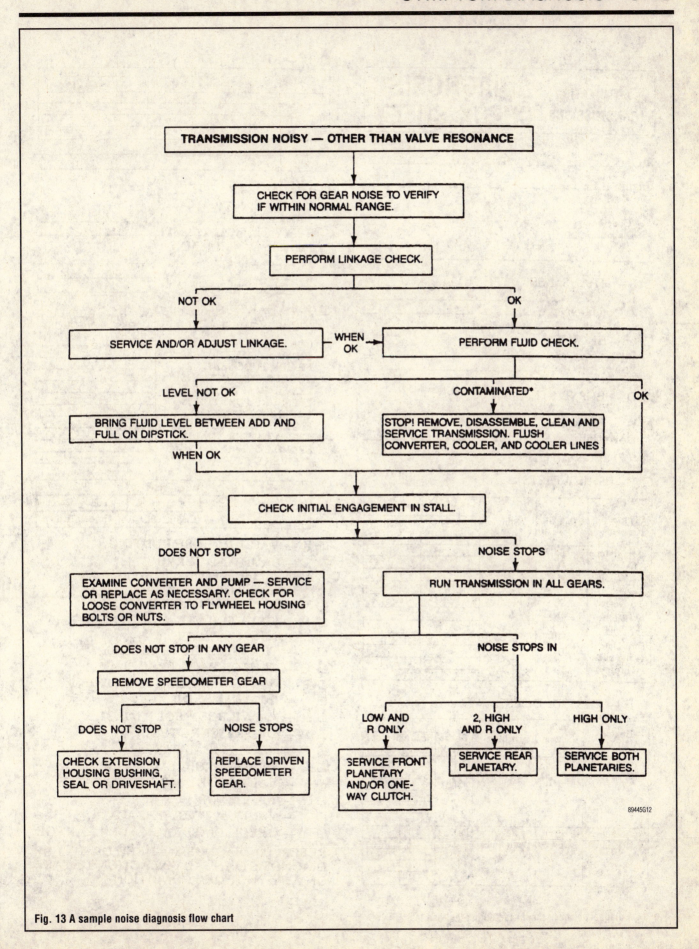

Fig. 13 A sample noise diagnosis flow chart

89445G12

AUTOMATIC TRANSMISSIONS	**DIAGNOSIS CHECK SHEET**

R.O. _____ Trans. _____

Engine _____

	Code on Diagnosis Wheel	Check/Test	Remarks

☐ **B — TRANSMISSION FLUID**
1. Level _____ _____
2. Condition _____ _____

☐ **C — ENGINE**
Idle _____ _____
Power _____ _____

☐ **D — EGR SYSTEM**

☐ **E — LINKAGE**
Downshift _____ _____
Manual _____ _____

☐ **F — SHIFT TESTS**

Throttle Opening	Range	Shift	Shift Points (MPH)	
			Record Actual	Record Spec.
Minimum (Above 12'' Vacuum)	D	1-2		
	D	2-3		
	D	3-1		
	1	2-1		
To Detent (Torque Demand)	D	1-2		
	D	2-3		
		3-2		
Thru Detent (wide open Throttle)	D	1-2		
	D	2-3		
	D	3-2		
	D	2-1 or 3-1		

☐ **G — PRESSURE TEST**

Engine RPM	Manifold Vacuum In-Hg	Throttle	Range	PSI	
				Record Actual	Record Spec.
Idle	Above 12	Closed	P		
			N		
			D		
			2		
			1		
			R		
As Required	10	As Required	D, 2, 1		
As Required	Below 3	Wide Open	D		
			2		
			1		
			R		

☐ **H — STALL TEST**

Range	Specified Engine RPM	Record Actual Engine RPM
D		
2		
1		
R		

Results _____

☐ **I — GOVERNOR TEST**
Cutback Speed (C3, C4, C6)
10'' Vacuum _____ MPH _____
0-2'' Vacuum _____ MPH _____
Pressure at MPH (FMX)
10 _____ PSI _____
20 _____ PSI _____
30 _____ PSI _____

☐ **J — LEAK TEST**

CHECK THESE	OK	OIL/FLUID *(COLOR)
CONVERTER AREA		
OIL PAN GASKET		
FILLER TUBE/SEAL		
COOLER/CONNECTIONS		
LEVER SHAFT SEALS		
PRESSURE PORT PLUGS		
EXTENSION/CASE GASKET		
EXTENSION SEAL/BUSHING		
SPEEDOMETER ADAPTER		
SERVO COVERS		
AIR VENT		

*Color Codes	Auto. Trans.	Red
	Power Steering	Yellow-Green
	Engine Oil	Golden Brown

☐ **K — VACUUM HOSE ROUTING**
☐ **L — BAND AND SERVO**
1. Intermediate Band Adj.
2. Reverse Band Adj.
3. Polished, Glazed Band, Drum
☐ **M — DRIVESHAFT, U-JOINTS, ENGINE MOUNTS**
☐ **N — TRANSMISSION END PLAY**
☐ **O — CLUTCH PACK FREE PLAY**
☐ **P — VALVE BODY DIRTY, STICKING**
☐ **Q — INTERNAL LINKAGE**
☐ **R — VALVE BODY BOLT TORQUE**
☐ **S — AIR PRESSURE TEST**
☐ **T — MECHANICAL PARTS**
☐ **U — VERIFY PROBLEM**
☐ **V — VALVE BODY MOUNTING FACES**
☐ **W — SPEEDO DRIVEN GEAR**
☐ **X — VACUUM TO DIAPHRAGM**
☐ **Y — CHECK DIAPHRAGM FOR LEAKAGE**

REFER TO DIAGNOSIS WHEEL OR TO CAR DIAGNOSIS MANUAL FOR ACTION TO TAKE ON ANY "NOT OK" CONDITION.

89445G14

Fig. 14 A diagnostic check sheet used during the road test

can be plugged into the cigar lighter or the 12V power outlet. It acts as an audio tachometer by picking up the generator rpm. The shift points and shift quality are easily detected. As an alternative technique, tune the local AM radio station to the off-beat high-frequency side of the broadcast.

The Garage Shift Feel is assessed before moving the vehicle onto the road. With the brake firmly applied, move the shift lever into each of the range positions. Wait for the gear to engage and analyze the results. You are checking for power transmission/transaxle and the quality of the engagement in the forward and reverse gears. There should be no delay, no slip, and no chatter, that indicates that the pump is turning, line pressure exists, and the torque converter is full.

Clutch and band slippage during the road test will cause the engine to flare. By process of elimination, any clutches/bands that slip can be verified, as well as those that are in proper working order. The idea is to correct the external condition before a total overhaul is needed to restore normal operation.

The key to clutch/band analysis is in comparing the results of the road test readings to a shift speed chart for the model of transmission/transaxle being serviced. Comparing the values on the chart to the values recorded will indicate between which gears the problems lie.

When diagnosing a noise while road testing, it is important that the driveline is in good condition. A worn driveshaft center bearing or U-joint could mimic a transmission/transaxle noise. A CV joint that is worn could produce enough noise to indicate major engine/transaxle failure. In rear wheel drive transmissions most of the noise problems are located in the torque converter and planetary gear set. Transaxles share those areas as well as the sprocket-and-chain assemblies and the final drive.

To identify what is making the noise, road test the vehicle and listen to the pitch of the noise, what gear it is present in, if it is load sensitive, and during what driving conditions it occurs. Many factory service manuals utilize flow charts to guide the technician through the diagnosis procedure. It is important to that you understand how to use a flow chart. Using the flow chart and the data gathered during the road test you should be able to identify the source of the noise. If the noise is difficult to pinpoint, try operating the vehicle on a lift, with all wheels off the ground, so that listening devices can be used on the transmission/transaxle to locate the noise. Three simple listening devices are a piece of steel rod, a piece of heater hose, and a stethoscope. By placing the listening device against the case and listening to the end the noise will enter your ear. Then, by moving the contact point around the case, you can listen to different areas of the transmission/transaxle . Extreme care must be taken when working around the revolving drive train.

Conditions/Cause/Checks

The following is a general list of common automatic transmission/transaxle fault conditions with a typical cause for that condition, then a possible direction to begin investigating. If there are multiple conditions, compare the causes to find a common cause, this may help in pinpointing the area of malfunction.

IF YOU FIND THAT THERE IS OIL COMING OUT OF THE VENT OR FOAMING:

- If the oil level is too low, check and fill transaxle as necessary
- The transmission/transaxle fluid could be overfilled, if so correct the oil level
- If there is contamination with antifreeze, engine overheating, or the lube pipes are leaking, the transmission/transaxle fluid should be checked for foaming
- If there is a damaged seal assembly or a cracked or damaged filter assembly, the oil filter or seal should be checked for foaming and replaced if necessary
- If the vent does not close when hot, is not insulated correctly, or has incorrect pin heights, the thermos element in the case should be checked
- If the gasket is damaged or incorrectly installed, then the upper channel plate gasket should be checked

- If there are plugged drain back holes, then the drive sprocket support should be checked
- If the bi-metallic element is stuck open, replace the element

IF YOU FIND THAT THERE IS HIGH OR LOW OIL PRESSURE:

- If the oil level is too low, check and fill transaxle as necessary
- If an electronic pressure control solenoid is inoperative or damaged, need to do further testing on electrical system controlling the
- If the oil pump assembly slide is stuck, the seals damaged, the vanes damaged or the pump driveshaft broken or damaged, the oil pump assembly should be checked
- The vacuum line could be leaking, pinched, disconnected, or cut, the line should be checked and replaced if necessary
- The diaphragm could be leaking or damaged, the modulator should be checked and replaced if necessary
- The modulator valve could be nicked, scored or cut, the modulator valve should be checked and replaced if necessary
- The pressure regulator valve or springs could be nicked or scored or the springs damaged, the pressure regulator valve or springs should be checked and source of damage determined
- If the throttle position sensor is out of range, adjust accordingly
- If there is a damaged spring or missing ball, the pressure relief valve should be checked

IF YOU FIND THAT THERE IS A DELAYED ENGAGEMENT:

- The converter is not seating and allowed draining back, check the cooler checkball
- If there is a cut or damaged seal, then reverse the servo assembly
- If there is a damaged retainer and ball assembly, then check the shaft and housing assembly
- If there is a damaged or mislocated seal or a damaged and cut seal, then forward the servo assembly
- If there is a damaged or cut/rolled seal, then input clutch order piston seal

IF YOU FIND THAT THERE IS SLIPPING IN DRIVE:

- The fluid level could be low, if so correct the fluid level
- If the vacuum line is pinched so that the vacuum response is slowed, then check the vacuum line
- If the modulator is damaged, then check the modulator and replace if necessary
- If there is low oil pressure, check the conditions for High/Low oil pressure
- If the modulator valves is stuck or binding, then check the modulator valve and replace if necessary
- If there is a damaged or missing cushion spring or retainer, then forward the servo assembly
- If there is a damaged seal, then forward the servo piston
- If there is a plugged oil filter, then check the oil filter and replace if necessary
- If there is a cut or damaged filter seal, then check the filter seal and replace if necessary
- If there is leaking, damaged or missing seal or low torque of the bolts, check the forward servo seal and replace if necessary
- If the stator roller clutch is not holding, check the torque converter
- If there is low torque that allows leakage for driven sprocket support, then check the bolt and tighten or replace as necessary
- If there are leaks at ball capsule or seals or the seals are damaged, check the input clutch assembly and replace as necessary
- If there is a sticking slide or leaking slide seals, then check the pump assembly and replace as necessary
- If the 1-2 Shift solenoid or torque converter clutch solenoid has malfunctioned or mislocated, check the 1-2 shift solenoid or torque converter clutch solenoid and replace as necessary

IF YOU FIND THAT THERE IS NO DRIVE IN DRIVE RANGE:

- The fluid level could be low, if so correct the fluid level
- If there is low oil pressure, check the conditions for High/Low oil pressure
- If the manual linkage is misadjusted or disconnected, then check the linkage and adjust or connect as necessary
- If there are damaged or missing O-rings or plugged oil filter, check and change as needed
- If there is a worn or damaged oil pump or oil pump drive shaft is damaged, check and change as necessary
- If the oil pump assembly slide is stuck, the seals damaged, the vanes damaged or the pump drive damaged, the oil pump assembly should be checked
- If the half shaft has damaged splines or is disengaged from transaxle, check and replace as necessary
- If the sprocket shaft to converter turbine spline is damaged, check and replace drive sprocket
- If the sprocket shaft to direct/intermediate clutch hub is damaged, check and replace drive sprocket
- If the 1-2 Shift solenoid or torque converter clutch solenoid has malfunctioned or mislocated, check the 1-2 shift solenoid or torque converter clutch solenoid and replace as necessary
- If there is a damaged seal or piston, then forward the servo assembly
- If there is a damaged roller clutch, check the clutch and adjust as necessary or replace
- If the drive axles are disengaged, check the axles and adjust as necessary
- If the drive chain assembly is damaged or broken, replace as needed
- If the checkball is missing, check and replace
- If because of the stator roller clutch the vehicle moves but is very sluggish, check the torque converter and adjust or change as necessary
- If there are damaged or broken drive link chain or damaged sprockets, check the drive link assembly and replace or adjust as necessary
- If there are burned or missing clutch plates, damaged piston or seals, leaking housing checkball assembly, damaged input shaft, or blocked feed passages on the input shaft, check the input clutch assembly and replace and adjust as needed
- If there is an improper sun gear assembly or a damaged sprag, check and input the sprag and sun gear assembly as needed
- If there are damaged pinions, internal gear or sun gear, input carrier and reaction carrier assembly as necessary
- If the output shaft is damaged or misassembled with axles, check the output shaft and adjust or replace as necessary
- If there is a burned forward band assembly or band apply pin is mislocated with the band, check the forward band assembly and replace as needed
- If there is a broken spring and the pawl remains engaged, check the parking prawl and replace parts as necessary
- If the side gear, gears, pinion, or internal gear are damaged, check the final drive assembly or final drive sun gear shaft and replace or repair as necessary

IF YOU FIND THAT THERE IS NO REVERSE IN REVERSE RANGE:

- If there are conditions for High/Low oil pressure, check the possible diagnostic trouble code set
- If there is a damaged or missing cushion spring or retainer, then reverse the servo assembly
- If there is a damaged seal or piston or misassembled servo, then reverse the servo
- If there are damaged or broken drive link chain or damaged sprockets, check the drive link assembly and replace or adjust as necessary
- If there are burned or missing clutch plates, damaged piston or seals, leaking housing checkball assembly, damaged input shaft, or blocked feed passages on the input shaft, check the input clutch assembly and replace and adjust as needed

- If there is an improper sun gear assembly or a damaged sprag, check and input the sprag and sun gear assembly as needed
- If there are damaged pinions, internal gear or sun gear, input carrier and reaction carrier assembly as necessary
- If the oil pump assembly slide is stuck, the seals damaged, the vanes damaged or the pump drive damaged, the oil pump assembly should be checked
- If there is a burned or damaged reverse band or band apply pin is mislocated, check the reverse band assembly and replace as needed
- If the 1-2 Shift solenoid or torque converter clutch solenoid has malfunctioned or walked out, check the 1-2 shift solenoid valve or torque converter clutch solenoid valve and replace as necessary
- If there are damaged splines, check the reverse reaction drum

IF YOU FIND THAT THERE IS FIRST GEAR ONLY:

- If the microprocessor is damaged, need to do further testing on electrical system controlling the automatic transmission/transaxle
- If the shift solenoid has a wiring short or open, need to do further testing on electrical system controlling the automatic transmission/transaxle
- If there is sticking or binding of 1-2 shift valve or mispositioned or damaged space plate or gaskets, then check the control valve assembly
- If there are damaged oil seals, check the driven sprocket support
- If there are damaged clutch plates, damaged piston or seals, or misassembled parts, check the second clutch assembly
- If the eave spring is damaged or missing, check and service as required
- If the clutch spring is damaged, check and service as required
- If the intermediate clutch tap plug is loose or missing, check and service as required
- If there are damaged splines, reverse reaction drum
- If the front carrier is damaged, inspect welds and service as needed
- If the check ball is missing or damaged, check and replace the check ball
- If the control assembly bolts are too loose or too tight, check the bolts and tighten
- If the clutch hub seals are damaged or missing or the holes are blocked, check and service as required

IF YOU FIND THAT THERE IS HARSH OR SOFT 1-2 SHIFT FEEL:

- The fluid level could be low, if so correct the fluid level
- If cover bolts are improperly tightened, the pistons or seals are damaged, the spring is damaged, the gaskets are damaged or mispositioned, the bolts are torqued low, or the piston and springs are installed upside down, check the 1-2 accumulator piston, cover and springs and adjust or replace as necessary
- If the 1-2 accumulator valve is stuck , nicked or damaged, or the 1-2 accumulator bushing retainer is mislocated or missing, check the control valve assembly
- If there is a checkball mislocated, check and relocate
- If the driven sprocket support ring has a rolled or twisted seal or a worn sleeve in the second clutch housing, check the driven sprocket support and change parts as required
- If there is low oil pressure, check the conditions for High/Low oil pressure
- If there is a cracked or damaged filter, check and change
- If the electronic pressure control solenoid is inoperative or damaged, need to de further testing on electrical system controlling the
- If the microprocessor is damaged, need to do further testing on electrical system controlling the automatic transmission/transaxle

IF YOU FIND THAT THERE IS HARSH OR SOFT 2-1 SHIFT FEEL:

- If there is a wrong spacer plate, check and replace
- If the seal is cut, check and replace 1-2 accumulator seal
- If there is a stuck valve, check the 1-2 accumulator valve

IF YOU FIND THAT THERE IS 1-2 SHIFT SPEED HIGH OR LOW:

- If there is an inoperative vehicle speed sensor, need to do further testing on electrical system controlling the automatic transmission/transaxle If there is wrong or missing speedometer gear, install correct speedometer driven gear
- If the 1-2 accumulator valve is stuck , nicked or damaged, or the 1-2 accumulator bushing retainer is mislocated or missing, check the control valve assembly

IF YOU FIND THAT THERE IS 1-2 SHIFT SHUDDER:

- If there is worn fiber plates, leaking clutch valve ball, cut seal, damaged steel plated, or mispositioned snap ring, check the second clutch
- If there are damaged seal rings, check the driven sprocket support
- If there is low oil pressure, check the conditions for High/Low oil pressure

IF YOU FIND THAT THERE IS HIGH OR LOW UPSHIFT OR DOWNSHIFT SPEED:

- If there is an incorrect tooth count on VS sensor, wrong PROM, or wrong final drive ratio or model, check the diagnostic trouble code set Also check service bulletins

IF YOU FIND THAT THERE IS FIRST AND SECOND GEAR ONLY (B SOLENOID STUCK ON):

- If there is foreign material in the solenoid, the PCM signal is grounded or the solenoid return wire to ground is pinched, check the 2-3 shift solenoid ON failure

IF YOU FIND THAT THERE IS FIRST AND FOURTH GEAR ONLY (A SOLENOID STUCK ON):

- If there is foreign material in the solenoid, the PCM signal is grounded or the solenoid return wire to ground is pinched, check the 1-2 shift solenoid ON failure

IF YOU FIND THAT THERE IS SECOND AND THIRD GEAR ONLY (A SOLENOID STUCK ON):

- If there is foreign material plugging the filter, the PCM signal is not grounded, the O-ring has failed, the solenoid does not sufficient force, no voltage is supplied to the 1-2 shift solenoid or the wires are not connected to the solenoid, then check the 1-2 shift solenoid OFF failure

IF YOU FIND THAT THERE IS NO 2-3 SHIFT:

- If the electronic pressure control solenoid is inoperative or damaged, need to de further testing on electrical system controlling the
- If the microprocessor is damaged, need to perform further testing on electrical system controlling the
- If the servo bore or piston is damaged, install correct apply rod if necessary
- If the case servo release passage is blocked, determine the source of blockage and service
- If the servo release tube is leaking or improperly installed, check and correct
- If the wrong apply rod for the low/intermediate servo is applied, install correct apply rod
- If the 2-3 shift valve or the 3-2 manual downshift valve is stuck, nicked, or damaged, check the control valve assembly
- If the bolts are too tight or too loose, tighten the control assembly as necessary
- If the check balls are missing, check and replace

- If the channel plate gasket is mispositioned or damaged, check and replace the channel plate gasket as necessary
- If the third clutch passage is blocked, check the driven sprocket support
- If there are damaged seals or blocked oil passages, check the input housing and shaft assembly
- If the clutch plates are burned or the piston, seals or checkball assembly is damaged, check the third clutch assembly
- If the cage or springs are damaged, the rollers are out of the cage, the input sun gear shaft is misassembled, no third gear or no engine braking in Manual Lo, check the third roller clutch assembly

IF YOU FIND THAT THERE IS HARSH OR SOFT 2-3/3-2 SHIFT FEEL:

- If the electronic pressure control solenoid is inoperative or damaged, need to de further testing on electrical system controlling the transmission/transaxle
- The microprocessor could be damaged, need to perform further testing on electrical system controlling the transmission/transaxle
- If the wrong apply rod for the low/intermediate servo is applied, install correct apply rod
- If there are conditions for High/Low oil pressure, check the possible diagnostic trouble code set
- If the checkball is mislocated for a soft shift, check the checkball and relocate
- If the checkball is mislocated for a harsh shift, check the checkball and relocate
- If the checkball is missing, check the control valve assembly
- If there are missing springs, cut seal, damaged gaskets, or low torque of bolts, check the 2-3 accumulator pistons, gaskets and seals and replace as necessary
- If there is a stuck valve, check the 2-3 Accumulator Valve
- If there is a valve stuck, nicked or damaged, check the 3-2 shift timing
- If there is damaged or missing springs or incorrect servo apply rod length, check for correct apply rod and replace if necessary

IF YOU FIND THAT THERE IS 2-3 SHIFT SPEED HIGH OR LOW:

- If there is an inoperative vehicle speed sensor, need to do further testing on electrical system controlling the automatic transmission/transaxle
- If there is wrong or missing speedometer gear, install correct speedometer driven gear
- If the 1-2 shift valve is stuck , nicked or damaged, check the control valve assembly
- If the shift solenoid is damaged, check the control assembly

IF YOU FIND THAT THERE IS THIRD GEAR ONLY:

- If there is a loose or corroded connector or defective PCM, check the PCM and transmission/transaxle wiring
- If there is a loose or corroded connector or defective generator, check the generator and wiring
- If there is a loose or corroded connector or defective speed sensor, check the speed sensor and wiring
- If there is a loose or corroded connector or defective quad driver, check the PCM and wiring 1-2- and 2-3 shift solenoid valves

IF YOU FIND THAT THERE IS FOURTH GEAR STARTS:

- The solenoid could be mislocated from the control valve assembly, check the 2-3 shift solenoid valve

IF YOU FIND THAT THERE IS THIRD AND FOURTH GEAR OPERATION:

- If there is foreign material and plugging in the filter, insufficient force of the solenoid, O-ring failure, an open wire from the shift solenoid to the PCM or the PCM is not grounding, no supply voltage to the solenoid, or the wires are not connected to the solenoid, check the 2-3 shift solenoid for OFF failure

IF YOU FIND THAT THERE IS NO 3-4 SHIFT:

- If there is a wiring short or open of shift solenoid, need to do further testing on electrical system controlling the transmission/transaxle
- The microprocessor could be damaged, need to do further testing on electrical system controlling the transmission/transaxle
- If the overdrive band assembly is not holding, perform air pressure test and service as necessary
- If the wrong apply rod for the overdrive servo assembly is applied, install correct apply rod
- If the servo bore or piston is damaged, the piston seals are damaged or missing, or return spring or retaining clip is missing or broken, install correct apply rod and service as necessary
- If the 1-2 shift valve is stuck, nicked or damaged or the spring is missing, check the control valve assembly
- If the 3-4 shift valve or the 4-3 manual downshift valve is stuck, nicked or damaged or the spring is damaged, check the control valve assembly
- If the spline is damaged, check the fourth clutch shaft
- If the clutch plate is burned, the pistons or seals damaged, or the clutch plates and pistons mislocated, check the forward clutch assembly
- If the shift cable does not shift, adjust the shift cable
- If the control assembly bolts are too loose or too tight, check and tighten bolts

IF YOU FIND THAT THERE IS HARSH OR SOFT 3-4/4-3 SHIFT FEEL:

- If the electronic pressure control solenoid is inoperative or damaged, need to de further testing on electrical system controlling the transmission/transaxle
- The microprocessor could be damaged, need to do further testing on electrical system controlling the transmission/transaxle
- If there are conditions for High/Low oil pressure, check the possible diagnostic trouble code set
- If the checkball is missing or damaged, check and replace the appropriate checkball
- If there are missing springs or damaged seals, check the accumulator cover and pistons and replace as necessary
- If there is a stuck 3-4 accumulator valve, check the control valve assembly
- If there is a cut seal, check the 3-4 accumulator seal
- If there is a stuck valve, check the 3-4 accumulator valve

IF YOU FIND THAT THERE IS 3-4 SHIFT SPEED HIGH OR LOW:

- If there is an inoperative vehicle speed sensor, need to do further testing on electrical system controlling the automatic transmission/transaxle
- If there is wrong or missing speedometer gear, install correct speedometer driven gear
- If the 1-2 shift valve/ or 3-4 shift valve is stuck , nicked or damaged, check the control valve assembly
- If the shift solenoid is damaged, check the control assembly

IF YOU FIND THAT YOU HAVE NO DRIVE BUT REVERSE IS OKAY:

- If the 2-3 servo regulator valve is stuck, check and replace as necessary
- If the band assembly is burned or has broken ends, check the low/intermediate band assembly
- If the wrong apply rod for the low/intermediate servo assembly is applied, install correct apply rod
- If the piston, seal, or rod for the low/intermediate servo assembly is damaged, check and replace as necessary

- If there is damaged oil tubes or damaged case bores, check and replace low/intermediate servo oil tubes

IF YOU FIND THAT YOU HAVE NO REVERSE BUT DRIVE IS OKAY:

- If there are burned or missing plates in the reverse clutch, check and replace as needed
- If there is leaking or improperly installed reverse apply tube, check the reverse apply tube and replace as needed

IF YOU FIND THAT YOU ARE LOCKED UP IN DRIVE/ REVERSE:

- If there is a deformed or dented parking prawl, check the final drive internal gear
- If there is a misalignment of pin to bore, check reverse servo pin and bore

IF YOU FIND THAT YOU SLIP IN REVERSE:

- If there are damaged splines, check the reverse reaction drum
- If there is a damaged seal, check reverse servo assembly
- If there are conditions for High/Low oil pressure, check the possible diagnostic trouble code set

IF YOU FIND THAT YOU HAVE NO PARK RANGE:

- If there is a problem with the park pawl spring, parking pawl, or park gear, check the final drive internal gear
- If there is a damaged spring, check the actuator assembly
- If there is an adjustment to be made to the shift cable, check and adjust the shift cable

IF YOU FIND THAT YOU HAVE HARSH NEUTRAL TO REVERSE:

- If there is a loss of vacuum due to damaged lines or modulator, check the modulator or lines
- If there is a broken or wrong spring, check the reverse servo cushion spring
- If there is a missing checkball that results in harsh reverse, check the control valve assembly

IF YOU FIND THAT YOU HAVE HARSH NEUTRAL TO DRIVE:

- If the engine idle speed is too high, check engine curb idle speed and correct as necessary
- If there is a loss of vacuum due to damaged lines or modulator, check the modulator or lines
- If there is a missing checkball that results in harsh drive, check the control valve assembly
- If there is a broken or wrong spring, check forward servo cushion spring
- If thermal element does not close when the element is warm, causing harsh neutral to drive, check the spacer plate
- If accumulator piston is stuck, accumulator seal or springs are damaged or missing, check the neutral-drive accumulator assembly and service as needed

IF YOU FIND THAT SECOND GEAR STARTS:

- If the 1-2 shift valve is stuck, check the control valve assembly
- If there is a scored drum surface or hot spots caused by band slipping, check the second clutch housing and drum assembly
- If there is burned fiber material, check the reverse band assembly

- If a checkball is missing or mislocated, check the checkball and adjust
- If there is a damaged or wrong spring, check the servo cushion spring

IF YOU FIND THAT THERE IS NO ENGINE BRAKING IN MANUAL SECOND OR LOW RANGE:

- If there is burned or glazed fiber material, check the 2-1 manual band
- If there is nor engaging of 2-1 manual band, check apply pin
- If the seals are cut or damaged or the filter is missing which allows foreign material to damage the seal and the bore, check seals and replace accordingly
- If the gaskets are damaged, check and replace accordingly
- If the the bolts are loose, check and tighten the bolts

IF YOU FIND THAT THE CONVERTER CLUTCH APPLY:

- If there is nor lock up signal, by-pass solenoid damaged or inoperative, bulkhead connector damaged or a pinched wire, need to do further testing on the transaxle electrical system or electronic engine control
- If there are damaged or missing seals, check the turbine shaft
- If the bypass clutch control valve is stuck or the bypass plunger stuck, check the bypass clutch control valve and service as necessary
- If there are missing or damaged seals or cup plugs, check the pump shaft
- If there are damaged or misaligned valve body pilot sleeve, check the sleeve and replace as necessary

IF YOU FIND THAT THE CONVERTER CLUTCH DOES NOT RELEASE:

- If the torque converter clutch solenoid valve does not exhaust, check the torque converter clutch solenoid valve
- If the converter clutch valve is stuck in apply position, the check the control valve assembly
- If there is a missing screen that allows foreign material to stick the torque converter clutch solenoid On, check the torque converter clutch screen and replace if necessary
- If there is no unlock signal, bypass solenoid is damaged or inoperative, or the bulkhead connector wires are damaged, check electronic engine control or transaxle electrical system

IF YOU FIND THE CONVERTER CLUTCH STUCK ON IN 3RD AND 4TH GEARS:

- If there is foreign material, pinched solenoid wire to ground or PCM signal is grounded, check the torque converter clutch solenoid valve for ON failure

IF YOU FIND THE 3-1, 2-1 DOWNSHIFTS HARSH:

- If there is an incorrect servo apply rod length, damaged servo piston or seal or missing springs, check the rod for correct installation and replace if necessary
- If the oil level is too low or high, check and fill transaxle as necessary
- If the checkball is missing, check and replace as needed

IF YOU FIND THE 4-3 DOWNSHIFTS HARSH:

- If there is an incorrect servo apply rod length, damaged servo piston or seal or missing springs, check the rod for correct installation and replace if necessary

- If there is no converter clutch release, need to do further testing on electrical system controlling the automatic transmission/transaxle

IF YOU FIND THAT THERE IS 3-2 DOWNSHIFT HARSH:

- If the transaxle electrical system is not operational, need to do further testing on electrical system controlling the automatic transmission/transaxle
- If the wrong apply rod for the low/intermediate servo is applied, install correct apply rod
- If there are conditions for High/Low oil pressure, check the possible diagnostic trouble code set
- If the checkball is missing, replace the checkball gaskets and seals and replace as necessary
- If there is a stuck valve, nicked or damaged, check the backout valve
- If there is a valve stuck, nicked or damaged, check the 3-2 shift timing
- If there is damaged or missing springs or incorrect servo apply rod length, check for correct apply rod and replace if necessary

IF YOU FIND THE HARSH CONVERTER CLUTCH (TORQUE CONVERTER CLUTCH) APPLY:

- If there is foreign material in torque converter clutch PWM solenoid valve, no PCM signal, insufficient force of PCM, failure of the O-rings, open wire, or no supply voltage to the torque converter clutch PWM solenoid valve, check the torque converter clutch PWM Solenoid Valve for OFF Failure

IF YOU FIND THE CONVERTER CLUTCH APPLY ROUGH, SLIPS, OR SHUDDERS:

- If the regulator valve in the converter clutch is stuck, check the control valve assembly
- If there are damaged or missing seals or damaged or unseated blow off checkball for the converter clutch, check the turbine shaft
- If there is a regulator valve in the converter clutch, check the channel plate
- If there is a worn bushing, check the drive sprocket support
- If there is worn or glazed fiber material, check and replace the torque converter if necessary

IF YOU FIND THAT THERE IS NO TORQUE CONVERTER CLUTCH APPLY:

- If there is improper operation or wiring, check to verify proper PCM operation and Vehicle wiring
- If there are damaged or loose connector, pinched wires, or inoperative torque converter clutch solenoid valve, check wiring harness
- If there is stuck converter clutch valve or converter clutch regulator valve, check the control valve assembly
- If the solenoid O-ring is leaking, check the torque converter clutch solenoid valve
- If the screen is blocked, check the solenoid screen
- If there are damaged seals, check the turbine shaft and the oil pump drive shaft
- If there is a damaged or unseated blow off checkball for the converter clutch, check the channel plate
- If there is foreign material and plugging in the filter, no PCM signal, insufficient force, failure of the O-rings, open wire, or no supply voltage to the torque converter clutch solenoid valve, check the torque converter clutch Solenoid Valve for OFF Failure

- If there is foreign material, pinched wire, or grounded PCM signal, check torque converter clutch Solenoid Valve for ON Failure
- If there is no supply voltage, corroded connector, corroded switch contact, loose connector, or pinched wire, check the brake switch for no torque converter clutch Apply
- If there is a loose or corroded connector, corroded switch contact, pinched wire, or PRNDL connector wires A and B are switched around, check the PRNDL Circuit problem
- If there is a loose or corroded connector, incorrect resistance, or pinched wire, check the coolant temperature sensor

IF YOU FIND THAT THE TRANSAXLE OVERHEATS:

- If there are excessive tow loads, check the guidelines for tow limits The fluid level could be low or high, if so correct the fluid level
- If there is incorrect engine idle or performance, check the owners manual for diagnosis
- If there is a dirty or sticking valve body, check , clean, service or replace valve body
- If there is a seized converter one-way clutch, replace converter
- If there is an improper clutch or band application or oil pressure control system, perform a control pressure test

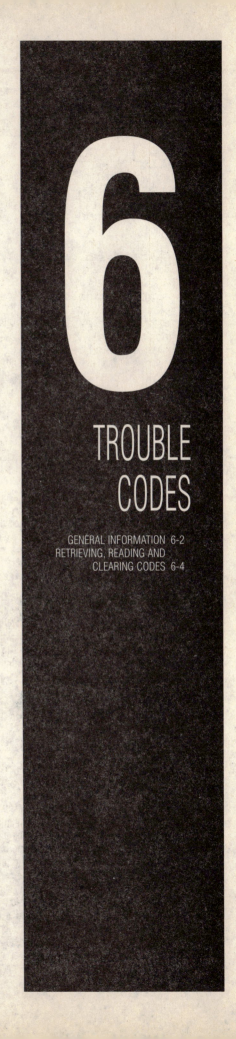

6

TROUBLE CODES

GENERAL INFORMATION

Vehicle Diagnostic Software

Starting in the mid-1980's, vehicle manufacturer's began to implement electronic control systems for automatic transmissions. Suddenly, electronic computers decided when the transmission/transaxle should up-shift, down-shift or when the torque converter should lock-up, among other things. To help with diagnosing these electronic control systems, each manufacturer came up with their own (also, often shared amongst manufacturers) diagnostic system.

These early diagnostic systems used Diagnostic Trouble Codes (DTC's) to inform the servicing technician which circuit/component was detected as having a problem or fault. As vehicles have become more and more advanced, DTC coverage has also expanded. Starting in the late 1980's, the Federal government implemented a limited set of standards for automotive diagnostics, referred to as On Board Diagnostics (OBD I). By 1996, the federal government decreed that every vehicle's diagnostic system and DTC's must conform to a standardized SAE format. This newer format is known as On Board Diagnostics 2nd Generation or OBD II.

Up until the implementation of OBD II standards (1995–96), DTC's were representative codes that, depending on the manufacturer or system, were interpreted in many different ways. DTC's are still representative codes for a corresponding circuit and component, but the interpretation has been standardized, which makes interpreting the codes easier. However, because of the complexity of the OBD II systems designed today, retrieving DTC's from the vehicle's computer has become much more complicated. Often it is impossible to retrieve codes without an expensive scan tool; unlike pre-OBD II systems, which usually only require a voltmeter to retrieve transmission/transaxle related codes.

Therefore, this section is aimed toward helping the Do-It-Yourselfer, using normal diagnostic tools (voltmeter, jumper wires, etc.) to retrieve, read (interpret) and clear transmission/transaxle -related DTC's from pre-OBD II vehicles. It would require a nicely sized tome (encyclopedic in size) to inform you how to do this with every production and aftermarket scan tool on every vehicle. Luckily, most scan tool manufacturer's include instructions on how to use their scan tools for code retrieval and reading.

➡**One point to keep in mind when working with DTC's is that they do not indicate that a component is bad. They only indicate that a malfunction or fault was detected within that component's circuit; the fault could lie within the component itself, within the wires that make up the circuit, or within both. Therefore, do not condemn the corresponding component without also inspecting it's circuit for faults.**

Vehicle Diagnostic Hardware

VEHICLE COMPUTERS

The actual hardware (computers, wires, sensors, etc.) used in the vehicles' control systems have changed over the years as well as the software used (DTC retrieval, clearing, etc.). However, despite the large number of manufacturers and vehicles designed between the mid-1980's and today, there are generally only three control computer configurations used. Back when the first computer-controlled vehicles were introduced, they used a simple feedback system consisting of one computer (usually known as the Engine Control Module or ECM) used exclusively for the engine systems (air/fuel and ignition primarily). As the systems grew and became more complicated, a transmission/transaxle control computer (often called the Transmission Control Module or TCM) was added to the vehicle. This computer communicated with the ECM. Together, based on predetermined parameters, they decided what was best for the entire powertrain (engine & transmission). Eventually, the ECM and the TCM were formed into one computer, most commonly referred to as the Powertrain Control Module (PCM). The PCM configuration is used on most, if not all, vehicles manufactured today.

DIAGNOSTIC TOOLS

▶ **See Figures 1 thru 7**

Normally, it is necessary to use a scan tool to retrieve, read and clear OBD II DTC's, but not for OBD I codes. This section is aimed toward helping you retrieve, read and clear OBD I codes using common diagnostics tools, such as voltmeters and jumper wires. Occasionally, when possible with common tools, I have explained how to retrieve, read and clear OBD II codes as well. So, grab your voltmeter and jumper wires (a paper clip or two also makes an appearance) and, after thoroughly reading all of the following precautions, jump right in.

Precautions

The electronic control systems used by the vehicles covered in this section are very delicate and complicated. Please, to save yourself aggravation, money and time, be sure to adhere to the following points when working on your vehicle's control system:

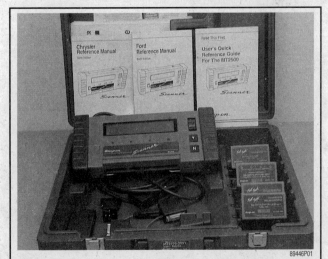

Fig. 1 Although you could go out and buy an expensive scan tool for transmission/transaxle diagnosis . . .

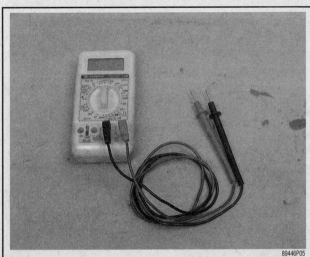

Fig. 2 . . . if all you need to do is read codes from pre-OBD II vehicles, only a voltmeter and jumper wires are needed

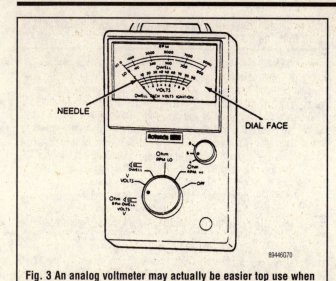

Fig. 3 An analog voltmeter may actually be easier top use when reading the DTC's

Fig. 6 . . . such as this OBD II connector

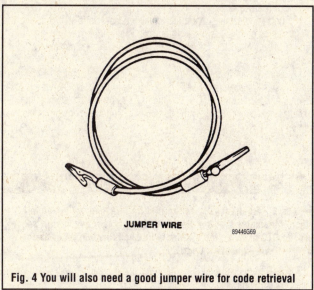

Fig. 4 You will also need a good jumper wire for code retrieval

Fig. 7 However, they also require software cartridges for each manufacturer

- Unless otherwise instructed, always disconnect the battery cables when servicing the electronic system.
- When disconnecting the battery, always be sure to detach the negative battery cable FIRST, then the positive cable. This simple practice will almost completely prevent the chance of arcing or shorting the system.
- Never pierce or cut the insulation off of a wire for testing purposes. Many of the control systems wires are designed to handle a precise amount of electrical resistance, and the computer expects to see a certain predetermined amount of resistance. If you pierce or cut the insulation off of a wire, corrosion can build up in the wiring, leading to decreased control system efficiency, DTC storing or possibly even component damage.
- Whenever handling (that includes just touching) the system's control computer, use a ground strap to prevent accidental static electricity charges from damaging the computer's electronic chips.
- Never subject any control computer to excessive jolts (such as dropping).
- If welding on the vehicle (don't worry, nothing in this section requires welding), always disconnect the computer from the vehicle's wiring harness.
- Always be careful when working around a running engine. Loose clothing or long hair can easily become tangled in moving components. Hot exhaust system parts can cause painful burns.

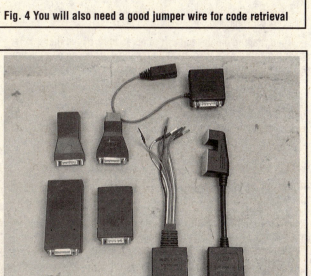

Fig. 5 So that they can be used with most makes and models, some scan tools come with multiple connectors . . .

- Always wear eye protection when servicing a vehicle.
- On vehicles equipped with an air bag system (make sure you check for this!), it is necessary to properly disarm the system before working on or around any air bag components, since the air bag system can often deploy even if the negative battery cable has been disconnected. This includes the steering wheel, front impact sensors, system wiring, and the instrument panel. If you do not know the proper disarming procedure, refer to the Chilton's Total Car Care manual for your specific vehicle.
- Never detach a wiring harness connector when the ignition switch is turned **ON**.
- Always operate the vehicle in a well-ventilated area. Also, never smoke around the battery or fuel system.
- Always keep a dry chemical (class B) fire extinguisher at hand.

Preliminary Inspection

Prior to trying to determine what components are the cause of DTC's, it is always a good idea to do a preliminary inspection. I have heard, and experienced, many examples when a loose or corroded wire or terminal has caused the control computer to store "ghost" DTC's. Time and sanity are usually sacrificed when skipping a general inspection before servicing or repairing components and circuits indicated by stored DTC's. Perform a general inspection, as follows:

1. Retrieve and write down the stored DTC's as described through-out this section, then clear them.
2. Open the hood.
3. Check the transmission/transaxle fluid level and fill it if necessary. While the engine is running, to heat the transmission/transaxle up, watch for fluid leaks. Any leaks must be repaired immediately, otherwise transmission/transaxle damage can occur.

✳✳ CAUTION

Checking the fluid level may require that the transmission/transaxle be hot. So take this into consideration when working around the vehicle during this inspection. Painful burns can result from the exhaust system.

4. Check the battery voltage with a voltmeter; it should be approximately 12.6 volts. If it lower than 11 volts, you should charge it with a battery charger. If the battery will not hold a full charge, it may be defective. Have it tested at a competent automotive service center.
5. Inspect every connector you can get at (especially the transmission/transaxle connectors) to ensure that they are tightly connected. Separate the connector halves and ensure that the terminals inside are corrosion-free and not loose. Re-engage all of the connector halves making sure they are properly attached. If any of the connectors are faulty, fix or replace them.
6. If possible, trace the wires to ensure that they are not cut, burned or similarly damaged. If any such damage is found, the wire must be repaired.
7. Check all vacuum hoses to ensure they are properly attached. Replace any vacuum line that is cracked or crumbling; this can cause a vacuum leak, which can adversely affect the engine and/or transmission/transaxle .
8. Inspect all engine and transmission-related linkage to ensure that it actuates smoothly and properly. Maladjusted linkage can cause early or late shifts, excessively hard shifts and other problems.
9. Check all of the transmission/transaxle fluid lines for kinks, cracks, or leaks. Fix any defects.
10. Raise and safely support the vehicle on jackstands so that access to the underside of the transmission/transaxle is gained. Inspect the under vehicle wires, connectors, lines, linkage and fluid lines as previously described.
11. Once you have satisfactorily inspected the engine and transmission/transaxle and fixed any problems found, drive the vehicle for ½ hour and recheck for DTC's. You might be surprised that some of the DTC's are no longer stored by the computer (if you found any problems, that is).

RETRIEVING, READING AND CLEARING CODES

Domestic Vehicles

CHRYSLER CORPORATION & JEEP

▶ See Figures 8 and 9

Reading and Retrieving Codes

Chrysler-built transmission/transaxle diagnostic systems use two-digit DTC's, as with most pre-OBD II vehicles, to inform to the mechanic or technician the problem circuit or component.

Unfortunately, it is necessary to use a diagnostic scan tool to retrieve DTC's from almost all of the Chrysler automatic transmissions and transaxles, with the exception of the 42RH and 46RH assemblies. The other transmissions/transaxles use a separate computer (TCM) to control the electronic transmission/transaxle , and require a scan tool with a model-specific cartridge to retrieve and read the DTC's. However, the 42RH and 46RH units are controlled by the Powertrain Control Module, and, therefore, their DTC's can be read along with the engine-related DTC's. Chrysler models which utilize either the 42RH or 46RH use a light, located in the instrument cluster and usually referred to as the Malfunction Indicator Lamp (MIL), CHECK ENGINE light or POWER LOSS light (on earlier models), to display the DTC's as a series of flashes or blinks.

The two-digit codes are flashed out in a logical series of blinks. For example, when first starting to retrieve the DTC's the MIL will flash eight times, pause briefly, then flash eight more times followed by a longer pause. This indicates a Code 88. Therefore, a flash pattern of two flashes-pause-three flashes-long pause would indicate a Code 23.

➡Code 88 should always be the first DTC flashed out by the MIL, since it indicates that the diagnostic system is ready to start showing the DTC's. When the system is finished flashing all of the DTC's, it will indicate a Code 55.

The DTC's are displayed in numerical sequence starting with the lowest code number and moving to the highest.

➡When reading the codes be sure not to miss any of them, otherwise the entire sequence must be repeated. This can be tiresome if a large number of codes was stored, and you missed the last one.

Unlike most other manufacturers, the Chrysler diagnostic computer system is very easy to actuate, simply turn the ignition key **ON**, **OFF**, **ON**, **OFF**, **ON** within 2 seconds. The DTC's will start to flash after turning the ignition key to the **ON** position for the third time. Code 55 indicates that all of the DTC's have been flashed out, and that the DTC cycle is done.

Clearing Codes

➡It is not recommended that the negative battery cable be disconnected to clear DTC's. Doing so will clear all settings from the vehicle's computer system, resulting in lost radio presets, seat memories, anti-theft codes, driveability parameters, etc., and there are better ways of spending your afternoon than resetting all of these.

There are two methods for clearing the DTC's from Chrysler models: the first is by using a scan tool, the second is slightly more time consuming. To erase the DTC's without a scan tool, you must turn the ignition switch **ON** and **OFF** 51 times! Yes, you read correctly, 51 times! Since the system is designed to erase codes if they have not been evident during the last 50

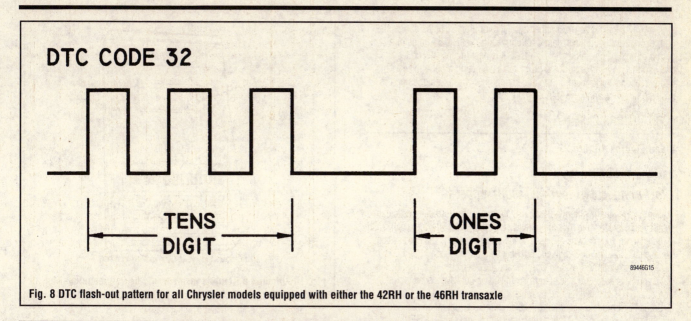

Fig. 8 DTC flash-out pattern for all Chrysler models equipped with either the 42RH or the 46RH transaxle

DIAGNOSTIC TROUBLE CODES—CHRYSLER 42RH AND 46RH TRANSMISSIONS

Code	Component/Circuit Fault
11	No distributor reference signal
12	Number of ignition key cycles since the last fault was erased
13	MAP sensor or circuit
14	MAP sensor or circuit
15	Vehicle Speed Sensor (VSS) or circuit
16	Battery Input Sense
17	Engine temperature too low
21	Oxygen sensor or circuit
22	Engine Coolant Temperature (ECT) sensor or circuit
23	Throttle body temperature sensor or circuit
24	TP sensor or circuit
25	AIS motor circuits
26	Fuel injectors or circuits
27	Fuel injectors or circuits
31	Purge solenoid or circuit
32	EGR system or solenoid circuit
33	A/C clutch relay or circuit
34	S/C servo solenoids or circuit
35	Idle switch
36	Air switch solenoid or circuit
37	PTU solenoid circuit
41	Alternator Field circuit
42	ASD relay or circuit
43	Ignition control circuit
44	FJ2 voltage sense
45	Overdrive solenoid or circuit
46	Battery voltage too high
47	Low charging output
51	Lean air/fuel mixture
52	Rich air/fuel mixture
53	Internal self-test
55	Not used
62	EMR miles not stored
63	EEPROM write denied

Fig. 9 Chrysler 42RH and 46RH transaxle DTC's

ignition key cycles, you must either cycle the ignition 51 times at one sitting, or wait until you have started and driven the vehicle 51 times or a period of time. (For diagnostic purposes, it is best to clear the codes right away so that you can ensure the problems have been properly fixed.) This is one reason (not to mention the added bonus of ease of diagnosing the systems) that a scan tool would be more advantageous.

FORD MOTOR COMPANY

▶ **See Figures 10 thru 27**

Reading and Retrieving Codes

EXCEPT PROBE AND 1991–92 ESCORT MODELS

Compared with the other domestic manufacturers, the Ford diagnostic computer system is either the most comprehensive or the most difficult to work with (depending on your outlook—glass half full or half empty?). The Ford system can use either 2-digit or 3-digit codes, depending on the year and/or model of the vehicle at hand. And, as with Chrysler and Jeep vehicles, it is most beneficial if you have access to a scan tool. But, if you do not have a scan tool, don't worry the DTC's can still be read by more conventional methods.

Some Ford models use a light, located in the instrument cluster and usually referred to as the Malfunction Indicator Lamp (MIL), CHECK ENGINE light or a similarly-named light, to display the DTC's as a series of flashes or blinks. Unfortunately, not all Ford vehicles are equipped with such a light. In these cases, you will need either an analog voltmeter (with an arm—not the digital read-out type), a test probe light, or any other device capable of reading voltage sweeps.

➡**Although a scan tool is not necessary to retrieve and read the DTC's, it can be very helpful, especially on vehicles which use 3-digit trouble codes. Reading Code 586 as a series of flashes (flash, flash, flash, flash, flash, pause, flash, flash, flash, flash, flash, flash, flash, flash, pause, flash, flash, flash, flash, flash, flash—you get the idea) can be irritating. Be sure you have enough paper and a pen/pencil that works!**

The MIL or analog voltmeter (analog voltmeter will be used as an example of any voltage measuring device you plan on using to read the codes for vehicles which do not utilize a MIL) will indicate the DTC's in a logical series of flashes, or sweeps. Individual digits (say, a 5) are displayed as a series of flashes with a ½ second pause between each flash. Therefore, the digit 5 would be five flashes separated by ½ second pauses. Individual digits belonging to the same 2- or 3-digit number (say the 2 and the 3 of digit 23) are separated by two second pauses. Therefore, Code 23 would be indicated as follows: flash, ½ second pause, flash, 2 second pause, flash,½ second pause, flash, ½ second pause, flash. In between each individual trouble code a 4 second pause is used.

To retrieve the codes, perform the following:

1. Perform the preliminary inspection, located earlier in this section. This is very important, since a loose or disconnected wire, or corroded connector terminals can cause a whole slew of unrelated DTC's to be stored by the computer; you will waste a lot of time performing a diagnostic "goose chase."

2. Grab some paper and a pencil or pen to write down the DTC's when they are flashed out.

3. Ensure that the ignition key is in the **OFF** position.

4. Locate the Self-Test Connector (STC), which is comprised of two separate connectors: the Self-Test Output (STO) connector and the Self-Test Input (STI) connector. The STC is usually located under hood near one of the strut towers or near the firewall. If you cannot locate the STC either ask someone at a local dealership for help with its location, or refer to the Ford model-specific Chilton's Total Car Care manual.

5. Once the STC is located, identify the two connectors. The smaller of the two, with only one terminal, is the STI connector. The larger connector is the STO.

6. If the vehicle at hand is not equipped with a MIL, the analog voltmeter must be connected to the computer system. To attach the voltmeter, perform the following:

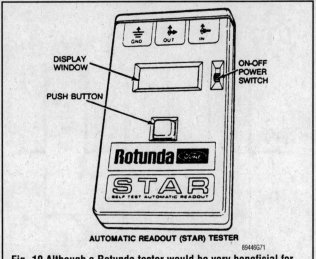

Fig. 10 Although a Rotunda tester would be very beneficial for reading DTC's, it is not necessary for pre-OBD II models

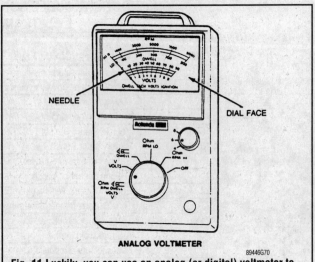

Fig. 11 Luckily, you can use an analog (or digital) voltmeter to retrieve the DTC's . . .

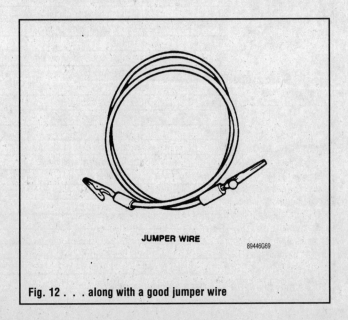

Fig. 12 . . . along with a good jumper wire

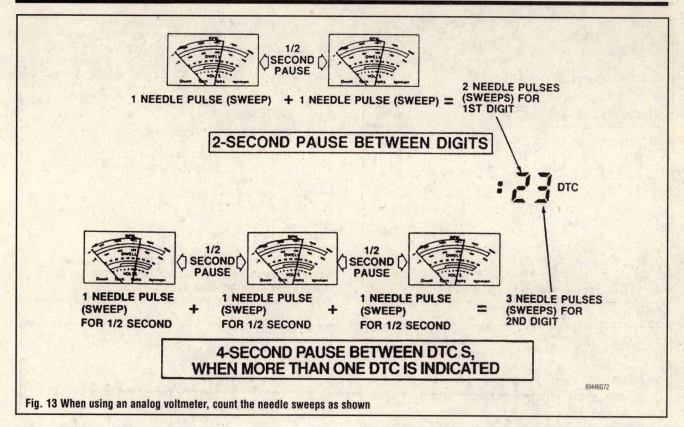

Fig. 13 When using an analog voltmeter, count the needle sweeps as shown

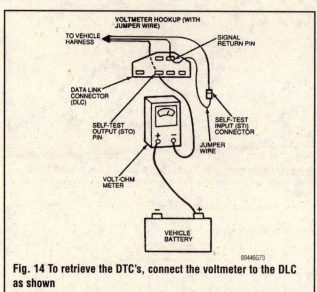

Fig. 14 To retrieve the DTC's, connect the voltmeter to the DLC as shown

a. Using a jumper wire, attach the STI terminal to a good engine ground or to the negative terminal of the battery.

b. Attach the negative voltmeter lead to the STO terminal and the positive lead to the battery positive terminal.

7. For models equipped with a MIL, simply jumper the STI terminal to a good engine ground or to the negative battery cable.

8. Turn the ignition switch **ON**, but do not start the engine. This condition is known as Key On—Engine Off (KOEO). After the ignition is turned **ON**, depending on the instrument used to read the codes, either the MIL will start flashing the DTC's as a series of blinks or the voltmeter needle will start moving between 0 and 12 volts in a series sweeps.

9. Count and note the number and sequence of the flashes/sweeps.

The DTC system will progress through two sets of codes at this time. First, all of the KOEO On-Demand codes will be displayed. These codes are not normally stored in the computer memory. Rather, the computer performs a quick self-diagnostic check once the ignition key is turned **ON** and displays any applicable codes for faults that are currently just detected. If there are no current trouble faults in the computer system, only Code 11 (Code 111 or vehicles which use 3-digit codes) will be displayed. Code 11 or Code 111 indicates that there are no "hard," or current malfunction, faults.

After the system displays all of the ON-Demand codes or Code 11 (111), it will display a Code 10, which appears as a single flash/sweep (the computer cannot display a zero). Code 10 indicates that the system will enter the second set of codes, which are referred to as the KOEO Continuous Memory Codes.

→Before and after Code 10 there is a long pause (5 to 10 seconds); be patient. Do not prematurely switch the ignition switch OFF, otherwise the Continuous Memory codes (and potential problems) will be missed.

The Continuous Memory codes are the codes which are stored in the computer's memory. These codes can indicate a constant or intermittent problem with the system. Which is one reason why comparing the On-Demand and Continuous Memory codes are helpful in diagnosing transmission/transaxle problems. The lack of the corresponding On-Demand code when a Continuous Memory code is present may indicate an intermittent fault.

As with the On-Demand codes, if there are no Continuous Memory codes, the computer will display a Code 11 (111).

10. Disconnect the ground wire from the STI terminal.

11. Position the transmission/transaxle gear selector in **P**.

12. Block the drive wheels and apply the parking brake.

13. Start the engine and allow it to reach normal operating temperature.

14. Turn the engine **OFF**, then reconnect the ground wire to the STI terminal.

15. Start the engine.

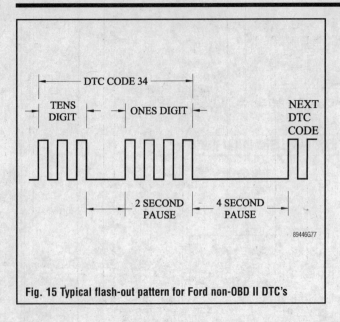

Fig. 15 Typical flash-out pattern for Ford non-OBD II DTC's

16. Watch for a series of flashes/sweeps, which are the engine ID code. 4-cylinder engines will display a two flash/sweep code, a 6-cylinder engine will display a three flash/sweep code, and an 8-cylinder engine will display a four flash/sweep code—the code is always half the total number of cylinders of the engine. As soon as the engine ID code is displayed, perform the following:

    a. Turn the steering wheel at least ½ turn.

    b. Depress the brake pedal.

    c. Cycle the transmission/transaxle control switch ON and OFF, if equipped.

17. On some models, after completing the previous three substeps, there should be a long pause (possibly as long as 20-30 seconds), then a single flash/sweep signal. When this signal occurs, depress the accelerator to the floor and release it quickly. This is a dynamic response test; the computer system uses it in its diagnostic self-test. After depressing the accelerator, there may be a small series of slight voltage changes, which is simply the computer sending its information to the scan tool, in case one is being used. In our case, just ignore these small signals.

On models which do not display the dynamic response test single flash/sweep, the system will go into the KOER On-Demand code display without you having to depress the accelerator pedal.

18. Six seconds after the small voltage signals are detected, the system

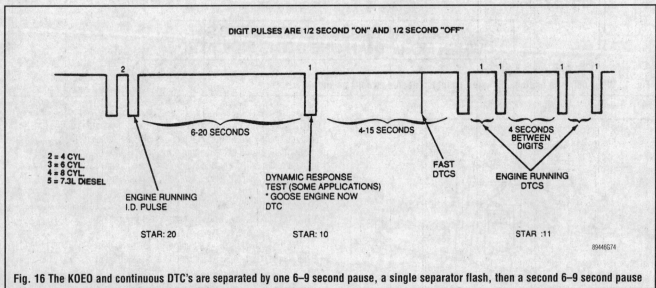

Fig. 16 The KOEO and continuous DTC's are separated by one 6–9 second pause, a single separator flash, then a second 6–9 second pause

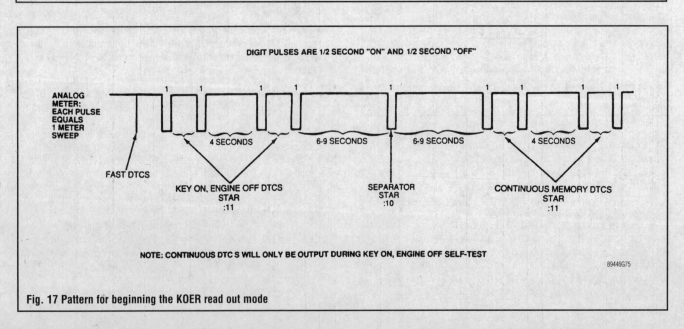

Fig. 17 Pattern for beginning the KOER read out mode

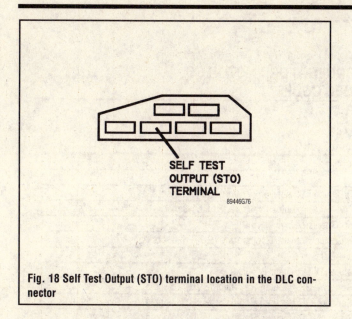

Fig. 18 Self Test Output (STO) terminal location in the DLC connector

will start to display the KOER On-Demand. As with the KOEO ON-Demand codes, these codes are not stored in the computer; they are on-the-spot codes and will be erased when the DTC retrieval cycle is finished. Therefore, be sure to pay attention as they are displayed to avoid having to perform this entire procedure again. As with the other two DTC sets, a Code 11 (111) will be displayed if no DTC's are found.

19. After noting all of the displayed DTC's, turn the ignition switch **OFF**, remove the STI terminal ground wire, and the voltmeter, test light, etc. (if used).

The Ford computer system is also capable of numerous self-diagnostic tests to aid in trouble-shooting system faults. Some of the tests include the Wiggle Test and the Output State Test. If you are interested in these more specific test subroutines, refer to the Chilton's Total Car Care manual specific to your vehicle or the Ford factory manual.

1991–92 ESCORT MODELS

Early Escort models equipped with a 1.8L or 1.9L engine were provided with the F4EAT transaxle, which were manufactured by Mazda. Although there is no outward way of knowing when or if the vehicle has detected a DTC, there is a way to retrieve codes. The 1991–92 Escort models use two-digit DTC's, not the more complex three-digit ones available in some Ford vehicles. An analog voltmeter or similar voltage measuring device is necessary to read the DTC's.

The two-digit codes are flashed out in a logical series of voltmeter arm sweeps. The tens digits are displayed as long sweeps (in duration of time, not distance the arm moves), and the ones digits are displayed as short sweeps. For example, when first starting to retrieve the DTC's the voltmeter sweeps one long time, pause briefly, then sweep two short times followed by a longer pause. This indicates a Code 12 (long sweep-pause-two short sweeps-long pause). Therefore, a pattern of five long sweeps-pause-five short sweeps-long pause would indicate a Code 55.

Each DTC is displayed three times, then the next code is displayed out (if more than one code has been stored). The codes are displayed in numerical sequence starting with the lowest code number and moving to the highest.

When reading the codes be sure not to miss any of them, otherwise the entire sequence must be repeated. Although the codes will continue to be read until the ignition key is turned **OFF**, this can be tiresome if a large number of codes was stored, and you missed the last one.

To retrieve the codes, perform the following:

1. Perform the preliminary inspection, located earlier in this section. This is very important, since a loose or disconnected wire, or corroded connector terminals can cause a whole slew of unrelated DTC's to be stored by the computer; you will waste a lot of time performing a diagnostic "goose chase."

2. Grab some paper and a pencil or pen to write down the DTC's when they are flashed out.

3. Ensure that the ignition key is in the **OFF** position.

4. Locate the diagnostic connector (STC). The diagnostic connector is usually located under hood near one of the strut towers or near the firewall. If you cannot locate the diagnostic connector either ask someone at a local dealership for help with its location, or refer to the Ford model-specific Chilton's Total Car Care manual.

5. Attach a negative voltmeter lead to the STO terminal and the positive lead to the battery positive terminal.

6. Turn the ignition switch **ON**, but do not start the engine. This condition is known as Key On—Engine Off (KOEO). After the ignition is turned **ON**, the voltmeter needle will start moving between 0 and 12 volts in a series sweeps.

7. Count and note the number, sequence and duration of the sweeps.

PROBE MODELS

As with the 1991–92 Escort models, the Probe vehicles also utilize Mazda-built transaxles. The computer diagnostic system also uses 2-digit trouble codes, which are displayed through flashing of the MANUAL SHIFT light or the CHECK ENGINE light.

The Probe DTC's are displayed in the same format (with the exception that they are flashes of a light rather than sweeps of an analog voltmeter needle), in that the tens digits are long flashes and the ones digits are short flashes. For example, when first starting to retrieve the DTC's the MANUAL SHIFT or CHECK ENGINE light flashes one long time, pauses briefly, then flash two short times followed by a longer pause. This indicates a Code 12 (long flash-pause-two short flash-long pause). Therefore, a pattern of five long flashes-pause-five short flashes-long pause would indicate a Code 55.

The codes are displayed from the lowest number on through to the highest number, and three times in a row. Therefore the first code is flashed three times, then the second code is flashed three times, etc., until the highest code number has also been flashed three times. After the highest code number is displayed, the system begins with the lowest again. The system will continue to cycle through all of the codes in this manner; so if you missed a code or two, you can wait until it starts the cycle over again.

To retrieve the codes for 1989–92 models, perform the following:

1. Perform the preliminary inspection, located earlier in this section. This is very important, since a loose or disconnected wire, or corroded connector terminals can cause a whole slew of unrelated DTC's to be stored by the computer; you will waste a lot of time performing a diagnostic "goose chase."

2. Grab some paper and a pencil or pen to write down the DTC's when they are flashed out.

3. Locate the Diagnostic Request (DR) terminal, which is usually located next to or near the transmission/transaxle control computer. If the DR terminal cannot be found, either call your local Ford dealer or refer to the applicable Chilton's Total Car Care manual for your vehicle.

4. Turn the ignition key **OFF**, if not already done.

5. Using a jumper wire, attach the DR terminal to a good chassis or engine ground.

6. Without starting the engine, turn the ignition switch **ON**.

7. The MANUAL SHIFT light should start to flash out any stored DTC's.

To retrieve the codes for 1993 and newer non-OBD II models, perform the following:

8. Perform the preliminary inspection, located earlier in this section. This is very important, since a loose or disconnected wire, or corroded connector terminals can cause a whole slew of unrelated DTC's to be stored by the computer; you will waste a lot of time performing a diagnostic "goose chase."

9. Grab some paper and a pencil or pen to write down the DTC's when they are flashed out.

10. Locate the Diagnostic Link Connector (DLC), which is usually located near the left-hand side of the engine compartment and the firewall. If the DLC cannot be found, either call your local Ford dealer or refer to the applicable Chilton's Total Car Care manual for your vehicle.

11. Turn the ignition key **OFF**, if not already done.

12. Using a jumper wire, attach the TAT terminal to a good chassis or engine ground.

DIAGNOSTIC TROUBLE CODES—FORD MOTOR CO.
2-DIGIT KOEO CODES

Codes	Ford Pinpoint Test Designation	System/Component/Circuit Fault
11	—	All Systems Functioning Normally
19	—	Faulty processor
21	DA	Intake Air Temperature (IAT)/Engine Coolant Temperature (ECT) Sensors
22	DF	Manifold Absolute Pressure (MAP)/Barometric Pressure (BARO) Sensor
23	DH	Throttle Position (TP) Sensor (Gasoline Engines)
24	DA	Intake Air Temperature (IAT)/Engine Coolant Temperature (ECT) Sensors
26	DC	Mass Air Flow (MAF) Sensor—Cars
26	TE	Transmission Fluid Temperature (TFT) Sensor—Trucks
31	DN	EGR Valve Position (EVP) Sensor/EGR Vacuum Regulator (EVR) Solenoid
32	DN	EGR Valve Position (EVP) Sensor/EGR Vacuum Regulator (EVR) Solenoid
34	DN	EGR Valve Position (EVP) Sensor/EGR Vacuum Regulator (EVR) Solenoid
35	DN	EGR Valve Position (EVP) Sensor/EGR Vacuum Regulator (EVR) Solenoid
47	TB	4x4 Low/Transmission Control Switch TCIL-TCS-TCSM
51	DA	Intake Air Temperature (IAT)/Engine Coolant Temperature (ECT) Sensors
52	FF	Power Steering Pressure (PSP) Switch
53	DH	Throttle Position (TP) Sensor (Gasoline Engines)
53	DQ	Throttle Position (TP) Sensor (7.3L Diesel)
54	DA	Intake Air Temperature (IAT)/Engine Coolant Temperature (ECT) Sensors
56	DC	Mass Air Flow (MAF) Sensor—Cars
56	TE	Transmission Fluid Temperature (TFT) Sensor—Trucks
57	KP	Octane Adjust (OCT ADJ)
61	DA	Intake Air Temperature (IAT)/Engine Coolant Temperature (ECT) Sensors
63	DH	Throttle Position (TP) Sensor—Gasoline Engines
63	DQ	Throttle Position (TP) Sensor—7.3L Diesel Engines
64	DA	Intake Air Temperature (IAT)/Engine Coolant Temperature (ECT) Sensors
66	TE	Transmission Fluid Temperature (TFT) Sensor—Trucks
67	KM	WOT A/C Cutout (WAC) A/C Demand
67	TA	Park/Neutral Position (PNP)/Clutch Pedal Position (CPP) Switches
79	KM	WOT A/C Cutout (WAC) A/C Demand
81	KC	Secondary Air Injection (AIRB)/(AIRD) Solenoids
82	KC	Secondary Air Injection (AIRB)/(AIRD) Solenoids
84	DN	EGR Valve Position (EVP) Sensor/EGR Vacuum Regulator (EVR) Solenoid
85	KD	Canister Purge (CANP) Solenoid
86	TC	Transmission Solenoids
87	J	Fuel Pump or Circuit
89	TC	Transmission Solenoids
91	TC	Transmission Solenoids
92	TC	Transmission Solenoids
93	TC	Transmission Solenoids
94	TC	Transmission Solenoids
98	TC	Transmission Solenoids
99	TC	Transmission Solenoids
All Other Codes Not Listed	QA	No Diagnostic Trouble Codes (DTC)/DTC Not Listed

89446G60

Fig. 19 Ford Diagnostic Trouble Codes—2-digit KOEO codes

DIAGNOSTIC TROUBLE CODES—FORD MOTOR CO.
2-DIGIT KOER CODES

Codes	Pinpoint Test Designation	System/Component/Circuit Fault
11	—	All Systems Functioning Normally
12	KE	Idle Air Control (IAC) Solenoid
13	KE	Idle Air Control (IAC) Solenoid
16	KE	Idle Air Control (IAC) Solenoid
18	PA	Spark Timing Check—Distributor Ignition (DI)
19	—	Faulty processor
21	DA	Intake Air Temperature (IAT)/Engine Coolant Temperature (ECT) Sensors
22	DF	Manifold Absolute Pressure (MAP)/Barometric Pressure (BARO) Sensor
23	DH	Throttle Position (TP) Sensor (Gasoline Engines)
24	DA	Intake Air Temperature (IAT)/Engine Coolant Temperature (ECT) Sensors
25	DG	Knock Sensor (KS)
26	DC	Mass Air Flow (MAF) Sensor—Cars
26	TE	Transmission Fluid Temperature (TFT) Sensor—Trucks
31	DN	EGR Valve Position (EVP) Sensor/EGR Vacuum Regulator (EVR) Solenoid
32	DN	EGR Valve Position (EVP) Sensor/EGR Vacuum Regulator (EVR) Solenoid
33	DN	EGR Valve Position (EVP) Sensor/EGR Vacuum Regulator (EVR) Solenoid
34	DN	EGR Valve Position (EVP) Sensor/EGR Vacuum Regulator (EVR) Solenoid
35	DN	EGR Valve Position (EVP) Sensor/EGR Vacuum Regulator (EVR) Solenoid
41	H	Fuel Control System
42	HA	Adaptive Fuel System
44	KC	Secondary Air Injection (AIRB)/(AIRD) Solenoids
45	KC	Secondary Air Injection (AIRB)/(AIRD) Solenoids
46	KC	Secondary Air Injection (AIRB)/(AIRD) Solenoids
52	FF	Power Steering Pressure (PSP) Switch
65	TB	4x4 Low/Transmission Control Switch TCIL-TCS-TCSM
72	DF	Manifold Absolute Pressure (MAP)/Barometric Pressure (BARO) Sensor
73	DH	Throttle Position (TP) Sensor (Gasoline Engines)
74	FD	Brake On/Off (BOO) switch
75	FD	Brake On/Off (BOO) switch
77	M	Dynamic Response Test
91	H	Fuel Control System
92	H	Fuel Control System
94	KC	Secondary Air Injection (AIRB)/(AIRD) Solenoids
All Other Codes Not Listed	QA	No Diagnostic Trouble Codes (DTC)/DTC Not Listed

KOER - Key On, Engine Running

89446G61

Fig. 20 Ford Diagnostic Trouble Codes—2-digit KOER codes

DIAGNOSTIC TROUBLE CODES—FORD MOTOR CO.
2-DIGIT CONTINUOUS CODES

Codes	Pinpoint Test Designation	System/Component/Circuit Fault
11	—	All Systems Functioning Normally
14	NA	Ignition Diagnostic Monitor (IDM)/Distributor Ignition(DI)
15	QB	Continuous Memory Diagnostic Trouble Code (DTC)
18	NA	Ignition Diagnostic Monitor (IDM)/Distributor Ignition(DI)
22	DF	Manifold Absolute Pressure (MAP)/Barometric Pressure (BARO) Sensor
29	DP	Vehicle Speed Sensor (VSS)—Cars
29	DS	Programmable Speedometer/Odometer Module (PSOM)—Trucks
31	DN	EGR Valve Position (EVP) Sensor/EGR Vacuum Regulator (EVR) Solenoid
32	DN	EGR Valve Position (EVP) Sensor/EGR Vacuum Regulator (EVR) Solenoid
33	DN	EGR Valve Position (EVP) Sensor/EGR Vacuum Regulator (EVR) Solenoid
34	DN	EGR Valve Position (EVP) Sensor/EGR Vacuum Regulator (EVR) Solenoid
35	DN	EGR Valve Position (EVP) Sensor/EGR Vacuum Regulator (EVR) Solenoid
41	H	Fuel Control System
49	TG	Electronic Transmission/Continuous Memory Diagnostic Trouble Codes (DTC)
51	DA	Intake Air Temperature (IAT)/Engine Coolant Temperature (ECT) Sensors
53	DH	Throttle Position (TP) Sensor (Gasoline Engines)
53	DQ	Throttle Position (TP) Sensor (7.3L Diesel)
54	DA	Intake Air Temperature (IAT)/Engine Coolant Temperature (ECT) Sensors
56	DC	Mass Air Flow (MAF) Sensor—Cars
56	TG	Electronic Transmission/Continuous Memory Diagnostic Trouble Codes (DTC)
59	TG	Electronic Transmission/Continuous Memory Diagnostic Trouble Codes (DTC)
61	DA	Intake Air Temperature (IAT)/Engine Coolant Temperature (ECT) Sensors
62	TG	Electronic Transmission/Continuous Memory Diagnostic Trouble Codes (DTC)
63	DH	Throttle Position (TP) Sensor (Gasoline Engines)
63	DQ	Throttle Position (TP) Sensor (7.3L Diesel)
64	DA	Intake Air Temperature (IAT)/Engine Coolant Temperature (ECT) Sensors
66	DC	Mass Air Flow (MAF) Sensor—Cars
66	TG	Electronic Transmission/Continuous Memory Diagnostic Trouble Codes (DTC)
67	TA	Park/Neutral Position (PNP)/Clutch Pedal Position (CPP) Switches
67	TG	Electronic Transmission/Continuous Memory Diagnostic Trouble Codes (DTC)
69	TG	Electronic Transmission/Continuous Memory Diagnostic Trouble Codes (DTC)
81	DF	Manifold Absolute Pressure (MAP)/Barometric Pressure (BARO) Sensor
87	J	Fuel Pump or Circuit
91	H	Fuel Control System
95	J	Fuel Pump or Circuit
96	J	Fuel Pump or Circuit
99	TG	Electronic Transmission/Continuous Memory Diagnostic Trouble Codes (DTC)

89446G62

Fig. 21 Ford Diagnostic Trouble Codes—2-digit continuous codes

DIAGNOSTIC TROUBLE CODES—FORD MOTOR CO.
3-DIGIT KOEO CODES

Codes	Pinpoint Test Designation	System/Component/Circuit Fault
111	—	All Systems Functioning Normally
112	DA	Intake Air Temperature (IAT)/Engine Coolant Temperature (ECT) Sensors
113	DA	Intake Air Temperature (IAT)/Engine Coolant Temperature (ECT) Sensors
114	DA	Intake Air Temperature (IAT)/Engine Coolant Temperature (ECT) Sensors
116	DA	Intake Air Temperature (IAT)/Engine Coolant Temperature (ECT) Sensors
117	DA	Intake Air Temperature (IAT)/Engine Coolant Temperature (ECT) Sensors
118	DA	Intake Air Temperature (IAT)/Engine Coolant Temperature (ECT) Sensors
121	DH	Throttle Position (TP) Sensor (Gasoline Engines)
122	DH	Throttle Position (TP) Sensor (Gasoline Engines)
123	DH	Throttle Position (TP) Sensor (Gasoline Engines)
126	DF	Manifold Absolute Pressure (MAP)/Barometric Pressure (BARO) Sensor
158	DC	Mass Air Flow (MAF) Sensor—Cars
159	DC	Mass Air Flow (MAF) Sensor—Cars
226	NC	Ignition Diagnostic Monitor (IDM)/Distributor Ignition(DI)—EDIS Ignition System
327	DL	Pressure Feedback EGR (PFE)/Differential PFE (DPFE) Sensor/EGR Vacuum Regulator (EVR) Solenoid—except Mustang, 5.0L Thunderbird/Cougar & Trucks
327	DN	EGR Valve Position (EVP) Sensor/EGR Vacuum Regulator (EVR) Solenoid—Mustang, 5.0L Thunderbird/Cougar & Trucks
328	DN	EGR Valve Position (EVP) Sensor/EGR Vacuum Regulator (EVR) Solenoid
334	DN	EGR Valve Position (EVP) Sensor/EGR Vacuum Regulator (EVR) Solenoid
335	DL	Pressure Feedback EGR (PFE)/Differential PFE (DPFE) Sensor/EGR Vacuum Regulator (EVR) Solenoid
336	DL	Pressure Feedback EGR (PFE)/Differential PFE (DPFE) Sensor/EGR Vacuum Regulator (EVR) Solenoid
337	DL	Pressure Feedback EGR (PFE)/Differential PFE (DPFE) Sensor/EGR Vacuum Regulator (EVR) Solenoid—except Mustang, 5.0L Thunderbird/Cougar & Trucks
337	DN	EGR Valve Position (EVP) Sensor/EGR Vacuum Regulator (EVR) Solenoid—Mustang, 5.0L Thunderbird/Cougar & Trucks
341	KP	Octane Adjust (OCT ADJ)
511	—	Faulty processor
513	—	Faulty processor
519	FF	Power Steering Pressure (PSP) Switch
524	X	Constant Control Relay Module (CCRM)
525	TA	Park/Neutral Position (PNP)/Clutch Pedal Position (CPP) Switches
539	KM	WOT A/C Cutout (WAC) A/C Demand
551	KT	Intake Manifold Runner Control (IMRC) System
552	KC	Secondary Air Injection (AIRB)/(AIRD) Solenoids
553	KC	Secondary Air Injection (AIRB)/(AIRD) Solenoids
554	KN	Fuel Pressure Regulator Control (FPRC) Solenoid
556	J	Fuel Pump or Circuit—except Tempo/Topaz, 2.3L Mustang, Probe, Taurus/Sable, Continental
556	X	Constant Control Relay Module (CCRM)—Tempo/Topaz, 2.3L Mustang, Probe, Taurus/Sable, Continental
557	X	Constant Control Relay Module (CCRM)
558	DL	Pressure Feedback EGR (PFE)/Differential PFE (DPFE) Sensor/EGR Vacuum Regulator (EVR) Solenoid—except Mustang, 5.0L Thunderbird/Cougar & Trucks
558	DN	EGR Valve Position (EVP) Sensor/EGR Vacuum Regulator (EVR) Solenoid—Mustang, 5.0L Thunderbird/Cougar & Trucks
559	KM	WOT A/C Cutout (WAC) A/C Demand

89446G67

Fig. 22 Ford Diagnostic Trouble Codes—3-digit KOEO codes

DIAGNOSTIC TROUBLE CODES—FORD MOTOR CO.
3-DIGIT KOEO CODES

Codes	Pinpoint Test Designation	System/Component/Circuit Fault
563	KF	Low Fan Control (LFC)/High Fan Control (HFC)—Excort/Tracer
563	X	Constant Control Relay Module (CCRM)—except Excort/Tracer
564	KF	Low Fan Control (LFC)/High Fan Control (HFC)—Excort/Tracer
564	X	Constant Control Relay Module (CCRM)—except Excort/Tracer
565	KD	Canister Purge (CANP) Solenoid
566	TC	Transmission Solenoids
569	KD	Canister Purge (CANP) Solenoid
621	TC	Transmission Solenoids
622	TC	Transmission Solenoids
624	TC	Transmission Solenoids
625	TC	Transmission Solenoids
626	TC	Transmission Solenoids
627	TC	Transmission Solenoids
629	TC	Transmission Solenoids
631	TB	4x4 Low/Transmission Control Switch TCIL-TCS-TCSM
633	TB	4x4 Low/Transmission Control Switch TCIL-TCS-TCSM
634	TD	Manual Lever Position (MLP) Sensor
636	TE	Transmission Fluid Temperature (TFT) Sensor—Trucks
637	TE	Transmission Fluid Temperature (TFT) Sensor
638	TE	Transmission Fluid Temperature (TFT) Sensor
641	TC	Transmission Solenoids
643	TC	Transmission Solenoids
652	TC	Transmission Solenoids
998	TC	Transmission Solenoids
Codes Not Listed	QA	No Diagnostic Trouble Codes (DTC)/DTC Not Listed

KOEO - Key (Ignition) On, Engine Off

89446G68

Fig. 23 Ford Diagnostic Trouble Codes—3-digit KOEO codes continued

DIAGNOSTIC TROUBLE CODES—FORD MOTOR CO.
3-DIGIT KOER CODES

Codes	Pinpoint Test Designation	System/Component/Circuit Fault
111	—	All Systems Functioning Normally
114	DA	Intake Air Temperature (IAT)/Engine Coolant Temperature (ECT) Sensors
116	DA	Intake Air Temperature (IAT)/Engine Coolant Temperature (ECT) Sensors
121	DH	Throttle Position (TP) Sensor (Gasoline Engines)
126	DF	Manifold Absolute Pressure (MAP)/Barometric Pressure (BARO) Sensor
129	DC	Mass Air Flow (MAF) Sensor—Cars, Ranger, Aerostar and Explorer
129	DF	Manifold Absolute Pressure (MAP)/Barometric Pressure (BARO) Sensor—Trucks except Ranger, Aerostar and Explorer
136	H	Fuel Control System
137	H	Fuel Control System
159	DC	Mass Air Flow (MAF) Sensor—Cars
167	DH	Throttle Position (TP) Sensor (Gasoline Engines)
172	H	Fuel Control System
173	H	Fuel Control System
213	PA	Spark Timing Check—Distributor Ignition (DI)
213	PB	Spark Timing Check—Electronic Ignition (Low Data Rate)
213	PC	Spark Timing Check—Electronic Ignition (High Date Rate)
225	DG	Knock Sensor (KS)
311	KC	Secondary Air Injection (AIRB)/(AIRD) Solenoids
312	KC	Secondary Air Injection (AIRB)/(AIRD) Solenoids
313	KC	Secondary Air Injection (AIRB)/(AIRD) Solenoids
314	KC	Secondary Air Injection (AIRB)/(AIRD) Solenoids
326	DL	Pressure Feedback EGR (PFE)/Differential PFE (DPFE) Sensor/EGR Vacuum Regulator (EVR) Solenoid
327	DL	Pressure Feedback EGR (PFE)/Differential PFE (DPFE) Sensor/EGR Vacuum Regulator (EVR) Solenoid—except Mustang, 5.0L Thunderbird/Cougar & Trucks
327	DN	EGR Valve Position (EVP) Sensor/EGR Vacuum Regulator (EVR) Solenoid—Mustang, 5.0L Thunderbird/Cougar & Trucks
328	DN	EGR Valve Position (EVP) Sensor/EGR Vacuum Regulator (EVR) Solenoid
332	DL	Pressure Feedback EGR (PFE)/Differential PFE (DPFE) Sensor/EGR Vacuum Regulator (EVR) Solenoid—except Mustang, 5.0L Thunderbird/Cougar & Trucks
332	DN	EGR Valve Position (EVP) Sensor/EGR Vacuum Regulator (EVR) Solenoid—Mustang, 5.0L Thunderbird/Cougar & Trucks
334	DN	EGR Valve Position (EVP) Sensor/EGR Vacuum Regulator (EVR) Solenoid
336	DL	Pressure Feedback EGR (PFE)/Differential PFE (DPFE) Sensor/EGR Vacuum Regulator (EVR) Solenoid
337	DL	Pressure Feedback EGR (PFE)/Differential PFE (DPFE) Sensor/EGR Vacuum Regulator (EVR) Solenoid—except Mustang, 5.0L Thunderbird/Cougar & Trucks
337	DN	EGR Valve Position (EVP) Sensor/EGR Vacuum Regulator (EVR) Solenoid—Mustang, 5.0L Thunderbird/Cougar & Trucks
411	KE	Idle Air Control (IAC) Solenoid
412	KE	Idle Air Control (IAC) Solenoid
511	—	Faulty processor
513	—	Faulty processor
521	FF	Power Steering Pressure (PSP) Switch
536	FD	Brake On/Off (BOO) switch
538	M	Dynamic Response Test
632	TB	4x4 Low/Transmission Control Switch TCIL-TCS-TCSM
636	TE	Transmission Fluid Temperature (TFT) Sensor—Trucks
639	TF	Transmission Shaft Speed Sensor (TSS)/Output Shaft Speed (OSS) Sensor

Fig. 24 Ford Diagnostic Trouble Codes—3-digit KOER codes

89446G63

DIAGNOSTIC TROUBLE CODES—FORD MOTOR CO.
3-DIGIT CONTINUOUS CODES

Codes	Pinpoint Test Designation	System/Component/Circuit Fault
111	—	All Systems Functioning Normally
112	DA	Intake Air Temperature (IAT)/Engine Coolant Temperature (ECT) Sensors
113	DA	Intake Air Temperature (IAT)/Engine Coolant Temperature (ECT) Sensors
117	DA	Intake Air Temperature (IAT)/Engine Coolant Temperature (ECT) Sensors
118	DA	Intake Air Temperature (IAT)/Engine Coolant Temperature (ECT) Sensors
121	G	In Range MAF/TP/Fuel Injector Pulse Width Test
122	DH	Throttle Position (TP) Sensor (Gasoline Engines)
123	DH	Throttle Position (TP) Sensor (Gasoline Engines)
124	G	In Range MAF/TP/Fuel Injector Pulse Width Test
125	G	In Range MAF/TP/Fuel Injector Pulse Width Test
126	DF	Manifold Absolute Pressure (MAP)/Barometric Pressure (BARO) Sensor
128	DF	Manifold Absolute Pressure (MAP)/Barometric Pressure (BARO) Sensor
139	H	Fuel Control System
144	H	Fuel Control System
157	DC	Mass Air Flow (MAF) Sensor—Cars, Ranger, Aerostar and Explorer
158	DC	Mass Air Flow (MAF) Sensor—Cars
171	H	Fuel Control System
172	H	Fuel Control System
173	H	Fuel Control System
174	H	Fuel Control System
175	H	Fuel Control System
176	H	Fuel Control System
177	H	Fuel Control System
178	H	Fuel Control System
179	HA	Adaptive Fuel System
181	HA	Adaptive Fuel System
182	HA	Adaptive Fuel System
183	HA	Adaptive Fuel System
184	G	In Range MAF/TP/Fuel Injector Pulse Width Test
185	G	In Range MAF/TP/Fuel Injector Pulse Width Test
186	G	In Range MAF/TP/Fuel Injector Pulse Width Test
187	G	In Range MAF/TP/Fuel Injector Pulse Width Test
188	HA	Adaptive Fuel System
189	HA	Adaptive Fuel System
191	HA	Adaptive Fuel System
192	HA	Adaptive Fuel System
194	H	Fuel Control System
195	H	Fuel Control System
211	NA	Ignition Diagnostic Monitor (IDM)/Distributor Ignition(DI)—TFI Ignition System
211	NB	Ignition Diagnostic Monitor (IDM)/Distributor Ignition(DI)—DIS Ignition System
211	NC	Ignition Diagnostic Monitor (IDM)/Distributor Ignition(DI)—EDIS Ignition System
212	NA	Ignition Diagnostic Monitor (IDM)/Distributor Ignition(DI)—TFI Ignition System
214	DR	Cylinder Identification (CID) Circuits
219	PB	Spark Timing Check—Electronic Ignition (Low Data Rate)
327	DN	EGR Valve Position (EVP) Sensor/EGR Vacuum Regulator (EVR) Solenoid—Mustang, 5.0L Thunderbird/Cougar & Trucks
327	DN	EGR Valve Position (EVP) Sensor/EGR Vacuum Regulator (EVR) Solenoid—except Mustang, 5.0L Thunderbird/Cougar & Trucks

89446G64

Fig. 25 Ford Diagnostic Trouble Codes—3-digit KOER codes continued

DIAGNOSTIC TROUBLE CODES—FORD MOTOR CO.
3-DIGIT CONTINUOUS CODES

Codes	Pinpoint Test Designation	System/Component/Circuit Fault
328	DN	EGR Valve Position (EVP) Sensor/EGR Vacuum Regulator (EVR) Solenoid
332	DL	Pressure Feedback EGR (PFE)/Differential PFE (DPFE) Sensor/EGR Vacuum Regulator (EVR) Solenoid—except Mustang, 5.0L Thunderbird/Cougar & Trucks
332	DN	EGR Valve Position (EVP) Sensor/EGR Vacuum Regulator (EVR) Solenoid—Mustang, 5.0L Thunderbird/Cougar & Trucks
334	DN	EGR Valve Position (EVP) Sensor/EGR Vacuum Regulator (EVR) Solenoid
336	DL	Pressure Feedback EGR (PFE)/Differential PFE (DPFE) Sensor/EGR Vacuum Regulator (EVR) Solenoid
337	DL	Pressure Feedback EGR (PFE)/Differential PFE (DPFE) Sensor/EGR Vacuum Regulator (EVR) Solenoid—except Mustang, 5.0L Thunderbird/Cougar & Trucks
337	DN	EGR Valve Position (EVP) Sensor/EGR Vacuum Regulator (EVR) Solenoid—Mustang, 5.0L Thunderbird/Cougar & Trucks
338	DA	Intake Air Temperature (IAT)/Engine Coolant Temperature (ECT) Sensors
339	DA	Intake Air Temperature (IAT)/Engine Coolant Temperature (ECT) Sensors
452	DP	Vehicle Speed Sensor (VSS)—Cars
452	DS	Programmable Speedometer/Odometer Module (PSOM)—Trucks
511	—	Faulty processor
512	QB	Continuous Memory Diagnostic Trouble Code (DTC)
513	—	Faulty processor
524	X	Constant Control Relay Module (CCRM)
525	TA	Park/Neutral Position (PNP)/Clutch Pedal Position (CPP) Switches
528	TA	Park/Neutral Position (PNP)/Clutch Pedal Position (CPP) Switches
529	ML	Self-Test Output (STO)/Malfunction Indicator Lamp (MIL)
533	ML	Self-Test Output (STO)/Malfunction Indicator Lamp (MIL)
536	FD	Brake On/Off (BOO) switch
556	J	Fuel Pump or Circuit—except Tempo/Topaz, 2.3L Mustang, Probe, Taurus/Sable, Continental
556	X	Constant Control Relay Module (CCRM)—Tempo/Topaz, 2.3L Mustang, Probe, Taurus/Sable, Continental
557	X	Constant Control Relay Module (CCRM)
617	TG	Electronic Transmission/Continuous Memory Diagnostic Trouble Codes (DTC)
618	TG	Electronic Transmission/Continuous Memory Diagnostic Trouble Codes (DTC)
619	TG	Electronic Transmission/Continuous Memory Diagnostic Trouble Codes (DTC)
621	TC	Transmission Solenoids
622	TC	Transmission Solenoids
624	TG	Electronic Transmission/Continuous Memory Diagnostic Trouble Codes (DTC)
625	TG	Electronic Transmission/Continuous Memory Diagnostic Trouble Codes (DTC)
628	TG	Electronic Transmission/Continuous Memory Diagnostic Trouble Codes (DTC)
634	TD	Manual Lever Position (MLP) Sensor—Escort/Tracer, E-Series/F-Series Trucks, Bronco
634	TG	Electronic Transmission/Continuous Memory Diagnostic Trouble Codes (DTC)—except Escort/Tracer, E-Series/F-Series Trucks, Bronco
637	TE	Transmission Fluid Temperature (TFT) Sensor—Escort/Tracer
637	TG	Electronic Transmission/Continuous Memory Diagnostic Trouble Codes (DTC)—except Escort/Tracer
638	TE	Transmission Fluid Temperature (TFT) Sensor—Escort/Tracer
638	TG	Electronic Transmission/Continuous Memory Diagnostic Trouble Codes (DTC)—except Escort/Tracer
639	TF	Transmission Shaft Speed Sensor (TSS)/Output Shaft Speed (OSS) Sensor
641	TC	Transmission Solenoids—Escort/Tracer

89446G65

Fig. 26 Ford Diagnostic Trouble Codes—3-digit continuous codes

DIAGNOSTIC TROUBLE CODES—FORD MOTOR CO.
3-DIGIT CONTINUOUS CODES

Codes	Pinpoint Test Designation	System/Component/Circuit Fault
643	TC	Transmission Solenoids
645	TG	Electronic Transmission/Continuous Memory Diagnostic Trouble Codes (DTC)
646	TG	Electronic Transmission/Continuous Memory Diagnostic Trouble Codes (DTC)
647	TG	Electronic Transmission/Continuous Memory Diagnostic Trouble Codes (DTC)
648	TG	Electronic Transmission/Continuous Memory Diagnostic Trouble Codes (DTC)
649	TG	Electronic Transmission/Continuous Memory Diagnostic Trouble Codes (DTC)
651	TG	Electronic Transmission/Continuous Memory Diagnostic Trouble Codes (DTC)
654	TD	Manual Lever Position (MLP) Sensor
656	TG	Electronic Transmission/Continuous Memory Diagnostic Trouble Codes (DTC)

89446G66

Fig. 27 Ford Diagnostic Trouble Codes—3-digit continuous codes continued

➡Some vehicles may not use the TAT terminal in the DLC. If this is the case with the vehicle at hand, ground the TEN terminal of the DLC. If this is done, however, watch the CHECK ENGINE light for the DTC's, not the MANUAL SHIFT light .

13. Without starting the engine, turn the ignition switch **ON**.
14. The MANUAL SHIFT light should start to flash out any stored DTC's.

Clearing Codes

EXCEPT PROBE AND 1991–92 ESCORT MODELS

To clear the DTC's from the vehicle's memory, simply disconnect the ground jumper wire from the STI terminal while the retrieval cycle is running. The computer will erase all applicable codes.

PROBE AND 1991–92 ESCORT MODELS

To clear the DTC's from the vehicle's memory, disconnect the negative battery cable for at least one minute. Reconnect the cable and all of the DTC's should be purged from the computer's memory.

GENERAL MOTORS

▶ See Figures 28 thru 33

Reading and Retrieving Codes

EXCEPT GEO

General Motors (GM) transmission/transaxle diagnostic systems use two-digit DTC's, as with most pre-OBD II vehicles, to inform to the mechanic or technician the problem circuit or component.

Fortunately, it is not necessary to use a diagnostic scan tool to retrieve DTC's from a pre-OBD II GM vehicle. General Motors models use a light, located in the instrument cluster and usually referred to as the Malfunction Indicator Lamp (MIL), CHECK ENGINE light or SERVICE ENGINE SOON light, to display the DTC's as a series of flashes or blinks.

The two-digit codes are flashed out in a logical series of blinks. For example, when first starting to retrieve the DTC's the MIL will flash once, pause briefly, then flash two more times followed by a longer pause. This indicates a Code 12 (one flash-pause-two flashes-long pause). Therefore, a flash pattern of two flashes-pause-three flashes-long pause would indicate a Code 23.

➡Code 12 should always be the first DTC flashed out by the MIL, since it indicates that the diagnostic system is functioning properly.

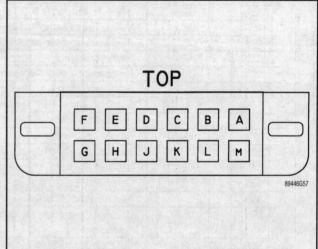

89446G57

Fig. 28 To enable the flash-out mode of the diagnostic computer system, connect terminals A and B of the DLC

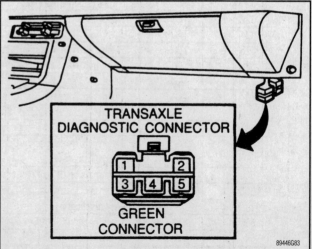

89446G83

Fig. 29 The transaxle diagnostic connector is located under the right side of the instrument panel, as shown

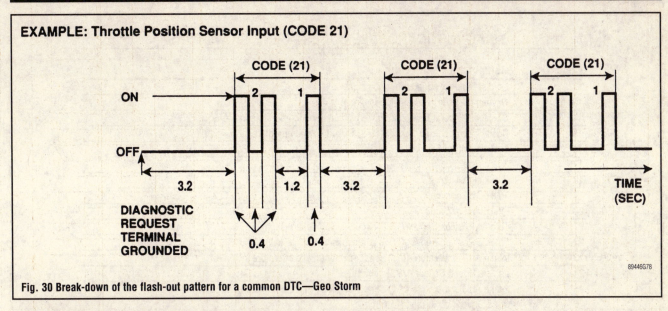

EXAMPLE: Throttle Position Sensor Input (CODE 21)

Fig. 30 Break-down of the flash-out pattern for a common DTC—Geo Storm

Each DTC is flashed out by the MIL three times, then the next code is flashed out (if more than one code has been stored). The codes are displayed in numerical sequence starting with the lowest code number and moving to the highest. Therefore, if Codes 21, 14, and 16 were stored by the vehicle computer, the MIL would display them in the following order: 12 (always the first code—it indicates that the ignition key is **ON**, but the engine is not running), 12, 12, 14, 14, 14, 16, 16, 16, 21, 21, 21. Remember that each code is flashed three times.

When reading the codes be sure not to miss any of them, otherwise the entire sequence must be repeated. This can be tiresome if a large number of codes was stored, and you missed the last one.

To retrieve the codes, perform the following:

1. Perform the preliminary inspection, located earlier in this section. This is very important, since a loose or disconnected wire, or corroded connector terminals can cause a whole slew of unrelated DTC's to be stored by the computer; you will waste a lot of time performing a diagnostic "goose chase."

2. Grab some paper and a pencil or pen to write down the DTC's when they are flashed out.

3. Locate the Diagnostic Link Connector (DLC) or Assembly Line Diagnostic Link (ALDL). This is the connector that provides a direct link with the vehicle's computer system. It is usually located underneath the instrument panel or in the glove box, but occasionally can be located in the engine compartment, near the firewall or relay box. If the DLC/ALDL cannot be found, either call your local GM dealer or refer to the applicable Chilton Total Car Care manual for your vehicle.

4. Using a jumper wire or paper clip (you have to bend the paper clip into a large U shape), connect DLC terminals **A** and **B**, which grounds the test terminal and switches the computer system into the diagnostic display mode.

5. Turn the ignition switch **ON**, but DO NOT start the engine.

➡**Remember that when the ignition is ON, but the engine is not running, a Code 12 is normal.**

6. At this point, write down the codes flashed out by the MIL.
7. Turn the ignition switch **OFF**.
8. Fix any problems indicated by the DTC's.

GEO STORM

The Geo Storm uses two-digit DTC's, which can be read by having the ECONOMY light flash the codes out as a series of flashes or blinks.

The two-digit codes are flashed out in a logical series of blinks. For example, when first starting to retrieve the DTC's the MIL will flash once, pause briefly, then flash three more times followed by a longer pause. This indicates a Code 13 (one flash-pause-three flashes-long pause). Therefore,

a flash pattern of two flashes-pause-four flashes-long pause would indicate a Code 24.

Each DTC is displayed from the lowest number on through to the highest number. After the highest code number is displayed, the system begins with the lowest again. The system will continue to cycle through all of the codes in this manner; so if you missed a code or two, you can wait until it starts the cycle over again.

To retrieve the codes, perform the following:

1. Perform the preliminary inspection, located earlier in this section. This is very important, since a loose or disconnected wire, or corroded connector terminals can cause a whole slew of unrelated DTC's to be stored by the computer; you will waste a lot of time performing a diagnostic "goose chase."

2. Grab some paper and a pencil or pen to write down the DTC's when they are flashed out.

3. Locate the transaxle diagnostic connector, which is a 5-terminal green connector and is usually located behind the right-hand lower kick-panel. If the transaxle diagnostic connector cannot be found, either call your local Geo dealer or refer to the applicable Chilton's Total Car Care manual for your vehicle.

4. Turn the ignition key **OFF**, if not already done.

5. Using a jumper wire, connect terminals 1 (black/white wire) and 2 (black/blue wire).

6. Ensure the transmission/transaxle mode selector is set to the NORMAL position.

7. Without starting the engine, turn the ignition switch **ON**.

8. The ECONOMY light should start to flash out any stored DTC's.

Clearing Codes

EXCEPT GEO

➡**It is not recommended that the negative battery cable be disconnected to clear DTC's. Doing so will clear all settings from the vehicle's computer system, resulting in lost radio presets, seat memories, anti-theft codes, driveability parameters, etc., and there are better ways of spending your afternoon than resetting all of these.**

1. Locate the vehicle's fuse box. The fuse box is usually located under the driver's side of the instrument panel or in the engine compartment.

2. If applicable, remove the fuse box cover.

3. Look for a PCM, ECM or similarly marked fuse (computer, battery +, etc.). This fuse is used for the battery voltage circuit that supplies the PCM with power even when the ignition is switched **OFF**.

4. Remove this fuse from the fuse box, and wait at least 10 seconds.

DIAGNOSTIC TROUBLE CODES—GENERAL MOTORS

Code	Component/Circuit Fault
12	No distributor or speed reference signal
13	Oxygen (O_2) sensor circuit (open circuit)
14	Coolant Temperature Sensor (CTS) circuit (high temperature indicated)
15	Coolant Temperature Sensor (CTS) Circuit (low temperature indicated)
16	System voltage high/low
17	Spark reference circuit
18	Crank signal circuit
19	Fuel pump or circuit
20	Fuel pump or circuit
21	Throttle Position Sensor (TPS) circuit (signal voltage high)
22	Throttle Position Sensor (TPS) circuit (signal voltage low)
23	Intake Air Temperature (IAT) sensor circuit (low temperature indicated)
24	Vehicle Speed Sensor (VSS) circuit
25	Intake Air Temperature (IAT) sensor or circuit (high temperature indicated)
26	Quad-Driver (QDM) fault (1 of 3)
27	2nd gear switch or circuit; 4th gear switch or circuit
28	3rd gear switch or circuit; 4th gear switch or circuit
29	4th gear switch or circuit; canister purge solenoid or circuit; 4th gear switch or circuit
30	ISC or circuit
31	PRNDL switch circuit
32	Barometric pressure sensor or circuit; EGR system; MAP sensor or circuit
33	MAP sensor or circuit; MAF sensor or circuit
34	Mass Air Flow (MAF) sensor or circuit
35	ISC or circuit; IAC or circuit; Barometric pressure sensor or circuit
36	Transaxle shift control problem
37	MAT sensor or circuit; Brake switch stuck on
38	TCC brake input circuit
39	TCC or VCC engagement fault
40	Power Steering Pressure (PSP) circuit
41	Camshaft sensor or circuit (Type 1 Ignition)
41	Camshaft sensor or circuit (Type 2 Ignition)
42	Electronic Spark Timing (EST) circuit
43	Electronic Spark Control (ESC) circuit
44	Oxygen (O_2) sensor or circuit (lean exhaust indicated)
45	Oxygen (O_2) sensor or circuit (rich exhaust indicated)
46	PSP switch or circuit; Anti-theft system; Left-to-right bank fuel delivery difference
47	ECM/PCM/BCM fault
48	Misfire detected
49	Air management system
50	Second gear pressure circuit
51	MEM-CAL error (faulty or incorrect MEM-CAL)
52	Mem-Cal CalPal ECM fault; ECM/PCM memory reset indicator; EOT sensor or circuit
53	System voltage too high; distributor signal interrupt; EGR system; Anti-theft system; Air switch solenoid or circuit; Reference voltage overload (Diesel)
54	M/C solenoid or circuit; Fuel pump or circuit
55	ECM fault; Air switch valve or circuit; TPS or circuit
56	4th gear switch or circuit; 3-4 shift solenoid or circuit; vacuum sensor or circuit
57	PCM/BCM fault; 4th gear switch or circuit; 3-4 shift solenoid or circuit
58	Personal Automotive Security System (PASS-Key) fuel enable circuit
59	Transmission temperature circuit
60	Cruise control—transmission not in drive

89446G86

Fig. 31 General Motors Diagnostic Trouble Codes, 1 of 2—except Geo Storm

DIAGNOSTIC TROUBLE CODES—GENERAL MOTORS

Code	Component/Circuit Fault
61	Cruise vent solenoid circuit
62	Cruise vacuum solenoid circuit
63	Cruise control—vehicle speed and set speed difference; MAP sensor or circuit; EGR flow check fault; Right oxygen sensor or circuit
64	Cruise control—vehicle acceleration too high; EGR flow check fault; Right oxygen sensor or circuit
65	Cruise servo position sensor or circuit
66	Cruise control—engine speed too high; A/C pressure sensor or circuit; Power switch or circuit; 3-2 control solenoid or circuit
67	Cruise control switches or circuits
68	Cruise control system fault
69	A/C head pressure switch or circuit
70	Intermittent TPS sensor or circuit fault
71	Intermittent MAP sensor or circuit fault
72	Vehicle speed control circuit
73	Force motor current error
74	Intermittent MAT sensor or circuit fault
75	Intermittent VSS sensor or circuit fault
79	Transmission fluid too hot
80	Fuel system overly rich
81	Shift solenoid B circuit
82	Shift solenoid A or circuit
83	TCC solenoid or circuit
85	Throttle angle incorrect range
86	Low ratio error—solenoid B closed
87	High ratio error—solenoid B open
90	VCC or circuit
91	PRNDL switch or circuit
92	Heated windshield request problem
96	Torque converter overstress
97	P/N to D/R at high throttle angle
98	High RPM from P/N to D/R in ISC range
99	Cruise control servo application fault

89446G56

Fig. 32 General Motors Diagnostic Trouble Codes, 2 of 2—except Geo Storm

CODE	DIAGNOSTIC AREA	MODE		FAILURE
–	*Normal/Fault Detected	ON OFF  ⊓⊔⊓⊔⊓⊔⊓⊔⊓⊔		–
11	Vehicle speed sensor (Transaxle mounting)			Open or Short
13	Rpm signal			Open or Short
15	ATF temperature sensor			Open
21	Throttle position sensor input			Open or Short
24	Vehicle speed sensor (I/P cluster mounting)			Open or Short
31	Shift A solenoid			Open or Short
32	Shift B solenoid			Open or Short
33	Overrun clutch solenoid			Open or Short
34	TCC solenoid			Open or Short
35	Line pressure duty solenoid			Open or Short

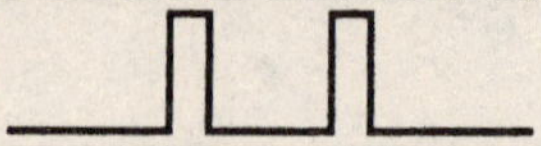

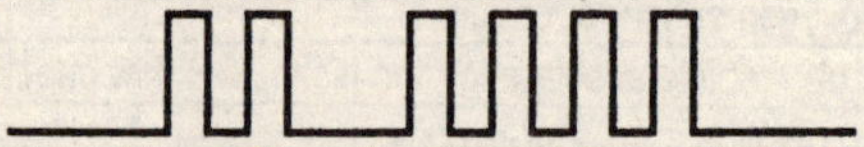

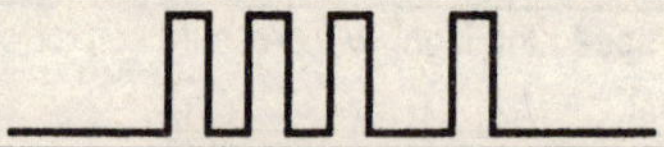

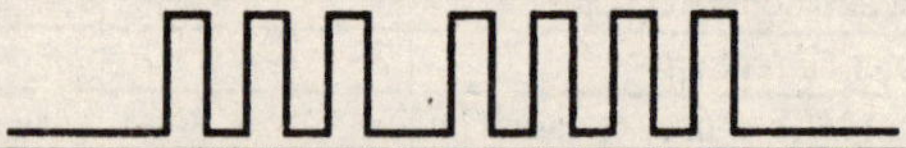

*Normal with diagnostic request terminal grounded.
Fault detected without diagnostic request terminal grounded.

89446G79

Fig. 33 Geo Storm Diagnostic Trouble Codes

5. Reinstall the fuse, and install the fuse box cover (if equipped).
The transmission/transaxle DTC's should now be cleared. To double check, perform the code retrieval procedure again. Only Code 12 should be displayed, if not either the codes were not cleared (Are the codes identical to those flashed out previously?) or the underlying problem is still there (Are only some of the codes the same as previously?)

GEO STORM

➡It is not recommended that the negative battery cable be disconnected to clear DTC's. Doing so will clear all settings from the vehicle's computer system, resulting in lost radio presets, seat memories, anti-theft codes, driveability parameters, etc., and there are better ways of spending your afternoon than resetting all of these.

To clear the DTC's from the vehicle's memory, locate the 4AT fuse in the fuse block. Remove the fuse for approximately 60 seconds, then reinstall it. All of the codes should now be cleared. To double check, perform the code retrieval procedure again. If there are still codes present, either the codes

were not cleared (Are the codes identical to those flashed out previously?) or the underlying problem is still there (Are only some of the codes the same as previously?)

Import Vehicles

ACURA & HONDA

◆ See Figures 34 thru 49

Reading and Retrieving Codes

EARLY MODELS

➡Early models refers to the following vehicles: 1989–91 Civic 4WD, 1990–93 Integra, 1988–90 Legend and 1988–91 Prelude models.

All of these models display stored DTC's through a Light Emitting Diode (LED) located on the transmission/transaxle control computer, which con-

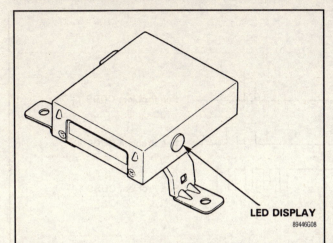

Fig. 34 The LED is usually located on the side of the transaxle control computer, and, if a fault is detected, flashes whenever the ignition key is ON—early models

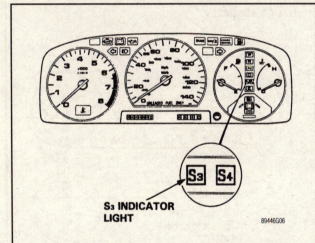

Fig. 35 Typical location of the S3 indicator light used to alert you when DTC's have been stored—1991 Prelude shown

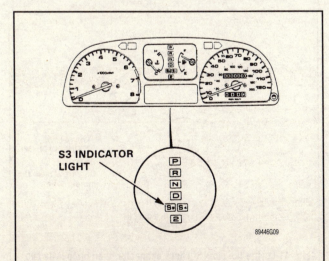

Fig. 36 Another common location of the S3 indicator light in the instrument cluster display—1989–91 Civic 4WD shown

stantly flashes out the DTC's whenever the ignition key is **ON**. When the computer stores a transmission-related DTC, the S3 light in the instrument cluster illuminates all the time, regardless of the gear selected, to let the driver know there is a stored DTC.

These vehicles use one and two-digit DTC's, which are displayed through the LED on the transmission/transaxle computer in a series of flashes. For example, if the LED flashes five times, pauses, then flashes six times and pauses, this indicates Codes 5 and 6 have been stored by the transmission/transaxle computer. For two-digit codes, the tens digits are displayed as long flashes and the ones digits are displayed as shorter flashes. Therefore, if the LED displays one long flash and four short flashes followed by a pause, this indicates a Code 14. The system displays the codes from the lowest to the highest, and when it reaches the last code it returns to the lowest and cycles through the codes again.

To retrieve the codes, perform the following:

1. Perform the preliminary inspection, located earlier in this section. This is very important, since a loose or disconnected wire, or corroded connector terminals can cause a whole slew of unrelated DTC's to be stored by the computer; you will waste a lot of time performing a diagnostic "goose chase."

2. Grab some paper and a pencil or pen to write down the DTC's when they are flashed out.

3. Locate the transmission/transaxle control computer. The computer is generally located in the following positions:
• 1987–90 Legend Sedan—under the driver's seat
• 1988–90 Legend Coupe—under the passenger's seat
• 1988–90 Prelude—under the carpet in the passenger's side front footwell
• 1989–91 Civic 4WD—under the driver's seat
• 1990–93 Integra—up under the dashboard to the left side of the instrument cluster
• 1991 Prelude with the 2.0L engine—behind the center console, below the radio (it may be visible without removing the center console)
• 1991 Prelude with the 2.1L engine—under the carpet in the passenger's side front footwell

4. Turn the ignition switch **ON** and observe the computer's LED. Note all of the DTC's.

5. Once all of the DTC's have been noted, turn the ignition switch **OFF**.

LATE MODELS

➥Late models refers to the following vehicles: 1990–96 Accord, 1996–97 Civic, 1992–96 Prelude, 1994–96 Integra, 1991–94 Legend, 1995–97 2.5TL, 1996–97 3.2TL, 1992–94 Vigor, 1996–97 3.5RL models.

Unlike the other Acura/Honda models described earlier in this section, these models display the transmission-related DTC's through either the S or D4 light on the instrument cluster. These models use 2-digit trouble codes, which are displayed in a series of flashes. The tens digits are displayed as long flashes and the ones digits are displayed as shorter flashes. Therefore, if the S or D4 light displays one long flash and four short flashes followed by a pause, this indicates a Code 14.

➥Some of the early models of these vehicles have the added benefit, in which you can read the DTC's through the S/D4 light and also from the LED on the computer. The LED located on the computer was slowly phased out during these model years.

To retrieve the DTC's, perform the following:

1. Perform the preliminary inspection, located earlier in this section. This is very important, since a loose or disconnected wire, or corroded connector terminals can cause a whole slew of unrelated DTC's to be stored by the computer; you will waste a lot of time performing a diagnostic "goose chase."

2. Grab some paper and a pencil or pen to write down the DTC's when they are flashed out.

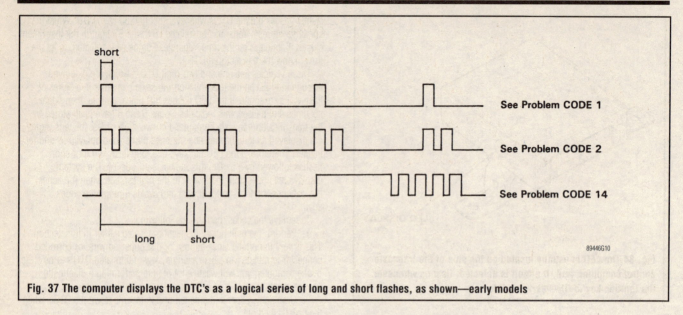

Fig. 37 The computer displays the DTC's as a logical series of long and short flashes, as shown—early models

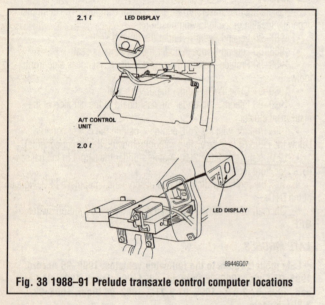

Fig. 38 1988–91 Prelude transaxle control computer locations

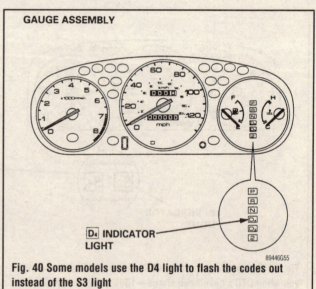

Fig. 40 Some models use the D4 light to flash the codes out instead of the S3 light

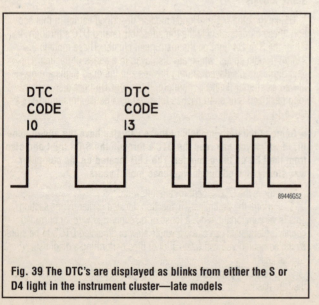

Fig. 39 The DTC's are displayed as blinks from either the S or D4 light in the instrument cluster—late models

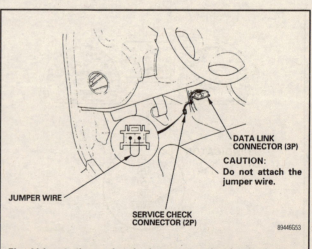

Fig. 41 Locate the service check connector, then jumper its two terminals to activate the diagnostic system—1995 Prelude shown

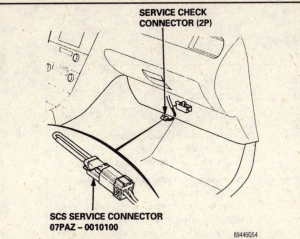

Fig. 42 Acura/Honda actually makes a service tool to connect the two terminals (but you can use a jumper wire just as easily)—1996–97 3.5RL shown

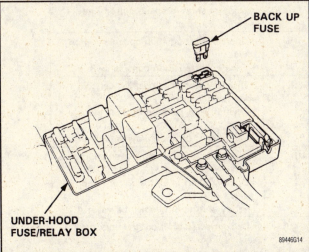

Fig. 43 To clear the DTC's from most Acura/Honda models' memory, remove the backup fuse from the under hood fuse/relay box

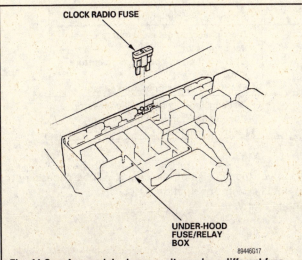

Fig. 44 On a few models, however, it may be a different fuse, such as with this 1995 Prelude

❊❊ WARNING

The Diagnostic Link Connector (DLC) is usually situated next to the service check connector; do not jump the terminals of the DLC by mistake. This could damage the vehicle's computer.

3. Locate the service check connector. The service check connector is either a single connector with two wires running to it or it is two separate one-wire connectors next to each other. The service check connector(s) are not attached to anything. It can be located according to the model of the vehicle at hand, as follows:

• 1990–96 Accord—beneath the far right-hand side of the instrument panel
• 1996–97 Civic—beneath the passenger's side instrument panel
• 1992–96 Prelude—behind the left-hand side of the center console
• 1994–97 Integra—behind the passenger's side lower right kick-panel
• 1991–94 Legend—beneath the passenger's side instrument panel
• 1995–97 2.5TL—beneath the passenger's side instrument panel
• 1996–97 3.2TL—beneath the passenger's side instrument panel
• 1992–94 Vigor—beneath the passenger's side instrument panel, against the firewall
• 1996–97 3.5RL—beneath the passenger's side instrument panel, against the firewall

4. Using a jumper wire or bent paper clip, connect the two terminals of the service check connector.

5. Turn the ignition switch **ON** and observe the S or D4 light in the instrument cluster. Note all of the DTC's.

6. Once all of the DTC's have been noted, turn the ignition switch **OFF** and remove the jumper wire/paper clip.

Clearing Codes

➡ It is not recommended that the negative battery cable be disconnected to clear DTC's. Doing so will clear all settings from the vehicle's computer system, resulting in lost radio presets, seat memories, anti-theft codes, driveability parameters, etc., and there are better ways of spending your afternoon than resetting all of these.

For most models, locate the backup fuse in the engine compartment fuse block. For the exceptions, refer to the following list:

• 1989–91 Civic 4WD—fuse No. 34
• 1989–91 Prelude—fuse No. 35
• 1992–96 Prelude—clock/radio fuse
• 1991–93 Integra—fuse No. 34
• 1988–90 Legend—fuse No. 22
• 1991–95 Legend—fuse No. 15

➡ The only fuse not located in the engine compartment fuse block is fuse No. 15 for the 1991–95 Legend, which is located in the under dash fuse box.

Remove the fuse for at least 10–15 seconds, then reinstall it. All of the codes should now be cleared. To double check, road test the vehicle and perform the code retrieval procedure again. If there are still codes present, either the codes were not cleared (Are the codes identical to those flashed out previously?) or the underlying problem is still there (Are only some of the codes the same as previously?)

EUROPEAN MODELS

A great number of European vehicles use transmissions/transaxles of the ZF transmission/transaxle family. These transmissions/transaxles are used in Audis, BMWs, Volvos and many other European vehicles. Unfortunately, as is the case with almost all of the European import vehicles, the DTC's cannot be retrieved without the use of an expensive, specific scan tool.

Although some of the codes on OBD II equipped models may be retrieved using a generic OBD II scan tool, not all codes can be read in this manner.

Number of LED display blinks	S3 indicator light	Symptom	Probable Cause
1	Blinks	• Lock-up clutch does not engage. • Lock-up-clutch does not disengage. • **Unstable idle speed.**	• Disconnected lock-up control solenoid valve A connector • Open or short in lock-up control solenoid valve A wire • Faulty lock-up control solenoid valve A
2	Blinks	• Lock-up clutch does not engage.	• Disconnected lock-up control solenoid valve B connector • Open or short in lock-up control solenoid valve B wire • Faulty lock-up control solenoid valve B
3	Blinks or OFF	• Lock-up clutch does not engage.	• Disconnected throttle angle sensor connector • Open or short in throttle angle sensor wire • Faulty throttle angle sensor
4	Blinks	• Lock-up clutch does not engage.	• Disconnected speed pulser connector • Open or short in speed pulser wire • Faulty speed pulser
5	Blinks	• Fails to shift other than 2nd ⟷ 4th gear. • Lock-up clutch does not engage.	• Short in shift position console switch wire • Faulty shift position console switch
6	OFF	• Fails to shift other than 2nd ⟷ 4th gear. • Lock-up clutch does not engage. • Lock-up clutch engages and disengages alternately.	• Disconnected shift position console switch connector • Open in shift position console switch wire • Faulty shift position console switch.
7	Blinks	• Fails to shift other than 1st ⟷ 4th, 2nd ⟷ 4th, or 2nd ⟷ 3rd gears. • Fails to shift (stuck in 4th gear).	• Disconnected shift control solenoid valve A connector • Open or short in shift control solenoid valve A wire • Faulty shift control solenoid valve A
8	Blinks	• Fails to shift (stuck in 1st gear or 4th gear).	• Disconnected shift control solenoid valve B connector • Open or short in shift control solenoid valve B wire • Faulty shift control solenoid valve B
9	Blinks	• Lock-up clutch does not engage.	• Disconnected A/T speed pulser • Open or short in A/T speed pulser wire • Faulty A/T speed pulser
10	Blinks	• Lock-up clutch does not engage.	• Disconnected coolant temperature sensor connector • Open or short in coolant temperature sensor wire • Faulty coolant temperature sensor
11	OFF	• Lock-up clutch does not engage.	• Disconnected ignition coil connector • Open or short in ignition coil wire • Faulty ignition coil

NOTE:
- If a customer describes the symptoms for codes 3, 6 or 11, yet the LED is not blinking, it will be necessary to recreate the symptom by test driving, and then checking the LED with the ignition STILL ON.
- If the LED display blinks 12 or more times, the control unit is faulty.

89446G01

Fig. 45 Acura/Honda DTC's—1989–91 Civic 4WD, 1990–93 Integra, 1988–90 Legend and 1988–91 Prelude models

Number of D_4 indicator light blinks while Service Check Connector is jumped.	D_4 indicator light	Possible Cause	Symptom
1	Blinks	• Disconnected lock-up control solenoid valve A connector • Short or open in lock-up control solenoid valve A wire • Faulty lock-up control solenoid valve A	• Lock-up clutch does not engage. • Lock-up clutch does not dis-engage. • Unstable idle speed.
2	Blinks	• Disconnected lock-up clontrol solenoid valve B connector • Short or open in lock-up control solenoid valve B wire • Faulty lock-up control solenoid valve B	• Lock-up clutch not engage.
3	Blinks or OFF	• Disconnected throttle position (TP) sensor connector • Short or open in TP sensor wire • Faulty TP sensor	• Lock-up clutch does not engage.
4	Blinks	• Disconnected vehicle speed sensor (VSS) connector • Short or open in VSS wire • Faulty VSS	• Lock-up clutch does not engage.
5	Blinks	• Short in A/T gear position switch wire • Faulty A/T gear position switch	• Fails to shift other than 2nd ↔ 4th gears. • Lock-up clutch does not engage.
6	OFF	• Disconnected A/T gear position switch connector • Open in A/T gear position switch wire • Faulty A/T gear position switch	• Fails to shift other than 2nd ↔ 4th gears. • Lock-up clutch does not engage. • Lock-up clutch engages and disengages alternately.
7	Blinks	• Disconnected shift control solenoid valve A connector • Short or open in shift control solenoid valve A wire • Faulty shift control solenoid vavle A	• Fails to shift (between 1st ↔ 4th, 2nd ↔ 4th or 2nd ↔ 3rd gears only). • Fails to shift (stuck in 4th gear)
8	Blinks	• Disconnected shift control solenoid valve B connector • Short or open in shift control solenoid valve B wire • Faulty shift control solenoid valve B	• Fails to shift (stuck in 1st or 4th gears).
9	Blinks	• Disconnected countershaft speed sensor connector • Short or open in the counter-shaft speed sensor wire • Faulty countershaft speed sensor	• Lock-up clutch does not engage.

89446G02

Fig. 46 Acura/Honda DTC's, 1 of 2—1990–95 Accord, 1992–95 Prelude, 1994–95 Integra, 1991–94 Legend, 1995 2.5TL, and 1992–94 Vigor models

Number of D4 indicator light blinks while Service Check Connector is jumped.	D4 indicator light	Possible Cause	Symptom
10	Blinks	•Disconnected engine coolant temperature (ECT) sensor connector •Short or open in the ECT sensor wire •Faulty ECT sensor	•Lock-up clutch does not engage.
11	OFF	•Disconnected ignition coil connector •Short or open in ignition coil wire •Faulty ignition coil	•Lock-up clutch does not engage.
14	OFF	•Short or open in FAS (BRN/WHT) wire between the D16 terminal and ECM •Trouble in ECM	•Transmission jerks hard when shifting.
15	OFF	•Disconnected mainshaft speed sensor connector •Short or open in mainshaft speed sensor wire •Faulty mainshaft speed sensor	•Transmission jerks hard when shifting.

89446G03

Fig. 47 Acura/Honda DTC's, 2 of 2—1990–95 Accord, 1992–95 Prelude, 1994–95 Integra, 1991–94 Legend, 1995 2.5TL, and 1992–94 Vigor models

Diagnostic Trouble Code (DTC)*	D4 Indicator Light	Symptom	Possible Cause
P1753 (1)	Blinks	• Lock-up clutch does not engage. • Lock-up clutch does not disengage. • Unstable idle speed.	• Disconnected lock-up control solenoid valve A connector • Short or open in lock-up control solenoid valve A wire • Faulty lock-up control solenoid valve A
P1758 (2)	Blinks	• Lock-up clutch does not engage.	• Disconnected lock-up control solenoid valve B connector • Short or open in lock-up control solenoid valve B wire • Faulty lock-up control solenoid valve B
P1791 (4)	Blinks	• Lock-up clutch does not engage.	• Disconnected vehicle speed sensor (VSS) connector • Short or open in VSS wire • Faulty VSS
P1705 (5)	Blinks	• Fails to shift other than 2nd – 4th gears. • Lock-up clutch does not engage.	• Short in A/T gear position switch wire • Faulty A/T gear position switch
P1706 (6)	OFF	• Fails to shift other than 2nd – 4th gears. • Lock-up clutch does not engage. • Lock-up clutch engages and disengages alternately.	• Disconnected A/T gear position switch connector • Open in A/T gear position switch wire • Faulty A/T gear position switch
P0753 (7)	Blinks	• Fails to shift (between 1st – 4th, 2nd – 4th or 2nd – 3rd gear only). • Fails to shift (stuck in 4th gear).	• Disconnected shift control solenoid valve A connector • Short or open in shift control solenoid valve A wire • Faulty shift control solenoid valve A
P0758 (8)	Blinks	• Fails to shift (stuck in 1st or 4th gears).	• Disconnected shift control solenoid valve B connector • Short or open in shift control solenoid valve B wire • Faulty shift control solenoid valve B
P0720 (9)	Blinks	• Lock-up clutch does not engage.	• Disconnected countershaft speed sensor connector • Short or open in countershaft speed sensor wire • Faulty countershaft speed sensor
P0715 (15)	OFF	• Transmission jerks hard when shifting.	• Disconnected mainshaft speed sensor connector • Short or open in mainshaft speed sensor wire • Faulty mainshaft speed sensor
P1768 (16)	Blinks	• Transmission jerks hard when shifting. • Lock-up clutch does not engage.	• Disconnected linear solenoid connector • Short or open in linear solenoid wire • Faulty linear solenoid

(DTC)*: The DTCs in parentheses are the number of the D4 indicator light blinks when the service check connector is connected with the special tool (SCS service connector).

89446G04

Fig. 48 Acura/Honda DTC's, 1 of 2—1996–97 Accord, 1996–97 Civic, 1996–97 Prelude, 1996–97 Integra, 1996–97 2.5TL, 1996–97 3.2TL and 1996–97 3.5RL models

Diagnostic Trouble Code (DTC)*	D4 Indicator Light	Symptom	Possible Cause
P0740 (40)	OFF	• Lock-up clutch does not engage. • Lock-up clutch does not disengage. • Unstable idle speed.	• Faulty lock-up control system
P0730 (41)	OFF	• Fails to shift (between 1st – 2nd, 2nd – 4th or 2nd – 3rd gears only). • Fails to shift (stuck in 1st or 4th gears).	• Faulty shift control system

(DTC)*: The DTCs in parentheses are the number of the D4 indicator light blinks when the service check connector is connected with the special tool (SCS service connector).

89446G05

Fig. 49 Acura/Honda DTC's, 2 of 2—1996–97 Accord, 1996–97 Civic, 1996–97 Prelude, 1996–97 Integra, 1996–97 2.5TL, 1996–97 3.2TL and 1996–97 3.5RL models

ISUZU

♦ **See Figures 50 thru 59**

Reading and Retrieving Codes

IMPULSE MODELS

The Isuzu Impulse utilizes two-digit DTC's to inform the technician in which circuit and/or component a fault was detected. The two-digit codes are displayed by the ECONOMY light, located in the instrument cluster. The light flashes the digits out by first flashing the tens digit, then after a slight pause, the ones digit. Therefore, if the ECONOMY light flashes twice, pauses slightly, then flashes once, a Code 21 was displayed. If there is more than one DTC stored in the computer's memory, the light will display the lowest number first and move upward through all of the stored codes in numerical order. After the progression reaches the highest code number, the cycle will repeat itself until the ignition switch is turned **OFF**.

To determine if the vehicle's computer has stored any DTC's, switch the transmission/transaxle mode switch to the NORMAL position and turn the ignition switch **ON**. The ECONOMY light should illuminate for three seconds, then extinguish. If the ECONOMY light stays ON in the NORMAL mode, the computer has stored DTC's.

To retrieve the stored DTC's, perform the following:

1. Perform the preliminary inspection, located earlier in this section. This is very important, since a loose or disconnected wire, or corroded connector terminals can cause a whole slew of unrelated DTC's to be stored by the computer; you will waste a lot of time performing a diagnostic "goose chase."

2. Grab some paper and a pencil or pen to write down the DTC's when they are flashed out.

3. Locate the five terminal, green transaxle diagnostic connector, which is usually located behind the passenger's kickpanel. If the transaxle diagnostic connector cannot be found, either call your local Isuzu dealer or refer to the applicable Chilton's Total Car Care manual for your vehicle.

4. Turn the ignition switch **OFF**.

5. Using a jumper wire or a bent paper clip, connect terminals 1 and 2 of the transaxle diagnostic connector. A black wire with a white tracer stripe is usually attached to terminal 1, and a black wire with a blue tracer stripe is usually attached to terminal 2 of the transaxle diagnostic connector.

6. Set the transaxle mode selector in the NORMAL position, then turn the ignition switch to the **ON** position **without starting the engine**.

7. Observe the ECONOMY light and note any stored DTC's flashed out by it.

8. After all of the codes have been noted, turn the ignition switch **OFF**, and remove the jumper wire/paper clip from the transaxle diagnostic connector.

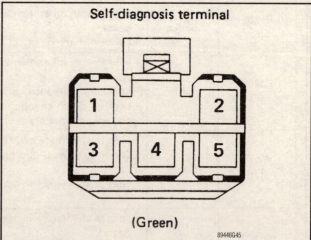

Self-diagnosis terminal

(Green)

89446G45

Fig. 50 To activate the diagnostic system's flash-out sequence, use a jumper wire to connect terminals 1 and 2 of the transaxle diagnostic connector

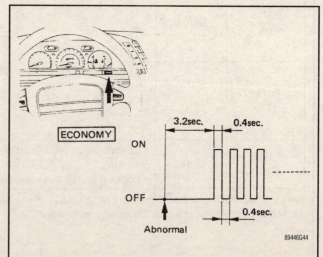

89446G44

Fig. 51 The DTC's are displayed by the computer system flashing the ECONOMY light—Impulse models

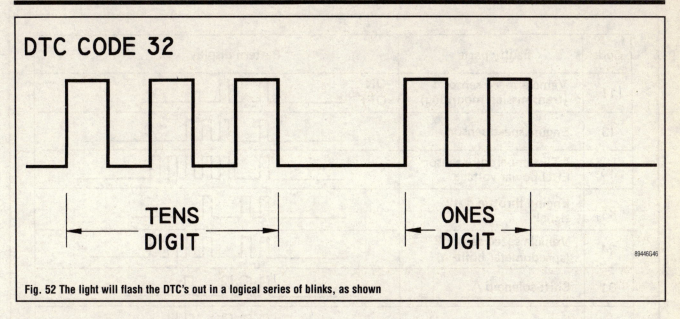

Fig. 52 The light will flash the DTC's out in a logical series of blinks, as shown

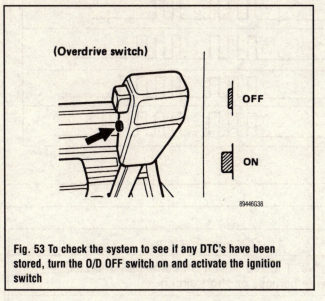

Fig. 53 To check the system to see if any DTC's have been stored, turn the O/D OFF switch on and activate the ignition switch

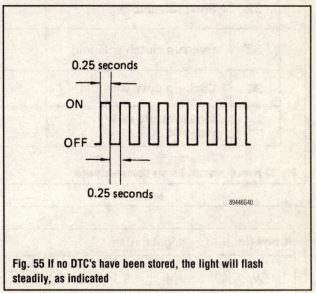

Fig. 55 If no DTC's have been stored, the light will flash steadily, as indicated

Fig. 54 The system will display codes by flashing the O/D OFF light, located in the center console—1988–90 Trooper

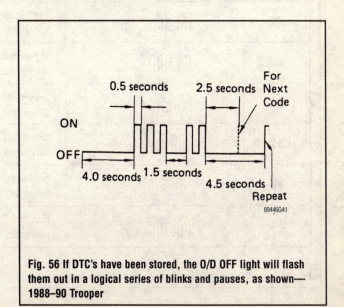

Fig. 56 If DTC's have been stored, the O/D OFF light will flash them out in a logical series of blinks and pauses, as shown—1988–90 Trooper

Code	Faulty parts	Pattern display
11	Vehicle speed sensor 1 (transmission mounting)	ON / OFF 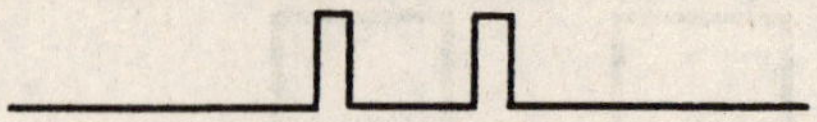
13	Engine speed sensor	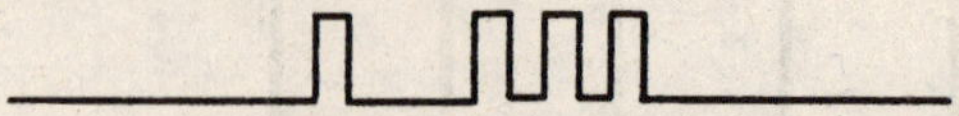
15	ATF temperature sensor ECU power voltage	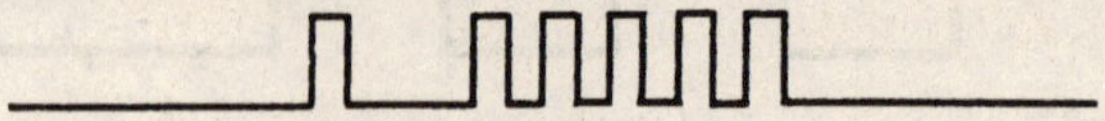
21	Engine throttle duty signal	
24	Vehicle speed sensor 2 (speedometer built-in)	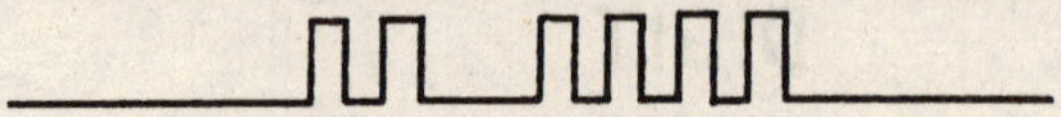
31	Shift solenoid A	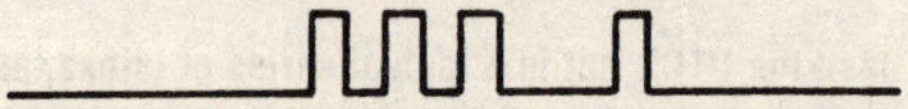
32	Shift solenoid B	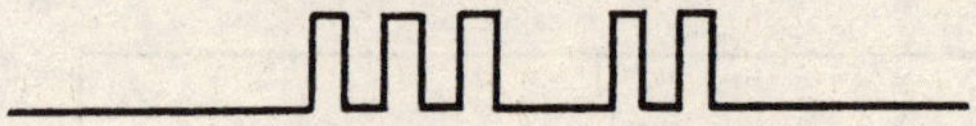
33	Overrun clutch solenoid	
34	Lock-up duty solenoid	
35	Line pressure duty solenoid	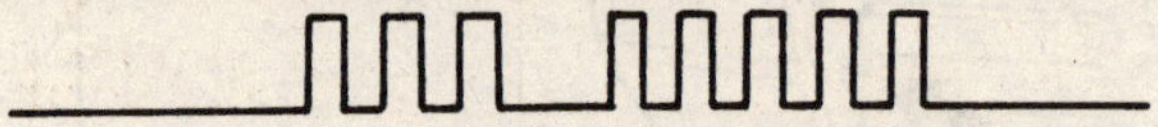

89446G43

Fig. 57 Isuzu Diagnostic Trouble Codes—Impulse

Code No.	Light Pattern	Diagnosis System
21	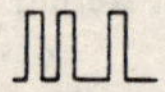	Defective No. 1 speed sensor (in combination meter) — severed wire harness or short circuit
22	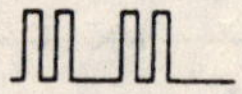	Defective No. 2 speed sensor (in Automatic transmission) — severed wire harness or short circuit
23	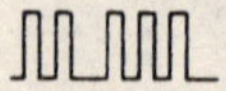	Severed throttle sensor or short circuit — Severed wire harness or short circuit
31	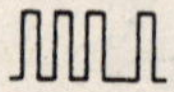	Severed No. 1 solenoid or short circuit — severed wire harness or short circuit
32	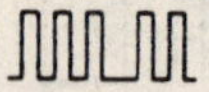	Severed No. 2 solenoid or short circuit — severed wire harness or short circuit
33	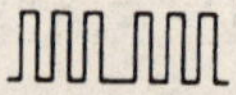	Severed No. 3 solenoid or short circuit — severed wire harness or short circuit
34	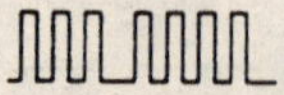	Severed No. 4 solenoid or short circuit — severed wire harness or short circuit

89446G42

Fig. 58 Isuzu Diagnostic Trouble Codes—1988–90 Trooper

1988–90 TROOPER MODELS

The 1988–90 Isuzu Trooper models utilize two-digit DTC's to inform the technician in which circuit and/or component a fault was detected. The two-digit codes are displayed by the O/D OFF light, located in the center console shifter display. The light flashes the digits out by first flashing the tens digit, then after a slight pause, the ones digit. Therefore, if the O/D OFF light flashes twice, pauses slightly, then flashes once, a Code 21 was displayed. If there is more than one DTC stored in the computer's memory, the light will display the lowest number first and move upward through all of the stored codes in numerical order. After the progression reaches the highest code number, the cycle will repeat itself until the ignition switch is turned OFF.

➡There will be a longer pause between the individual codes than there is between the tens and ones digits of each code.

To determine if the vehicle's computer has stored any DTC's, switch the transmission/transaxle mode switch so that transmission/transaxle overdrive is enabled (the O/D OFF switch in the out position) and turn the ignition switch ON. The O/D OFF light should illuminate for three seconds, then extinguish. If the light flashes ON and OFF, the computer has stored DTC's.

➡The O/D OFF light should stay illuminated when the overdrive is turned off.

To retrieve the stored DTC's, perform the following:
1. Perform the preliminary inspection, located earlier in this section. This is very important, since a loose or disconnected wire, or corroded connector terminals can cause a whole slew of unrelated DTC's to be stored by the computer; you will waste a lot of time performing a diagnostic "goose chase."
2. Grab some paper and a pencil or pen to write down the DTC's when they are flashed out.
3. Locate the two terminal diagnostic request connector, which is usually located up and behind the left-hand side of the instrument panel (near the hood latch release handle). If the diagnostic request connector cannot be found, either call your local Isuzu dealer or refer to the applicable Chilton's Total Car Care manual for your vehicle.
4. Turn the ignition switch OFF.
5. Using a jumper wire or a bent paper clip, connect the two terminals of the diagnostic request connector.
6. Set the transaxle mode selector to the overdrive enabled position (O/D OFF button in the out position), then turn the ignition switch to the ON position without starting the engine.
7. Observe the O/D OFF light and note any stored DTC's flashed out by it.
8. After all of the codes have been noted, turn the ignition switch OFF, and remove the jumper wire/paper clip from the diagnostic request connector.

1991 AND NEWER PRE-OBD II TROOPER MODELS

The 1991 and newer pre-OBD II Isuzu Trooper models, as with the earlier models, utilize two-digit DTC's to inform the technician in which circuit and/or component a fault was detected. The two-digit codes are displayed by the CHECK TRANS light, located in the instrument cluster display. The light flashes the digits out in the same manner as with the 1988–90 mod-

---

DIAGNOSTIC TROUBLE CODES—ISUZU 1991 AND NEWER TROOPER

Code	Component/Circuit Fault
17	1-2/3-4 shift solenoid or circuit
21	Throttle Position (TP) sensor or circuit
22	Throttle Position (TP) sensor or circuit
23	Engine Coolant Temperature (ECT) sensor or circuit
25	1-2/3-4 shift solenoid or circuit
26	2-3 shift solenoid or circuit
28	2-3 shift solenoid or circuit
29	TCC solenoid or circuit
31	No engine speed signal from engine speed sensor
32	Line pressure solenoid or circuit
33	Line pressure solenoid or circuit
34	Band apply solenoid or circuit
35	Band apply solenoid or circuit
36	TCC solenoid or circuit
39	Vehicle Speed Sensor (VSS) or circuit
41	Gear error
43	Computer ground control circuit turned off, because of fault in a solenoid circuit
46	Downshift error
48	Low voltage or bad computer ground
55	Computer failure
56	Shift lever position switch or TP sensor circuit
65	ATF temperature sensor or circuit
66	ATF temperature sensor or circuit
77	Kickdown switch or circuit, or TP sensor or circuit
82	Shift lever position switch

89446G47

Fig. 59 Isuzu Diagnostic Trouble Codes—1991 and newer pre-OBD II Trooper models

els. Therefore, if the CHECK TRANS light flashes twice, pauses slightly, then flashes once, a Code 21 was displayed. If there is more than one DTC stored in the computer's memory, the light will display the lowest number first and move upward through all of the stored codes in numerical order. After the progression reaches the highest code number, the cycle will repeat itself until the ignition switch is turned **OFF**.

➡**There will be a longer pause between the individual codes than there is between the tens and ones digits of each code.**

To determine if the vehicle's computer has stored any DTC's, turn the ignition switch **ON** and observe the CHECK TRANS light. The light should illuminate for three seconds, then extinguish. If the light flashes ON and OFF continuously, the computer has stored DTC's.

To retrieve the stored DTC's for 1991 vehicles, perform the following:

1. Perform the preliminary inspection, located earlier in this section. This is very important, since a loose or disconnected wire, or corroded connector terminals can cause a whole slew of unrelated DTC's to be stored by the computer; you will waste a lot of time performing a diagnostic "goose chase."

2. Grab some paper and a pencil or pen to write down the DTC's when they are flashed out.

3. Locate the **two terminal** diagnostic request connector (not the three terminal connector) attached to the transmission/transaxle computer connector harness, which is usually located up and behind the left-hand side of the instrument panel (near the hood latch release handle). If the diagnostic request connector cannot be found, either call your local Isuzu dealer or refer to the applicable Chilton's Total Car Care manual for your vehicle.

4. Turn the ignition switch **OFF**.

5. Using a jumper wire or a bent paper clip, connect the two terminals of the diagnostic request connector.

6. Turn the ignition switch to the **ON** position **without starting the engine**.

7. Observe the CHECK TRANS light and note any stored DTC's flashed out by it.

8. After all of the codes have been noted, turn the ignition switch **OFF**, and remove the jumper wire/paper clip from the diagnostic request connector.

To retrieve the stored DTC's for 1992 and newer models, perform the following:

9. Perform the preliminary inspection, located earlier in this section. This is very important, since a loose or disconnected wire, or corroded connector terminals can cause a whole slew of unrelated DTC's to be stored by the computer; you will waste a lot of time performing a diagnostic "goose chase."

10. Grab some paper and a pencil or pen to write down the DTC's when they are flashed out.

11. Locate the **three terminal** diagnostic request connector, which may be located either in the same spot as the 1991 models (up and behind the left-hand side of the instrument panel, near the hood latch release handle), or behind the left-hand side of the center console. If the diagnostic request connector cannot be found, either call your local Isuzu dealer or refer to the applicable Chilton's Total Car Care manual for your vehicle.

12. Turn the ignition switch **OFF**.

13. Using a jumper wire or a bent paper clip, connect the two OUTER terminals of the diagnostic request connector.

14. Turn the ignition switch to the **ON** position **without starting the engine**.

15. Observe the CHECK TRANS light and note any stored DTC's flashed out by it.

16. After all of the codes have been noted, turn the ignition switch **OFF**, and remove the jumper wire/paper clip from the diagnostic request connector.

Clearing Codes

IMPULSE MODELS

➡**It is not recommended that the negative battery cable be disconnected to clear DTC's. Doing so will clear all settings from the vehicle's computer system, resulting in lost radio presets, seat memories, anti-theft codes, driveability parameters, etc., and there are better ways of spending your afternoon than resetting all of these.**

To clear the stored DTC's from the vehicle's computer memory, remove the 4AT fuse from the fuse block for a minimum of one minute. Reinstall the 4AT fuse and all of the stored DTC's should be deleted. If there are still codes present, either the codes were not properly cleared (Are the codes identical to those flashed out previously?) or the underlying problem is still there (Are only some of the codes the same as previously?)

TROOPER MODELS

➡**It is not recommended that the negative battery cable be disconnected to clear DTC's. Doing so will clear all settings from the vehicle's computer system, resulting in lost radio presets, seat memories, anti-theft codes, driveability parameters, etc., and there are better ways of spending your afternoon than resetting all of these.**

To clear the stored DTC's from the vehicle's computer memory, remove the CLOCK fuse (1988–90 Trooper II models), the STOP/AT CONT fuse (1988–90 Trooper models) or the ROOM LAMP fuse (1991 and newer pre-OBD II Trooper models) from the fuse block for a minimum of one minute. Reinstall the applicable fuse and all of the stored DTC's should be deleted. If there are still codes present, either the codes were not properly cleared (Are the codes identical to those flashed out previously?) or the underlying problem is still there (Are only some of the codes the same as previously?)

MAZDA

◗ **See Figures 60, 61, 62, 63 and 64**

Reading and Retrieving Codes

Generally, the transaxle related DTC's are easy to retrieve and read on all pre-OBD II Mazda vehicles.

➡**The 1994 and newer pre-OBD II 626 and MX6 models with the 2.0L engine use a Ford transaxle. Please refer to the Ford procedures in this section for these models.**

All Mazda models utilize two-digit DTC's to inform the technician in which circuit and/or component a fault was detected. The two-digit codes are displayed by either the HOLD light or the CHECK ENGINE light, both of which are located in the instrument cluster display. The light flashes the digits out by first flashing the tens digits as long flashes, then after a slight pause, the ones digits as shorter flashes. Therefore, if the HOLD or CHECK ENGINE light flashes once (long flash), pauses slightly, then flashes twice (short flashes), a Code 12 was displayed. If there is more than one DTC stored in the computer's memory, the light will display the lowest number first and move upward through all of the stored codes in numerical order. After the progression reaches the highest code number, the cycle will repeat itself until the ignition switch is turned **OFF**.

➡**There will be a longer pause between the individual codes than there is between the tens and ones digits of each code.**

To determine if the vehicle's computer has stored any DTC's, turn the ignition switch **ON** and observe the HOLD or CHECK ENGINE light. The light should illuminate for three seconds, then extinguish. If the light flashes ON and OFF continuously, the computer has stored DTC's.

To retrieve the stored DTC's, perform the following:

1. Perform the preliminary inspection, located earlier in this section. This is very important, since a loose or disconnected wire, or corroded

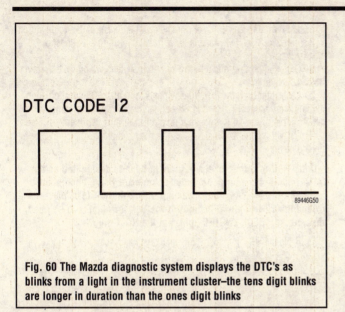

DTC CODE 12

Fig. 60 The Mazda diagnostic system displays the DTC's as blinks from a light in the instrument cluster–the tens digit blinks are longer in duration than the ones digit blinks

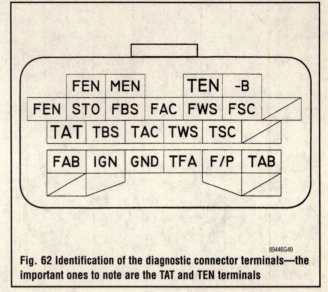

Fig. 62 Identification of the diagnostic connector terminals—the important ones to note are the TAT and TEN terminals

Code No.	Buzzer pattern	Diagnosed circuit	Condition	Point
06		Speedome-ter sensor	No input signal from speedometer sensor while driving at drum speed above 600 rpm in D, S, or L ranges	• Speedometer sensor connector • Wiring from speedometer sensor to instrument cluster • Wiring from instrument cluster to EC-AT CU • Speedometer sensor resistance
12		Throttle sensor	Open or short circuit	• Throttle sensor connector • Wiring from throttle sensor to EC-AT CU • Throttle sensor resistance.
55		Pulse generator	No input signal from pulse generator while driving at 40 km/h (25 mph) or higher in D, S, and L ranges	• Pulse generator connector • Wiring from pulse generator to EC-AT CU • Pulse generator connector
57		Reduce torque sig-nal 1	Open or short circiut of reduce torque signal 1 wire harness	• Wiring from ECU to EC-AT CU
58		Reduce torque sig-nal 2	Open or short circuit of reduce torque signal 2 and/or torque reduced sig-nal/water thermo signal wire harness	• Wiring from ECU to EC-AT CU

Fig. 61 A few examples of some of the possible DTC's displayed by Mazda vehicles

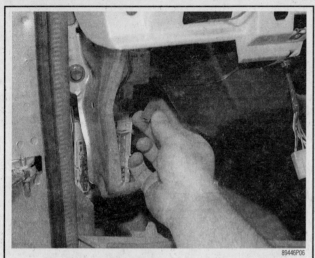

Fig. 63 The DRC for MPV models is a single wire connector, located up, under the left side of the instrument panel

connector terminals can cause a whole slew of unrelated DTC's to be stored by the computer; you will waste a lot of time performing a diagnostic "goose chase."

2. Grab some paper and a pencil or pen to write down the DTC's when they are flashed out.

3. Locate the Diagnostic Request Connector (DRC). The location and configuration of the DRC depends largely on the model of the particular vehicle at hand. The DRC will either be a green or blue two-terminal connector, or will be one terminal as a part of the large diagnostic connector in the engine compartment.

➡**If the DRC terminal for the vehicle at hand is designated as the TAT terminal of the large diagnostic connector in the engine compartment, but the connector does not utilize the TAT terminal, ground the TEN terminal instead. If the TEN terminal is to be used, observe the CHECK ENGINE light, not the HOLD light.**

To locate the DRC in the specific vehicle at hand, refer to the following general list:
- 1990–91 323 and MX3—TAT terminal, located in the diagnostic connector near the battery, actuates the HOLD light
- 1992–95 323 and MX3—TAT terminal, located in the diagnostic connector near the battery, actuates the CHECK ENGINE light

DIAGNOSTIC TROUBLE CODES—MAZDA

Code	Component/Circuit Fault
1	No engine speed signal from ECM
6	Vehicle Speed Sensor (VSS) or circuit
7	Output speed sensor or circuit
12	Throttle Position (TP) sensor or circuit
55	Input speed sensor or circuit
56	ATF temperature sensor or circuit
57 ①	Lost communication between engine computer terminal 1S and transmission and computer terminal 1J
58 ①	Lost communication between engine computer terminal 1B and transmission and computer terminal 1L
59 ①	Lost communication between engine and transmission computer terminals 1K
60	Shift solenoid A (1-2) or circuit
61	Shift solenoid B (2-3) or circuit
62	Shift solenoid C (3-4) or circuit
63	Lockup solenoid or circuit
64 ②	3-2 control solenoid or circuit
64 ③	Line pressure control solenoid or circuit
65	Lock-up engagement control solenoid or circuit
66 ②	Line pressure control solenoid or circuit
46	Downshift error
48	Low voltage or bad computer ground
55	Computer failure
56	Shift lever position switch or TP sensor circuit
65	ATF temperature sensor or circuit
66	ATF temperature sensor or circuit
77	Kickdown switch or circuit, or TP sensor or circuit
82	Shift lever position switch

① GF4AEL only
② Except R4AEL
③ R4AEL

89446G51

Fig. 64 Mazda Diagnostic Trouble Codes

• 1988–89 626 and MX6—blue, single connector, located under the far left-hand side of the instrument panel, near the transaxle computer, actuates the HOLD light

• 1990–92 323 and MX3 non-Turbo—green, single connector, located near the left, front shock tower, actuates the CHECK ENGINE light

• 1990–92 626 and MX6 Turbo—blue, single connector, located under the far left-hand side of the instrument panel, near the transaxle computer, actuates the HOLD light

• 1993 626 and MX6—TAT terminal, located in the diagnostic connector near the left, front shock tower, actuates the HOLD light

• 1994–95 626 and MX6 models with 2.5L engine—TAT terminal, located in the diagnostic connector near the left, front shock tower, actuates the HOLD light

• 1988–91 929—blue, single connector, located near the right, front shock tower, actuates the HOLD light

• 1992–94 929—TAT terminal, located in the diagnostic connector near the left, front shock tower, actuates the HOLD light

• 1989–91 RX7—blue, single connector, located near the left, front shock tower, actuates the HOLD light

• 1992–93 RX7—TAT terminal, located in the diagnostic connector near the left, front shock tower, actuates the HOLD light

• All MPV—blue, single connector, located under the far left-hand side of the instrument panel, near the transaxle computer, actuates the HOLD light

If the DRC cannot be found, either call your local Mazda dealer or refer to the applicable Chilton's Total Car Care manual for your vehicle.

4. Ensure the ignition switch is in the **OFF** position.

5. Using a jumper wire, attach the applicable DRC terminal to a good engine or chassis ground. If the DRC is located in the diagnostic connector, you can even attach it to the GND terminal of the connector.

6. Without starting the engine, turn the ignition to the **ON** position.

7. Observe the applicable light and note the DTC's.

8. After all of the codes have been noted, turn the ignition switch **OFF** and remove the jumper wire.

Clearing Codes

To clear the DTC's from the vehicle's computer memory, disconnect the negative battery cable for at least one minute, then reattach it. Reattach the negative battery cable and all of the stored DTC's should be deleted. If there are still codes present, either the codes were not properly cleared (Are the codes identical to those flashed out previously?) or the underlying problem is still there (Are only some of the codes the same as previously?)

MITSUBISHI & HYUNDAI

◆ **See Figures 65 thru 85**

Reading and Retrieving Codes

GENERAL INFORMATION

Most Mitsubishi computerized transaxle control systems have no way of indicating a problem in the system. If the computer detects a problem, it begins pulsing a code at the diagnostic link connector. The only way you'll know there's a code is to look for it; nothing lights up to tell you there's a problem.

The only exception to that is 1996 and newer Hyundai models, which do have a Malfunction Indicator Lamp (MIL) in the instrument cluster display. The MIL illuminates when the diagnostic system detects a problem in the transaxle or control system.

Mitsubishi/Hyundai vehicles use Diagnostic Link Connectors (DLC) in a possible three different configurations, four different sets of diagnostic codes (types 1 through 4), and three methods for displaying DTC's. Mitsubishi transaxles appear in Mitsubishi (of course), Hyundai and some Chrysler vehicles.

The Diagnostic Link Connector (DLC) must be located to retrieve any DTC's. Luckily, unlike some of the other import models covered in this manual, there are only four various locations for the DLC. The four locations are as follows: 1985–88 Galant, inside the glove box above the com-

partment itself; 1989 and newer pre-OBD II Mitsubishi and Chrysler vehicles, either in the fuse box, or under the far left-hand side of the instrument panel; 1989–95 Hyundai models, next to the fuse box inside the left side kick panel; all OBD II vehicles, under the left-hand side of the instrument panel.

→**Remember that only OBD II models can let the driver know that codes have been stored; all other models do not alert the driver that any faults have been detected. Therefore, the only way to ascertain whether codes have been stored is to actually perform the retrieval procedure.**

Over the years, the transaxle computer control system used with Mitsubishi vehicles has been designed to display the transaxle DTC's in three different formats. All of these formats consist of changes to the on- and off-time of the signal. Also, four different sets of DTC's have been utilized to inform the person servicing the transaxle of any detected problems. The Mitsubishi models and their corresponding code types are presented in the following list:

→**The precise date when models switched from one code type to another wasn't always consistent, and using this list as a Bible for the breaking points between code types could have you chasing down the wrong component for the wrong code.**

Type 1 codes

• 1985–88 Galant models
• 1989 Sonata 4-cylinder models

Type 2 codes

• 1989–90 ½ Galant models
• 1989–93 Precis models
• 1989–90 except Galant and Precis models
• 1989–90 Chrysler models equipped with Mitsubishi transaxles
• 1989–93 Hyundai models, except for 1989 Sonata 4-cylinder models

Type 3 codes

• 1990½ and newer Galant models
• 1994 and newer Precis models
• 1991 and newer Galant and Precis models
• 1991 and newer Chrysler models equipped with Mitsubishi transaxles
• 1994–95 Hyundai models
• 1994 and newer Hyundai Sonata models

Type 4 codes

• 1996 and newer Hyundai models, except for Sonota

Since the precise date when models switched from one code type to another listed in the preceding list is not set in stone, record any codes in memory, clear the codes, then create a fault in the transaxle computer system. For example, if the shift solenoid code was not stored by the transaxle computer, disconnect the shift solenoid transmission/transaxle connector and restart the engine, after clearing the previous codes. The diagnostic system should store the appropriate code for a shift solenoid/circuit fault, which will allow you to ascertain precisely what set of codes the system being worked on uses.

TYPE 1 CODES

The first format only appeared in 1985–88 models, with the addition of the 4-cyl;inder 1989 Sonata. The Type 1 system was designed with what is known as the "Type 1" computer system. The Type 1 DTC's are made up of a series of signal pulses that varied in on-time and off-time. There is no regular pattern to these signals, except for a 4 second start and end signal, to indicate when each DTC starts and finishes. To identify the code, you must compare the signal to the accompanying code chart.

The Type 1 Mitsubishi transaxle diagnostic system functions so that once the system detects a malfunction, it pulses the DTC continuously through the DLC. There's no wire to ground, no special procedure to perform; just connect a DVOM to the respective DLC terminals and read the DTC's.

Because of the lack of a regular pattern, Type 1 codes can be difficult to read and interpret. One way to make these early codes easier to read is to use an oscilloscope or Digital Multi-Meter (DMM) with a graphic display instead of a voltmeter. By capturing the signal as a single waveform, you can compare the waveform to the accompanying code chart.

The Type 1 code system utilized volatile code memory, which means that whenever the ignition key is turned **OFF**, the codes are cleared. Also, because of the simplistic design of the system, any code will force the vehicle to run in a failsafe mode.

➡**The Type 1 code system can only display one DTC at a time. Therefore, you must repair the problem for the first DTC before the second DTC can be read. So do not be surprised if you must fix several problems before there are no DTC's left.**

To retrieve the codes, perform the following:

1. Perform the preliminary inspection, located earlier in this section. This is very important, since a loose or disconnected wire, or corroded connector terminals can cause a whole slew of unrelated DTC's to be stored by the computer; you will waste a lot of time performing a diagnostic "goose chase."

2. Grab some paper and a pencil or pen to write down the DTC's when they are flashed out.

3. Locate the Diagnostic Link Connector (DLC), which, depending on the vehicle at hand, may be in one of two locations: inside the glove box above the compartment itself for 1985–88 Galant models or next to the fuse box inside the left side kick panel for 1989 Sonata 4-cylinder models. If the DLC cannot be found, either call your local Mitsubishi/Hyundai dealer or refer to the applicable Chilton's Total Car Care manual for your vehicle.

4. Start the engine and drive the vehicle until the transaxle goes into the failsafe mode.

5. Park the vehicle, but do not turn the ignition **OFF**. Allow it to idle.

➡**It is important to allow the vehicle to idle while reading the DTC's, otherwise a Code 12—lack of an ignition signal—will be detected by the computer, leading to misdirected troubleshooting and wasted time.**

6. Attach a Digital Volt-Ohmmeter (DVOM), a Digital Multi-Meter (DMM)—expensive versions are often equipped with a scope display function capable of showing waveforms—or an oscilloscope to the test terminals on the Diagnostic Link Connector (DLC). For 1985–86 Galant models, attach the instrument's (DVOM/DMM/oscilloscope) negative lead to terminal 9 of the DLC and the positive lead to terminal 6. For 1987–88 Galant and 1989 Sonata 4-cylinder models, attach the negative lead to terminal 6 of the DLC and the positive lead to terminal 12.

7. Observe the DVOM/DMM/oscilloscope and note the wave pattern. Compare the pattern to the accompanying chart to determine which code it corresponds to.

➡**Be sure of which system the vehicle at hand uses, since Type 1 Code 12 looks the same as Types 2 and 3 Code 15.**

8. After all of the DTC(s) have been retrieved, fix the applicable problems, clear the codes, drive the vehicle, and perform the retrieval procedure again to ensure that all of the codes are gone.

TYPE 2 AND 3 CODES

➡**Reading and retrieving Type 2 and Type 3 codes is identical; the only difference between Type 2 and 3 codes is what each of the codes stands for.**

Unlike the Type 1 diagnostic system, the system used for Type 2 and 3 codes was designed with the capability of storing up to 10 DTC's. Therefore, you will not have to fix each problem one at a time. However, unlike other import manufacturers, the codes are not displayed from lowest number to highest number. Rather, they are flashed out in the order in which they detected. This means that if the same fault was detected numerous times, the same code will be flashed out the same number of times as detected. Also, since only 10 codes can be stored at one time, if the computer detects an eleventh code, the first code stored will be deleted to make room for the latest code. This makes it important to check for codes whenever you suspect a problem with the transaxle.

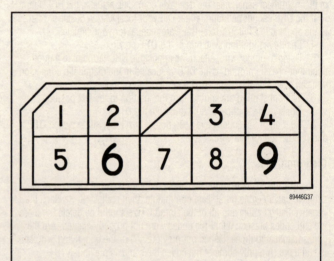

Fig. 66 To read the DTC's on 1985–86 Galant models, attach the voltmeter leads to terminals 6 and 9 of the DLC

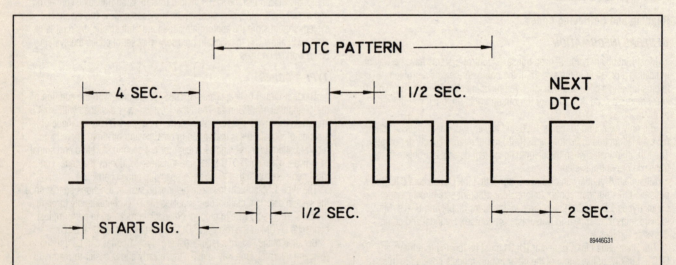

Fig. 65 The Type 1 codes are comprised of one 4 second start pulse followed by five individual pulses, lasting either 1½ seconds (ON) or ½ second (OFF) in duration

➡Remember that only Type 4 (OBD II) models can let the driver know that codes have been stored; all other models do not alert the driver that any faults have been detected. Therefore, the only way to ascertain whether codes have been stored is to actually perform the retrieval procedure.

With the introduction of the Type 2 code system, Mitsubishi also developed Failsafe Codes. Failsafe codes are DTC's which, if detected, will switch the vehicle's management system into a failsafe mode (thus the name). The failsafe mode sets the control computer to run (albeit not as efficiently) on predetermined parameters, so that erroneous information from the vehicle's sensors will be ignored until the problem at hand can be repaired. Other manufacturers call the failsafe mode the Limp In mode.

➡Remember that since many codes are redundant, the system may store a non-failsafe code and a failsafe code for the same malfunction. In which case, the system will go into failsafe mode.

Mitsubishi also implemented an easier way for reading the DTC's in Type 2 and 3 codes. The DTC's are 2-digit codes represented by a logical series of flashes (a DVOM, DMM or voltmeter can be used). The system displays the codes on the voltmeter by first flashing out the tens digit, which is a series of long flashes, followed by a short pause, then flashing out the ones digits, which is a series of short flashes. Therefore, two long flashes followed by three short flashes would indicate a Code 23. Between each complete DTC is a longer separator code.

To retrieve the codes, perform the following:

1. Perform the preliminary inspection, located earlier in this section. This is very important, since a loose or disconnected wire, or corroded connector terminals can cause a whole slew of unrelated DTC's to be stored by the computer; you will waste a lot of time performing a diagnostic "goose chase."

2. Grab some paper and a pencil or pen to write down the DTC's when they are flashed out.

3. Locate the Diagnostic Link Connector (DLC), which, depending on the vehicle at hand, may be in one of three locations: either in the fuse box, or under the far left-hand side of the instrument panel for 1989 and newer pre-OBD II Mitsubishi an Chrysler vehicles or next to the fuse box inside the left side kick panel for 1989–95 Hyundai models (except 1989 Sonata 4-cylinder models). If the DLC cannot be found, either call your local Mitsubishi/Hyundai dealer or refer to the applicable Chilton's Total Car Care manual for your vehicle.

4. Start the engine and drive the vehicle until the transaxle goes into the failsafe mode.

5. Park the vehicle, but do not turn the ignition **OFF**. Allow it to idle.

6. Attach a voltmeter (analog or digital) to the test terminals on the Diagnostic Link Connector (DLC). The negative lead should be attached to terminal 6 and the positive lead to terminal 12.

7. Observe the voltmeter and count the flashes (or arm sweeps if using an analog voltmeter); note the applicable codes.

8. After all of the DTC(s) have been retrieved, fix the applicable problems, clear the codes, drive the vehicle, and perform the retrieval procedure again to ensure that all of the codes are gone.

Fig. 67 When retrieving Type 2 and 3 codes, connect the DVOM leads to terminals 6 and 12 of the DLC, as indicated

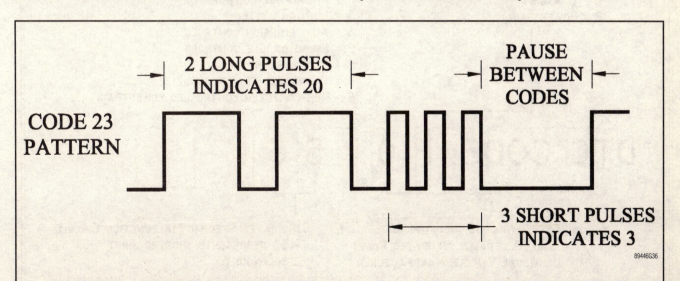

Fig. 68 Representation of Type 2 and 3 codes, as flashed out by a voltmeter

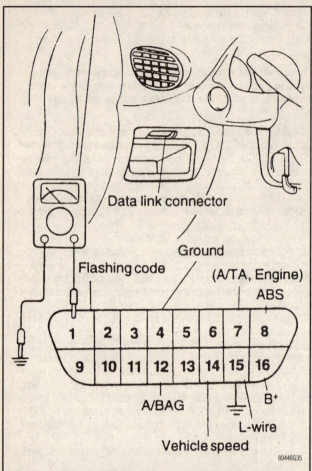

Fig. 69 For Type 4 (OBD II) code retrieval on Hyundai models, connect the DVOM and jumper wire as shown—for OBD II Mitsubishi and Chrysler models, attach the DVOM to terminals 4 and 6

TYPE 4 CODES (OBD-II CODES)

In 1996, all Hyundai (except the Sonata) switched from an arbitrary code listing and format, to the federally regulated On Board Diagnostics 2nd Generation (OBD II) code system. Normally, OBD II equipped vehicles do not have the option of allowing the person servicing the vehicle to flash the codes out with a voltmeter; usually a scan tool is necessary to retrieve OBD II codes. Hyundai, however, does provide this option.

The Federal government decided that it was time to create a standard for vehicle diagnostic systems codes for ease of servicing and to insure that certain of the vehicle's systems were being monitored for emissions purposes. Since OBD II codes are standardized (they all contain one letter and four numbers), they are easy to decipher. For a breakdown of a typical transaxle-related OBD II code, refer to the accompanying illustration.

The OBD II system in the Hyundai models is designed so that it will flash the DTC's out on a voltmeter (even though a scan tool is better). However, the first two characters of the code are not used. This is because the transaxle is a part of the powertrain, so all transaxle related codes will begin with a P. Also, since there are no overlapping numbers between SAE and Hyundai codes, the second digit is also not necessary.

The system flashes the codes out in a series of flashes in three groups, each group corresponding to one of the three last digits of the OBD II code. Therefore, Code P0753 would be flashed out in seven flashes, followed by five flashes, then by three flashes. Each group of flashes is separated by a brief pause. All of the flashes are of the same duration, with the only exception being zero. Zero is represented by a long flash. Therefore, seven flashes, one long flash, two flashes would indicate a P0702 code (shorted TP sensor circuit).

To retrieve the codes, perform the following:

1. Perform the preliminary inspection, located earlier in this section. This is very important, since a loose or disconnected wire, or corroded connector terminals can cause a whole slew of unrelated DTC's to be stored by the computer; you will waste a lot of time performing a diagnostic "goose chase."

2. Grab some paper and a pencil or pen to write down the DTC's when they are flashed out.

3. Locate the Diagnostic Link Connector (DLC), which is located under the left-hand side of the instrument panel. If the DLC cannot be found, either call your local Mitsubishi/Hyundai dealer or refer to the applicable Chilton's Total Car Care manual for your vehicle.

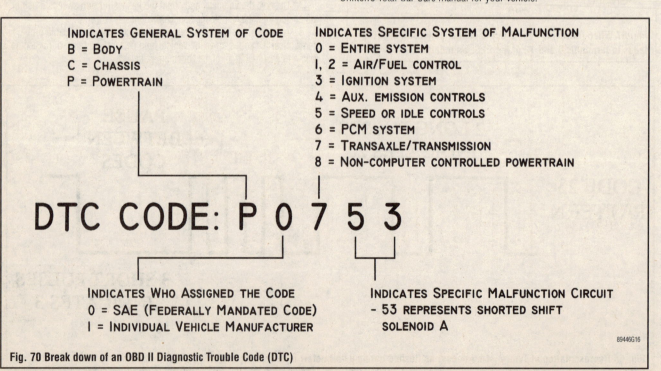

INDICATES GENERAL SYSTEM OF CODE
B = BODY
C = CHASSIS
P = POWERTRAIN

INDICATES SPECIFIC SYSTEM OF MALFUNCTION
0 = ENTIRE SYSTEM
1, 2 = AIR/FUEL CONTROL
3 = IGNITION SYSTEM
4 = AUX. EMISSION CONTROLS
5 = SPEED OR IDLE CONTROLS
6 = PCM SYSTEM
7 = TRANSAXLE/TRANSMISSION
8 = NON-COMPUTER CONTROLLED POWERTRAIN

DTC CODE: P 0 7 5 3

INDICATES WHO ASSIGNED THE CODE
0 = SAE (FEDERALLY MANDATED CODE)
1 = INDIVIDUAL VEHICLE MANUFACTURER

INDICATES SPECIFIC MALFUNCTION CIRCUIT
- 53 REPRESENTS SHORTED SHIFT SOLENOID A

Fig. 70 Break down of an OBD II Diagnostic Trouble Code (DTC)

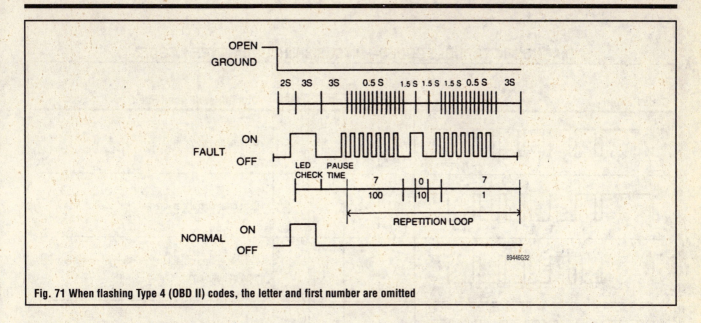

Fig. 71 When flashing Type 4 (OBD II) codes, the letter and first number are omitted

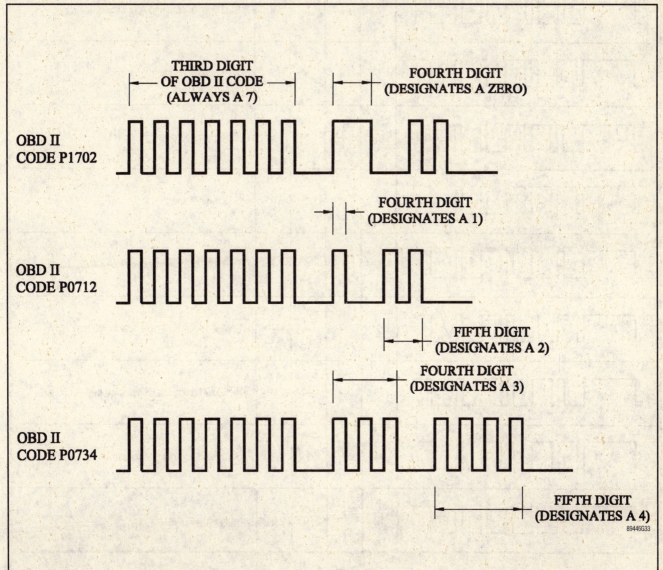

Fig. 72 Flash-out representation of three common OBD II codes—remember that first two digits of the OBD II code are not flashed out

DIAGNOSTIC TROUBLE CODES—MITSUBISHI/HYUNDAI TYPE 1 CODES

Code Pattern	Codes				Component/Circuit Fault
	1985	1986	1987	1988	
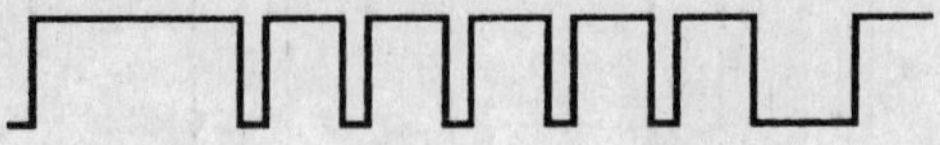	1	1	1	1	Control computer fault
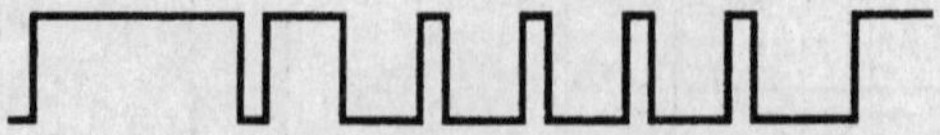	2	2	2	2	First gear signal detected at high speed
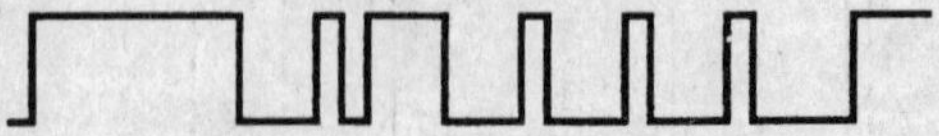	3	3	3	3	Vehicle speed detected by pulse generator B lower than actual vehicle speed
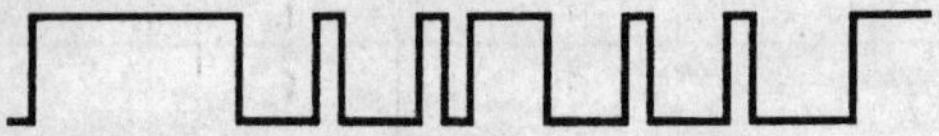	4	4	4	4	Shift control solenoid A or circuit
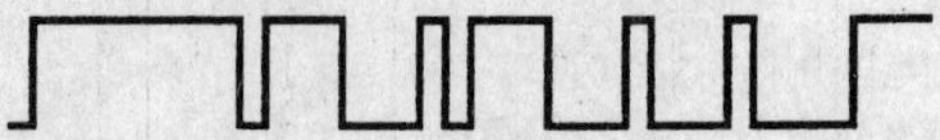	5	5	5	5	Shift control solenoid B or circuit
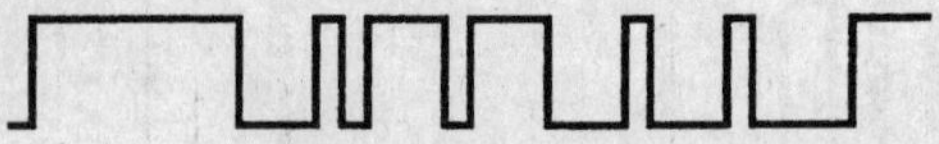	6	—	6	6	Kickdown servo switch or circuit
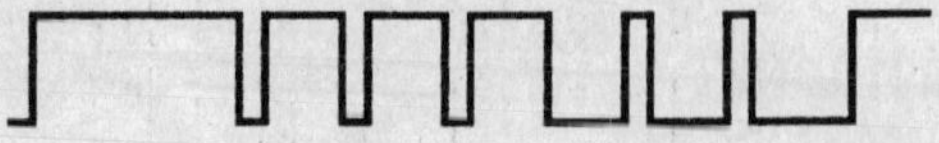	7	6	7	7	Slipping detected
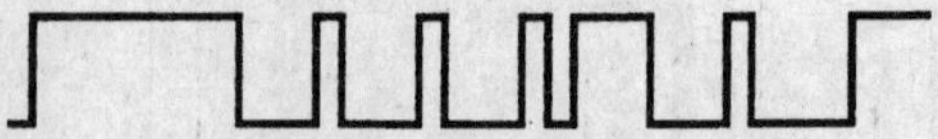	8	7	8	8	Pressure control solenoid or circuit
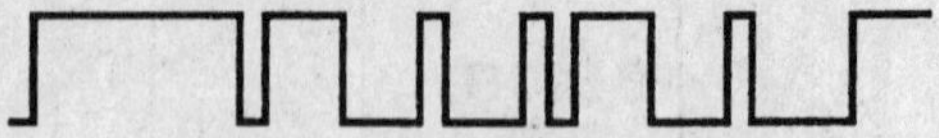	9	—	—	—	Engine speed detected over 6,500 rpm
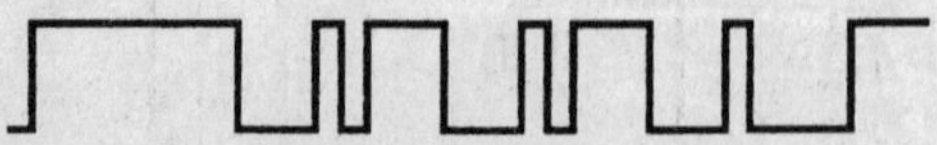	10	—	—	—	Kickdown drum speed detected over 6,500 rpm
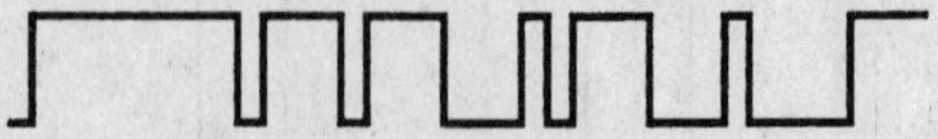	11	8	9	—	Damper clutch control system
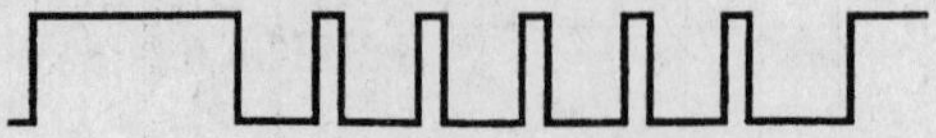	12	9	10	9	No ignition signal

89446G18

Fig. 73 Mitsubishi/Hyundai DTC's—Type 1 Codes

Fault code	Fault code (for voltmeter)	Cause	Remedy
21	 5V ----- ‾‾‾‾ ‾‾ 0V ----- ___	Abnormal increase of TPS output	o Check the throttle position sensor connector. o Check the throttle position sensor itself.
22	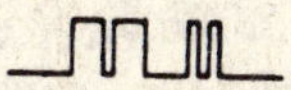	Abnormal decrease of TPS output	o Adjust the throttle position sensor. o Check the accelerator switch (No.28: output or not).
23		Incorrect adjustment of the throttle-position sensor system	o Check the throttle position sensor output circuit harness.
24	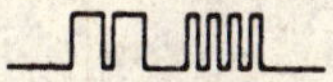	Damaged or disconnected wiring of the oil temperature sensor system	o Check the oil temperature sensor circuit harness. o Check the oil temperature sensor connector. o Check the oil temperature sensor itself.
25		Damaged or disconnected wiring of the kickdown servo switch system, or improper contact	o Check the kickdown servo switch output circuit harness. o Check the kickdown servo switch connector. o Check the kickdown servo switch itself.
26		Short circuit of the kickdown servo switch system	
27		Damaged or disconnected wiring of the ignition pulse pick-up cable system	o Check the ignition pulse signal line.
28		Short circuit of the accelerator switch system or improper adjustment	o Check the accelerator switch output circuit harness. o Check the accelerator switch connector. o Check the accelerator switch itself. o Adjust the accelerator switch.
31	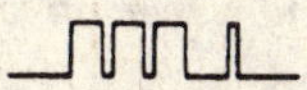	Malfunction of the microprocessor	o Replace the control unit.

89446G19

Fig. 74 Mitsubishi/Hyundai DTC's, 1 of 4—Type 2 Codes

Fault code	Fault code (for voltmeter)	Cause	Remedy
32	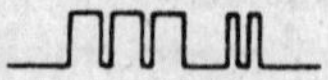	First gear command during high-speed driving	o Replace the control unit
33	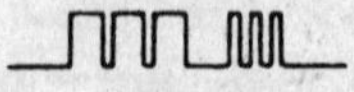	Damaged or disconnected wiring of the pulse generator B system	o Check the pulse generator B output circuit harness. o Check pulse generator B itself. o Check the vehicle speed reed switch (for chattering).
41	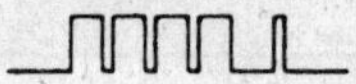	Damaged or disconnected wiring of the shift control solenoid valve A system	o Check the solenoid valve connector. o Check shift control solenoid valve A itself.
42	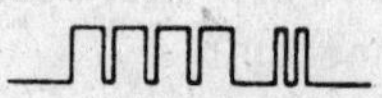	Short circuit of the shift-control solenoid valve A system	o Check the shift control solenoid valve A drive circuit harness
43	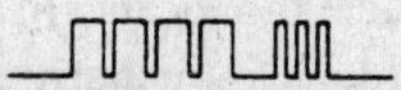	Damaged or disconnected wiring of the shift control solenoid valve B system	o Check the solenoid valve connector. o Check shift control solenoid valve B itself.
44	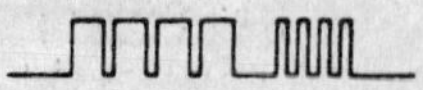	Short circuit of the shift control solenoid valve B system	o Check the shift control solenoid valve B drive circuit harness.
45	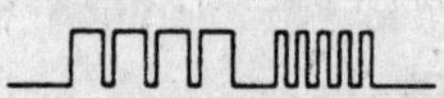	Damaged or disconnected wiring of the pressure control solenoid valve system	o Check the solenoid valve connector. o Check the pressure control solenoid valve itself.
46	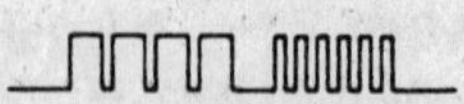	Short circuit of the pressure control solenoid valve system	o Check the pressure control solenoid valve drive circuit harness.
47	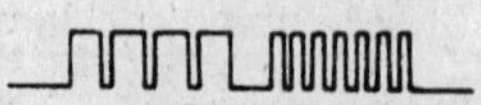	Damaged or disconnected wiring of the damper clutch control solenoid valve system	o Check the solenoid valve connector. o Check the damper clutch control solenoid valve itself.
48		Short circuit of the damper clutch control solenoid valve system	o Check the damper clutch control solenoid valve drive circuit harness.

89446G20

Fig. 75 Mitsubishi/Hyundai DTC's, 2 of 4—Type 2 Codes

FAIL-SAFE ITEM

Code No.	Output pattern (for voltmeter)	Description	Fail-safe	Note (relation to fault code)
	Output code			
11	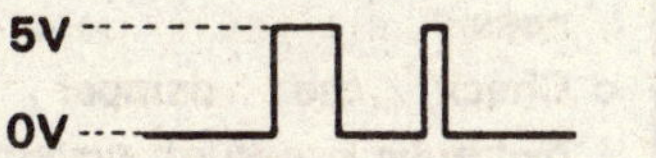 5V ······· 0V ····	Malfunction of the microprocessor	3rd gear hold	When code No. 31 is generated a 4th time.
12	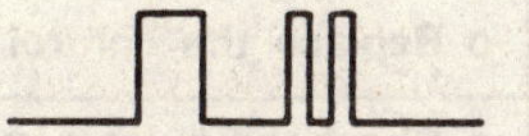	First gear command during high speed driving	3rd gear (D) or 2nd gear (2, L) hold	When code No. 32 is generated a 4th time.
13		Damaged or disconnected wiring of the pulse generator B system	3rd gear (D) or 2nd gear (2, L) hold	When code No. 33 is generated a 4th time.
14		Damaged or disconnected wiring, or short circuit, of shift control solenoid valve A	3rd gear hold	When code No. 41 or 42 is generated a 4th time.
15		Damaged or disconnected wiring, or short circuit, of shift control solenoid valve B	3rd gear hold	When code No. 43 or 44 is generated a 4th time.
16		Damaged or disconnected wiring, or short circuit, of the pressure control solenoid valve	3rd gear (D) or 2nd gear (2, L) hold	When code No. 45 or 46 is generated a 4th time.
17		Shift steps nonsynchronous	3rd gear (D) or 2nd gear (2, L) hold	When code No. 51, 52 53 or 54 is generated a 4th time.

89446G21

Fig. 76 Mitsubishi/Hyundai DTC's, 3 of 4—Type 2 Codes

Fault code	Fault code (for voltmeter)	Cause	Remedy
49	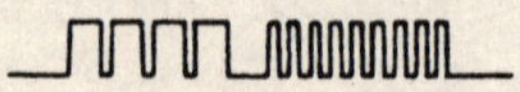	Malfunction of the damper clutch system	o Check the damper clutch control solenoid valve drive circuit harness. o Check the damper clutch hydraulic pressure system. o Check the damper clutch control solenoid valve itself. o Replace the control unit.
51	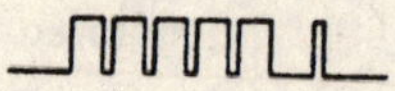	First gear non-synchronous	o Check the pulse generator output circuit harness. o Check the pulse generator connector. o Check pulse generator A and pulse generator B themselves. o Kickdown brake slippage.
52	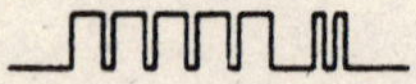	Second gear non-synchronous	o Check the pulse generator A output circuit harness. o Check the pulse generator A connector. o Check pulse generator A itself. o Kickdown brake slippage.
53	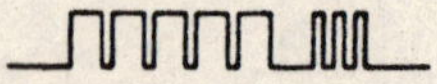	Third gear non-synchronous	o Check the pulse generator A output circuit harness. o Check the pulse generator connector. o Check pulse generator A and pulse generator B themselves. o Front clutch slippage. o Rear clutch slippage.
54	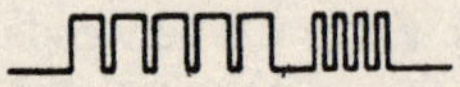	Fourth gear non-synchronous	o Check the pulse generator A output circuit harness. o Check the pulse generator A connector. o Check pulse generator A itself. o Kickdown brake slippage.

89446G22

Fig. 77 Mitsubishi/Hyundai DTC's, 4 of 4—Type 2 Codes

code	Diagnostic trouble code (for voltmeter)	Cause	Remedy
11		o Throttle position sensor malfunction o Open throttle position sensor circuit o Shorted throttle position sensor circuit	o Check the throttle position sensor connector. o Check the throttle position sensor itself. o Check the idle switch o Check the throttle position sensor wiring harness o Check the wiring between ECM and throttle position sensor
12			
13	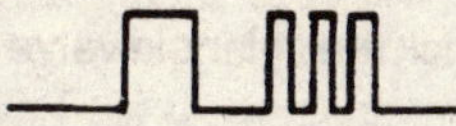		
14			
15		Oil temperature sensor damaged or disconnected wiring	o Oil temperature sensor connector inspection o Oil temperature sensor inspection o Oil temperature sensor wiring harness inspection
21		Open kickdown servo switch circuit.	o Check the kickdown servo switch connector. o Check the kickdown servo switch. o Check the kickdown serro switch wiring harness
22		Shorted kickdown servo switch circuit.	
23		Open ignition pulse pickup cable circuit	o Check the ignition pulse signal line. o Check the wiring between ECM and ignition system
24		Open-circuited or improperly adjusted idle switch	o Check the idle switch connector o Check the idle switch itself o Adjust the idle switch. o Check the idle switch wiring harness

89446G23

Fig. 78 Mitsubishi/Hyundai DTC's, 1 of 4—Type 3 Codes

code	Diagnostic trouble code (for voltmeter)	Cause	Remedy
31	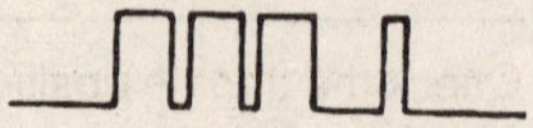	Pulse generator A damaged or disconnectorted wiring	o Check the pulse generator A and pulse generator B.
32	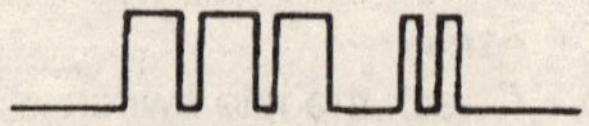	Pluse generator B damaged or disconnected wiring	o Check the vehicle speed reed switch (for chattering) o Check the pulse generator A and B wiring harness
41	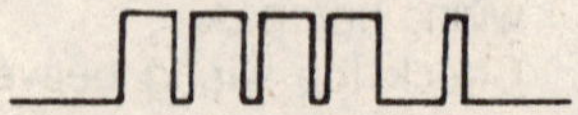	Open shift control solenoid valve A circuit	o Check the solenoid valve connector o Check the shift control solenoid valve A.
42		Shorted shift control solenoid valve A circuit	o Check the shift control solenoid valve A wiring harness
43	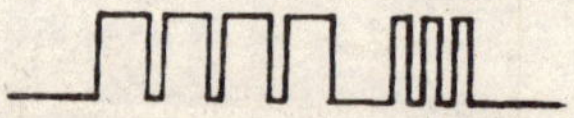	Open shift control solenoid valve B circuit	o Check the solenoid valve connector. o Check the shift control solenoid valve B wiring harness
44	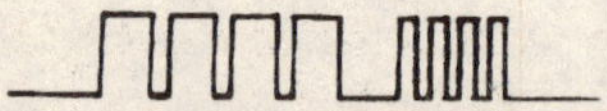	Shorted shift control solenoid valve B circuit	o Check the solenoid valve connector.
45	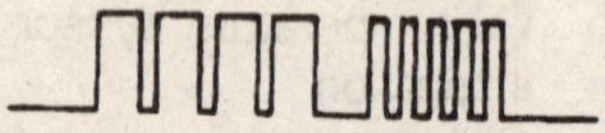	Open pressure control solenoid valve circuit	o Check the pressure control solenoid valve. o Check the pressure control solenoid valve wiring harness
46		Shorted pressure control solenoid valve circuit	
47	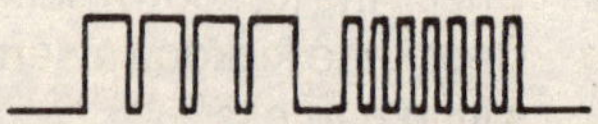	Open circuit in damper clutch control control solenoid valve Short circuit in damper clutch control solenoid valve	o Inspection of solenoid valve connector. o Individual inspection of damper clutch control solenoid valve
48		Defect in the damper clutch system	o Check the damper clutch control solenoid valve wiring harness o Check the TCM
49			o Inspection of damper clutch hydraulic system
51		Shifting to first gear does not match the engine speed.	o Check the pulse generator A and pulse generator B connector o Check the pulse generator A and pulse generator B o Check the one way clutch or rear clutch o Check the pulse generator wiring harness

89446G24

Fig. 79 Mitsubishi/Hyundai DTC's, 2 of 4—Type 3 Codes

code	Diagnostic trouble code (for voltmeter)	Cause	Remedy
52		Shifting to second gear does not match the engine speed.	o Check the pulse generator A and pulse generator B connector o Check the pulse generator A and pulse generator B. o Check the one way clutch or rear clutch o Check the pulse generator wiring harness
53		Shifting to third gear does not match the engine speed	o Check the rear clutch or control system o Check the pulse generator A and pulse generator B connector o Check the pulse generator A and pulse generator B. o Check the front clutch slippage or control system o Check the rear clutch slippage or control system o Check the pulse generator wiring harness
54		Shifting to fourth gear does not match the engine speed.	o Check the pulse generator A and B connector o Check the pulse generator A and B o Kickdown brake slippage. o Check the end clutch or control system o Check the pulse generator wiring harness
		Normal	
		Defective transaxle control module (TCM)	o TCM power supply inspection o TCM earth inspection o TCM replacement

89446G25

Fig. 80 Mitsubishi/Hyundai DTC's, 3 of 4—Type 3 Codes

FAIL-SAFE ITEM

Output code		Description	Fail-safe	Note (relation to diagnostic trouble code)
Code No.	Output pattern (for voltmeter)			
81		Open-circuited or damaged pulse generator A	Locked in third (D) or second (2,L)	When code No. 31 is generated fourth time
82		Open-circuited or damaged pulse generator B	Locked in third (D) or second (2, L)	When code No. 32 is generated fourth time
83		Open-circuited, shorted or damaged shift control solenoid valve A	Lock in third	When code No. 41 or 42 is generated fourth time
84		Open-circuited, shorted or damaged shift control solenoid valve B	Lock in third gear	When code No. 43 or 44 is generated fourth time
85		Open-circuited, shorted or damage pressure control solenoid valve	Locked in third (D) or second (2, L)	When code No. 45 or 46 is generated fourth time.
86		Gear shifting does not match the engine speed	Locked in third (D) or second (2, L)	When either code No. 51, 52, 53 or 54 is generated fourth time.
-	Constant output (or 0V)	Defective transaxle control module (TCM)	Fixed for third speed	-

89446G26

Fig. 81 Mitsubishi/Hyundai DTC's, 4 of 4—Type 3 Codes

Code	Output pattern (for voltmeter)	Cause	Remedy
P1702	 A5AT005E	Shorted throttle position sensor circuit	o Check the throttle position sensor connector o check the throttle position sensor itself o Check the closed throttle position switch o Check the throttle position sensor wiring harness o Check the wiring between ECM and throttle position sensor
P1701	 A5AT005F	Open throttle position sensor circuit	
P1704	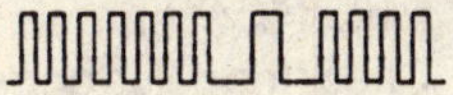 A5AT005H	Throttle position sensor malfunction Improperly adjusted throttle position sensor	
P0712	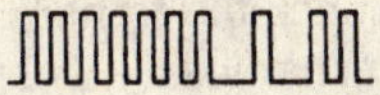 A5AT005I	Open fluid temperature sensor circuit	o Fluid temperature sensor connector inspection o Fluid temperature sensor inspection o Fluid temperature sensor wiring harness inspection
P0713	 A5AT005J	Shorted fluid temperature sensor circuit	
P1709	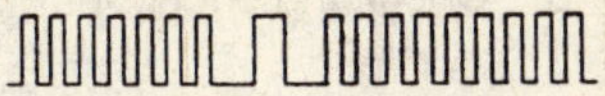 A5AT005K	Open kickdown servo switch circuit Shorted kickdown servo switch circuit	o Check the kickdown servo switch connector o Check the kickdown servo switch o Check the kickdown servo switch wiring harness

89446G27

Fig. 82 Mitsubishi/Hyundai DTC's, 1 of 4—Type 4 (OBD II) Codes

Code	Output pattern (for voltmeter)	Cause	Remedy
P0727	 A5AT005L	Open ignition pulse pickup cable circuit	o Check the ignition pulse signal line o Check the wiring between ECM and ignisiton system
P1714	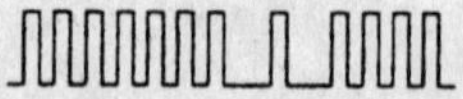 A5AT005M	Short-circuited or improperly adjusted closed throttle position switch	o Check the closed throttle position switch connector o Check the closed throttle position switch itself o Adjust the closed throttle position switch o Check the closed throttle position switch wiring harness
P0717	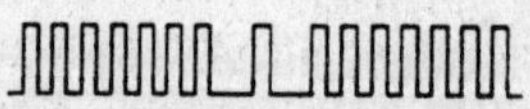 A5AT005N	Open-circuited pulse generator A	o Check the pulse generator A and pulse generator B o Check the vehicle speed reed switch (for chattering) o Check the pulse generator A and B wiring harness
P0722	 A5AT005O	Open-circuited pulse generator B	
P0707	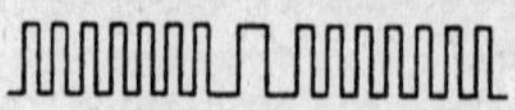 A5AT005Z	No input signal	o Check the transaxle range switch o Check the transaxle range wiring harness o Check the manual control cable
P0708	 A5ATA05A	More than two input signals	
P0752	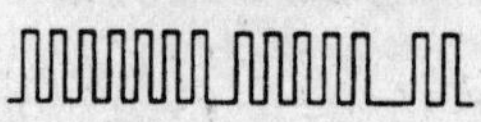 A5AT005P	Open shift control solenoid vave A circuit	o Check the solenoid valve connector o Check the shift control solenoid valve A o Check the shift control solenoid valve A wiring harness
P0753	 A5AT005Q	Shorted shift control solenoid valve A circuit	
P0757	 A5AT005R	Open shift control solenoid valve B circuit	o Check the shift control solenoid valve connector o Check the shift control solenoid valve B wiring harness o Check the shift control solenoid valve B
P0758	 A5AT005S	Short shift control solenoid valve B circuit	
P0747	 A5AT005T	Open pressure control solenoid valve circuit	o Check the pressure control solenoid valve o Check the pressure control solenoid valve wiring harness
P0748	 A5AT005U	Shorted pressure control solenoid valve circuit	

89446G28

Fig. 83 Mitsubishi/Hyundai DTC's, 2 of 4—Type 4 (OBD II) Codes

Code	Output pattern (for voltmeter)	Cause	Remedy
P0743	A5AT005V	Open circuit in damper clutch control solenoid valve	o Inspection of solenoid valve connector
P0742	A5AT005W	Short circuit in damper clutch control solenoid valve	o Individual inspection of damper clutch control solenoid valve
P0740	A5AT005X	Defect in the damper clutch system	o Check the damper clutch control solenoid valve wiring harness
P1744	A5AT005Y		o Chck the TCM
			o Inspection of damper clutch hydraulic system
P0731	A5ATA05B	Shifting to first gear does not match the engine speed	o Check the pulse generator A and pulse generator B connector
P0732	A5ATA05C	Shifting to second gear does not match the engine speed	o Check the pulse generator A and puls generator B o Check the one way clutch or rear clutch o Check the pulse generator wiring harness o Kickdown brake slippage
P0733	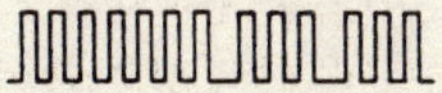 A5ATA05D	Shifting to third gear does not match the engine speed	o Check the rear clutch or control system o Check the pulse generator A and pulse generator B connector o Check the pulse generator A and pulse generator B o Check the pulse generator wiring harness o Check the rear clutch slippage or control system o Check the front clutch slippage or control system
P0734	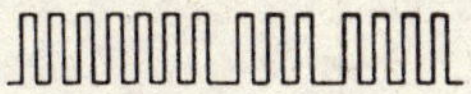 A5ATA05E	Shifting to fourth gear does not match the engine speed	o Check the pulse generator A and B connector o Check the pulse generator A and B o Kickdown brake slippage o Check the end clutch or control system o Check the pulse generator wiring harness
-		Normal	

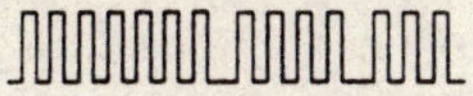

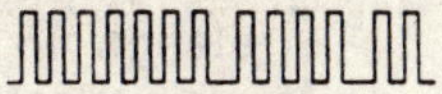

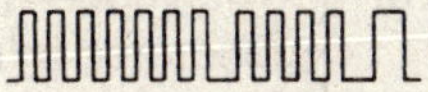

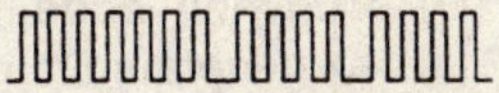

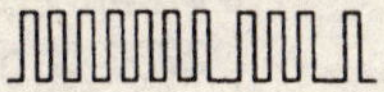

89446G29

Fig. 84 Mitsubishi/Hyundai DTC's, 3 of 4—Type 4 (OBD II) Codes

FAIL-SAFE ITEM

Code No.	Output code — Output pattern (for voltmeter)	Description	Fail-safe	Note (relation to diagnostic trouble code)
P0717	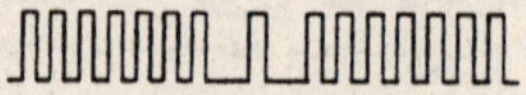 A5AT005N	Open-circuited pulse generator A	Locked in third (D) or second (2,L)	When code No.0717 is generated fourth time
P0722	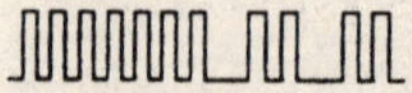 A5AT005O	Open-circuited pulse generator B	Locked in third (D) or second (2,L)	When code No.0722 is generated fourth time
P0752	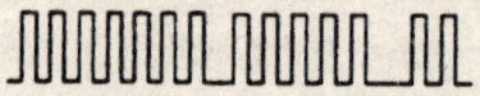 A5AT005P	Open-circuited or shorted shift control solenoid valve A	Lock in third	When code No.0752 or 0753 is generated fourth time
P0753	 A5AT005Q			
P0757	 A5AT005R	Open-circuited or shorted shift control solenoid valve B	Lock in third gear	When code No.0757 or 0758 is generated fourth time
P0758	 A5AT005S			
P0747	 A5AT005T	Open-circuited or shorted pressure control solenoid valve	Locked in third (D) or second (2,L)	When code No.0747 or 0748 is generated fourth time
P0748	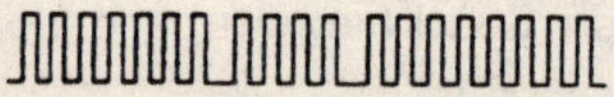 A5AT005U			
P0731	 A5ATA05B	Gear shifting does not match the engine speed	Locked in third (D) or second (2,L)	When either code No.0731, 0732, 0733 or 0734 is generated fourth time
P0732	 A5ATA05C			
P0733	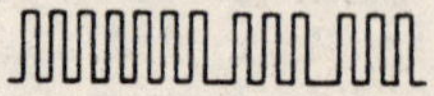 A5ATA05D			
P0734	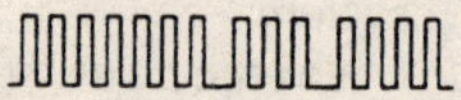 A5ATA05E			

89446G30

Fig. 85 Mitsubishi/Hyundai DTC's, 4 of 4—Type 4 (OBD II) Codes

4. Start the engine and drive the vehicle until the transaxle goes into the failsafe mode.

5. Park the vehicle, but do not turn the ignition **OFF**. Allow it to idle.

6. Attach a voltmeter (analog or digital) to the test terminals on the Diagnostic Link Connector (DLC). The negative lead should be attached to terminal 4 and the positive lead to terminal 1.

7. Observe the voltmeter and count the flashes (or arm sweeps if using an analog voltmeter); note the applicable codes.

8. After all of the DTC(s) have been retrieved, fix the applicable problems, clear the codes, drive the vehicle, and perform the retrieval procedure again to ensure that all of the codes are gone.

Clearing Codes

TYPE 1 CODES

It is not necessary to clear Type 1 codes, because they are stored in volatile memory and automatically erase whenever the ignition key is turned **OFF**.

TYPE 2, 3 AND 4 CODES

➡It is not recommended that the negative battery cable be disconnected to clear DTC's. Doing so will clear all settings from the vehicle's computer system, resulting in lost radio presets, seat memories, anti-theft codes, driveability parameters, etc., and there are better ways of spending your afternoon than resetting all of these.

1. Turn the ignition switch **OFF**.
2. Disconnect the wiring harness connector from the control system computer.
3. Wait at least one minute, then reconnect the computer. The codes should now be cleared. If there are still codes present, either the codes were not properly cleared (Are the codes identical to those flashed out previously?) or the underlying problem is still there (Are only some of the codes the same as previously?)

NISSAN & INFINITI

◆ **See Figures 86, 87, 88 and 89**

Reading and Retrieving Codes

Nissan and Infiniti vehicles provide diagnostic trouble codes through one of four different lights, depending on the specific model at hand: Power Shift lamp, O/D Off lamp, A/T Check lamp (300ZX and J30 models), digital readout in the diagnostic information display (Q45 models). The vehicle will flash the applicable lamp 16 times whenever the ignition key is turned **ON** when the system has stored DTC's.

Nissans and M30 Infiniti display DTC's using an 11-flash sequence. The light flashes 11 times in a row. The sequence always starts with a long start flash (approximately two seconds in duration), and is followed by 10 shorter flashes (approximately ¼ second in duration). If there are no malfunctions detected by the system, all ten flashes will be short, but if the computer recognizes a fault in the system, one of the 10 flashes will be longer (approximately one second in duration)—the long flash identifies the code stored in memory. For example, if the third flash after the two second start flash is the long one, you're looking at a Code 3.

If the computer has detected more than one code, it displays all of the codes in the same pass. For an example, please refer to the accompanying illustration. After the computer displays the code(s), the light will remain off for approximately 2½ seconds, then repeats the code(s). If the light flashes on and off, in regular, one-second intervals, it indicates the battery is low or was disconnected long enough to interfere with computer memory.

Infiniti J30 models display diagnostic trouble codes in the same manner as with the other models, the only difference is that they use a 13-flash sequence.

➡**If the light remains on or off, try performing the sequence again. You may have missed one of the steps in the preliminary or retrieval procedure.**

Infiniti Q45 models display DTC's in a hexadecimal format through a digital display, which also serves as the odometer display. Therefore, it is possible to read codes 1 through 9 and A through D. Refer to the DTC chart at the end of this section for the applicable codes.

➡**For Q45 models, if there are no codes in memory, the odometer will display OK.**

PRELIMINARY PROCEDURE

To prepare the vehicle for code retrieval, perform the following:

➡**It is important to perform all of the following steps; some of the steps are a preliminary inspection, which prepares the system for fault code retrieval.**

1. Start the engine and allow it to reach normal operating temperature.
2. Turn the ignition key **OFF** and apply the parking brake.
3. Without starting the engine, turn the ignition key to the **ON** position.

➡**The next 7 steps do not apply for Q45 models.**

4. For vehicles equipped with an O/D Off button, activate the switch so that the O/D Off light illuminates. Then, deactivate the switch so that the light turns off.
5. For vehicles equipped with a Mode switch, activate the switch so that the Power Shift lamp illuminates. Then, deactivate the switch so that the light turns off.
6. Turn the ignition key **OFF**, and wait a few seconds.
7. Once again turn the ignition switch **ON**, without starting the engine.

The indicator light on the instrument panel should illuminate for a few seconds (lamp bulb check),

then extinguish. This is to check the light circuit, to make sure it's working satisfactorily for code flashing. If the lamp does not illuminate, inspect the bulb and circuit for a malfunction.

8. Turn the ignition switch **OFF**.
9. Move the transmission/transaxle gear selector to the D position.
10. If equipped, turn the O/D Off switch off.

CODE RETREIVAL—EXCEPT QUEST, J30 AND Q45 MODELS

1. Perform the preliminary procedure before commencing with this procedure.
2. Turn the ignition key **ON**, without starting the engine, and wait for a few seconds.
3. Move the transmission/transaxle gear selector to position 2.
4. Turn the O/D switch on (lamp off).
5. Move the transmission/transaxle gear selector to position 1.
6. Turn the O/D switch off (lamp on).
7. Depress the accelerator pedal to the floor and release it.
8. On models which use the O/D Off light for code display, turn the O/D switch on.
9. Observe the appropriate light in the instrument cluster and note the DTC's.

CODE RETREIVAL—QUEST MODELS

1. Perform the preliminary procedure before commencing with this procedure.
2. Depress the O/D Off button and turn the ignition switch **ON**, without starting the engine.
3. Wait a few seconds, and release the O/D Off button.

➡**At this time, the O/D Off light should be illuminated.**

4. Move the transmission/transaxle gear selector to position 2.
5. Depress and release the O/D Off switch (the O/D Off lamp should extinguish).
6. Move the transmission/transaxle gear selector to position 1.
7. Depress and release the O/D Off switch (the O/D Off lamp should illuminate again).
8. Depress the accelerator pedal to the floor and release it.

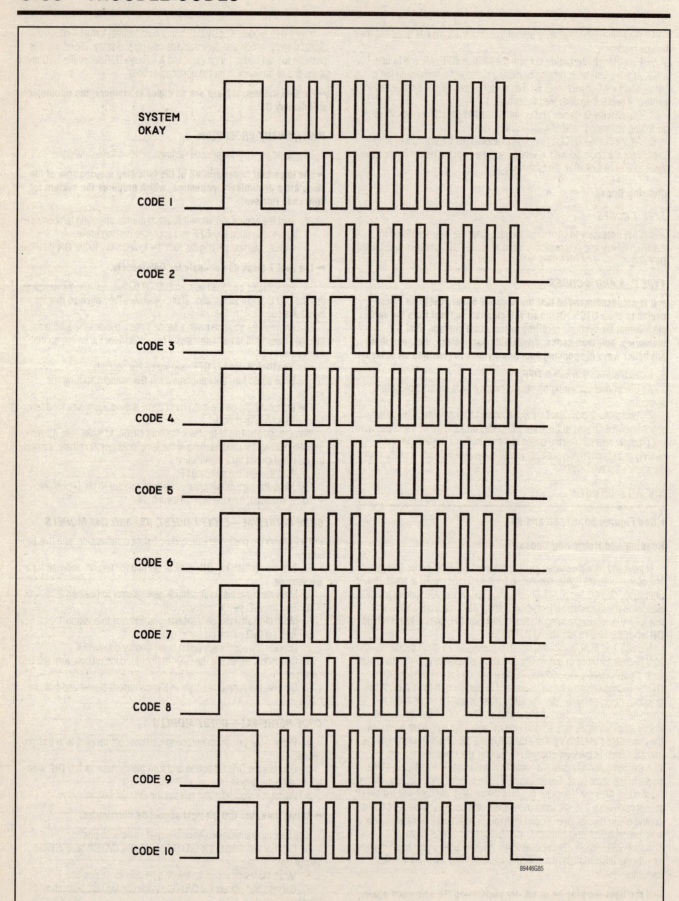

Fig. 86 Diagnostic trouble code patterns for codes 1 through 10—Nissans and M30 Infiniti

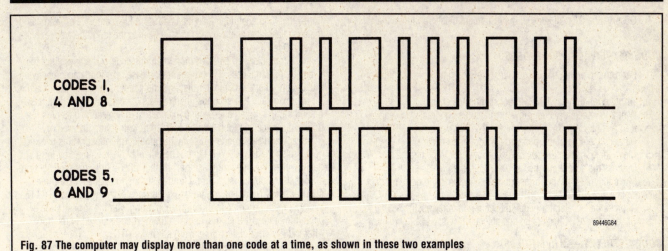

Fig. 87 The computer may display more than one code at a time, as shown in these two examples

DIAGNOSTIC TROUBLE CODES—NISSAN, EXCEPT J30 AND Q45

Code	Component/Circuit Fault
1	Vehicle Speed Sensor (VSS) or circuit
2	Speedometer VSS or circuit
3	Throttle Position (TP) sensor or circuit
4	Shift solenoid A or circuit
5	Shift solenoid B or circuit
6	Timing solenoid or circuit, or over-run clutch solenoid or circuit
7	Lock-up solenoid or circuit
8	Transaxle fluid temperature sensor or circuit, or computer power insufficient
9	Engine speed signal circuit
10	Line pressure solenoid or circuit
Regular Flashing	Battery voltage low, or power was disconnected from the computer

Fig. 88 Nissan and Infiniti Diagnostic Trouble Codes—except J30 and Q45 models

DIAGNOSTIC TROUBLE CODES—NISSAN J30 AND Q45

Codes J30	Q45	Component/Circuit Fault
1	1	Vehicle Speed Sensor (VSS) or circuit
2	2	Speedometer VSS or circuit
3	3	Throttle Position (TP) sensor or circuit
4	4	Shift solenoid A or circuit
5	5	Shift solenoid B or circuit
6	6	Timing solenoid or circuit, or over-run clutch solenoid or circuit
7	7	Lock-up solenoid or circuit
8	8	Transaxle fluid temperature sensor or circuit, or computer power insufficient
9	9	Engine speed signal circuit
10	A	Turbine shaft speed sensor or circuit
11	B	Line pressure solenoid or circuit
12	C	Engine/Transmission communication circuit
Regular Flashing	D	Battery voltage low, or power was disconnected from the computer

Fig. 89 Nissan and Infiniti Diagnostic Trouble Codes—J30 and Q45 models

9. Observe the O/D Off lamp in the instrument cluster and note the DTC's.

CODE RETREIVAL—J30 MODELS

1. Perform the preliminary procedure before commencing with this procedure.
2. Turn the ignition key **ON**, without starting the engine.
3. Wait a few seconds, then move the transmission/transaxle gear selector to position 3.
4. Depress the accelerator pedal to the floor, then release it.
5. Move the transmission/transaxle gear selector to position 2.
6. Depress the accelerator pedal to the floor, then release it.
7. Move the transmission/transaxle gear selector to the right, which positions the selector in Manual 1.
8. Depress the accelerator pedal to the floor, then release it.
9. Observe the A/T Check lamp and note the DTC's.

CODE RETREIVAL—Q45 MODELS

1. Perform the preliminary procedure before commencing with this procedure.
2. Turn the odometer reset counter knob counterclockwise, and hold it there during the next step.
3. Turn the ignition switch **ON**, without starting the engine, then release the odometer reset knob. At this time the odometer display should display **AT CHECK**.
4. Move the transmission/transaxle gear selector to position 3.
5. Depress the accelerator pedal to the floor, then release it.
6. Move the transmission/transaxle gear selector to position 2.
7. Depress the accelerator pedal to the floor, then release it.
8. Move the transmission/transaxle gear selector to the right, which positions the selector in Manual 1.
9. Depress the accelerator pedal to the floor, then release it.
10. Observe the digital odometer display and note the DTC's.

Clearing Codes

➡**It is not recommended that the negative battery cable be disconnected to clear DTC's. Doing so will clear all settings from the vehicle's computer system, resulting in lost radio presets, seat memories, anti-theft codes, driveability parameters, etc., and there are better ways of spending your afternoon than resetting all of these.**

Nissan and Infiniti vehicles will automatically clear any DTC's when the corresponding problem has been repaired and the ignition key has been turned **ON** twice.

SUBARU

◆ **See Figures 90 thru 92**

Reading and Retrieving Codes

EXCEPT JUSTY

Subaru indicates malfunctions in the transmission/transaxle system and displays DTC's via the POWER lamp in the instrument cluster display. This lamp should illuminate when the ignition key is turned **ON**, then extinguish when the system is functioning normally. However, when the system detects a malfunction, the POWER light will blink on and off for a couple of seconds during start up. This isn't a trouble code; it's the computer's way of letting you know that there's a problem in the system. To retrieve the DTC's themselves, you must perform the retrieval procedure.

Subaru models, except Justy, Legacy and SVX, display DTC's using an 12-flash sequence. The light flashes 12 times in a row. The sequence always starts with a long start flash (approximately two seconds in duration), and is followed by 11 shorter flashes (approximately ¼ second in duration). If there are no malfunctions detected by the system, all 11 flashes will be short, but if the computer recognizes a fault in the system, one of the 11 flashes will be longer (approximately one second in dura-

tion)—the long flash identifies the code stored in memory. For example, if the third flash after the two second start flash is the long one, you're looking at a Code 3.

If the computer has detected more than one code, it displays all of the codes in the same pass. For an example, please refer to the accompanying illustration. After the computer displays the code(s), the light will remain off for approximately 2½ seconds, then repeats the code(s). If the light flashes on and off, in regular, one-second intervals, it indicates the battery is low or was disconnected long enough to interfere with computer memory.

Legacy and SVX models display diagnostic trouble codes as two digit codes, also via the POWER lamp. They display the tens digit as long flashes, then the ones digit as short flashes after a brief pause. Therefore, a Code 13 would consist of one long flash, a short pause, then three short flashes. There is a longer pause between each individual DTC. The system repeats each code three times, starting with the lowest number code in memory. It then moves on to the next code stored in memory. After the system finishes displaying all of the DTC's, the process repeats itself. The system continues displaying DTC's until the ignition key is turned **OFF**.

If there are no DTC's stored in memory, the POWER light will flash on and off, in short flashes. The POWER light flashing very rapidly indicates low battery voltage.

If the lamp worked prior to the diagnostic procedure, but will not display any DTC's, there may be a problem in one or more of the following:
- Throttle position idle switch
- Shift lever position switch
- Hold switch

To retrieve codes on Subaru models except Justy, Legacy or SVX, perform the following:
1. If equipped with 4WD, lock the four-wheel drive system into two wheel drive, as follows:
   a. Locate the special plug, marked FWD. It is usually located around the left front shock tower.

➡**This is the 4WD lockout connector.**

   b. Open the connector cover.
   c. Insert a 15A spade-type fuse into the connector, thereby locking the system out of 4WD.
2. Start the engine and allow it reach normal operating temperature.
3. Turn the ignition switch **OFF**.
4. Without starting the engine, turn the ignition key to the **ON** position. At this time the POWER lamp should come on.
5. Turn the ignition key **OFF** and apply the parking brake.

➡**From this point on, do not disturb or touch either the accelerator or brake pedal unless specifically instructed to.**

6. Move the transmission/transaxle gear selector to D.
7. Turn the 1-Hold switch on (down position).
8. Turn the ignition key **ON**, without starting the engine.
9. Move the transmission/transaxle gear selector to position 3.
10. Turn the 1-Hold switch off (up position).
11. Move the transmission/transaxle gear selector to position 2.
12. Turn the 1-Hold switch on (down position).
13. Depress the accelerator pedal halfway to the floor.
14. Observe the POWER lamp and note the DTC's.

Legacy and SVX models are designed with two types of DTC's: existing codes and history codes.

Existing codes indicate problems that are currently detected by the diagnostic system, also known as hard codes. History codes indicate problems that were there in the past; you won't find the problem there now. Intermittent problems will often set history codes. If a problem sets a history code, but not an existing code, look for conditions like loose connections or bad grounds; anything that can cause a failure that comes and goes. The procedure for setting the system to display existing codes is different from the procedure for displaying history codes.

To retrieve existing codes on Subaru Legacy and SVX models, perform the following:
15. Start the engine and allow it reach normal operating temperature.

DIAGNOSTIC TROUBLE CODES—SUBARU EXCEPT JUSTY, LEGACY & SVX

Code	Component/Circuit Fault
1	No. 1 speed sensor
2	No. 2 speed sensor
3	Throttle Position (TP) sensor
4	No. 1 shift solenoid
5	No. 2 shift solenoid
6	No. 3 shift solenoid
7	Duty cycle solenoid B
8	Duty cycle solenoid C
9	ATF temperature sensor
10	No ignition signal
11	Duty cycle solenoid A

89446GAA

Fig. 90 Subaru Diagnostic Trouble Codes—except Justy, Legacy and SVX

DIAGNOSTIC TROUBLE CODES—SUBARU LEGACY & SVX

Code	Component/Circuit Fault
11	Duty solenoid A or circuit
12	Duty solenoid B or circuit
13	Shift solenoid No. 3 or circuit
14	Shift solenoid No. 2 or circuit
15	Shift solenoid No. 1 or circuit
21	ATF temperature sensor or circuit
22	Atmospheric pressure (Baro) sensor
23	Engine speed signal circuit
24	Duty solenoid C or circuit
31	Throttle Position (TP) sensor or circuit
32	No. 1 Vehicle Speed Sensor (VSS) or circuit
33	No. 2 Vehicle Speed Sensor (VSS) or circuit

89446GAB

Fig. 91 Subaru Diagnostic Trouble Codes—Legacy and SVX

16. If the transmission/transaxle is functioning well enough to road test the car, drive it at least 12 mph (20 km/h).

17. Apply the parking brake, then turn the ignition switch **OFF**.

18. Move the transmission/transaxle gear selector to D.

19. Turn the Hold or Manual switch on (down or in position).

20. Turn the ignition key **ON**, without starting the engine.

21. Move the transmission/transaxle gear selector to position 3.

22. Turn the Hold or Manual switch off (up or out position).

23. Move the transmission/transaxle gear selector to position 2.

24. Turn the Hold or Manual switch on (down or in position).

25. Move the transmission/transaxle gear selector to position 1.

26. Turn the Hold or Manual switch off (up or out position).

27. Depress the accelerator pedal halfway to the floor.

28. Observe the POWER lamp and note the DTC's.

To retrieve history codes on Subaru Legacy and SVX models, perform the following:

29. Apply the parking brake, then turn the ignition switch **OFF**.

30. Move the transmission/transaxle gear selector to 1.

31. Turn the Hold or Manual switch on (down or in position).

32. Turn the ignition key **ON**, without starting the engine.

33. Move the transmission/transaxle gear selector to position 2.

34. Turn the Hold or Manual switch off (up or out position).

35. Move the transmission/transaxle gear selector to position 3.

36. Turn the Hold or Manual switch on (down or in position).

37. Move the transmission/transaxle gear selector to position D.

38. Turn the Hold or Manual switch off (up or out position).

39. Depress the accelerator pedal halfway to the floor.

40. Observe the POWER lamp and note the DTC's.

JUSTY

The Subaru Justy diagnostic system indicates detected faults in the transmission/transaxle system and displays DTC's CHECK ECVT lamp, located in the instrument cluster display. This lamp should illuminate when the ignition key is turned **ON**, then extinguish when the system is functioning normally. However, when the system detects a malfunction, the CHECK ECVT light will illuminate during vehicle operation. This isn't a trouble code; it's the computer's way of letting you know that there's a problem in the system. To retrieve the DTC's themselves, you must perform the retrieval procedure.

Justy models display diagnostic trouble codes as two digit codes, also via the POWER lamp. They display the tens digit as long flashes, then the ones digit as short flashes after a brief pause. Therefore, a Code 13 would consist of one long flash, a short pause, then three short flashes. There is a

DIAGNOSTIC TROUBLE CODES—SUBARU JUSTY

Code	Component/Circuit Fault
13	D-range switch or circuit
14	Ds-range switch or circuit
15	R-range switch or circuit
21	ECM communication circuit
22	Coolant temperature signal from ECM
25 ①	Slow Cut solenoid, CFC solenoid or circuit
31	Throttle pedal switch or circuit
32	Throttle position sensor or circuit
33	No vehicle speed sensor signal
34	Clutch coil system current out of specifications
35	Line pressure solenoid or circuit
41, 42	Improper high altitude signal from ECM
45	Brake switch or circuit

① Coolant temperature sensor problems may also set a Code 25.

89446GAC

Fig. 92 Subaru Diagnostic Trouble Codes—Justy

longer pause between each individual DTC. The system repeats each code three times, starting with the lowest number code in memory. It then moves on to the next code stored in memory. After the system finishes displaying all of the DTC's, the process repeats itself. The system continues displaying DTC's until the ignition key is turned **OFF**.

If there are no DTC's stored in memory, the POWER light will flash on and off, in short flashes. The POWER light flashing very rapidly indicates low battery voltage.

To retrieve the codes, perform the following:

1. Perform the preliminary inspection, located earlier in this section. This is very important, since a loose or disconnected wire, or corroded connector terminals can cause a whole slew of unrelated DTC's to be stored by the computer; you will waste a lot of time performing a diagnostic "goose chase."

2. Grab some paper and a pencil or pen to write down the DTC's when they are flashed out.

3. Locate the Transaxle Check Mode Connectors and the Engine Check Connectors. The Transaxle Check Mode Connectors are single terminal, white, disconnected connector halves hanging from the transaxle computer harness, just before the computer connector. Don't confuse this with the two-terminal memory connector that's already connected. The Engine Check Connectors are usually located near the Transaxle Check Mode Connectors, and are disconnected, green single-terminal connector halves. If neither of these can be found, either call your local Subaru dealer or refer to the applicable Chilton's Total Car Care manual for your vehicle.

➡**The code retrieval procedure must be performed on the road. Take the vehicle to an open stretch of roadway, and pull off to the side to begin. You will need to accelerate the vehicle for ⅛ to ¼ mile (0.2–0.4 km) during the procedure. Be sure there is enough clear road to complete the diagnostic procedure.**

4. Start the engine and allow it reach normal operating temperature with the A/C off.

5. Turn the ignition key **OFF**.

6. Connect the two Transaxle Check Mode connector halves together.

7. Start the engine, with the transaxle in P.

➡**From this point on, do not disturb or touch either the accelerator or brake pedal unless specifically instructed to. If you must use one of these pedals to avoid an accident, after the problem is dealt with shut the ignition key OFF and start over, beginning with Step 4.**

8. Engage the two green Engine Check Connector halves for 10 seconds, then separate them.

➡**If the CHECK ECVT light starts to blink or display codes, ignore it for now.**

9. Apply the parking brake.

10. Move the transmission/transaxle gear selector to R.

11. Move the transmission/transaxle gear selector to N.

12. Move the transmission/transaxle gear selector to D.

13. Move the transmission/transaxle gear selector to Ds.

14. Move the transmission/transaxle gear selector to D.

15. When the road is clear, disengage the parking brake.

16. Floor the throttle, and accelerate to approximately 25–30 mph (40–48 km/h), and hold the vehicle at this speed constantly for at least 5 seconds.

17. Release the throttle, and COAST to a stop. Pull off the road where it's safe to sit for a few minutes.

➡**Don't touch the brake pedal to stop; If necessary, use the parking brake to help stop the vehicle.**

18. Apply the parking brake.

19. Depress and release the brake pedal three times.

➡**The CHECK ECVT light will blink three sets of short flashes. These are normal system identification flashes, not DTC's.**

20. Observe the CHECK ECVT lamp and note the DTC's.

Clearing Codes

EXCEPT JUSTY

To clear the codes from memory, disconnect the negative battery cable for at least one minute. Then, reconnect the cable. The codes should now be cleared. If there are still codes present, either the codes were not properly cleared (Are the codes identical to those flashed out previously?) or the underlying problem is still there (Are only some of the codes the same as previously?)

JUSTY

To clear the DTC's from the system's memory, detach the transmission/transaxle Check Mode connector.

TOYOTA

◆ **See Figures 93, 94, 95, 96 and 97**

Reading and Retrieving Codes

Like most other manufacturers, Toyota vehicles built since 1985 are capable of displaying DTC's. The Toyota systems display the DTC's via the O/D OFF lamp (as with many other import manufacturers).

The DTC's displayed by Toyota models consist of two digits. They are flashed out so that first the tens digit is displayed, then the ones digit. And,

Model	Year	Design	Location
Camry	1985-94	2	Left Front Shock Tower
	1992-94	3	Under the Left Side of the Instrument Panel
Celica	1986-94	2	Left Front Shock Tower
Corolla	1987	1	Right Front Shock Tower
	1988-94	2	Left Front Shock Tower
Cressida	1985-91	2	Left Front Shock Tower
	1989-91	3	Under Dash; Left of Steering
MR2	1985-89	2	Left Front of Engine Compartment
	1991	2	Left Rear of the Engine Compartment
	1992-94	2	Right Rear Shock Tower
Paseo	1992-94	2	Left Side of the Engine Compartment, Near the No. 2 Fuse Box
Pick-up and 4Runner	1985-89	1	Left Front Shock Tower
	1990-94	2	Left Side of the Engine Compartment, Near the No. 2 Fuse Box
Previa Van	1991-94	2	Under Front of the Driver's Side Seat
Supra	1985-86	1	Left Side of the Engine Compartment
	1986-92	2	Left Front Shock Tower
	1993-94	2	Right Side of the Firewall

89446G59

Fig. 93 Before starting the diagnostic retrieval procedure, locate the connector

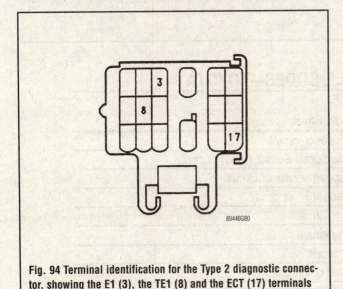

89446G80

Fig. 94 Terminal identification for the Type 2 diagnostic connector, showing the E1 (3), the TE1 (8) and the ECT (17) terminals

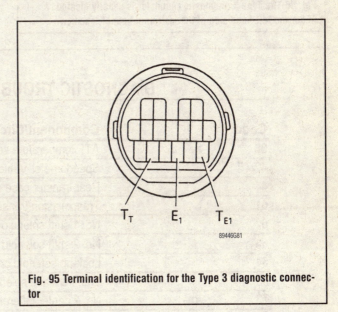

89446G81

Fig. 95 Terminal identification for the Type 3 diagnostic connector

since all of the O/D OFF lamp blinks are the same length of duration, Code 38 would be displayed as three flashes, then a short pause followed by eight more flashes. After which, the O/D OFF light would remain off for a few seconds, then the code would be repeated. The system repeats each DTC three times, then goes on to the next code stored in memory. After the system finishes displaying all of the codes stored in memory, it returns to the first DTC and starts all over. The system will continue to display the DTC's until the ignition key is turned **OFF**.

Before the codes can be retrieved, the diagnostic connector link must be located. Over the years, however, there have been three different configurations of diagnostic connector links. The earliest connector appears in 1985–88 vehicles. It is a single terminal (ECT terminal) connector. The next design was introduced on some models in 1985. It is an irregularly-shaped, multi-terminal connector and usually always appears in the engine compartment. On vehicles with a single, integrated (engine & transmission) Powertrain Control Module (PCM) the ECT terminal is changed to the

T_T terminal. The third configuration is referred to as the Toyota Diagnostic Communication Link (TDCL). This connector is used on 1989–91 Cressida and 1992 and newer pre-OBD II Camry that use a single, integrated (engine & transmission) Powertrain Control Module (PCM). Model equipped with this third type of connector also have the second design connector in the engine compartment. The third type connector is usually located under the driver's side of the instrument panel.

To retrieve DTC's, perform the following:
1. Locate the diagnostic connector location.
2. Turn the ignition key **OFF**.
3. Using a jumper wire, ground the ECT or TE1 terminal in the diagnostic connector.
4. Turn the ignition key **ON**, without starting the engine.
5. Observe the O/D OFF light and note all DTC's. If the O/D OFF keeps flashing off and on, in even, regular cycles, there are no codes stored in memory.

Clearing Codes

➡It is not recommended that the negative battery cable be disconnected to clear DTC's. Doing so will clear all settings from the vehicle's computer system, resulting in lost radio presets, seat memories, anti-theft codes, driveability parameters, etc., and there are better ways of spending your afternoon than resetting all of these.

To clear the codes from memory, remove the appropriate fuse from the fuse block for at least one minute, as follows:

1985–86 Camry—Turn the ignition switch **OFF**
1986 Camry—Fuse: ECU B 15A
1987–88 Camry with A140E—Fuse: radio No. 1 15A
1989 Camry except A540E—Fuse: dome 20A
1988–89 Camry with A540E—Fuse: EFI 15A
1986–87 Celica—Fuse: No. 1 radio 15A
1988–89 Celica—Fuse: dome 20A
1985–86 Cressida—Fuse: No. 1 radio 15A
1987–88 Cressida—Fuse: EFI 15A
1989–92 Cressida—Fuse: dome 7.5A
1987–88 FX—Fuse: stop lamp 15A
1986–89 MR2—Fuse: AM 7.5A
1985–86 Supra with A43DE—Fuse: ECU B 15A
1986–89 Supra with A340E—Fuse: No. 1 radio 15A
1985–88 Truck—Fuse: stop lamp 15A
1989–92 Truck—Fuse: EFI 15A

Then reinstall the fuse. The codes should now be cleared. If there are still codes present, either the codes were not properly cleared (Are the codes identical to those flashed out previously?) or the underlying problem is still there (Are only some of the codes the same as previously?).

To clear codes on vehicles that don't appear in the accompanying chart, you'll have to disconnect the computer connector, and leave it disconnected for one minute. Then reconnect the computer.

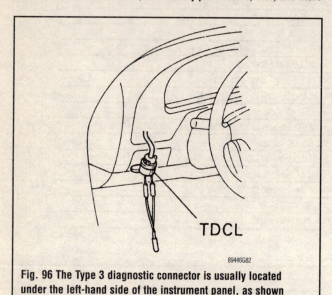

TDCL

89446G82

Fig. 96 The Type 3 diagnostic connector is usually located under the left-hand side of the instrument panel, as shown

DIAGNOSTIC TROUBLE CODES—TOYOTA

Code	Component/Circuit Fault
38	ATF temperature sensor or circuit
42	Speedometer Vehicle Speed Sensor (VSS) or circuit
44	Rear transfer case speed sensor or circuit
61	Transmission/transaxle VSS or circuit
62	No. 1 shift solenoid or circuit
63	No. 2 shift solenoid or circuit
64	Lock-up solenoid or circuit
65	Transfer case solenoid or circuit
73	No. 1 center differential control solenoid or circuit
74	No. 2 center differential control solenoid or circuit

89446G58

Fig. 97 Toyota Diagnostic Trouble Codes

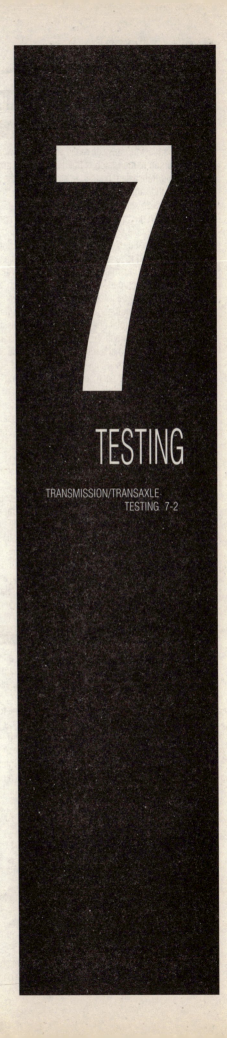

7

TESTING

TRANSMISSION/TRANSAXLE TESTING

Introduction

Any transmission/transaxle repair shop owner will tell you that the most valuable person in their shop is the diagnostician and the transmission/transaxle rebuilder. In some cases, the rebuilder and the diagnostician is the same person. Without this person, their would be no transmission/transaxle shop. The diagnostician is capable of pinpointing malfunctions. In our application, the diagnostician pinpoints problems and road tests the vehicle before and after the repair work has been completed. Sometimes the malfunction can be determined during the road test, otherwise further testing is required. A good diagnostician is familiar with the internal parts of a transmission/transaxle and fully understands the design and operation of each part. Without this knowledge, testing the transmission/transaxle would be totally useless. By now, you should be somewhat familiar with the internal parts of the transmission/transaxle and the contribution of each to the total operation of the transmission.

Their have been many horror stories told by unsatisfied customers about money and time wasted at a transmission/transaxle shop only to have the same problem occur when they drove their vehicle home. In many cases, the problem could have been corrected by replacing a vacuum hose or adjusting the linkage or throttle cable. On customer I know had taken his vehicle to no less than three different shops in order to repair an intermittent improper shift problem. His transmission/transaxle would shift erratically at times and be fine at other times. The problem turned out to be a crack in the vacuum hose between the intake manifold and the vacuum modulator. This problem was repaired by installing a new vacuum hose at a cost of $1.99 plus tax. Of course, not all transmission/transaxle problems are this inexpensive to repair.

A point to remember is that transmission/transaxle problems must be diagnosed dynamically while still in the vehicle. Once the unit is removed, the only way to find a problem is to disassembly the transmission/transaxle and carefully measure and examine each check-ball, seal, piston, clutch disc, solenoid, modulator and the list goes on. This method may take about a week which would prove very unproductive. On the other hand, if during the road test, the technician noticed that second gear was slipping, he would investigate the intermediate clutch and the related hydraulic circuits. A you can understand, this method would save time and money.

Safety Precautions

The same rules apply when testing and repairing automatic transmission/transaxle and transaxles as when working on or around any other automotive equipment. Proper clothes, work habits, and shop practices should be observed. Tools and test equipment should be used correctly and only in the manner that they were designed for as outlined by the tool manufacturer.

Special attention should be given when jacking or hoisting a vehicle. No one should ever be under a vehicle while it is being raised or lowered. Never go under a vehicle supported only by a jack. A jack is only designed to raise a vehicle, it is not a stable support if you intend to push or pull on components in the vehicle as during diagnosis or repair. Always place jackstands under the correct support points on the vehicle as described by the vehicle manufacturer.

Be sure the work area is clean and uncluttered. Always clean grease and oil off the floor immediately with oil dry absorbent or degreaser. Jack handles, hoses and tools should not be left around the work area carelessly. This is just inviting trouble to occur.

Other points to remember whenever working on a automatic transmission/transaxle are:

• Always turn the engine **OFF** when connecting test equipment such as tachometers and timing lights.

• Take care when working near hot engine, exhaust and transmission/transaxle parts. These parts are very hot especially after a road test.

• Automatic transmission/transaxle fluid can become very hot. Observe caution when disconnecting lines or removing the pan.

• Transmission/transaxle s are very heavy. Use the correct equipment when removing or installing them.

• Handle transmission/transaxle parts carefully. Many parts are machined to close tolerances and may be damaged easily and require replacement.

• Never drop transmission/transaxle parts. Carefully place them on a clean workbench and be sure they will not roll off.

• Absolute cleanliness of all parts is essential when working on automatic transmissions and transaxles. A small particle of dirt, lint or metal may require the removal and repair of the unit.

• The best way to avoid personal injury and damage to equipment is to use common sense, if you think you are doing something you shouldn't, you probably are.

➡Although this book contains many valuable and useful procedures, it is not meant to be a substitute for the service manual written specifically for your vehicle. The service manual will provide you with the specifications and exact procedures needed to diagnose and repair your specific transmission/transaxle .

Vacuum Testing

DESCRIPTION

▶ **See Figures 1 and 2**

As you remember, the throttle valve circuit provides the control pressure that represents engine load. Throttle valve pressure is low at engine idle or under light load. As the throttle is opened and engine load increases, the throttle valve pressure rises. If the throttle valve pressure is not correct, the transmission/transaxle will not shift properly and the main line pressure could be either high or low.

There are two types of throttle valve actuating systems in currently in use. One design uses a cable that connects the throttle linkage to the throttle valve in the transmission. The other system uses engine vacuum to power a vacuum modulator which moves the throttle valve. Some transmissions use both throttle cable and modular valve connected to engine vacuum.

Good engine vacuum is highly dependent upon the mechanical condition of the engine. An engine with burnt valves, worn camshaft, bad rings, broken rocker arms and/or air (vacuum) leaks through the intake manifold or throttle body gaskets, will not produce a good vacuum signal. In addition, worn or malfunctioning ignition system components will also contribute to a bad vacuum signal.

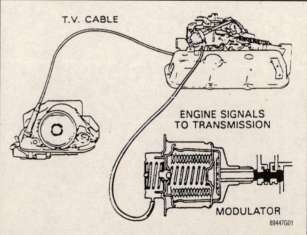

T.V. CABLE

ENGINE SIGNALS TO TRANSMISSION

MODULATOR

89447G01

Fig. 1 Some transmissions/transaxles such as the GM THM 4T60 use both modulator and throttle valve cable to control shift points

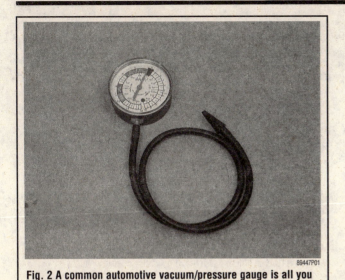

Fig. 2 A common automotive vacuum/pressure gauge is all you need to test for correct vacuum to the modulator assembly

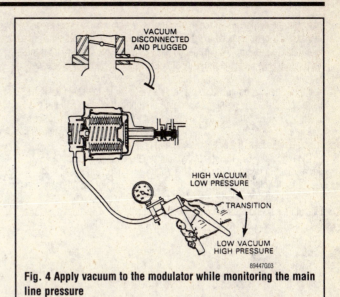

Fig. 4 Apply vacuum to the modulator while monitoring the main line pressure

This test should not be conducted until engine performance has been restored to the factory specifications.

TESTING

▶ See Figures 3, 4, 5 and 6

If the vehicle is equipped with a vacuum modulator, the first thing to check is the vacuum at the modulator. Tee a vacuum gauge into the vacuum line between the vacuum source and the modulator, then take a reading. If necessary, to gain access to the vacuum line, raise the vehicle and support it with jackstands so that a creeper can be used to reach the underside of the transmission. Avoid going under the vehicle when the engine is running but if you must, use extreme caution and be aware of all moving parts. The vacuum reading should be the same as a reading taken directly from the intake manifold. The standard vacuum readings should read between 14–22 in. Hg if taken at sea level. For every 1000 ft of elevation above sea level, subtract 1 in. of vacuum from the standard amount. If the vacuum is low, correct the problem and proceed with the test.

Disconnect the vacuum line from the modulator and plug the vacuum line. Attach the hand vacuum pump to the modulator. Pump the vacuum pump until the gauge reads the same amount of vacuum as was indicated during the first part of the test. Block the wheels and set the parking brake,

MODULATOR SYSTEM EVALUATION CHART		
VACUUM @ MOD (IN/Hg)	LINE PRESSURES	
	*KPA	*PSI
0	1145	166
2	1020	148
4	895	130
6	770	112
8	645	94
10	520	75
12	450	65
14	450	65
*(+ or − 35 KPA)/(+ or − 5 PSI)		

Fig. 5 Typical mainline pressure changes as vacuum is applied to the modulator

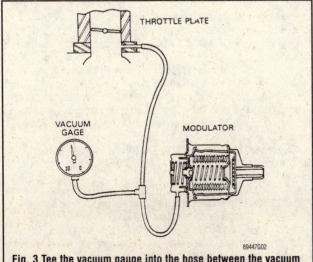

Fig. 3 Tee the vacuum gauge into the hose between the vacuum source and the modulator assembly as shown

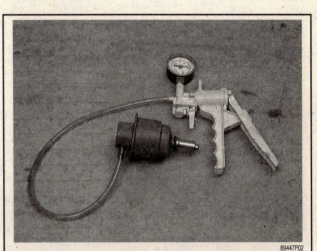

Fig. 6 The modulator can be bench tested by using a hand held vacuum pump as show. Replace the unit if it does not hold the vacuum.

then start the engine and place the transmission/transaxle in drive range and raise the engine speed to 1000 rpm. while watching the mainline pressure (refer to pressure testing) gauge, slowly release the vacuum from the hand pump. The mainline pressure should rise as the vacuum is released and then lower as the vacuum is applied to the modulator again.

If the vehicle is equipped with a throttle valve cable, the same basic test can be performed by running the engine at around 1000 rpm and pulling on the throttle valve cable to force the valve open. This rise in throttle valve pressure should cause the mainline pressure to rise also. If the mainline pressure increases, the system is working properly.

The modulator can also be tested individually by applying about 15 in. Hg. of vacuum to it with the hand held pump and checking to see if the vacuum will hold. If, while testing the vacuum modulator with the vacuum pump, you found that the modulator would not hold a vacuum, the unit has a vacuum leak and must be replaced. If the inside of the vacuum hose and its connection on the vacuum modulator are wet with ATF, the modulator must be replaced.

On vehicles using cables, the free operation of the cable should be checked and any problems corrected before adjusting the cable. The adjustment procedure is covered in Chapter 4.

Road Test

Refer to Section 5 for road testing procedures

Stall Test

DESCRIPTION

▶ See Figure 7

The stall test is used to evaluate the torque converter and the holding power of the bands and clutch packs within the transmission/transaxle . As stated earlier, engine performance must be within factory specifications in order to get the proper results from this test. This test will be performed with the engine running and the transmission/transaxle in gear. Be sure to set the parking brake and chock the wheels to prevent the vehicle from moving.

✳✳ WARNING

Some vehicle manufacturers do not recommend this test due to the high amount of stress placed of the transmission/transaxle. Damage to the unit may occur as a result of this test.

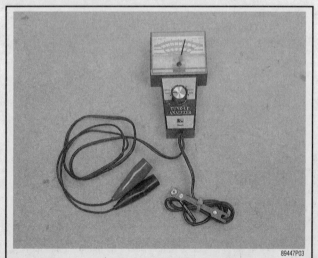

Fig. 7 Mark the maximum and minimum, if available, rpm ranges on the face of an analog meter as seen here

89447P03

The only test equipment needed to perform a stall test is the proper rpm specifications and tachometer. The tachometer should be installed according to instructions and placed where it can be seen from the driver's seat. The specifications for the stall test are usually stated as an rpm range with minimum and maximum amounts. After finding the specification, the maximum rpm value should be marked on the face of the tachometer with a grease pencil for easy reference. If a minimum rpm level is given, mark it also on the face of the meter.

Depending on the make and model of the transmission/transaxle being tested, the following tests may be made during this procedure.

• **D** position—forward clutch, one-way clutch in the transmission/transaxle , stator one-way clutch and engine power output.

• **2** position—forward clutch, one-way clutch in the transmission/transaxle , stator one-way clutch, intermediate band and servo (some models) and engine power output.

• **1** position—forward clutch, one-way clutch in the torque converter and engine power output.

• **R** position—reverse and high clutch, low and reverse band and servo or low and reverse holding clutch and the stator one-way clutch.

TESTING

▶ See Figures 8, 9 and 10

All safety precautions should be adhered to since this test is performed at Wide Open Throttle (WOT) with the gas pedal to the floor of the vehicle.

1. Check the level of the engine oil, coolant and automatic transmission/transaxle fluid. Add fluids as needed.

2. Connect the tachometer to the engine following the manufacturers instructions. Be sure to mark the face of the meter for easy reference.

3. Start the engine and allow it to reach normal operating temperature.

4. Apply the parking brake and chock the front and rear wheels of the vehicle being tested. Don't allow anyone to stand behind or in front of the vehicle.

5. Set the selector lever to the range (gear) to be tested.

✳✳ WARNING

If during the test, the engine speed exceeds the specified rpm, release the pedal immediately as transmission/transaxle slippage is indicated,

6. Press the accelerator pedal to the floor and note the maximum rpm on the tachometer. Release the pedal as soon as the reading is taken. Do not hold the pedal down for more than five seconds or damage to the transmission/transaxle may result due to overheating the fluid and components.

7. After each individual test, shift the transmission/transaxle into neutral and run the engine at about 1000rpm for at least 15 seconds to allow the torque converter to cool down before the next test.

8. Place the selector lever in each range and repeat the test using the previous steps. Record the rpm readings in each range using a chart as illustrated.

Compare the results to the specifications for your specific transmission/transaxle . Use the following as a guide to help isolate the problem.

• Stall speed is higher than normal—the transmission/transaxle is not holding due to a slipping friction element such as clutches or bands. Aerated transmission/transaxle fluid and low operating pressure can also be suspected.

• Stall speed is extremely high and runs away—suspect a slipping clutch and/or low operating pressure again. Don't rule out the possibility of sheared turbine splines on the input shaft.

• Stall speed is 100–200 rpm lower than normal—This condition may indicate that the engine is not operating to its full potential. Inspect the ignition system, install new spark plugs and ignition wires if necessary. If the results are still the same, inspect the mechanical condition of the engine by performing a compression test and/or a cylinder leak-down test the find the cause of low power output from the engine.

• Stall speed is about ⅓ lower than normal—Suspect a free-wheeling torque converter stator. During a stall test, the torque converter should be

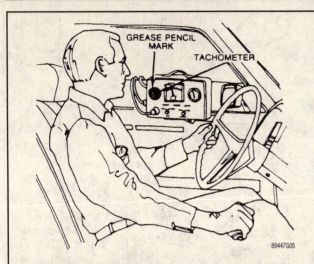

Fig. 8 Observe the tachometer from inside the vehicle while performing the stall test

Range	Specified Engine RPM	Record Actual Engine RPM
D		
2		
1		
R		

Results _____

Fig. 9 Use a chart like this one to record the results of the stall test

SELECTOR POSITIONS	STALL SPEED(S) HIGH (SLIP)	STALL SPEEDS LOW
D only	Low (planetary) one-way clutch	
D, 2 and 1	Forward clutch	1. Does engine misfire or bog down under load? Check engine for tune-up. If OK . . .
All driving ranges	Perform control pressure test	2. Remove torque converter and bench test for reactor one-way clutch slip.
R Only	Reverse and high clutch or low-reverse band or servo	

Fig. 10 Common stall test speed results

producing full torque transfer and the stator should be in the locked position. If the one-way roller clutch fails to take hold, the fluid action works against the impeller and engine rotation. The engine power transfer is drastically reduced and vehicle performance will be very poor during acceleration from a standing start position.

• Stall speed is normal—This condition indicates that the engine power is normal and the transmission/transaxle and torque converter is holding. However, it does not prove that the stator can release and permit the torque converter to act as a fluid coupling.

If the top speed of the vehicle at Wide Open Throttle (WOT) is greatly reduced, usually by about 1/3, a frozen stator can be suspected. Low speed acceleration will be normal, but cruising speed will be hampered by the stator if it does not release. The frozen stator acts as a dam or brake against the rotating force of the fluid. Within about 15–20 miles (24–32km), the fluid will overheat and the components in the transmission/transaxle may begin to slip and loose holding power.

To further test the torque converter stator, place the transmission/transaxle in **Neutral** and press the accelerator to WOT. If the engine speed easily passes 3,000 rpm, the stator is not frozen. If the engine can't exceed 3,000 rpm, the stator is frozen and the torque converter should be replaced.

Hydraulic Pressure Testing

DESCRIPTION

▸ **See Figures 11 and 12**

Although the road testing provides the means to identify which area in the transmission/transaxle is causing a problem, hydraulic pressure testing may be required to verify the diagnosis and isolate the problem further. If the road test yields conclusive results, pressure testing is not required. The number of hydraulic circuits that can be tested varies between different makes and models of transmissions/transaxles. Generally, all transmissions can be tested for line (control) pressure, including pressure increase and decrease during the various operating modes. Depending of the model, the usual circuits that may are measured are:

• Mainline pressure
• Governor pressure
• Throttle valve pressure
• Cooler line pressure
• Individual clutch or band applied pressure

Since hydraulic pressure is the power source used to apply the transmissions clutches and bands, much information can be gained about trans-

Fig. 11 Left side of Ford AOD transmission/transaxle showing the hydraulic fluid test ports

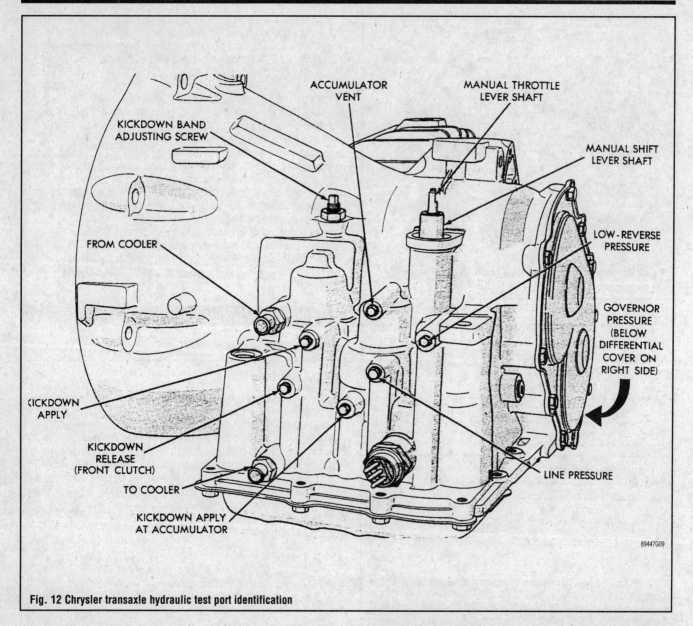

KICKDOWN BAND
ADJUSTING SCREW

ACCUMULATOR
VENT

MANUAL THROTTLE
LEVER SHAFT

MANUAL SHIFT
LEVER SHAFT

LOW-REVERSE
PRESSURE

FROM COOLER

GOVERNOR
PRESSURE
(BELOW
DIFFERENTIAL
COVER ON
RIGHT SIDE)

KICKDOWN
APPLY

KICKDOWN
RELEASE
(FRONT CLUTCH)

TO COOLER

LINE PRESSURE

KICKDOWN APPLY
AT ACCUMULATOR

89447G09

Fig. 12 Chrysler transaxle hydraulic test port identification

mission/transaxle condition and operation by measuring different circuit pressures.

The normal pressures that should be found in each circuit are listed in a pressure. All automatic transmissions have pressure test points, or taps, located on the case of the transmission. These test points are normally plugged with a small pipe plug which can be removed so that a pressure gauge can be attached. Each test point has an internal passageway connecting to a specific hydraulic circuit inside the transmission.

In most cases you do not need to take pressure readings at all the test points, but only at the circuits that are involved in the problem. The main line pressure should be checked first since it is the hydraulic power source. If the main line pressure is low, then all the other pressures will be low also. Remember that low pressure reduces the holding force on the clutches and bands which will allow them to slip.

TESTING

▶ **See Figures 13 thru 29**

The equipment that is needed to pressure test a transmission/transaxle will vary according to transmission/transaxle design. The basic test equipment that is used is a 0 to 50-psi pressure gauge, a 0 to 100-psi pressure

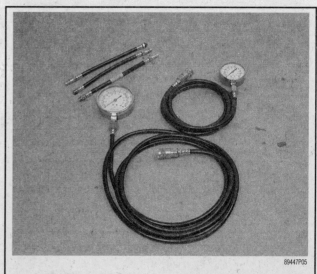

89447P05

Fig. 13 Typical fluid pressure test equipment including adapters

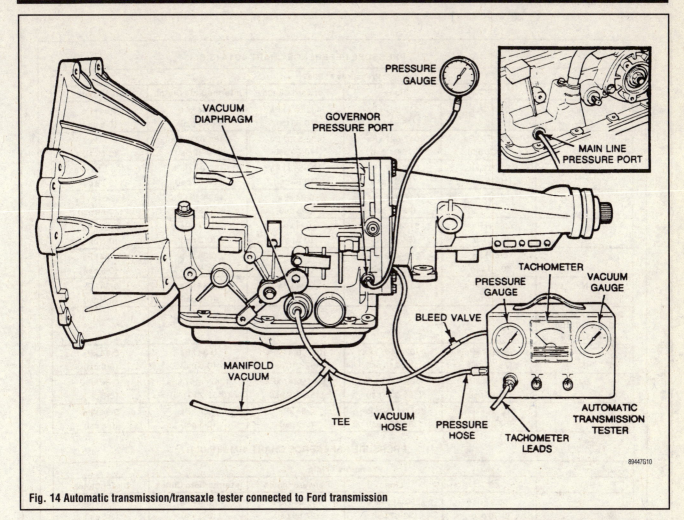

Fig. 14 Automatic transmission/transaxle tester connected to Ford transmission

gauge, a 0 to 350-psi pressure gauge, a hand-held engine tachometer, a vacuum gauge, and a hand held vacuum pump. The various gauges need the proper-length hoses and proper adapters to connect to the transmission.

To simplify the job of transmission/transaxle pressure testing, special test equipment is available. The testers include the same individual gauges mentioned earlier all contained in one unit. Instructions for using the tester are given by the manufacturer. Connecting the gauges to the proper test points is critical, so follow the manufacturer's instructions to the letter. A typical installation of a transmission/transaxle tester on a Ford C3 transmission is shown in the illustration.

Test procedures for all automatic transmissions/transaxles are not the same. Always refer to specifications for your transmission/transaxle . The following is only a general guide to be followed than can be used for all transmissions/transaxles.

    a. Be sure the fluid level is within the full range on the dipstick.

    b. Run the engine and transmission/transaxle until normal operating temperature is reached before testing.

    c. Always check the test equipment for leaks.

    d. Note the exact specifications and test port locations.

    e. Copy the test pressure specification onto the diagnostic check-sheet.

    f.   Install the test equipment according to the manufacturers instructions.

    g. Compare the test results with the specification chart.

Manufacturer diagnostic charts are very useful in interpreting the test results and should always be used when available. As an example, the following pages will illustrate the hydraulic circuits and pressures of the Chrysler 42LE transaxle. If the main line pressure is below specification,

a pump or pressure regulator problem is indicated. The throttle valve operation could also be at fault. The full list of possible problem areas listed in the diagnostic charts must be checked out. A low-pressure reading in only one hydraulic circuit usually indicates an internal leak in that circuit from a torn piston seal or worn sealing ring. During disassembly special attention should be paid to that circuit to find the failed part.

Govenor Testing

DESCRIPTION

Since the governor circuit provides the control pressure that reflects the speed of the vehicle, this test must be performed either during a road test or with the vehicle raised on jack stands.

✳✳ WARNING

Some manufacturers do not recommend testing their FWD transaxles on jack stands because of the extreme U-joint angles present under this condition. Always position the jackstands under the lower control arms to prevent the front wheels and axles from hanging down while rotating.

The only gauge necessary is a pressure gauge with enough capacity to handle the maximum pressure. This gauge is connected to the governor test port on the transmission/transaxle case. Position the gauge so that it is easily visible from the driver's seat, then proceed with the test.

PRESSURE REFERENCE CHART 401A (3.8L)

Gear	EPC	Line	Forward Clutch	Intermediate Clutch	Direct Clutch
Pressures At Idle					
1M 1D	103-172 kPa (15-25 psi)	517-627 kPa (75-91 psi)	482-613 kPa (70-89 psi)	0-34 kPa (0-5 psi)	0-34 kPa (0-5 psi)
2M 2D	103-172 kPa (15-25 psi)	517-627 kPa (75-91 psi)	482-613 kPa (70-89 psi)	482-613 kPa (70-89 psi)	0-34 kPa (0-5 psi)
3	103-172 kPa (15-25 psi)	517-627 kPa (75-91 psi)	482-613 kPa (70-89 psi)	482-613 kPa (70-89 psi)	482-613 kPa (70-89 psi)
4	103-172 kPa (15-25 psi)	517-627 kPa (75-91 psi)	0-34 kPa (0-5 psi)	482-613 kPa (70-89 psi)	482-613 kPa (70-89 psi)
R	0-69 kPa (0-10 psi)	372-661 kPa (54-96 psi)	0-34 kPa (0-5 psi)	0-34 kPa (0-5 psi)	0-34 kPa (0-5 psi)
P	0-69 kPa (0-10 psi)	220-462 kPa (32-67 psi)	0-34 kPa (0-5 psi)	0-34 kPa (0-5 psi)	0-34 kPa (0-5 psi)
N	103-172 kPa (15-25 psi)	517-627 kPa (75-91 psi)	0-34 kPa (0-5 psi)	0-34 kPa (0-5 psi)	0-34 kPa (0-5 psi)
Pressures at Wide Open Throttle (WOT) Stall					
1M 1D	572-641 kPa (83-93 psi)	1102-1447 kPa (160-210 psi)	1047-1447 kPa (152-210 psi)	0-34 kPa (0-5 psi)	0-34 kPa (0-5 psi)
2M	572-641 kPa (83-93 psi)	1102-1447 kPa (160-210 psi)	1047-1447 kPa (152-210 psi)	1047-1447 kPa (152-210 psi)	0-34 kPa (0-5 psi)
R	572-641 kPa (83-93 psi)	1337-1750 kPa (194-254 psi)	0-34 kPa (0-5 psi)	0-34 kPa (0-5 psi)	0-34 kPa (0-5 psi)

PRESSURE REFERENCE CHART 401B (5.0L)

Gear	EPC	Line	Forward Clutch	Intermediate Clutch	Direct Clutch
Pressures At Idle					
1M 1D	0-69 kPa (0-10 psi)	220-462 kPa (32-67 psi)	186-448 kPa (27-65 psi)	0-34 kPa (0-5 psi)	0-34 kPa (0-5 psi)
2M 2D	0-69 kPa (0-10 psi)	220-462 kPa (32-67 psi)	186-448 kPa (27-65 psi)	186-448 kPa (27-65 psi)	0-34 kPa (0-5 psi)
3	0-69 kPa (0-10 psi)	220-462 kPa (32-67 psi)	186-448 kPa (27-65 psi)	186-448 kPa (27-65 psi)	186-448 kPa (27-65 psi)
4	0-69 kPa (0-10 psi)	220-462 kPa (32-67 psi)	0-34 kPa (0-5 psi)	186-448 kPa (27-65 psi)	186-448 kPa (27-65 psi)
R	0-69 kPa (0-10 psi)	372-661 kPa (54-96 psi)	0-34 kPa (0-5 psi)	0-34 kPa (0-5 psi)	0-34 kPa (0-5 psi)
P	0-69 kPa (0-10 psi)	220-462 kPa (32-67 psi)	0-34 kPa (0-5 psi)	0-34 kPa (0-5 psi)	0-34 kPa (0-5 psi)
N	0-69 kPa (0-10 psi)	220-462 kPa (32-67 psi)	0-34 kPa (0-5 psi)	0-34 kPa (0-5 psi)	0-34 kPa (0-5 psi)
Pressures at Wide Open Throttle (WOT) Stall					
1M 1D	572-641 kPa (83-93 psi)	1102-1447 kPa (160-210 psi)	1047-1447 kPa (152-210 psi)	0-34 kPa (0-5 psi)	0-34 kPa (0-5 psi)
2M	572-641 kPa (83-93 psi)	1102-1447 kPa (160-210 psi)	1047-1447 kPa (152-210 psi)	1047-1447 kPa (152-210 psi)	0-34 kPa (0-5 psi)
R	572-641 kPa (83-93 psi)	1337-1750 kPa (194-254 psi)	0-34 kPa (0-5 psi)	0-34 kPa (0-5 psi)	0-34 kPa (0-5 psi)

89447G08

Fig. 15 Sample pressure reference chart for a 1995 Ford Mustang with either the 3.8L or the 5.0L engine

	Judgement	Suspected parts
At idle	Line pressure is low in all position.	• Oil pump wear • Control piston damage • Pressure regulator valve or plug sticking • Spring for pressure regulator valve damaged • Fluid pressure leakage between oil strainer and pressure regulator valve • Clogged strainer
	Line pressure is low in particular position.	• Fluid pressure leakage between manual valve and particular clutch. • For example, line pressure is: — Low in "R" and "1" positions, but — Normal in "D" and "2" positions. Then, fluid leakage exists at or around low and reverse brake circuit. Refer to "OPERATION OF CLUTCH AND BRAKE", AT-9.
	Line pressure is high.	• Mal-adjustment of throttle position sensor • Fluid temperature sensor damaged • Line pressure solenoid valve sticking • Short circuit of line pressure solenoid valve circuit • Pressure modifier valve sticking • Pressure regulator valve or plug sticking • Open in dropping resistor circuit
At stall speed	Line pressure is low.	• Mal-adjustment of throttle position sensor • Control piston damaged • Line pressure solenoid valve sticking • Short circuit of line pressure solenoid valve circuit • Pressure regulator valve or plug sticking • Pressure modifier valve sticking • Pilot valve sticking

89447G12

Fig. 16 Possible causes for abnormal hydraulic line pressure

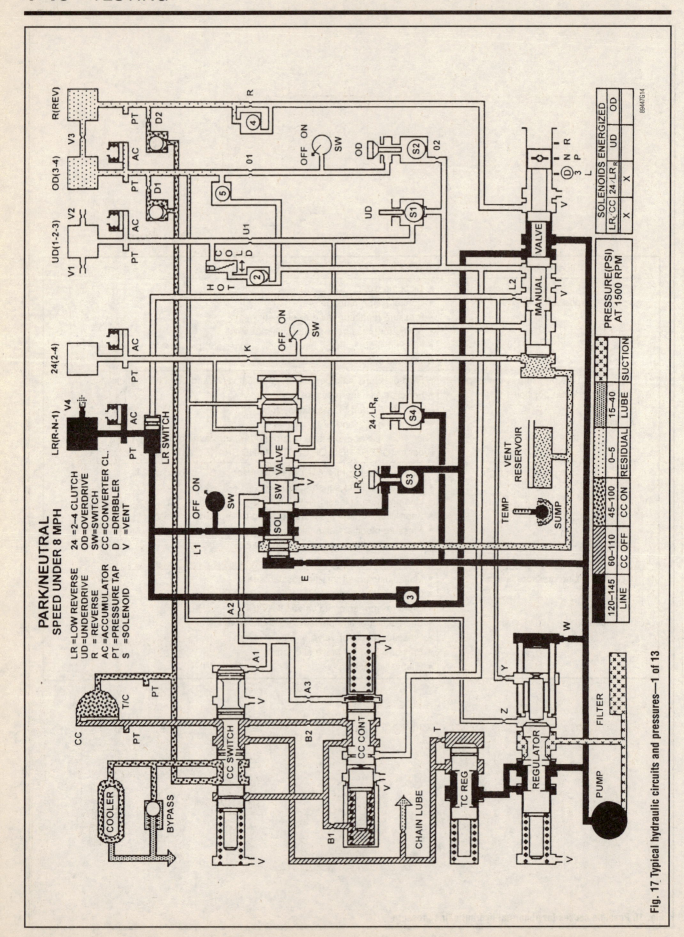

Fig. 17 Typical hydraulic circuits and pressures—1 of 13

PARK/NEUTRAL
SPEED UNDER 8 MPH

LR =LOW REVERSE   24 =2-4 CLUTCH
UD =UNDERDRIVE   OD =OVERDRIVE
R =REVERSE   SW =SWITCH
AC =ACCUMULATOR   CC =CONVERTER CL.
PT =PRESSURE TAP   D =DRIBBLER
S =SOLENOID   V =VENT

SOLENOIDS ENERGIZED			
LR/CC	24/LR$_R$	UD	OD
X	X		
	X	X	

PRESSURE(PSI) AT 1500 RPM					
120–145	60–110	45–100	0–5	15–40	
LINE	CC OFF	CC ON	RESIDUAL	LUBE	SUCTION

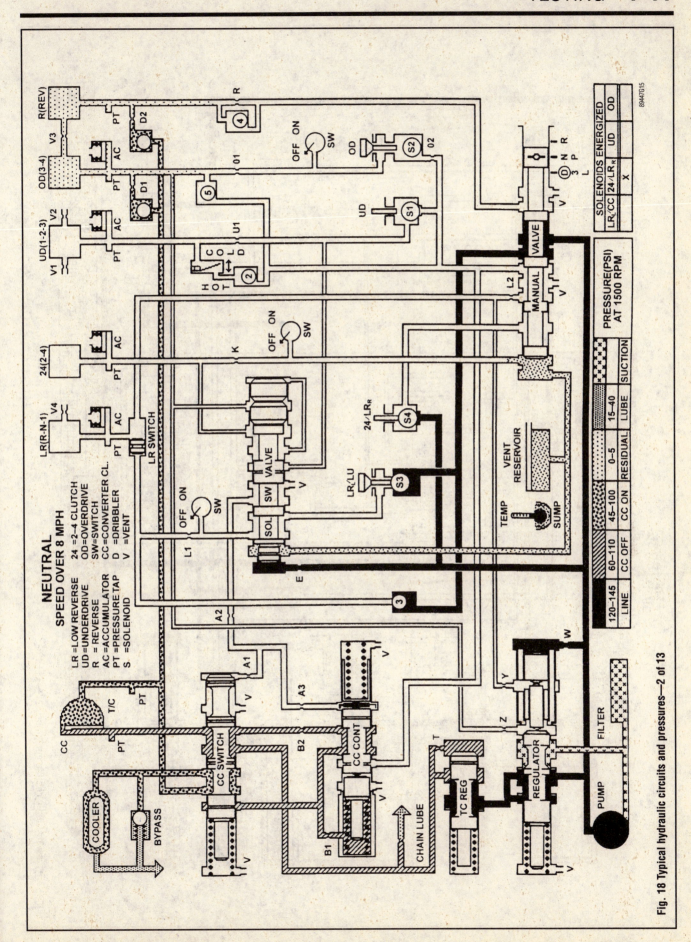

Fig. 18 Typical hydraulic circuits and pressures—2 of 13

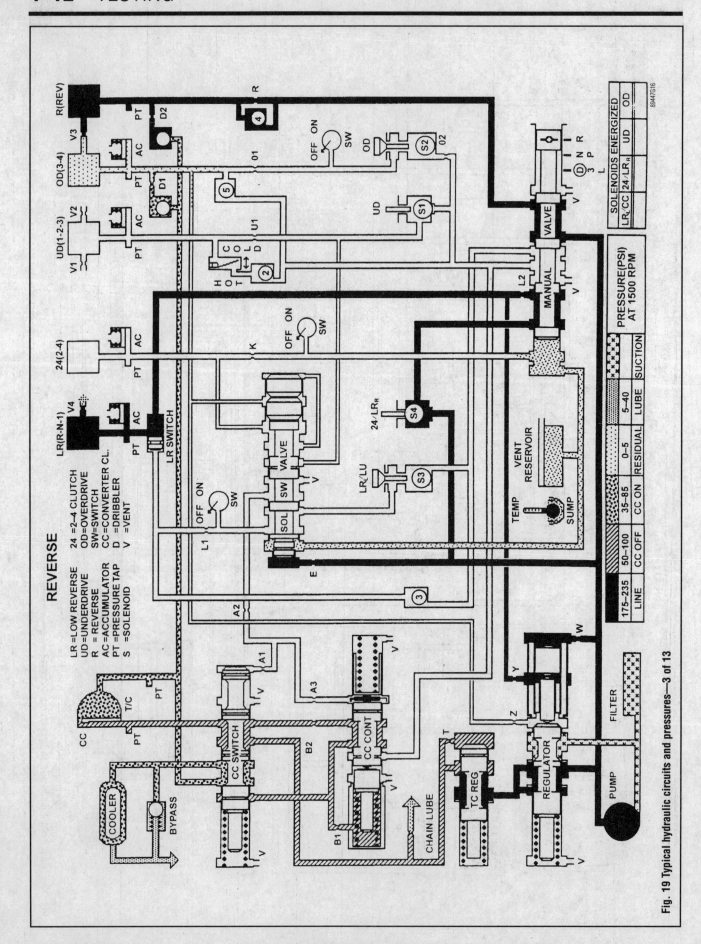

Fig. 19 Typical hydraulic circuits and pressures—3 of 13

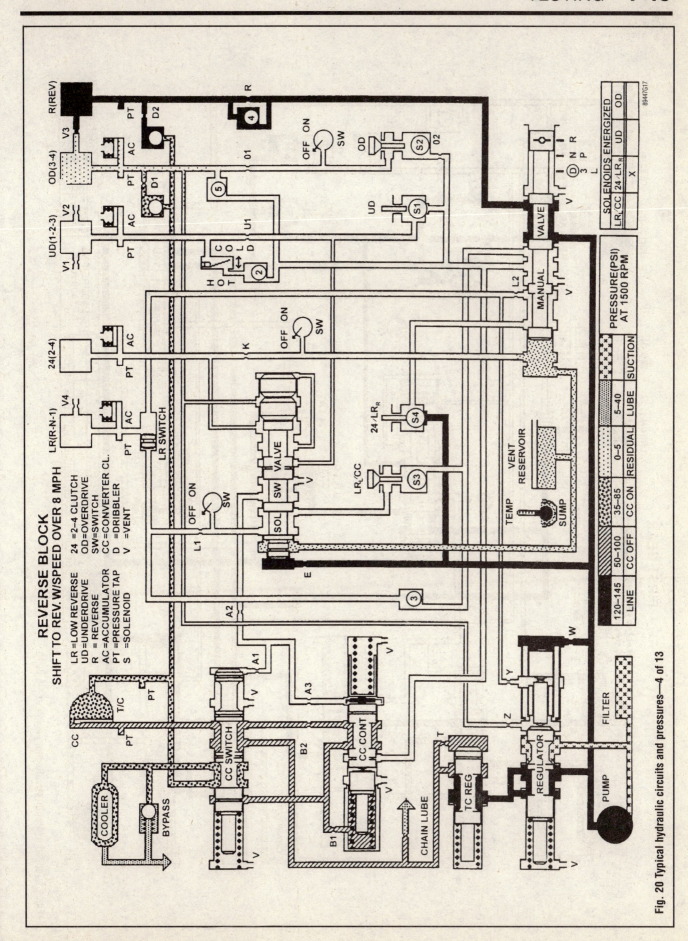

Fig. 20 Typical hydraulic circuits and pressures—4 of 13

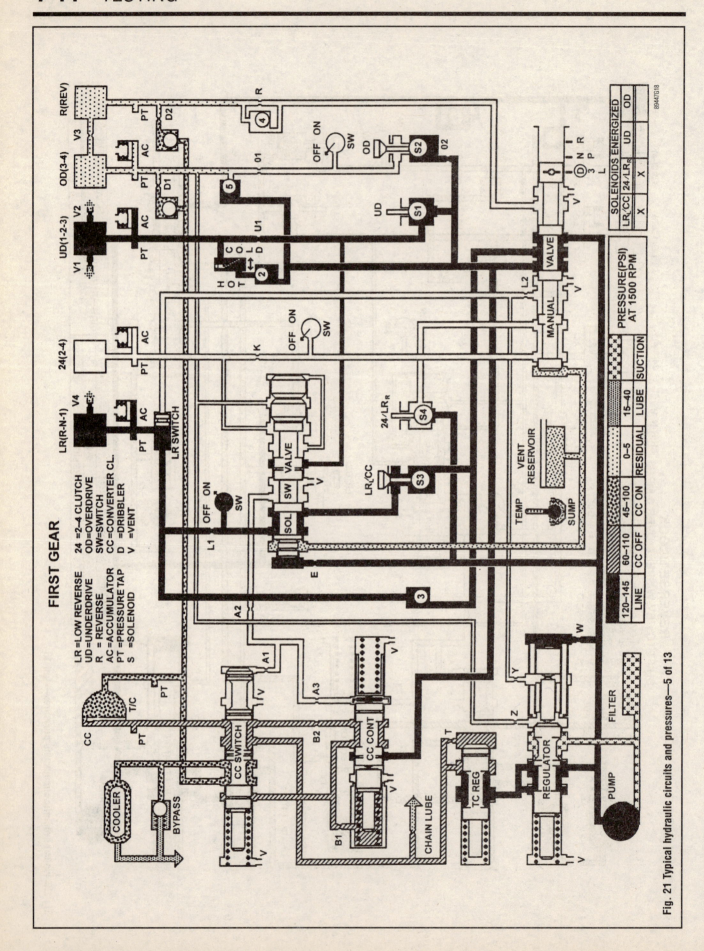

FIRST GEAR

LR =LOW REVERSE
UD =UNDERDRIVE
R = REVERSE
AC =ACCUMULATOR
PT =PRESSURE TAP
S =SOLENOID

24 =2-4 CLUTCH
OD=OVERDRIVE
SW=SWITCH
CC=CONVERTER CL.
D =DRIBBLER
V =VENT

	120-145	60-110	45-100	0-5	15-40		
	LINE	CC OFF	CC ON	RESIDUAL	LUBE	SUCTION	

	PRESSURE(PSI) AT 1500 RPM		

SOLENOIDS ENERGIZED			
LR/CC	24/LR$_R$	UD	OD
X		X	X

8947G18

Fig. 21 Typical hydraulic circuits and pressures—5 of 13

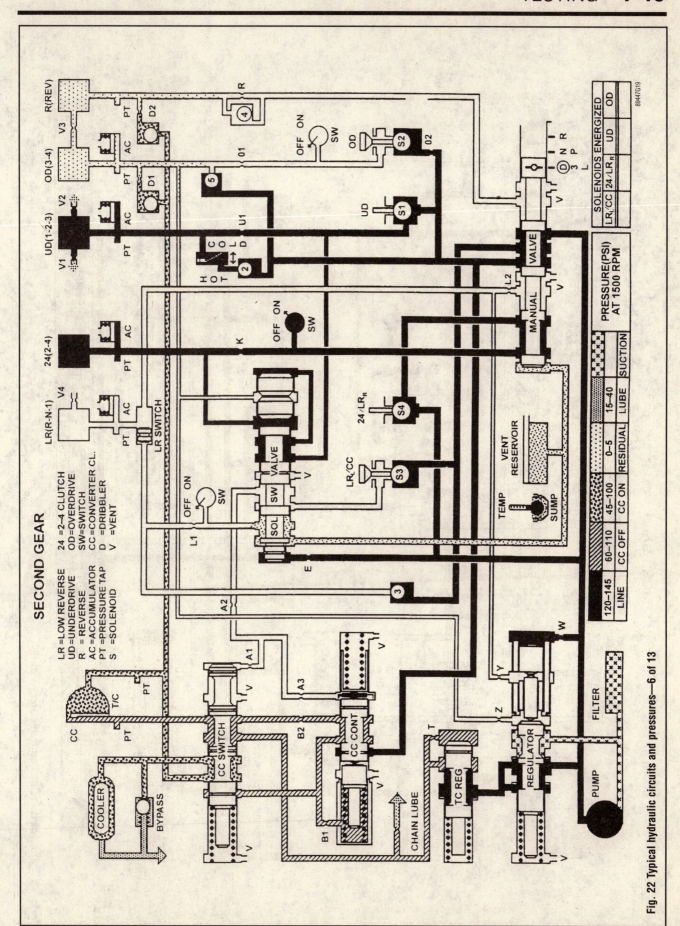

Fig. 22 Typical hydraulic circuits and pressures—6 of 13

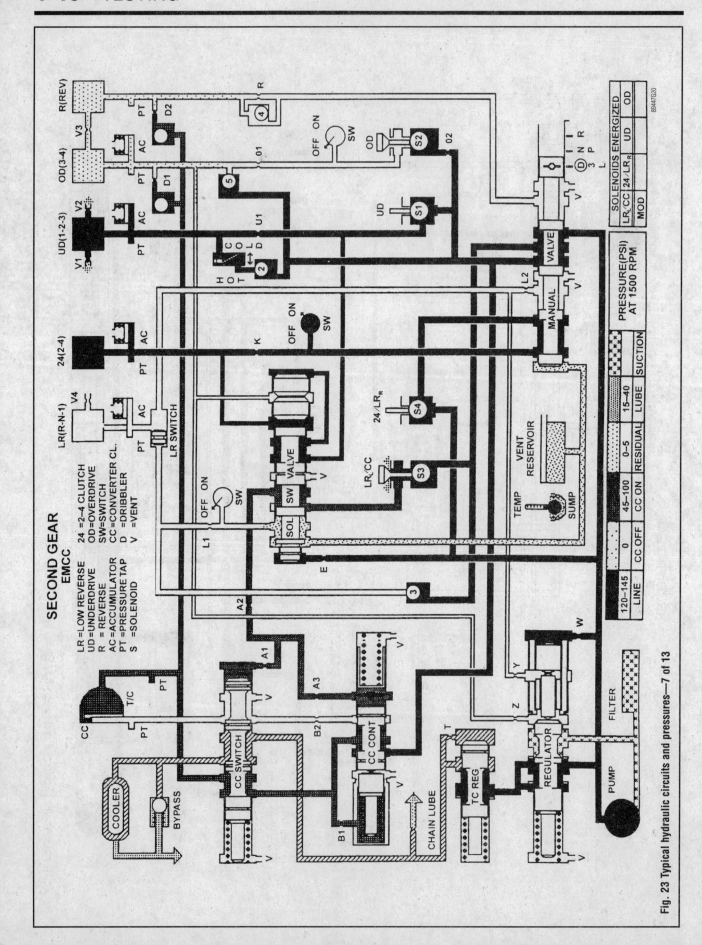

Fig. 23 Typical hydraulic circuits and pressures—7 of 13

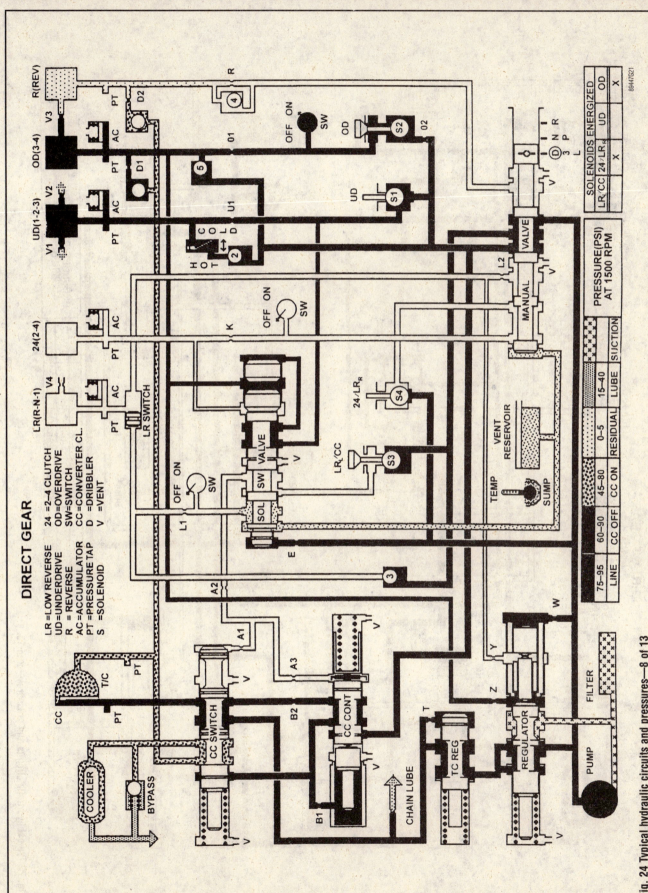

Fig. 24 Typical hydraulic circuits and pressures—8 of 13

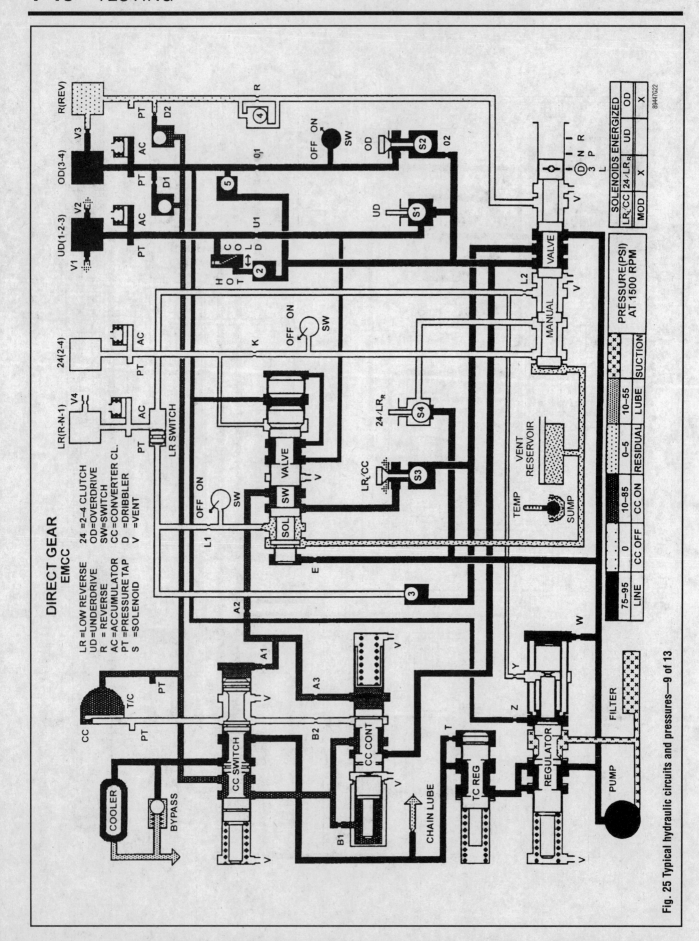

Fig. 25 Typical hydraulic circuits and pressures—9 of 13

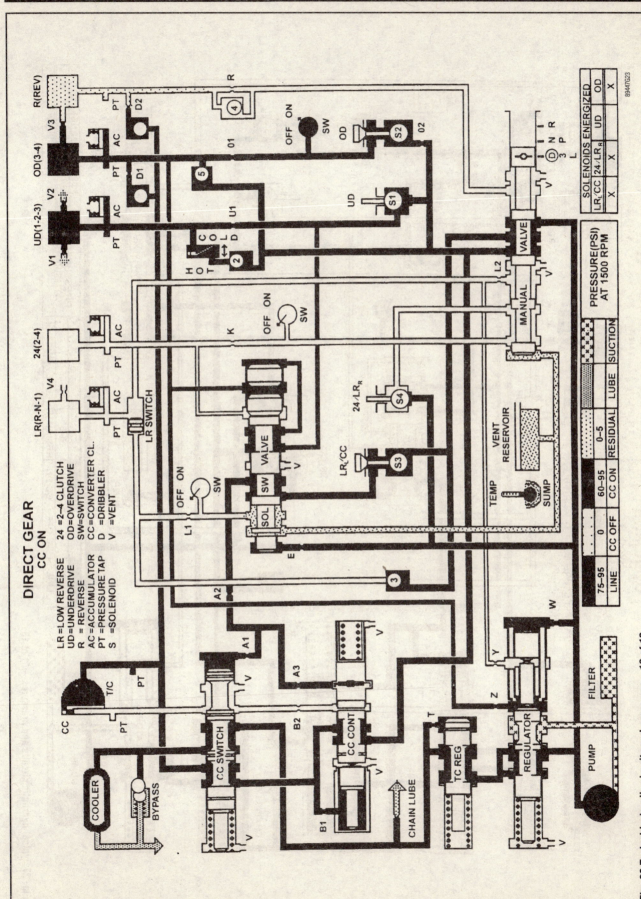

Fig. 26 Typical hydraulic circuits and pressures—10 of 13

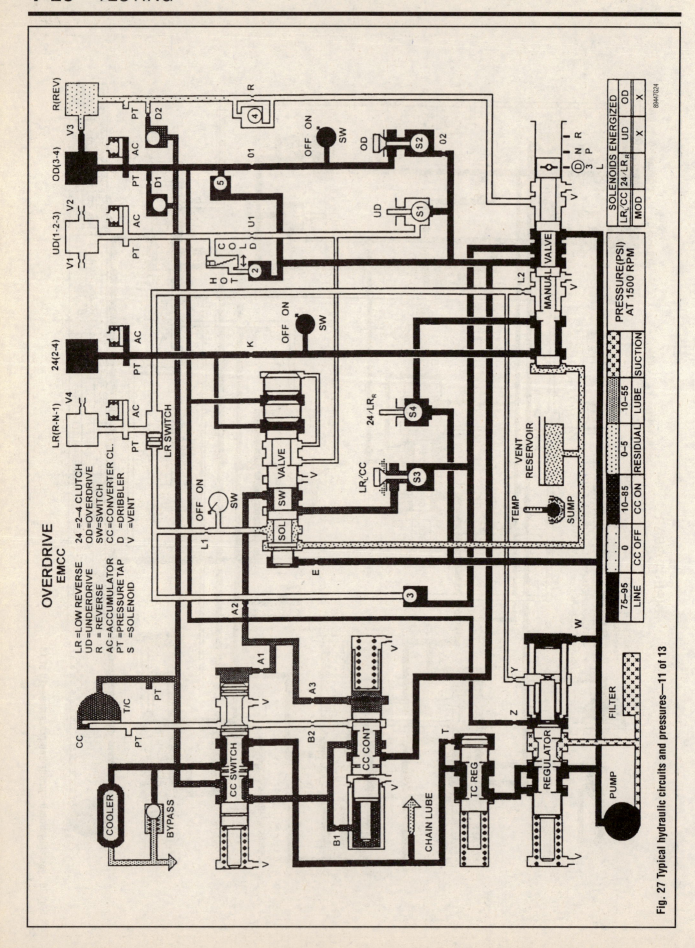

Fig. 27 Typical hydraulic circuits and pressures—11 of 13

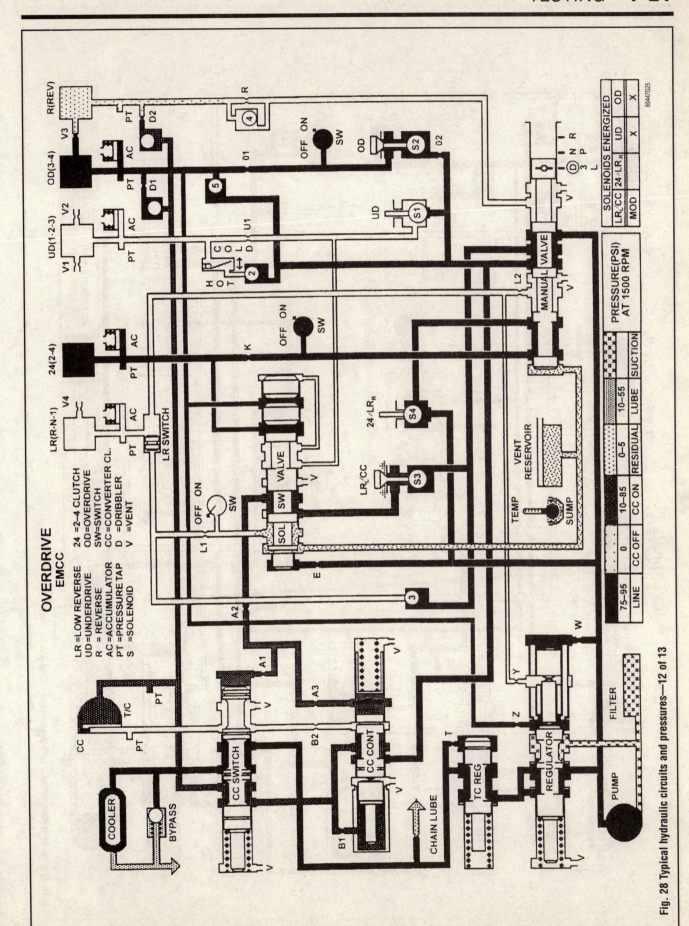

Fig. 28 Typical hydraulic circuits and pressures—12 of 13

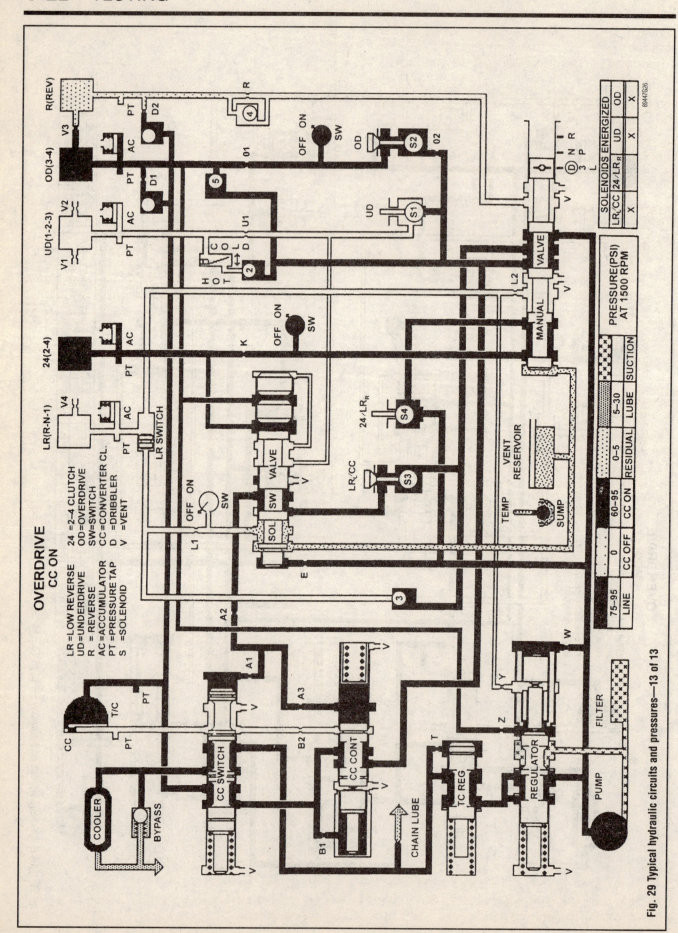

Fig. 29 Typical hydraulic circuits and pressures—13 of 13

TESTING

Slowly increase the vehicle speed while watching the pressure gauge. Look for a smooth increase in pressure as the speed increases and also a smooth decrease as the speed is reduced. Check the manufacturer for exact specifications and procedure for the transmission/transaxle you are testing.

If the governor pressure readings are in specifications with a smooth rise and fall in pressure, the system is operating properly. If the pressure is high or low or has a rough or erratic increase or decrease, the governor must be examined and repaired so that it can move freely. Refer to section 8 for governor service.

Air Pressure Testing

DESCRIPTION

▶ **See Figures 30 and 31**

Even with correct hydraulic fluid pressure, a NO DRIVE condition can exist. Air pressure testing allows the technician to test individual hydraulic circuits by substituting air pressure for hydraulic pressure. Turning the compressed air on and off will cause the piston or servo to cycle back and forth between application and release. By repeating a few of these cycles on each piston and servo, you will test the return spring action as well as the seals. If the piston or seals are in good, a thud will be heard as the piston or servo is applied. It is normal to head a small amount of air escaping. Any loud hiss of air escaping while the air is applied indicates a hydraulic leak in that particular circuit.

➡ **Use only clean, filtered and dried compressed air regulated to no more than 40psi (276kPa) during air pressure testing.**

A number of different types of air nozzles can be used for air testing. The first type uses a piece of curved tubing with the end cut at a 45° angle and the other type of nozzle uses a cone-shaped rubber tip. An air pressure regulator is also needed to control the maximum air pressure used for the tests.

Some transmissions require the use of an air test plate, which bolts to the transmission/transaxle case in the same position as the valve body. This plate seals the hydraulic circuits in the case, which are exposed when the valve body is removed. An air test plate for a Ford AOD-E transmission is shown in the illustration. Air test plates are not interchangeable. You must obtain the correct test plate for the transmission/transaxle you are testing.

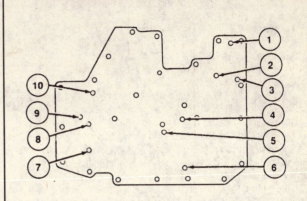

Item	Description
1	Converter Bypass
2	1-2 Accumulator Apply
3	Intermediate Clutch
4	Overdrive Servo Release
5	Reverse Servo
6	Overdrive Servo Apply
7	2-3 Accumulator Bottom
8	2-3 Accumulator Top
9	Forward Clutch
10	Direct Clutch

89447G11

Fig. 31 Air test plate and port identification designed specifically for the Ford AOD-E transmission

TESTING

▶ **See Figure 32**

The procedure and port locations will be different depending on the model of transmission/transaxle being tested. Check with the manufacturer for specific test procedures for the transmission/transaxle being tested. For most transmissions the air test procedures should include the following sub-steps:

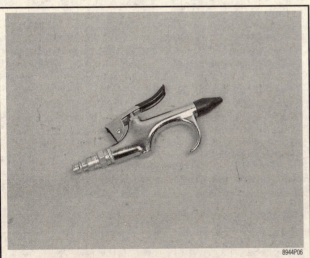

8944P06

Fig. 30 Air nozzle with a cone-shaped rubber tip—the tip seals the nozzle against the port or test plate

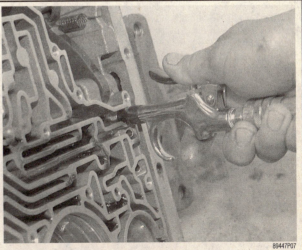

89447P07

Fig. 32 Air pressure is being applied to a clutch circuit in a Ford AOD transmission

1. Note the manufacturer specifications and procedures.
2. Support the vehicle on jack stands.
3. Drain the fluid and remove the pan.
4. Remove the valve body and separator plate assembly.
5. If necessary, install the air test plate and gasket.
6. Adjust the pressure regulator to the proper pressure.
7. Apply air pressure to the test point.
8. Listen for the "thud" of the piston as the air pressure is applied.
9. Listen for the hissing noise of an air leak.
10. Cycle the air on and off a few times to test the return spring action.
11. Write down the test results.

A successful air pressure test will apply the piston with a dull thud and there will be very little air loss with no loud hissing sounds. A servo will not make the same thud, but its operation can be watched as the band is applied. Any loud hissing sound indicates a leak in that particular hydraulic circuit. All of the seals, sealing rings, check balls and gaskets in that circuit must be inspected to locate the leak.

Fluid Cooler Flow Testing

DESCRIPTION

There are two basic tests used to determine if the ATF cooler is functioning properly. A flow test and a pressure test are used to provide information about cooler and transmission/transaxle operation. If the ATF cooler becomes clogged for any reason, the result will be reduced flow and less pressure in the lubrication circuit of the transmission. The reduced flow will also raise the operating temperature of the fluid and greatly decrease the life of the fluid and the transmission/transaxle.

With the development of four speed transmissions/transaxles, good fluid flow is even more critical due to the reduced speed of the fluid pump. Modern vehicles usually cruise at about 1500–1800 rpm at highway speeds. The speed of the pump is in direct relation to the speed of the engine. In other words, the speed of the fluid pump in today's vehicles is considerably slower than the speed of the pumps in vehicles a decade ago.

TESTING

To measure the transmission/transaxle cooler flow rate, you will need a quart container, a stopwatch, and a short piece of tubing. To perform the pressure test, a low pressure gauge and the necessary fittings to tee into the cooler lines will be needed. Check with the manufacturer to help select the gauge with the correct range.

The flow test is performed by disconnecting the cooler return line fitting at the transmission. The short piece of hose is slipped over the end of the tubing so that the flow can be directed into the container. First add two quarts of ATF to the transmission, let the engine run for 30 seconds, and then turn the engine **OFF**. Next, check the amount of fluid in the container. The normal amount for General Motors transmissions is approximately three quarts, with Chrysler and Ford transmissions flowing about two quarts. The minimum acceptable flow is one-half the normal flow rate. As always, check with the manufacturer for specific information about the vehicle that is being tested. Once the flow test is completed, the adapter fitting for the low-pressure (0 to 50) gauge is installed between the transmission/transaxle and return cooler line. With the gauge in place, you can read the cooler return line pressure.

If the flow test shows insufficient flow, the cooler is partially blocked and must be back flushed and blown out with low-pressure (30 psi) compressed air. After cleaning the cooler, another flow test should be performed to verify the repair. If the cooler remains blocked, it should be repaired at a radiator shop or replaced. If desired, an auxiliary cooler can be purchased and installed at home for no more than 50 dollars.

The pressure test can have a wide range of results, but as a general rule if the pressure is below 10 psi, the cooler could be restricted. The manufacturer's specifications must be used to provide the normal pressure range.

Any fluid restriction through the cooler will cause the fluid temperature to increase subsequently shortening the service life of the transmission. All fluid cooler problems must be repaired as soon as possible after they are found.

Torqe Converter Clutch Testing

DESCRIPTION

Diagnosing a torque converter clutch problem can be a big task. Some of the common complaints that customers have are total lack of converter clutch application, converter clutch will not disengage and stalls the engine when stopping, early or late application of the converter clutch, early or late release of the converter clutch, converter clutch shutter during application, or under acceleration. Because the torque converter clutch systems are so complex, there are many different areas that must be tested. Four of the most common areas to check are:

- Computerized engine controls
- Transmission/transaxle electrical/electronic system
- Torque converter hydraulic system
- Mechanical condition of the engine

Some of the more common equipment necessary for diagnosis and repair of the computerized engine control system are:

- Vacuum gauge
- Scantool
- Digital volt-ohmmeter (high impedance)
- Tachometer
- Test light
- Jumper wires
- Vacuum pump

Basic mechanical engine evaluation tools may includes the following

- Electronic engine analyzer
- Compression gauge
- Cylinder leakage gauge
- Vacuum gauge

To diagnose the electronic transmission/transaxle control system, the following tool may be needed:

- Analog volt-ohmmeter (low impedance)
- Testlight
- Jumper wires
- Special electrical test equipment

TESTING

Testing the torque converter clutch hydraulic system will require a 0 to 50-psi pressure gauge and adapter fittings. The sheer amount and type of test equipment listed above should indicate the complexity of diagnosing the torque converter clutch system.

Because of the many different types and designs of converter clutch systems, it is best to rely on the manufacturer for the exact tests and procedures recommended for each type of transmission/transaxle . A thorough knowledge of the operation of the particular converter clutch system being tested is also essential.

The test data collected from the four test areas are used with the diagnostic to pinpoint the malfunction. Because of the design differences between systems, only the recommended test procedures and diagnostic charts should be used for diagnosis. By following the factory recommendations closely, the difficult job of diagnosing converter clutch problems accurately is simplified.

Electronic/Electrical Testing

UNDERSTANDING ELECTRICITY

For any electrical system to operate, there must be a complete circuit. This simply means that the power flow from the battery must make a full circle. When an electrical component is operating, power flows from the battery to the component, passes through the component (load) causing it to function, and returns to the battery through the ground path of the circuit. This ground may be either another wire or a metal part of the vehicle (depending upon how the electrical circuit is designed).

Basic Circuits

▶ See Figure 33

Perhaps the easiest way to visualize a circuit is to think of connecting a light bulb (with two wires attached to it) to the battery. If one of the two wires was attached to the negative post (-) of the battery and the other wire to the positive post (+), the circuit would be complete and the light bulb would illuminate. Electricity could follow a path from the battery to the bulb and back to the battery. It's not hard to see that with longer wires on our light bulb, it could be mounted anywhere on the vehicle. Further, one wire could be fitted with a switch so that the light could be turned on and off. Various other items could be added to our primitive circuit to make the light flash, become brighter or dimmer under certain conditions, or advise the user that it's burned out.

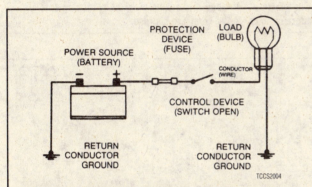

Fig. 33 Here is an example of a simple automotive circuit. When the switch is closed, power from the positive battery terminal flows through the fuse, then the switch and to the load (light bulb), the light illuminates and then, the circuit is completed through the return conductor and the vehicle ground. If the light did not work, the tests could be made with a voltmeter or test light at the battery, fuse, switch or bulb socket

GROUND

Some automotive components are grounded through their mounting points. The electrical current runs through the wiring harness, then into the component and back through the chassis of the vehicle. If you look, you'll see that the battery ground cable connects between the battery and the body of the vehicle.

LOAD

▶ See Figure 34

Every complete circuit must include a "load" (something to use the electricity coming from the source). If you were to connect a wire between the two terminals of the battery (DON'T do this, but take our word for it) without the light bulb, the battery would attempt to deliver its entire power supply from one pole to another almost instantly. This is a short circuit. The electricity is taking a short cut to get to ground and is not being used by any load in the circuit. This sudden and uncontrolled electrical flow can cause great damage to other components in the circuit and can develop a tremendous amount of heat. A short in an automotive wiring harness can develop sufficient heat to melt the insulation on all the surrounding wires and reduce a multiple wire cable to one sad lump of plastic and copper. Two common causes of shorts are broken insulation (thereby exposing the wire to contact with surrounding metal surfaces or other wires) or a failed switch (the pins inside the switch come out of place and touch each other).

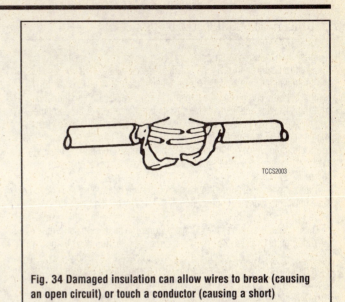

Fig. 34 Damaged insulation can allow wires to break (causing an open circuit) or touch a conductor (causing a short)

Switches And Relays

Some electrical components which require a large amount of current to operate also have a relay in their circuit. Since these circuits carry a large amount of current (amperage or amps), the thickness of the wire in the circuit (wire gauge) is also greater. If this large wire were connected from the load to the control switch on the dash, the switch would have to carry the high amperage load and the dash would be twice as large to accommodate wiring harnesses as thick as your wrist. To prevent these problems, a relay is used. The large wires in the circuit are connected from the battery to one side of the relay and from the opposite side of the relay to the load. The relay is normally open, preventing current from passing through the circuit. An additional, smaller wire is connected from the relay to the control switch for the circuit. When the control switch is turned on, it grounds the smaller wire to the relay and completes its circuit. The main switch inside the relay closes, sending power to the component without routing the main power through the inside of the vehicle. Some common circuits which may use relays are the horn, headlights, starter and rear window defogger systems.

Protective Devices

It is possible for larger surges of current to pass through the electrical system of your vehicle. If this surge of current were to reach the load in the circuit, it could burn it out or severely damage it. To prevent this, fuses, circuit breakers and/or fusible links are connected into the supply wires of the electrical system. These items are nothing more than a built-in weak spot in the system. It's much easier to go to a known location (the fusebox) to see why a circuit is inoperative than to dissect 15 feet of wiring under the dashboard, looking for what happened.

When an electrical current of excessive power passes through the fuse, the fuse blows (the conductor melts) and breaks the circuit, preventing the passage of current and protecting the components.

A circuit breaker is basically a self repairing fuse. It will open the circuit in the same fashion as a fuse, but when either the short is removed or the surge subsides, the circuit breaker resets itself and does not need replacement.

A fuse link (fusible link or main link) is a wire that acts as a fuse. One of these is normally connected between the starter relay and the main wiring harness under the hood. Since the starter is usually the highest electrical draw on the vehicle, an internal short during starting could direct about 130 amps into the wrong places. Consider the damage potential of introducing this current into a system whose wiring is rated at 15 amps and you'll understand the need for protection. Since this link is very early in the electrical path, it's the first place to look if nothing on the vehicle works, but the battery seems to be charged and is properly connected.

ELECTRONICALLY CONTROLLED UNITS

Today, transmissions/transaxles are being developed and manufactured with electronically controlled components. The demand for lighter, smaller and more fuel efficient vehicles has resulted in the use of electronics to control both the engine and the transmission/transaxle . Certain transmission/transaxle assemblies have become an integrated part of the computerized engine control system. Certain sensors send electronic signals such as engine speed, engine temperature, vehicle speed and throttle opening to the control module which in turn computes these signals and returns other signals to the transmission/transaxle for selection of the appropriate gear range.

Automatic transmissions/transaxles with microcomputers to determine gear selections are now commonplace. Sensors are used for engine and road speed, engine load, gear selector lever position. kickdown switch and states of the driving program to send signals to the microcomputer to determine the optimum gear selection according to a preset program. The gear shifting is accomplished by solenoid valves in the hydraulic system. The electronics also control the modulated hydraulic system during shifting. along with regulating engine torque to provide smooth shifts between gear ratio changes. This type of system can be designed for different driving programs, such as giving the operator the choice of operating the vehicle for either economy of performance.

GENERAL TROUBLESHOOTING

▶ **See Figures 35, 36, 37 and 38**

Electrical problems generally fall into one of three areas:
- The component that is not functioning is not receiving current.
- The component is receiving power but is not using it or is using it incorrectly (component failure).
- The component is improperly grounded.

The circuit can be can be checked with a test light and a jumper wire. The test light is a device that looks like a pointed screwdriver with a wire on one end and a bulb in its handle. A jumper wire is simply a piece of wire with alligator clips or special terminals on each end. If a component is not working, you must follow a systematic plan to determine which of the three is the cause of the problem.

1. Turn **ON** the switch that controls the item not working.

➡**Some items work only when the ignition switch is turned ON.**

2. Disconnect the power supply wire from the component.

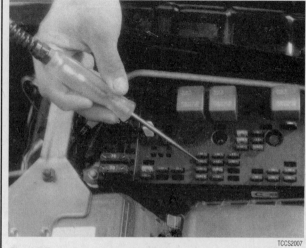

Fig. 36 Here, someone is checking a circuit by making sure there is power to the component's fuse

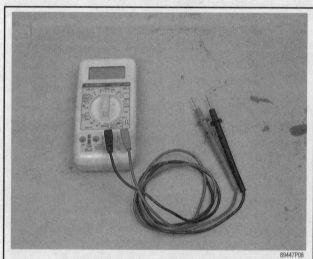

Fig. 37 A good quality digital multimeter is essential for testing sensitive computer circuits

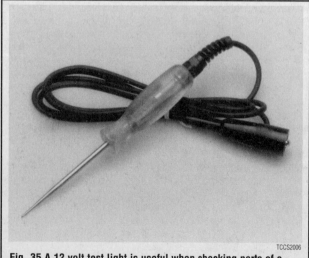

Fig. 35 A 12 volt test light is useful when checking parts of a circuit for power

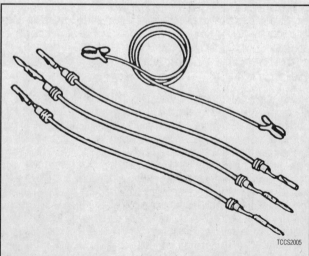

Fig. 38 Jumper wires with various connectors are handy for quick electrical testing

3. Attach the ground wire of a test light or a voltmeter to a good metal ground.

4. Touch the end probe of the test light (or the positive lead of the voltmeter) to the power wire; if there is current in the wire, the light in the test light will come on (or the voltmeter will indicate the amount of voltage). You have now established that current is getting to the component.

5. Turn the ignition or dash switch **OFF** and reconnect the wire to the component.

If there was no power, then the problem is between the battery and the component. This includes all the switches, fuses, relays and the battery itself. The next place to look is the fusebox; check carefully either by eye or by using the test light across the fuse clips. The easiest way to check is to simply replace the fuse. If the fuse is blown, and upon replacement, immediately blows again, there is a short between the fuse and the component. This is generally (not always) a sign of an internal short in the component. Disconnect the power wire at the component again and replace the fuse; if the fuse holds, the component is the problem.

✳✳ WARNING

DO NOT test a component by running a jumper wire from the battery UNLESS you are certain that it operates on 12 volts. Many electronic components are designed to operate with less voltage and connecting them to 12 volts could destroy them. Jumper wires are best used to bypass a portion of the circuit (such as a stretch of wire or a switch) that DOES NOT contain a resistor and is suspected to be bad.

If all the fuses are good and the component is not receiving power, find the switch for the circuit. Bypass the switch with the jumper wire. This is done by connecting one end of the jumper to the power wire coming into the switch and the other end to the wire leaving the switch. If the component comes to life, the switch has failed.

✳✳ WARNING

Never substitute the jumper for the component. The circuit needs the electrical load of the component. If you bypass it, you will cause a short circuit.

Checking the ground for any circuit can mean tracing wires to the body, cleaning connections or tightening mounting bolts for the component itself. If the jumper wire can be connected to the case of the component or the ground connector, you can ground the other end to a piece of clean, solid metal on the vehicle. Again, if the component starts working, you've found the problem.

A systematic search through the fuse, connectors, switches and the component itself will almost always yield an answer. Loose and/or corroded connectors, particularly in ground circuits, are becoming a larger problem in modern vehicles. The computers and on-board electronic (solid state) systems are highly sensitive to improper grounds and will change their function drastically if one occurs.

Remember that for any electrical circuit to work, ALL the connections must be clean and tight.

Use of the on-board computer to control the shifting of the transmission/transaxle has added an entirely new area of diagnosis. Before computer control, only the hydraulic and mechanical systems needed to be diagnosed, but now the electronic system must also be checked.

The electronic control system uses shift solenoids to control the pressure in the circuit, which moves the shift valves. Most electronic transmissions (as they are called) have two shift solenoids, which are turned on and off by the control module in the proper sequence to provide the next gear in the upshift or downshift. The design and operation of the computer control systems varies between vehicle manufacturers, so consequently their test procedures are also different.

Some of the most advanced electronic transmissions include a self-diagnosis program in the control module. Whenever one or more of the transmission/transaxle sensors is producing a signal that is out of the normal operational range, the computer will store a fault code. This fault code can be retrieved in the same manner as engine trouble codes. The fault code diagnostic chart is used after the code is retrieved to find the problem area. One electronic transmission/transaxle also includes a fail-safe system that will lock the transmission/transaxle in second or third gear after the computer has counted certain fault codes occurring for the fourth time. This will alert the driver that a problem exists and service needs to be performed.

Solenoids

DESCRIPTION

♦ See Figures 39 and 40

Basically, a solenoid is the same as the electromagnets we used to make as children for science experiments. A long piece of insulated/laminated wire is wrapped around a iron core, then each end of the wire is connected to a different polarity of the battery or power source. As long as the power is connected, the iron core is magnetic.

The modern automatic transmission/transaxle makes extensive use of solenoids to control the hydraulic fluid within the valve body. One common use of a solenoid in a transmission/transaxle is for engaging the Lock Up Torque Converter. The solenoid controls the hydraulic fluid pressure that causes the torque converter to lock up. The latest method of engaging the torque converter involves pulsing the solenoid rapidly to provide a gradual and smooth engagement. This is known as Pulse Width Modulation (PWM). Other solenoids are used in the valve body to control shifting and regulate fluid pressure.

A shift solenoid is usually an ON/OFF device, which means that it is either on or off. When voltage is applied, the solenoid is energized. As the vehicle speed increases and an upshift is required, the control module will either energize or de-energize the solenoid to cause the transmission/

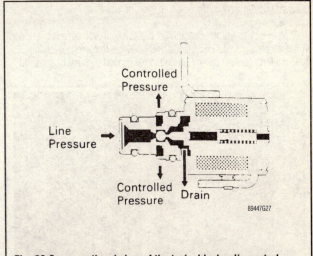

Fig. 39 Cross sectional view of the typical hydraulic control solenoid

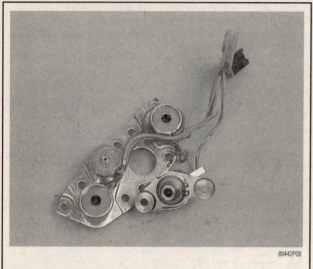

Fig. 40 Common hydraulic solenoid pack

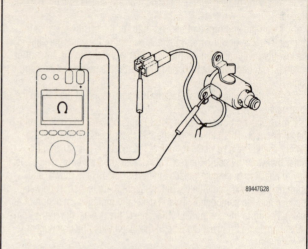

Fig. 42 Measuring the resistance of a typical hydraulic solenoid used in automatic transmissions/transaxles

transaxle to upshift or change gears. Solenoids that regulate pressure usually operate on a duty cycle. A duty cycle can be thought of as high frequency voltage pulses which cause the solenoid to operate somewhere between either on or off.

TESTING

▶ **See Figures 41 thru 46**

As we stated earlier, a solenoid is constructed of a long piece of insulated/laminated wire. As long as the wire in the solenoid is not broken, it will conduct electricity. In this case, we can test the solenoid for continuity with an ohmmeter.

Practically all solenoids will have two terminals. Each terminal represents one end of the coil of wire. In some cases, one of the terminals will be connected to ground. Ground may be the body of the solenoid or the bracket it is mounted to. Using an ohmmeter, measure the resistance of the coil of wire. If the resistance is not within the specification for the particular solenoid you are testing, the solenoid must be replaced. Typical solenoids may measure 5–30 ohms.

Another way to test the operation of the solenoid is to bypass the normal wiring and energize the coil manually. Disconnect the solenoid. Using jumper wires, connect battery ground to the ground terminal or solenoid

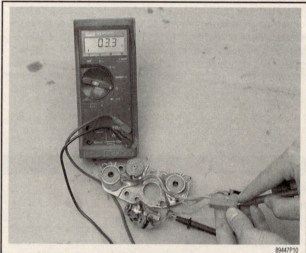

Fig. 43 Connect one meter lead to the common ground and the other lead to the solenoid to measure the resistance

Fig. 41 Testing the torque converter lock up solenoid with an ohmmeter—440T4 Transaxle shown

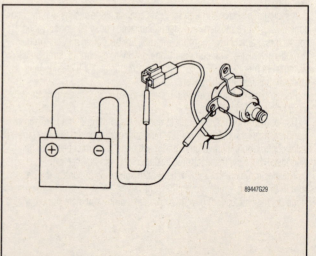

Fig. 44 Listen for a clicking sound while applying battery voltage to the solenoid

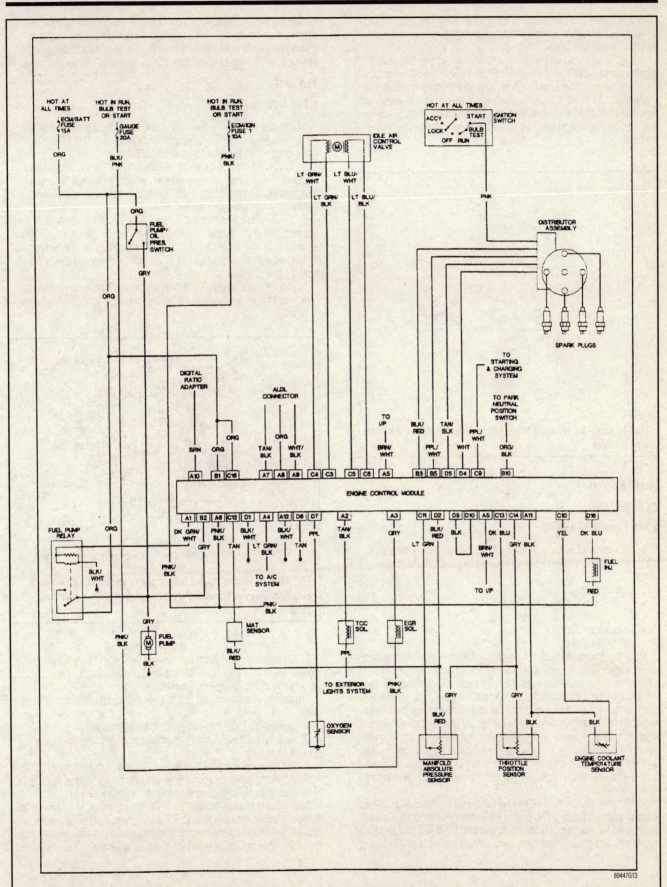

Fig. 45 Typical general motors schematic diagram showing the torque converter lock up solenoid circuit from terminal A2—the exact terminal will vary depending on the type of vehicle

body and battery positive voltage to the other terminal. You should be able to head the solenoid click when the battery voltage is applied. If not, the solenoid should be replace.

A third test that can be performed on the solenoid may just be the most important one, that is the ability of the valve in the solenoid to fully close and open. Compressed air is required to perform this test. Apply compressed air to the valve while connecting the solenoid to battery voltage. The valve should not leak while there is no battery voltage applied. Apply battery voltage to the solenoid and confirm that the air will flow through.

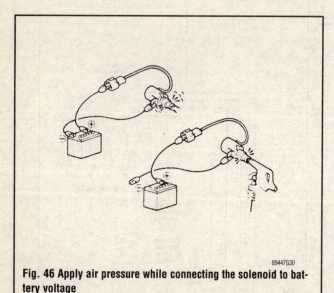

Fig. 46 Apply air pressure while connecting the solenoid to battery voltage

Speed Sensors

DESCRIPTION

▶ **See Figures 47 and 48**

Speed sensors or pulse generators can be used on today's vehicles to inform the control module of several vehicle conditions. Some of then include:

- Engine speed
- Transmission/transaxle input speed
- Transmission/transaxle output speed
- Speedometer operation
- Trouble diagnosis

Most speed sensors consist of a permanent magnet which rotates past a coil of wire. Usually the permanent magnet is engaged to the output (or input) shaft of the transmission/transaxle by a gear. The speed of the magnet is directly proportional to the speed of the shaft in the transmission. An electrical pulse is generated each time the magnet rotates past the coil of wire. That is why the sensor may also be referred to as a pulse generator. Magnets have an invisible field around them. If you place a magnet under a piece of paper and sprinkle iron fillings on the paper, you can see the flux lines that are normally invisible. As these lines pass through the conductor (coil of wire), an electric voltage is induced. This voltage can be measured with a volt meter.

Some speed sensors have no moving parts, they consist of a coil of wire wrapped around a magnet. This type of speed sensor does not have a gear on the end of it. The sensor is usually screwed or bolted into the case of the transmission/transaxle or transaxle. These sensors are located very close to

a rotating component that has notches cut in it. As the notches pass close to magnet, it causes the flux lines (magnetic field) of the magnet to change. Voltage is induced in the coil of wire as the flux lines pass through it. The voltage from this type of sensor can also be measured using a volt meter.

TESTING

▶ **See Figures 49 and 50**

The sensor that has the gear on the end can be tested in or out of the transmission/transaxle . The type with no moving parts should be tested while installed. To test either type in the vehicle:

1. Raise and safely support the vehicle if necessary. On front wheel drive vehicles, support the lower control arms so the halfshafts are near their normal operating position. The joints may be damaged if they are allowed to rotate while they are hanging down.

2. Disconnect the sensor and connect the leads of the voltmeter to the two terminals of the sensor. One lead to each terminal. Jumper wires with alligator clips may be needed to make a good connection. The leads should be long enough for the meter to be in the vehicle or where it can be seen from the driver's seat.

3. Set the meter to read AC voltage on the low (about five volts) scale.

4. Start the engine, apply the brake and place the transmission/transaxle in Drive.

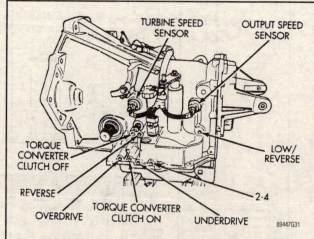

Fig. 47 Turbine (input) and output speed sensor locations on a common Chrysler transaxle—these sensors have no moving parts

5. Release the brake/clutch and observe the meter. The meter should indicate about 1–4 volts. It should not be necessary to apply the gas pedal. If a small voltage is generated, the sensor is good.

The sensor with the rotating gear may be tested out of the vehicle as follows:

6. Remove the sensor from the transmission/transaxle .

7. Connect the meter leads to the sensor terminals. One lead to each terminal as before.

8. Set the meter to read AC voltage on the low (about five volts) scale.

9. Rotate the gear with your fingers and observe the voltage reading. If the sensor is good if it is generating voltage. If not, replace it.

Both types of sensors can also be tested for continuity. There should be continuity between the two terminals. This indicates that the coil of fine wire inside is not broken. Some amount of resistance should be measured, this is normal. If the resistance is infinite, replace the sensor.

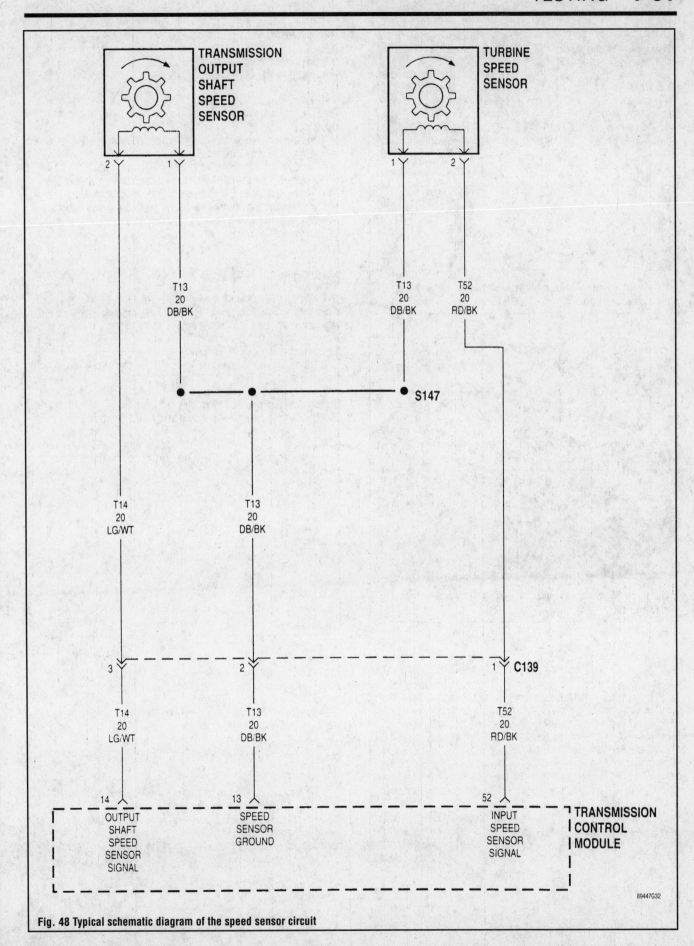

Fig. 48 Typical schematic diagram of the speed sensor circuit

89447G32

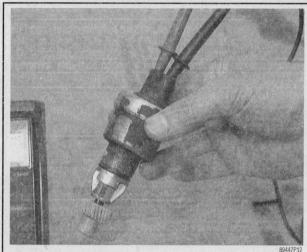

Fig. 49 Connect the leads of the voltmeter to the terminals of the sensor as shown

Fig. 50 Rotate the sensor gear with your fingers while observing the voltmeter

Range Sensor

DESCRIPTION

▶ See Figure 51

The range sensor, also known as the manual lever position sensor is a multifunction device which informs the control module of the following conditions:

• Weather the transmission/transaxle is in park or neutral
• What gear the transmission/transaxle is in
• Weather or not to illuminate the reverse (back-up) lights

The range sensor may be located on the outside of the unit or on the valve body inside the unit. The sensors that are mounted on the outside of the transmission/transaxle are mechanical switches. The range sensor mounted on the valve body consists of fluid pressure switches and possibly a transmission/transaxle fluid temperature sensor.

TESTING

▶ See Figure 52

Due to the many different varieties of range sensors, exact testing procedures cannot be explained here. Use the following as a guide when diagnosing the range sensor. A scan tool will be necessary to troubleshoot the internally mounted range sensors.

1. Observe all safety precautions.
2. Raise and safely support the vehicle if necessary to gain access to the range sensor.
3. Do not push the meter leads into the connectors. This may damage the connector and cause an intermittent problem.
4. Allow the transmission/transaxle to reach normal operating temperature on transmissions with internally mounted sensors before testing.
5. Identify the terminals for the temperature sensor. Measure the resistance of the temperature sensor and use chart in this manual for a general guideline for correct resistance.

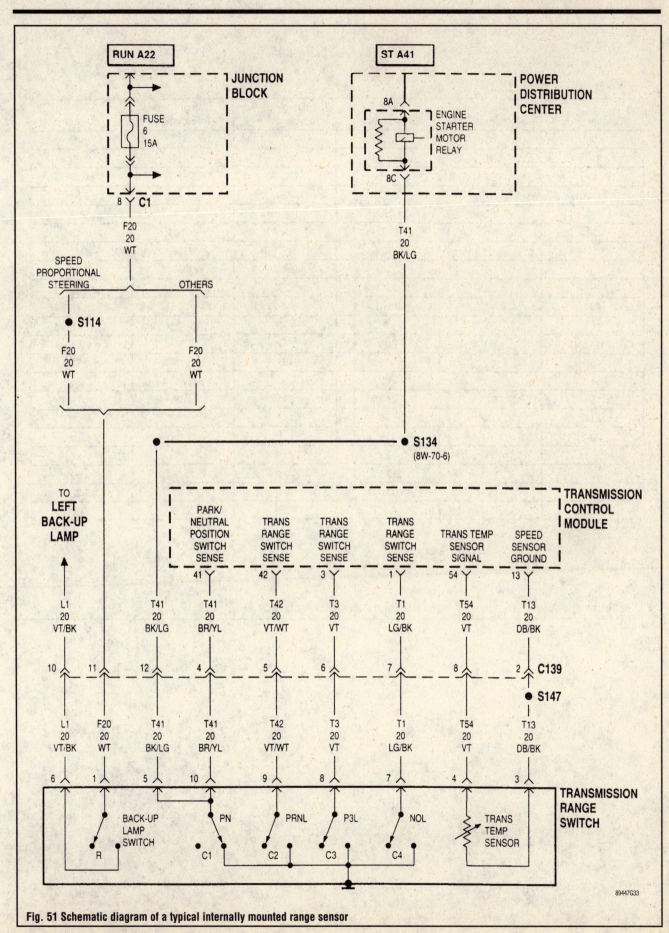

Fig. 51 Schematic diagram of a typical internally mounted range sensor

89447G33

Temperature	Temperature	Minimum Resistance	Nominal Resistance	Maximum Resistance	Signal
°F	°C	Ω	Ω	Ω	volts
-40	-40	90636	100707	110778	5.00
-22	-30	47416	52684	57952	4.78
-4	-20	25809	28677	31545	4.34
14	-10	14558	16176	17794	3.89
32	0	8481	9423	10365	3.45
50	10	5104	5671	6238	3.01
68	50	3164	3515	3867	2.56
86	30	2013	2237	2461	1.80
104	40	1313	1459	1605	1.10
122	50	876	973	1070	3.25
140	60	600	667	734	2.88
158	70	420	467	514	2.56
176	80	299	332	365	2.24
194	90	217	241	265	1.70
212	100	159	177	195	1.42
230	110	119	132	145	1.15
248	120	89.9	99.9	109.9	0.87
266	130	69.1	76.8	84.5	0.60
284	140	53.8	59.8	65.8	0.32
302	150	42.5	47.2	51.9	0.00

89447G34

Fig. 52 Transmission/transaxle temperature sensor resistance verses temperature chart

8
REPAIR

ON VEHICLE REPAIRS

Special Tools

Normally, the use of special factory tools is avoided for repair procedures, since these are not readily available for the do-it-yourself mechanic. When it is possible to perform the job with more commonly available tools, it will be pointed out, but occasionally, a special tool was designed to per-form a specific function and should be used. Before substituting another tool, you should be convinced that neither your safety nor the performance of the vehicle will be compromised.

Special tools can usually be purchased from an automotive parts store or from your dealer. In some cases special tools may be available directly from the tool manufacturer.

SERVICING YOUR VEHICLE SAFELY

♦ **See Figures 1, 2, 3 and 4**

It is virtually impossible to anticipate all of the hazards involved with automotive maintenance and service, but care and common sense will prevent most accidents.

The rules of safety for mechanics range from "don't smoke around gasoline," to "use the proper tool(s) for the job." The trick to avoiding injuries is to develop safe work habits and to take every possible precaution.

Do's

- Do keep a fire extinguisher and first aid kit handy.
- Do wear safety glasses or goggles when cutting, drilling, grinding or prying, even if you have 20–20 vision. If you wear glasses for the sake of vision, wear safety goggles over your regular glasses.
- Do shield your eyes whenever you work around the battery. Batteries contain sulfuric acid. In case of contact with the eyes or skin, flush the area with water or a mixture of water and baking soda, then seek immediate medical attention.

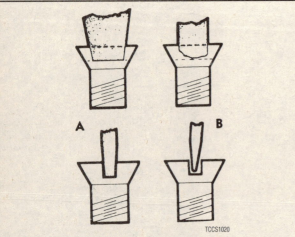

TCCS1020

Fig. 1 Screwdrivers should be kept in good condition to prevent injury or damage, which could result if the blade slips from the screw

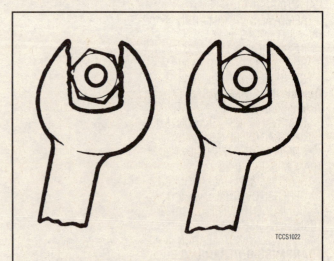

TCCS1022

Fig. 3 Using the correct size wrench will help prevent the possibility of rounding off a nut

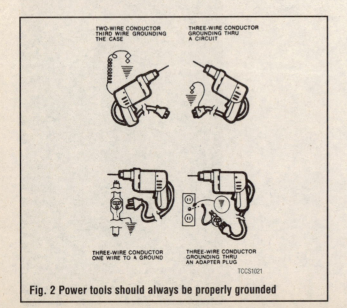

TWO-WIRE CONDUCTOR THIRD WIRE GROUNDING THE CASE

THREE-WIRE CONDUCTOR GROUNDING THRU A CIRCUIT

THREE-WIRE CONDUCTOR ONE WIRE TO A GROUND

THREE-WIRE CONDUCTOR GROUNDING THRU AN ADAPTER PLUG

TCCS1021

Fig. 2 Power tools should always be properly grounded

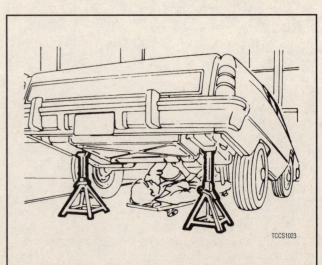

TCCS1023

Fig. 4 NEVER work under a vehicle unless it is supported using safety stands (jackstands)

• Do use safety stands (jackstands) for any under vehicle service. Jacks are for raising vehicles; jackstands are for making sure the vehicle stays raised until you want it to come down. Whenever the vehicle is raised, block the wheels remaining on the ground and set the parking brake.

• Do use adequate ventilation when working with any chemicals or hazardous materials. Like carbon monoxide, the asbestos dust resulting from some brake lining wear can be hazardous in sufficient quantities.

• Do disconnect the negative battery cable when working on the electrical system. The secondary ignition system contains EXTREMELY HIGH VOLTAGE. In some cases it can even exceed 50,000 volts.

• Do follow manufacturer's directions whenever working with potentially hazardous materials. Most chemicals and fluids are poisonous if taken internally.

• Do properly maintain your tools. Loose hammerheads, mushroomed punches and chisels, frayed or poorly grounded electrical cords, excessively worn screwdrivers, spread wrenches (open end), cracked sockets, slipping ratchets, or faulty droplight sockets can cause accidents.

• Likewise, keep your tools clean; a greasy wrench can slip off a bolt head, ruining the bolt and often harming your knuckles in the process.

• Do use the proper size and type of tool for the job at hand. Do select a wrench or socket that fits the nut or bolt. The wrench or socket should sit straight, not cocked.

• Do, when possible, pull on a wrench handle rather than push on it, and adjust your stance to prevent a fall.

• Do be sure that adjustable wrenches are tightly closed on the nut or bolt and pulled so that the force is on the side of the fixed jaw.

• Do strike squarely with a hammer; avoid glancing blows.

• Do set the parking brake and block the drive wheels if the work requires a running engine.

Don'ts

• Don't run the engine in a garage or anywhere else without proper ventilation—EVER! Carbon monoxide is poisonous; it takes a long time to leave the human body and you can build up a deadly supply of it in your system by simply breathing in a little every day. You may not realize you are slowly poisoning yourself. Always use power vents, windows, and fans and/or open the garage door.

• Don't work around moving parts while wearing loose clothing. Short sleeves are much safer than long, loose sleeves. Hard-toed shoes with neoprene soles protect your toes and give a better grip on slippery surfaces. Jewelry such as watches, fancy belt buckles, beads or body adornment of any kind is not safe working around a vehicle. Long hair should be tied back under a hat or cap.

• Don't use pockets for toolboxes. A fall or bump can drive a screwdriver deep into your body. Even a rag hanging from your back pocket can wrap around a spinning shaft or fan.

• Don't smoke when working around gasoline, cleaning solvent or other flammable material.

• Don't smoke when working around the battery. When the battery is being charged, it gives off explosive hydrogen gas.

• Don't use gasoline to wash your hands; there are excellent soaps available. Gasoline contains dangerous additives, which can enter the body through a cut or through your pores. Gasoline also removes all the natural oils from the skin so that bone dry hands will suck up oil and grease.

• Don't service the air conditioning system unless you are equipped with the necessary tools and training. When liquid or compressed gas refrigerant is released to atmospheric pressure it will absorb heat from whatever it contacts. This will chill or freeze anything it touches.

• Don't use screwdrivers for anything other than driving screws! A screwdriver used as a prying tool can snap when you least expect it, causing injuries. At the very least, you'll ruin a good screwdriver.

• Don't use an emergency jack (that little ratchet, scissors, or pantograph jack supplied with the vehicle) for anything other than changing a flat! These jacks are only intended for emergency use out on the road; they are NOT designed as a maintenance tool. If you are serious about maintaining your vehicle yourself, invest in a hydraulic floor jack of at least a 1½ ton capacity, and at least two sturdy jackstands.

Seal Replacement

The following are general procedures for typical external transmission/transaxle seals.

PUMP SEAL

▶ See Figure 5

1. Remove the transmission/transaxle from the vehicle.
2. On some transmissions/transaxles, it may be necessary to remove the fluid pump in order to remove the pump seal. As required, remove the fluid pump.

➡The transmission/transaxle fluid pump is usually found in the bell housing, but in some cases, the pump may be located under a side cover, at the rear of the case or under the valve body.

3. Using an appropriate tool, remove the pump housing bushing and/or seal.
4. Remove any remaining pump seals.

To install:
5. Carefully inspect the seal bore in the transmission pump for damage. Replace the pump if the seal bore is heavily damaged.
6. If the seal was previously coated with sealant, coat the outside of the pump seal case with a suitable sealer to prevent fluid leakage.
7. Lubricate the pump seals with transmission fluid and install.
8. Using a hammer and an appropriate seal driver tool, tap pump housing seal into it bore until it bottoms out.
9. If the fluid pump was removed, proceed as follows:
10. Clean the old pump gasket surfaces.
11. Place a new pump gasket into position. If the gasket will not stay in place, use some petroleum jelly in four or five locations on the gasket to hold it in place.
12. If necessary, install alignment dowels in the pump holes on the transmission/transaxle housing. If no dowels are available, you can make your own by finding long bolts with the same thread as the pump bolts and cutting the bolt heads off with a hacksaw.
13. After the alignment dowels are installed, slide the pump onto the dowels and into place on the transmission/transaxle housing.
14. Tighten the pump retainers.
15. Install the transmission/transaxle in the vehicle.

89448P10

Fig. 5 In some cases a seal puller may be used to remove the input shaft seal without removing the pump

TAILHOUSING SEAL

▶ **See Figures 6, 7 and 8**

The extension seal is located at the rear of the transmission case.
1. Raise and support the vehicle safely.
2. Drain the transmission of lubricant.
3. Matchmark the driveshaft to the yoke for reassembly and remove the driveshaft.
4. Remove the old seal using a seal puller or appropriate prytool.

To install:
5. Install a new seal, coated with sealing compound, using an appropriate seal installation tool.
6. Install the driveshaft, making certain to align the matchmark.
7. Fill the transmission to the proper level.
8. Lower the vehicle.
9. Road test the vehicle and check for proper operation and fluid leaks.

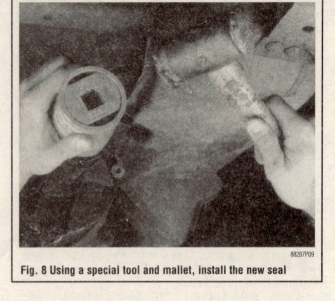

Fig. 8 Using a special tool and mallet, install the new seal

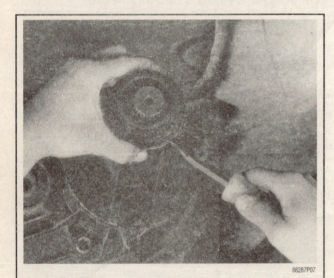

Fig. 6 Removing the extension housing seal

DIPSTICK TUBE SEAL

▶ **See Figures 9, 10 and 11**

1. Remove the dipstick.
2. Raise and support the vehicle with safety stands.
3. Unbolt any dipstick tube bracket retainers.
4. Remove the dipstick tube using a twisting and pulling motion.
5. Use a suitable puller or prytool to remove the dipstick seal.

To install:
6. Coat the seal with a suitable assembly lubricant, then install a new seal making sure it fully and evenly seated.
7. Install the dipstick tube using a twisting and pushing motion.
8. Install any dipstick tube bracket retainers.
9. Fill the transmission to the proper level.
10. Road test the vehicle and check for proper operation and fluid leaks.

Fig. 7 Always replace the old seal with a new one

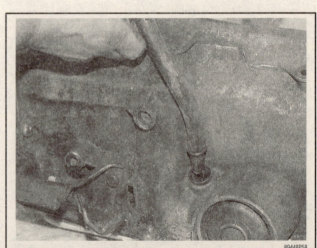

Fig. 9 Remove the dipstick tube from the transmission/transaxle (the transaxle shown has been removed from the vehicle for clarity)

Fig. 10 Remove the seal from the transmission/transaxle case

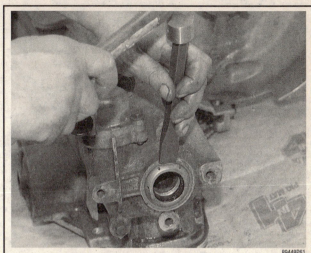

Fig. 12 A chisel may be needed to loosen the transaxle halfshaft seal

Fig. 11 When installing the new seal, coat it with a assembly lubrication before installation

Fig. 13 Use an appropriate seal puller to remove the transaxle halfshaft seal

TRANSAXLE HALFSHAFT SEAL

▶ **See Figures 12 and 13**

1. Raise and support the vehicle with safety stands.
2. Remove the halfshaft.
3. Use a suitable puller or prytool to remove the transaxle halfshaft seal.
4. In some cases it may be necessary to use a chisel and a hammer to unseat the seal before using the puller to remove the seal.

To install:

5. Install a new seal, coated with sealing compound, using an appropriate seal installation tool.
6. Install the halfshaft and lower the vehicle.
7. Fill the transmission to the proper level.
8. Road test the vehicle and check for proper operation and fluid leaks.

Driveshafts

➡The term driveshaft refers to the connection between a transmission and a rear differential on a rear wheel drive vehicle. It does not refer to halfshafts (often termed driveshafts by various manufac-turers), that connect the transaxle to the stub shafts on front wheel drive vehicles.

GENERAL INFORMATION

▶ **See Figures 14 and 15**

The driveshaft is a long steel tube used to transmit power from the transmission to the rear differential. Located at either end of the driveshaft is a universal joint (U-joint), which allows the driveshaft to move up and down (within designed limits) in order to match the motion of the rear axle. A slip joint is often attached to the U-joint closest to the transmission. The shaft is designed with yokes at each end that are in-line with each other in order to produce the smoothest possible running shaft.

➡**ALWAYS matchmark the shaft ends to the yoke or flange before removal.**

At the front of the driveshaft, the U-joint usually connects the driveshaft to a slip-jointed yoke. This yoke is internally splined and allows the driveshaft to move in and out on the transmission output shaft, which is externally splined. At the rear of the driveshaft, the U-joint is clamped to the yoke attached to the rear axle pinion flange. The rear yoke may also be a

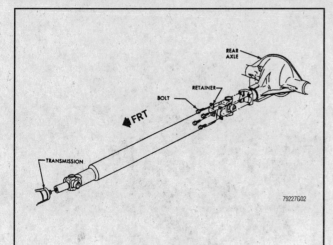

Fig. 14 Exploded view of a typical driveshaft assembly and attachment points

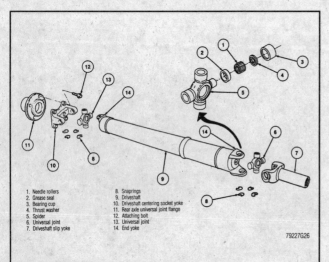

1. Needle rollers
2. Grease seal
3. Bearing cup
4. Thrust washer
5. Spider
6. Universal joint
7. Driveshaft slip yoke
8. Snaprings
9. Driveshaft
10. Driveshaft centering socket yoke
11. Rear axle universal joint flange
12. Attaching bolt
13. Universal joint
14. End yoke

Fig. 15 Exploded view of a typical driveshaft where the U-joint is pressed into the rear yoke, rather than bolted to it

flange that mates to the pinion flange on the differential. It is usually secured in the yoke by a bracket with two small bolts, one at either end.

Some rear U-joints are pressed into the yoke which is then bolted to the pinion flange on the differential.

On some production U-joints, nylon (plastic) is injected through a small hole in the yoke during manufacture and flows along a circular groove between the U-joint and the yoke, creating a non-metallic snapring.

➡Since plastic retaining rings must be sheared for removal and no snapring grooves are supplied, the production joints must be replaced with service U-joints with a snapring groove whenever they are removed from the shaft.

Bad U-joints, requiring replacement, will produce a clunking sound when the vehicle is put into gear and when the transmission shifts from gear-to-gear. This is due to worn needle bearings or scored trunnion ends. Most U-joints are permanently lubricated at the factory and require no periodic maintenance. Those that do have grease fittings should be lubricated at every oil change. Clean the fitting with a shop rag before pumping grease to avoid forcing dirt into the joint.

REMOVAL & INSTALLATION

▶ **See Figure 16**

1. Mark the relationship of the driveshaft-to-pinion flange or yoke and disconnect the rear universal joint by removing the bolts. If the bearing cups are loose, tape them together to prevent dropping them and losing the bearing rollers.
2. Slide the driveshaft forward to disengage it from the rear axle flange or yoke.
3. Move the driveshaft rearward to disengage it from the transmission slip-joint, passing it under the axle housing.

➡**DO NOT allow the driveshaft to hang by the U-joint or bend to extreme angles, as damage to the U-joint may occur. Support the driveshaft shaft during removal.**

To install:
4. Inspect the slip-joint and transmission output shaft for damage, burrs or wear, for this will damage the transmission seal or make installation difficult. Apply engine oil to all splined driveshaft yokes.

✳✳ WARNING

DO NOT use a hammer to force the driveshaft into place. Check for burrs on the transmission output shaft spline, twisted slip yoke splines or possibly the wrong U-joint. Make sure the splines agree in number and fit. To prevent trunnion seal damage, DO NOT place any tool between the yoke and splines.

5. Slide the driveshaft into the transmission.
6. Align the rear universal joint to the rear axle pinion flange, making sure the bearings are properly seated in the pinion flange yoke, if so equipped.
7. Install the retaining bolts and tighten them bolts securely.
8. Road test the vehicle.

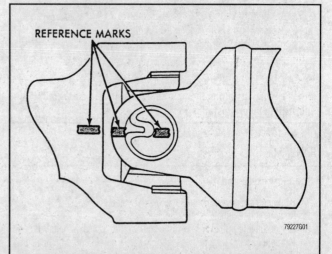

Fig. 16 Make alignment marks on the U-joint and shaft before disassembly to prevent possible vibration when assembled

BALANCING

▶ **See Figures 17 and 18**

The following procedure is used to help eliminate minor driveshaft vibration of an otherwise good driveshaft.

Before attempting this, carefully examine the driveshaft for damage such as dents and deformations. Driveshafts are subjected to large amounts of twisting force that can literally twist the driveshaft. Also check for missing weights that may have been knocked off of the shaft. If the driveshaft is deformed, replace it. If any weights appear to be missing, take the drive-shaft to a machine shop that is equipped to balance the shaft and have it repaired. Driveshafts typically turn at speeds 2½to 4 or more times faster than the rear axle, don't use a damaged driveshaft.

This type of balancing is performed by installing one or two hose clamps near the end of the driveshaft closest to the drive axle. The trial and error method is used to determine the best position of the clamp(s).

➡**Removing and turning the driveshaft 180°relative to the yoke may reduce some vibration. This should be done prior to the hose clamp method.**

1. Mark the rear of the driveshaft in four equal sections. Number the marks 1 through 4.
2. Install a hose clamp with the screw portion of the clamp on the No. 1 mark.

3. Test drive the vehicle to see if the vibration condition has improved.
4. Recheck the vibration with the clamp positioned at the remaining three positions. If the vibration is equally reduced at, for example, position number two and position number three, then position the screw portion of the clamp halfway between the marks.
5. Test drive the vehicle. If the vibration is still apparent, install another clamp in the same position as the first.
6. Test drive the vehicle. If the vibration is the same, move both clamps an equal distance from the point determined to be the best position. At first, position the clamps approximately ½ in. (12mm) apart.
7. Continue the process until the vibration is reduced to an acceptable level.

➡**If the vibration cannot be reduced to an acceptable level, take the driveshaft to a qualified machine shop for balancing.**

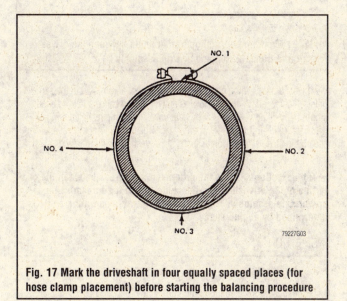

Fig. 17 Mark the driveshaft in four equally spaced places (for hose clamp placement) before starting the balancing procedure

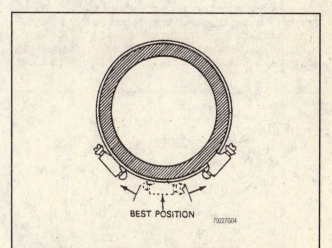

Fig. 18 Move the hose clamp heads an equal distance from the best position a little at a time until the vibration is reduced to an acceptable level

Troubleshooting Basic Driveshaft and Rear Axle Problems

When abnormal vibrations or noises are detected in the driveshaft area, this chart can be used to help diagnose possible causes. Remember that other components such as wheels, tires, rear axle and suspension can also produce similar conditions.

BASIC DRIVESHAFT PROBLEMS

Problem	Cause	Solution
Shudder as car accelerates from stop or low speed	• Loose U-joint • Defective center bearing	• Replace U-joint • Replace center bearing
Loud clunk in driveshaft when shifting gears	• Worn U-joints	• Replace U-joints
Roughness or vibration at any speed	• Out-of-balance, bent or dented driveshaft • Worn U-joints • U-joint clamp bolts loose	• Balance or replace driveshaft • Replace U-joints • Tighten U-joint clamp bolts
Squeaking noise at low speeds	• Lack of U-joint lubrication	• Lubricate U-joint; if problem persists, replace U-joint
Knock or clicking noise	• U-joint or driveshaft hitting frame tunnel • Worn CV joint	• Correct overloaded condition • Replace CV joint

79227C01

Universal Joints (U-Joints)

GENERAL INFORMATION

▶ **See Figures 19, 20 and 21**

The universal joint (U-Joint) is used to provide a strong and flexible connection between the driveshaft and axle assembly. A flexible joint is necessary because of the constant movement of the axle assembly relative to the body of the vehicle. A U-Joint consists of the spider (trunnion), needle (roller) bearings, bearing cups, seals and snaprings. In most cases, U-Joints will last the life of the vehicle. The life of the U-Joint may decrease significantly if the operating angle has been changed or exceeded. This occurs when the vehicle ride height is changed. Vehicles that have been lifted will benefit by using a Double Cardon type joint. The Double Cardon type joint has a greater operating angle than the single U-joint.

When two components are connected by a conventional U-joint, the bend that is formed is called the operating angle. The larger the angle, the larger the amount of angular acceleration and deceleration of the joint. In other words, when the driveshaft is turning at a steady speed, the pinion gear in the differential will actually speed up and slow down. This takes place as long as

PROPELLER SHAFT R.P.M.	MAX. NORMAL OPERATING ANGLES
5000	3°
4500	3°
4000	4°
3500	5°
3000	5°
2500	7°
2000	8°
1500	11°

79227G07

Fig. 21 Maximum normal operating angle between the driveshaft and transmission and/or axle assembly

the driveshaft and pinion gear shaft are at different angles (not in the same plane). The speeding up and slowing down must be canceled out to ensure a smooth flow of power. This is why both yokes on the driveshaft are in line with each other. For example, whereas the transmission output is at a steady speed, the angle at the U-joint causes the driveshaft speed to vary. In such a case, the rear U-joint cancels the fluctuations caused by the front U-joint.

➡ **The operating angle is the difference in degrees between the centerline of the driveshaft and the centerline of the transmission and/or axle assembly. The maximum allowable operating angle is determined by engine speed.**

INSPECTION

▶ **See Figure 22**

Remove and replace the U-joint if any of the following conditions are present:
- Knocking or clunking noise from the driveshaft when the vehicle is put into gear, or when coasting at 10 mph (16 km/h) in neutral.
- Squeaking noise from the U-joint that increases in frequency as the speed of the vehicle increases.
- Roughness in the U-joint bearing when felt by hand. The U-joint should turn smoothly.

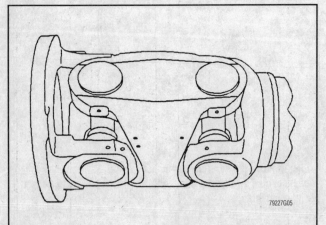

Fig. 19 A Double Cardon universal joint has a greater operating angle than a single joint. This joint has been punch marked before disassembly so the components can be reassembled in their original positions

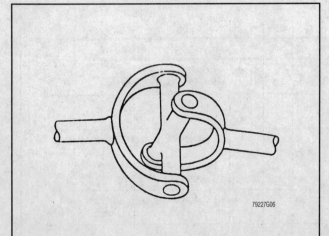

Fig. 20 This simplified version of a universal joint shows how the angles can change while still transmitting power

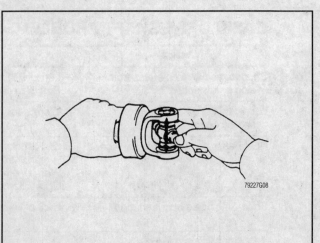

Fig. 22 Inspect the U-joint for excessive axial movement—replace the joint if the play is more than 0.002 in. (0.05mm).

• Axial play (up and down movement). Replace the U-joint if the axial play is more than 0.002 in. (0.05mm).

OVERHAUL

▶ **See Figures 23 thru 28**

1. Position the driveshaft assembly in a sturdy soft-jawed vise, BUT DO NOT place a significant clamp load on the shaft or you will risk deforming and ruining it.

➡Some original equipment U-joints are secured in the yoke by nylon (plastic) that has been injected at the factory. To remove this type of U-joint from the yoke, press the bearing cup until the plastic retaining ring breaks. The replacement U-joint will have a snapring groove like a conventional joint.

2. Remove the snaprings that retain the bearings in the yoke.

➡A U-joint removal and installation tool (which looks like a large C-clamp) is available to significantly ease the task, but it is very possible to replace the U-joints using an arbor press or a large vise and a variety of sockets.

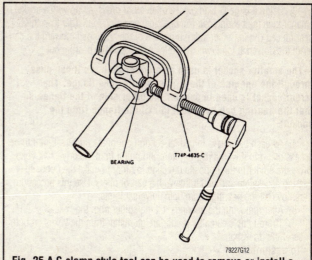

Fig. 25 A C-clamp style tool can be used to remove or install a U-joint successfully

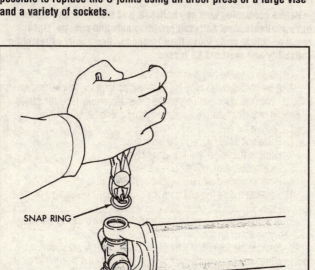

Fig. 23 Use a pair of snapring pliers, or similar tool, to remove the outer snapring that retains the bearing in the yoke

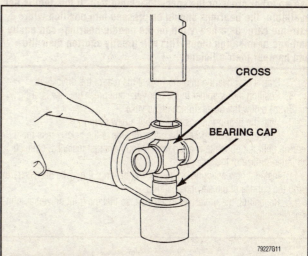

Fig. 26 Arbor press method of removing the U-joint from the yoke—the U-joint can also be installed in a similar fashion

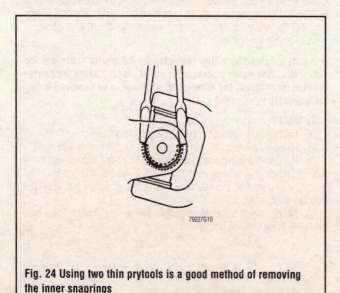

Fig. 24 Using two thin prytools is a good method of removing the inner snaprings

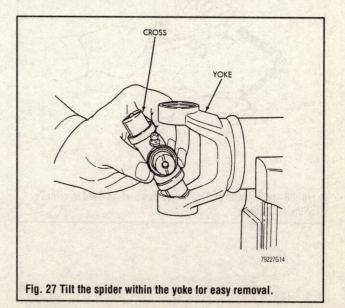

Fig. 27 Tilt the spider within the yoke for easy removal.

3. Using a large C-clamp, vise or an arbor press, along with a socket smaller than the bearing cap (on one side) and a socket larger than the bearing cap (on the other side), drive one of the bearings in toward the center of the universal joint, which will force the opposite bearing out.

➡The smaller socket is used as a driver here, as it can pass through the opening of the U-joint or slip yoke flange. The larger socket is used to support the other side of the flange so that the bearing cap has room to exit the flange (into the socket).

4. As each bearing is forced far enough out of the universal joint to be accessible, grip it with a pair of pliers and pull it from the driveshaft yoke. Drive the spider in the opposite direction in order to make the opposite bearing accessible and pull it free with a pair of pliers. Use this procedure to remove all the bearings from both universal joints.

5. After removing the bearings, lift the spider from the yoke.

6. Thoroughly clean all dirt and foreign matter from the yokes on both ends of the driveshaft.

To assemble:

❊❊ WARNING

When installing new bearings in the yokes, it is advisable to use an arbor press or the special C-clamp tool. If this tool is not available, the bearings should be pressed into position with extreme care, as a heavy jolt on the needle bearings can easily damage or misalign them. This will greatly shorten their life and hamper their efficiency.

7. Start a new bearing into the yoke at the rear of the driveshaft.

8. Position a new spider in the rear yoke and press the new bearing ¼ in. (6mm) below the outer surface of the yoke.

9. With the bearing in position, install a new snapring.

10. Start a new bearing into the opposite side of the yoke. Press the bearing until the opposite bearing, which you have just installed, contacts the inner surface of the snapring.

11. Install a new snapring on the second bearing. It may be necessary to grind the surface of the second snapring.

12. Reposition the driveshaft in the vise, so that the front universal joint is accessible.

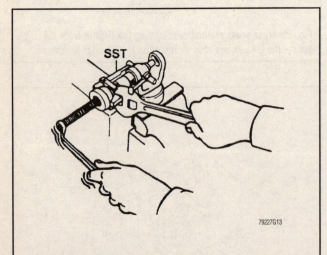

Fig. 28 The 2-jawed puller method can also be used to remove or install U-joints

13. Install the new bearings, new spider and new snaprings in the same manner as for the previously assembled rear joint.

14. Position the slip yoke on the spider. Install new bearings, nylon thrust bearings (if applicable) and snaprings.

15. Check both reassembled joints for freedom of movement, If misalignment of any part is causing a bind, a sharp rap on the side of the yoke with a brass hammer should seat the needle bearings and provide the desired freedom of movement. Care should be exercised to firmly support the shaft end during this operation, as well as to prevent blows to the bearings themselves. Under no circumstance should the driveshaft be installed in a car if there is any binding in the universal joints.

16. Grease the fittings, if equipped.

HalfShaft

REMOVAL & INSTALLATION

➡The term halfshaft refers to the connection between the transaxle and the stub shafts on front wheel drive vehicles.

➡Before continuing with any halfshaft procedure, make sure to have available, new halfshaft retaining nuts and circlips. Once removed, these parts loose their torque holding ability or retention capability and must not be reused.

1. Disconnect the negative battery cable.

2. If the halfshaft retaining nut is staked perform the following step:

   a. With the vehicle sitting on the ground, carefully raise the staked portion of the halfshaft retaining nut using a suitable small chisel.

3. Loosen the nut.

4. Raise and safely support the vehicle.

5. Remove the wheel and tire assembly.

6. If equipped, remove the splash shield(s).

7. Remove and discard the halfshaft retaining nut.

8. Separate the tie rod end from the steering knuckle using a suitable removal tool. Discard the cotter pin.

9. In some cases, it may be necessary to remove the sway bar nut before prying down on the control arm.

10. Remove the lower ball joint pinch bolt. Carefully pry down on the lower control arm to separate the ball joint stud from the steering knuckle.

11. Pull outward on the steering knuckle/brake assembly. Carefully pull the halfshaft from the hub and position it aside.

12. In some cases, a crossmember or sub-frame may have to be removed to facilitate halfshaft removal.

13. Position a drain pan under the transaxle.

14. Insert a prybar between the halfshaft and the transaxle case. Gently pry outward to release the halfshaft from the differential side gear. Be careful not to damage the transaxle case, oil seal, CV-joint or CV-joint boot.

15. Remove the halfshaft.

➡Install suitable plugs after removing the halfshafts to prevent the differential side gears from moving out of place. Should the gears become misaligned, the differential will have to be removed from the transaxle to align the gears.

To install:

16. Lubricate the splines lightly with a suitable grease.

17. Remove the plugs that were installed in the differential side gears.

18. Position the halfshaft so the CV-joint splines are aligned with the differential side gear splines. Push the halfshaft into the differential.

19. Pull outward on the steering knuckle/brake assembly and insert the halfshaft into the hub.

20. Pry downward on the lower control arm and position the lower ball joint stud in the steering knuckle.

21. If removed, install the crossmember or sub-frame that was removed to facilitate halfshaft removal.

22. Install the lower ball joint pinch bolt.

23. Attach the tie rod end to the steering knuckle. Install the castellated nut and tighten to specification. Install a new cotter pin.

24. If removed, attach the sway bar nut and tighten to specification.

25. If removed, install the splash shield(s).

26. Install the wheel and tire assembly.

27. Install a new halfshaft retaining nut and tighten to specification.

28. If the original halfshaft nut was staked, stake the halfshaft retaining nut using a suitable chisel with a rounded cutting edge.

➡**If the nut splits or cracks after staking, replace it with a new nut.**

29. Check and refill the transaxle with the proper type and quantity of fluid.

30. Connect the negative battery cable.

31. Road test the vehicle and check for proper operation.

CV-Joint Boots

INSPECTION

▸ **See Figures 29 and 30**

Whenever undercarriage work is performed, the Constant Velocity (CV) joint boots should be inspected for breaks and tears. The first sign of boot damage will be dark spots (grease) on the inside of the tire and wheel. If boot damage is caught early enough, cleaning, greasing and replacing the boot can save the joint. If the boot is left unrepaired, damage to the bearing will occur and replacement of the CV-joint is required. In most cases, it may be more economical to replace the entire halfshaft with a remanufactured one.

➡**Check with your parts supplier for price and availability to determine weather you should replace the entire halfshaft or separate components.**

BOOT REPLACEMENT

▸ **See Figures 31, 32 and 33**

Always follow the instructions included in the CV-joint boot kit. There are several variations and methods of boot replacement. Use the following

Fig. 29 Remove, clean and inspect this CV joint—it may be possible to save the joint by replacing the boot as long as the joint is not beyond repair

TCCS1011

Fig. 30 Push apart the bellows to inspect for tears or cracks that may be developing

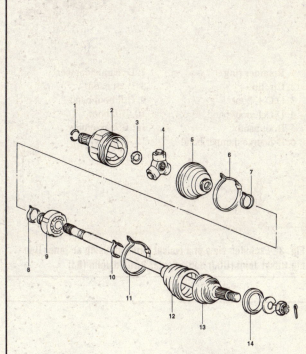

1. Circlip	8. Dynamic damper band
2. T.J. case	9. Dynamic damper
3. Snapring	10. Boot band
4. Spider assembly	11. B.J. boot band
5. T.J. boot	12. B.J. boot
6. T.J. boot band	13. B.J. assembly
7. Boot band	14. Dust cover

79247G15

Fig. 31 Exploded view of a typical halfshaft using an inner Tripot Joint (TJ) and an outer Birfield Joint (BJ)

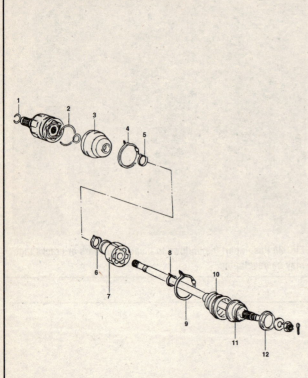

1. Retainer ring
2. Circlip
3. D.O.J. boot
4. D.O.J. boot band
5. Boot band
6. Dynamic damper band
7. Dynamic damper
8. Boot band
9. B.J. boot band
10. B.J. boot
11. B.J. assembly
12. Dust cover

79247G16

Fig. 32 Exploded view of a typical halfshaft using an inner Double Offset Joint (DOJ) and an outer Birfield Joint (BJ)

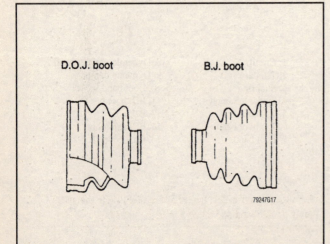

D.O.J. boot   B.J. boot

79247G17

Fig. 33 Typically, the boot on the Birfield Joint (BJ) has one extra valley compared to the boot on a Double Offset Joint (DOJ)

TCCS7031

Fig. 34 Pry under the hook to remove this type of band from the CV-joint

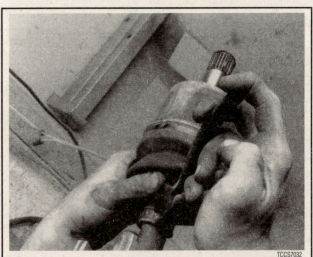

TCCS7032

Fig. 35 This type of band is crimped and must be cut before it can be removed

procedures as a general guide and in case the kit may not contain specific instructions.

➡Most outer CV joints on Asian vehicles, including Chrysler imports, use a Birfield joint, which should not be disassembled. To replace the outer boot, disassemble the inner joint, then slide the outer boot off the inner end of the shaft.

Outer Boot

▶ See Figures 34 thru 41

Generally the Double Offset Joint (DOJ) is used as the outer CV-joint.

A Tripot Joint (TJ) may also be referred to as a Tulip Joint because of the physical shape of it that resembles a tulip.

1. Remove the halfshaft and carefully place it in a vise using a protective covering on the vise jaws.

✳✳ WARNING

Some halfshafts may utilize hollow shafts between the CV-joints. Do not tighten the vise more than necessary.

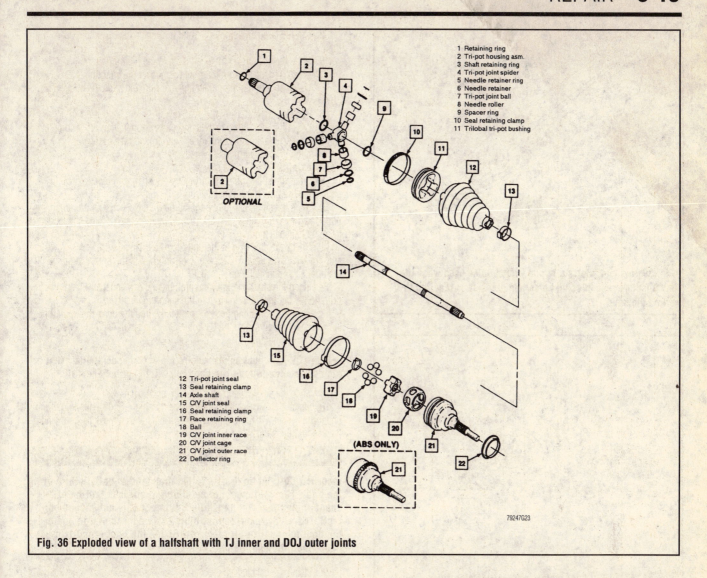

1 Retaining ring
2 Tri-pot housing asm.
3 Shaft retaining ring
4 Tri-pot joint spider
5 Needle retainer ring
6 Needle retainer
7 Tri-pot joint ball
8 Needle roller
9 Spacer ring
10 Seal retaining clamp
11 Trilobal tri-pot bushing

OPTIONAL

12 Tri-pot joint seal
13 Seal retaining clamp
14 Axle shaft
15 C/V joint seal
16 Seal retaining clamp
17 Race retaining ring
18 Ball
19 C/V joint inner race
20 C/V joint cage
21 C/V joint outer race
22 Deflector ring

(ABS ONLY)

79247G23

Fig. 36 Exploded view of a halfshaft with TJ inner and DOJ outer joints

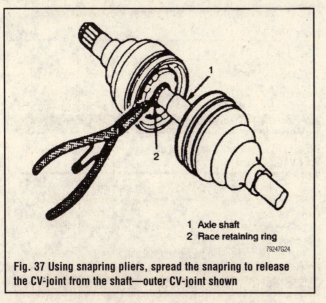

1 Axle shaft
2 Race retaining ring

79247G24

Fig. 37 Using snapring pliers, spread the snapring to release the CV-joint from the shaft—outer CV-joint shown

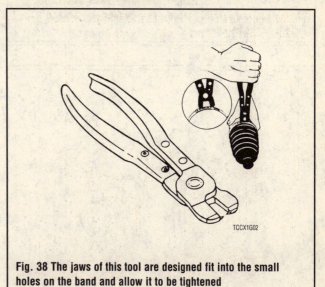

TCCX1G02

Fig. 38 The jaws of this tool are designed fit into the small holes on the band and allow it to be tightened

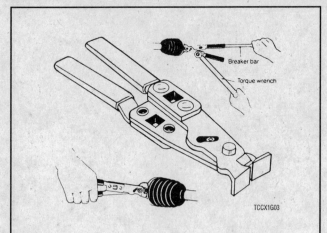

Fig. 39 This tool allows a torque wrench to be used when the manufacturer specifies that a certain pressure is required to crimp the band

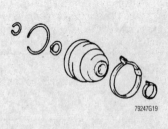

Fig. 40 The typical boot replacement kit contains a new boot, two clamps and special grease—new circlips may also be included is some kits

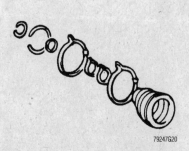

Fig. 41 The boot kit for the Birfield joint should contain an extra set of bands, because the inner joint must be removed in order to install a new boot on the outer (Birfield) joint

2. Cut the large and small CV-joint boot band clamps and discard them.
3. Slide the boot down the shaft uncovering the outer joint.
4. Clean the grease from the joint to uncover the snapring.
5. Using snapring pliers, open the snapring and slide the outer joint off the shaft.
6. Remove the boot from the shaft.
7. Clean the joint thoroughly using parts cleaner, then dry it completely with compressed air. Inspect the inner bearing and race assembly.
8. If the joint is worn or damaged, replace it.

To install:

9. Wrap the splines on the end of the halfshaft with tape to prevent damage to the boot during installation.
10. Slide the small CV-joint boot clamp onto the halfshaft and push the boot down several inches past the seal mounting area. Remove the tape from the halfshaft splines.
11. Check the snapring in the outer joint for damage or excessive wear and replace as necessary. Pack the joint with half of the grease supplied in the boot kit and install it on the shaft.
12. Insert the shaft into the joint until the splines engage. With a brass drift, lightly tap the joint down until the snapring engages.
13. Pack the remaining grease from the kit into the boot, then pull the large side of the boot over the CV-joint. Seat the small end of the boot on the seal mounting area.

➡Some CV-joint boot bands require the use of special pliers that are designed to grip the band and allow it to be tightened.

14. Slide the small clamp into position and secure it.
15. Install the large clamp in the proper position. Slide a small, dull tool under the lip of the boot to equalize the air pressure, remove the tool, and then secure the band.

✶✶ WARNING

Incorrect CV-joint boot installation may lead to early failure of the boot. The boot must not be dimpled, stretched or out of shape in any way when installed. If the boot is not shaped correctly, carefully insert a thin, blunt tool under the large end of the boot to equalize air pressure. Shape the boot properly by hand, then remove the tool.

16. Install the halfshaft.
17. Road test the vehicle to check for abnormal noise or vibration.

Inner Boot

▶ **See Figures 42 and 43**

1. Remove the halfshaft and carefully place it in a vise using a protective covering on the vise jaws.
2. Cut the large and small boot clamps, remove and discard the clamps.

✶✶ WARNING

Do not cut through the boot and damage the sealing surface of the CV-joint housing.

3. Pull the boot down the shaft to expose the joint.
4. If equipped, remove the large circlip from the inner edge of the outer bearing race.
5. Matchmark the bearing and outer case so they can be installed in their original positions.
6. Remove the housing from the spider and axle. Clean and dry all components thoroughly. Replace any parts that show signs of wear.
7. Push the spider assembly down the shaft to uncover the snapring on the end of the shaft. Remove the snapring and slide the spider assembly off the end of the shaft.
8. If equipped, remove the spacer ring from the shaft.

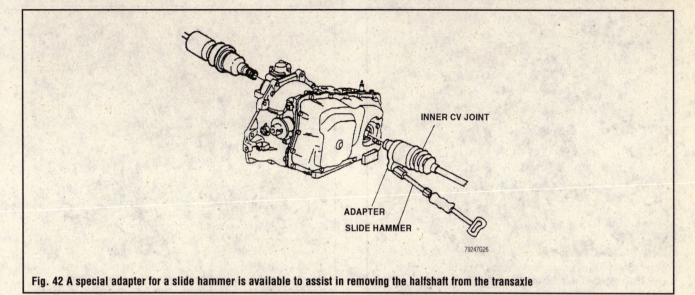

Fig. 42 A special adapter for a slide hammer is available to assist in removing the halfshaft from the transaxle

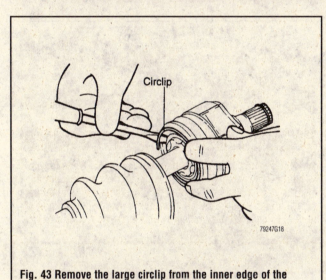

Fig. 43 Remove the large circlip from the inner edge of the outer race to release the bearing assembly

9. Remove the remaining circlip and slide the boot off the shaft.

10. Clean and dry all components thoroughly. Replace any parts that show signs of wear.

To install:

11. Slide the small clamp onto the halfshaft.

12. Slide the boot onto the shaft until the small end of the boot is in the original groove that it was removed from.

13. Install the applicable circlip.

14. If equipped, install the spacer ring on the shaft, several inches below the second spacer ring groove.

15. Install the spider assembly far enough down the shaft to expose the top snapring groove. Make sure the counterbored face of the spider faces the end of the shaft.

16. Install the top snapring and pull the spider assembly back up into position.

17. If equipped, lock the spacer ring in the spacer ring groove.

18. Pack the housing with half of the grease supplied in the kit and put the rest of the grease in the boot.

19. Slide the larger clamp over the boot.

20. Push the housing over the spider assembly.

21. If equipped, install the large circlip in the outer race.

22. Slide the larger diameter of the boot into position. Slide a small dull tool under the lip of the boot to equalize the air pressure, remove the tool, and then secure the band in position.

❋❋ WARNING

The boot must not be dimpled, stretched or out of shape in any way. If boot is not shaped correctly, carefully insert a thin flat blunt tool at the large end of the boot to equalize pressure. Shape the boot properly by hand and then remove the tool.

23. Install the halfshaft.

24. Road test the vehicle, check for abnormal road noise and vibrations.

CV-JOINT OVERHAUL

◆ See Figures 44 thru 57

Most manufacturers use several different types of joints. Engine size, transaxle type, whether the joint is an inboard or outboard joint, even which side of the vehicle is being serviced could make a difference in joint type. Be sure to properly identify the joint before attempting joint or boot replacement. Look for identification numbers at the large end of the boots and/or on the end of the metal retainer bands.

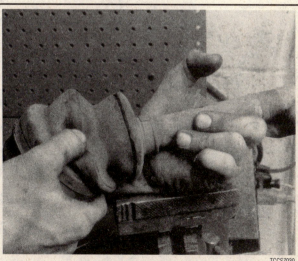

Fig. 44 Check the CV-boot for wear

Fig. 45 Removing the outer band from the CV-boot

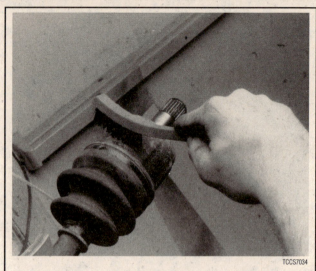

Fig. 48 Clean the CV-joint housing prior to removing boot

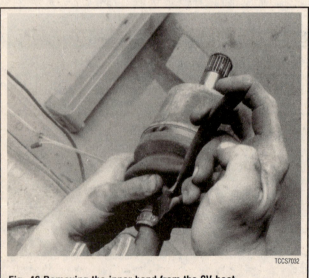

Fig. 46 Removing the inner band from the CV-boot

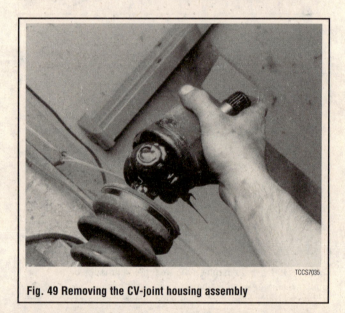

Fig. 49 Removing the CV-joint housing assembly

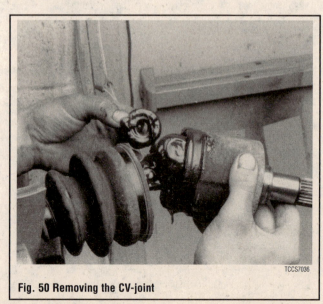

Fig. 47 Removing the CV-boot from the joint housing

Fig. 50 Removing the CV-joint

TCCS7037

Fig. 51 Inspecting the CV-joint housing

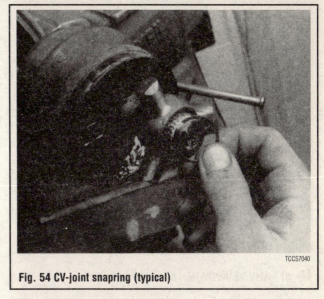

TCCS7040

Fig. 54 CV-joint snapring (typical)

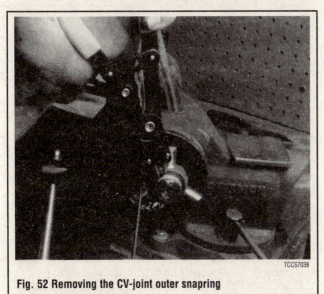

TCCS7038

Fig. 52 Removing the CV-joint outer snapring

TCCS7041

Fig. 55 Removing the CV-joint assembly

TCCS7039

Fig. 53 Checking the CV-joint snapring for wear

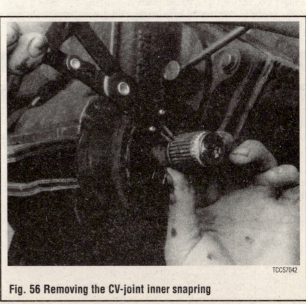

TCCS7042

Fig. 56 Removing the CV-joint inner snapring

Fig. 57 Installing the CV-joint assembly (typical)

The 3 most common types of joints used are the Birfield Joint, (B.J.), the Tripod Joint (T.J.) and the Double Offset Joint (D.O.J.).

➡ **Do not disassemble a Birfield joint. Service with a new joint or clean and repack using a new boot kit.**

In addition, some shafts may have dynamic damper installed on the shaft. Special grease is generally used with these joints and is often supplied with the replacement joint and/or boot. Do not use regular chassis grease.

The distance between the large and small boot bands is important and should be checked prior to and after boot service. This is so the boot will not be installed either too loose or too tight, which could cause early wear and cracking, allowing the grease to get out and water and dirt in, leading to early joint failure.

➡ **The driveshaft joints use special grease, do not add any grease other than that supplied with the kit.**

Double Offset Joint

The Double Offset Joint (D.O.J.) is bigger than other joints and in these applications, is normally used as an inboard joint.
1. Remove the halfshaft from the vehicle.
2. Side cutter pliers can be used to cut the metal retaining bands. Remove the boot from the joint outer race.
3. Locate and remove the large circlip at the base of the joint. Remove the outer race (the body of the joint).

4. Remove the small snapring and take off the inner race, cage and balls as an assembly. Clean the inner race, cage and balls without disassembling.
5. If the boot is to be reused, wipe the grease from the splines and wrap the splines in vinyl tape before sliding the boot from the shaft.
6. Remove the inner (D.O.J.) boot from the shaft. If the outer (B.J.) boot is to be replaced, remove the boot retainer rings and slide the boot down and off of the shaft at this time.
   To install:
7. Be sure to tape the shaft splines before installing the boots. Fill the inside of the boot with the specified grease. Often the grease supplied in the replacement parts kit is meant to be divided in half, with half being used to lubricate the joint and half being used inside the boot.
8. Install the cage onto the halfshaft so the small diameter side of the cage is installed first. With a brass drift pin, tap lightly and evenly around the inner race to install the race until it comes into contact with the rib of the shaft. Apply the specified grease to the inner race and cage and fit them together. Insert the balls into the cage.
9. Install the outer race (the body of the joint) after filling with the specified grease. The outer race should be filled with this grease.
10. Tighten the boot bands securely. Make sure the distance between the boot bands is correct.
11. Install the halfshaft to the vehicle.

Except Double Offset Joint

1. Disconnect the negative battery cable. Remove the halfshaft.
2. Use side cutter pliers to remove the metal retaining bands from the boot(s) that will be removed. Slide the boot from the T.J. case.
3. Remove the snapring and the tripod joint spider assembly from the halfshaft. Do not disassemble the spider and use care in handling.
4. If the boot is be reused, wrap vinyl tape around the spline part of the shaft so the boot(s) will not be damaged when removed. Remove the dynamic damper, if used, and the boots from the shaft.
   To install:
5. Double check that the correct replacement parts are being installed. Wrap vinyl tape around the splines to protect the boot and install the boots and damper, if used, in the correct order.
6. Install the joint spider assembly to the shaft and install the snapring.
7. Fill the inside of the boot with the specified grease. Often the grease supplied in the replacement parts kit is meant to be divided in half, with half being used to lubricate the joint and half being used inside the boot. Keep grease off the rubber part of the dynamic damper (if used).
8. Secure the boot bands with the halfshaft in a horizontal position. Make sure distance between boot bands is correct.
9. Install the halfshaft to the vehicle and reconnect the negative battery cable.

OFF VEHICLE REPAIRS

Transaxle

REMOVAL & INSTALLATION

1. Disconnect the negative battery cable.
2. Open and support the hood.
3. If equipped, tag and disconnect the throttle kickdown cable from the throttle linkage.
4. Install an engine support bar tool to support the weight of the engine. The engine must be properly supported for transaxle removal.
5. Remove the upper transaxle housing bolts.
6. Raise and safely support the vehicle.
7. Tag and disconnect the shift linkage. The linkage may be attached by means of a bolt, nut or some sort of a retaining clip.

8. Tag and disengage any electrical connectors, cables and hoses that would interfere with transaxle removal.
9. If the transaxle fluid pan has a drain plug, drain the fluid as follows:
   a. Place a suitable drain pan under the transaxle.
   b. Loosen the plug and allow the fluid to completely drain from the transaxle.
   c. Loosen all of the pan attaching bolts to within a few turns of complete removal, then carefully break the gasket seal and allow what is left of the fluid to drain over the edge of the pan.

✳✳ WARNING

DO NOT force the pan while breaking the gasket seal. DO NOT allow the pan flange to become bent or otherwise damaged.

d. When the fluid has drained, remove the pan bolts and carefully lower the pan, doing your best to drain the rest of the fluid into the drain pan.

10. If the transaxle fluid pan does not have a drain plug, drain the fluid as follows:

a. Place a large drain pan under the transaxle.

b. Loosen all the pan bolts except the bolts on the four corners.

c. Loosen the front two bolts about three turns and the rear two bolts about six turns.

d. Use a prytool to gently separate the pan from the transaxle.

e. As the fluid drains from the pan, keep loosening the bolts in the same two-to-one ratio allowing all the fluid to completely drain.

f. When fluid has drained, remove the pan bolts and the pan, doing your best to drain the rest of the fluid into the drain pan.

11. Install the pan and tighten the bolts until they are snug.

12. Unfasten the transaxle cooler line fittings at the radiator and disconnect the lines. Cap the lines to avoid fluid leakage.

13. If necessary, remove the starter motor.

14. Remove the torque converter cover.

15. Separate the halfshafts from both steering knuckles.

16. Remove both halfshafts from the transaxle. Install two transaxle plugs into the differential side gears.

17. Position a suitable transmission jack under the transaxle. Secure the transaxle to the jack.

18. Mark the retainers securing the torque converter to the flexplate so that they can be installed in their original positions.

19. Remove the retainers.

20. Unfasten the transaxle mount(s) retainers. Roll the mount(s) out of the way to allow clearance for transaxle removal, or remove the mount(s).

21. Remove any crossmembers that would interfere with transaxle removal.

22. Remove the remaining transaxle housing bolts.

23. Check that all mounts, wires, cables etc., that would hinder the removal of the transaxle are unbolted, unplugged, removed or moved aside.

24. Disengage the transaxle from the engine. Be careful not to drop the torque converter, then as soon as access is possible, install a torque converter retaining strap.

25. Lower the transaxle jack and remove the transaxle from the vehicle.

26. Remove the transaxle from the jack and place it on a work bench or a suitable holding device.

27. If your transaxle was removed because of internal damage, flush the transaxle fluid cooling system lines, fluid compartment (part of the radiator) or if equipped, an external cooler. Refer to your factory information for this procedure.

To install:

28. Make sure the torque converter is properly seated and a retaining strap is installed.

29. Position the transaxle on the jack. Secure the transaxle to the jack.

30. Raise the jack until the transaxle is in position.

31. Install the lower transaxle housing bolts.

32. Install any crossmembers that were removed.

33. Install the transaxle mount(s) and tighten the retainers.

34. Align and install the transaxle-to-flexplate retainers in their original positions.

35. Remove the transaxle plugs and install the halfshafts to the transaxle.

36. Attach the halfshafts to the steering knuckles.

37. Remove the transaxle jack.

38. Install the torque converter cover.

39. If removed, install the starter motor.

40. Unplug and attach the transaxle cooler lines to the radiator and tighten the fittings.

41. Tighten the pan bolts to specification. Refer to the Total Car Care manual for your vehicle for all torque specifications.

42. Connect the shift linkage and install the linkage retainer.

43. Attach any electrical connectors, cables and hoses that were removed.

44. Lower the vehicle.

45. Install and tighten the upper transaxle housing bolts.

46. If equipped, connect the throttle kickdown cable to the throttle linkage.

47. Connect the negative battery cable.

48. Add the proper type and quantity of transmission fluid as follows:

a. Add fluid to the transaxle until you reach the **COLD** or **LOW** mark on the dipstick.

b. Start the vehicle and let it reach normal operating temperature.

c. Move the gear selector through all gears in the shift pattern.

d. Check the transaxle fluid level. Add fluid, as necessary, to obtain the correct level. Be careful not to overfill the transaxle with fluid as this could cause internal damage to the assembly.

49. Check the transaxle for leaks and for proper operation.

50. Road test the vehicle and check for proper operation.

GENERAL OVERHAUL

▶ **See Figures 58 thru 101**

The following is a general description of automatic transaxle overhaul. After consulting a number of transmission specialty shops we chose a GM 440-T4 transaxle for photographic purposes as this unit contains components which are common to most automatic transaxles, though the components in your transaxle may use different names and terminology, the function of the components are basically the same. The photographs used should give you a basic idea of what to expect when you perform a teardown and overhaul of your transaxle assembly.

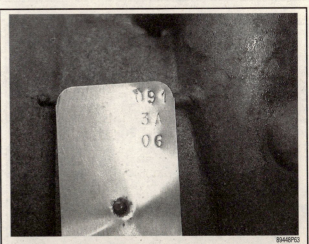

89448P63

Fig. 58 A identification tag is usually attached to the transaxle, this tag can help you identify the transaxle, and also can be of assistance when ordering parts

When overhauling a transaxle, the transaxle should be completely disassembled and cleaned. All seals, friction materials and bushings should be replaced and all adjustments made to ensure proper transaxle performance.

1. Remove the transaxle from the vehicle.
2. Place the transaxle on a suitable work bench or holding fixture.
3. Check the input shaft and output shaft end-play using a dial indicator and note the readings. Thrust washers in the transaxle assembly control the end-play.
4. When disassembling a transaxle assembly, note the locations of the bolts, washers, thrust washers, snaprings and seals. It is a good idea to make a template from Styrofoam or a similar material of your valve body so that the check balls and springs may be cleaned, stored and if being reused, replaced in their original position.
5. If your transaxle/transmission uses a chain, check the chain for wear and tear.
6. Tag all wires and hoses before unplugging or removing them so that they can be installed or attached in their original positions.
7. Prepare a large clean work surface on which to lay the components.
8. It is a good idea to clean and rebuild the components as they are removed, then set the components in the order they were removed to avoid confusion.

9. You should also draw or write down the component order of disassembly, this will make installation easier.
10. The transaxle components should be cleaned in a suitable parts washer and air dried. Do not clean the parts with a shop rag as this could leave lint deposits that could cause internal component damage.
11. Remove the torque converter by pulling it from the converter housing.
12. Remove the oil pan and/or any side pans also remove the transaxle fluid filter.
13. Remove the transaxle fluid pump. If necessary use a suitable pulling device to remove the pump.

➡️**The transaxle fluid pump is usually removed from the case bell housing, but in some cases the pump may be located under a side cover, at the rear of the case or under the valve body.**

14. Unfasten the valve body retainers and remove the valve body
15. Remove the accumulator piston assemblies and if equipped, governor assemblies from the transaxle by removing their covers and withdrawing them from the case.
16. If your transaxle uses band(s), loosen the band adjusters and remove the band(s).

Fig. 59 Unfasten the shift linkage retainers and remove the linkage

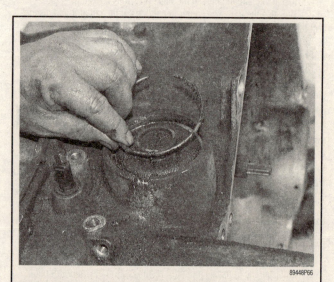

Fig. 61 Remove the accumulator snapring

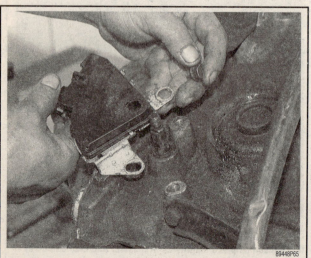

Fig. 60 Unfasten the neutral safety switch retainers and remove the switch

Fig. 62 A pair of pliers may be necessary to loosen the accumulator cover

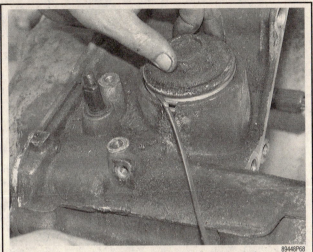

Fig. 63 If the cover is hard to remove, cut the cover seal and pull the seal to remove it with the cap still in place

Fig. 66 After all the retainers have been removed, separate the side cover from the transaxle case

Fig. 64 After the seal has been removed. Remove the cover and the accumulator assembly

Fig. 67 Tag and unplug all electrical connections from the valve body

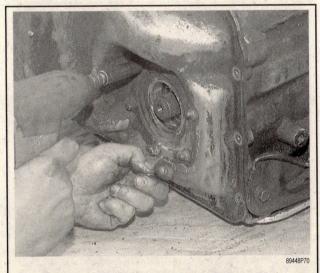

Fig. 65 Unfasten the transaxle side cover retainers

Fig. 68 Unfasten the valve body retainers and remove the valve body and pump as an assembly

Fig. 69 Unfasten the pump-to-valve body retainers and separate the pump from the valve body

Fig. 72 Remove the valve body-to-case gasket

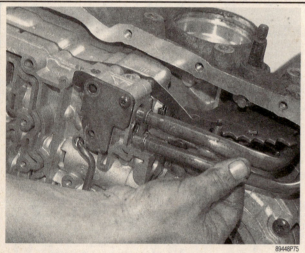

Fig. 70 Separate the fluid pipes from the lower valve body assembly

Fig. 73 Check the transaxle case and its components for cracks and damage

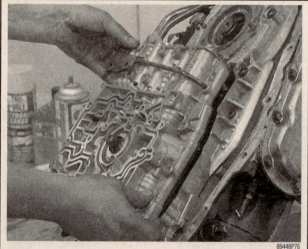

Fig. 71 Unfasten the lower valve body bolts and remove the valve body assembly

Fig. 74 Remove the oil pump shaft

Fig. 75 Remove the lower transaxle oil pan . . .

Fig. 78 Remove the accumulator gasket

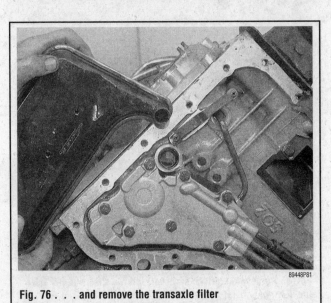

Fig. 76 . . . and remove the transaxle filter

Fig. 79 Remove the fluid pipe, seal, retainer and spring assemblies

Fig. 77 Remove the accumulator cover and the fluid pipes

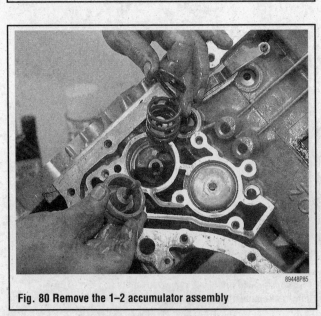

Fig. 80 Remove the 1–2 accumulator assembly

Fig. 81 Remove the 3– accumulator assembly from its housing in the case

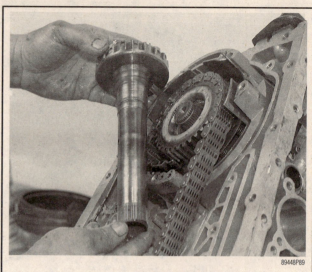

Fig. 84 Remove the fourth clutch hub and shaft assembly

Fig. 82 Unfasten the channel plate retainers and remove the channel plate from the transaxle

Fig. 85 Remove the driven sprocket, turbine shaft and the drive link assembly (chain)

Fig. 83 Remove the fourth clutch assembly

Fig. 86 Remove the driven sprocket support . . .

Fig. 87 . . . and the second clutch and input clutch assemblies

Fig. 90 Remove the input sun gear from the shaft

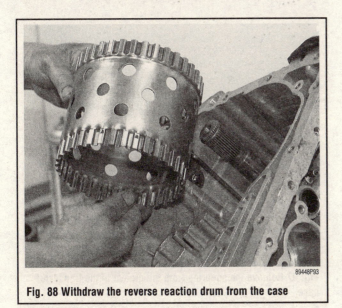

Fig. 88 Withdraw the reverse reaction drum from the case

Fig. 91 Remove the input carrier assembly

Fig. 89 Remove but do not disassemble the roller clutch assembly, check that the clutch turns smoothly in one direction only

Fig. 92 Remove the reaction carrier assembly, bearings and washers

Fig. 93 Gently slide the planetary gear set off the shaft

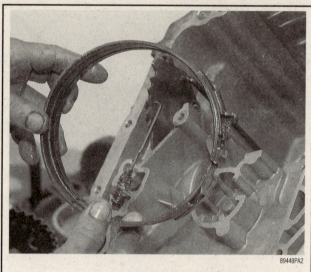

Fig. 96 . . . and the 1–2 band from the transaxle case

Fig. 94 Remove and inspect the planetary gear set bearing, replace as necessary

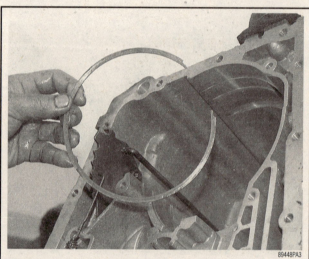

Fig. 97 Remove the snapring from the groove in the transaxle case

Fig. 95 Remove the reaction sun gear drum . . .

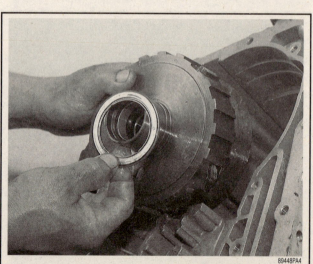

Fig. 98 Remove the internal final drive gear and bearing, inspect and replace the bearing as necessary

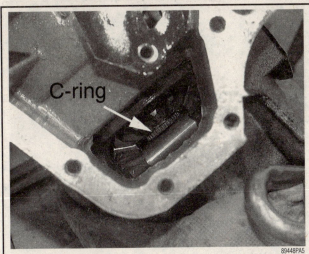

Fig. 99 The output shaft C-ring (arrow) must be removed in order to remove the output shaft

Fig. 100 After the C-ring has been removed, withdraw the output shaft from the case

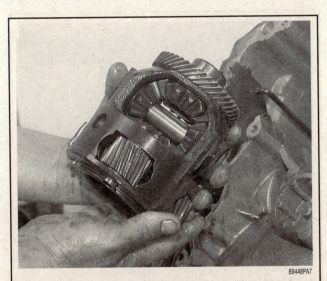

Fig. 101 Remove the final drive carrier from the transaxle case

➡The components of the transaxle are usually retained by a series of snaprings. The snaprings can be of varying sizes and installed in various locations. Some snaprings are located on the shafts, in the clutch drums or they may be on the transaxle case. Never try to force any of the components if they are hard to remove.

17. Remove the snaprings, thrust bearings, thrust washers, clutch packs, overrunning clutches and planetary gear sets from the transaxle/transmission case.

18. Remove all the oil rings, O-rings and seals.

19. Remove all of the transaxle case seals.

To install:

20. Disassemble, clean and rebuild all the transaxle/transmission components.

21. Thoroughly clean the transaxle case.

22. Install new external seals on the transaxle case.

23. By either consulting your own notes or using the correct factory information, install the clutch packs, overrun clutches, planetary gear sets, snaprings, seals, O-rings and gaskets in the reverse order of removal.

24. Install all the kickdown band(s), if equipped.

25. Install the accumulators and if equipped, the governor valve.

26. Air check the components using reduced compressed air and test plates. The test plates are designed by each manufacturer for a specific transaxle and can be purchased at an auto parts store or your vehicle's dealership.

27. Install the valve body.

28. Using your factory information, make any adjustments required by the transaxle manufacturer.

29. Install a new filter and the oil pan.

30. Install the transaxle pump.

31. Check the input and output shaft end-play. Compare your readings to the factory specifications. If the readings are not within specifications you may have to install thicker or thinner thrust washers.

32. Install the torque converter making sure that it is properly seated.

33. Install the transaxle/transmission in the vehicle.

34. Add the proper type and quantity of transmission fluid as follows:

    a. Add fluid to the transmission until you reach the **COLD** or **LOW** mark on the dipstick.

    b. Start the vehicle and let it reach normal operating temperature.

    c. Move the gear selector through all gears in the shift pattern.

    d. Check the transmission fluid level. Add fluid, as necessary, to obtain the correct level. Be careful not to overfill the transmission with fluid as this could cause internal damage to the assembly.

35. Road test the vehicle and check for proper operation.

COMPONENT OVERHAUL

♦ **See Figures 102 thru 108**

The following is a general description of automatic transaxle overhaul. After consulting a number of transmission specialty shops we chose a GM 440-T4 transaxle for photographic purposes as this unit contains components which are common to most automatic transaxles, though the components in your transaxle may use different names and terminology, but the function of the components are basically the same. The photographs used should give you a basic idea of what to expect when you perform a teardown and overhaul of your transaxle assembly. This procedure is for a typical overhaul of transaxle. It is intended to be used with factory materials by factory trained technicians. It is supplied here to provide the do-it-yourselfer with a general idea of the steps involved in overhauling a transaxle.

1. Remove the components from the transaxle assembly.

2. Throughly clean and dry all the components.

3. Once all components are cleaned and dried, they should be inspected for damage and if necessary, replaced.

4. Check that all bearings and thrust washers are damage free.

5. Check the planetary gear sets for damage such as worn gears, cracks and chipping.

6. Check that all oil passages are clean and free from any obstructions.

Fig. 102 Clean the pump case passages and if equipped, valves, with a suitable cleaner

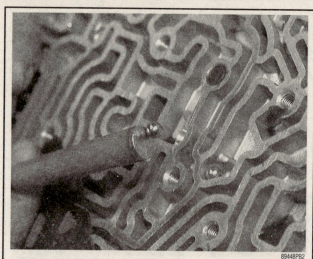

Fig. 105 If the valve body check balls are metal, a small magnet may be used to remove them.

Fig. 103 Unfasten the valve body separator plate retainers and remove the separator plate from the valve body

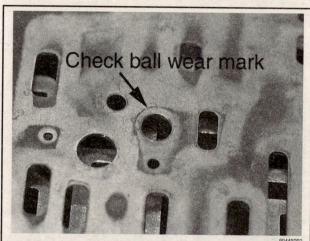

Fig. 106 Look at the back of the separator plate for check ball wear marks, this can help indicate the location of the check balls during valve body assembly

Fig. 104 Use a suitable cleaner to clean the valve body components

Fig. 107 Check the 1–2 band friction surface for wear

Fig. 108 Check the clutch pack components for wear and damage

7. Soak the new friction discs in unused transmission fluid for about half an hour before installation.

8. Clean all the metal oil rings grooves.

9. Install the new oil rings, make sure they fit snugly and there is no movement

10. Replace the O-rings with an O-ring of the same type and size. If the O-ring will not stay in place, coat it liberally with an assembly lube or petroleum jelly, this will usually hold the O-ring in place.

11. Check the condition of the overrun clutches, theses clutches are also referred to sprag clutches and one way clutches.

12. Make sure the overrun clutches turn smoothly in one direction and do not turn in the other direction. Always follow the factory instructions when installing the overrun clutch.

13. Check the condition of the kick-down band(s) friction material and replace if cracked, worn or otherwise damaged.

14. Separate the clutch and hub assemblies. Note the location of each component as it is removed, especially the thrust washers

15. Inspect the drums for cracks or damage and replace as necessary. Use a straightedge and check the outer drum surface for straightness where it may have been contacted by a band. If the drum is curved it may need to be replaced.

16. Remove the clutch assembly snapring and if applicable the clutch pressure plate.

17. Remove the clutch pack from the drum. Set the clutch pack aside in the order that they were removed, you can then use these for reference when installing the new clutch pack.

18. If the drum contains a piston return spring retainer and spring, proceed as follows:

a. Using a suitable tool, compress the clutch piston return spring and release the return spring snapring.

b. Remove the tool used to compress the clutch piston return spring.

c. Remove the snapring, spring retainer, return spring and the piston.

d. Clean all parts thoroughly.

e. Inspect the clutch piston for damage and replace as necessary.

f. Inspect the clutch plate steel plates for damage and flatness. Replace as necessary.

g. If the clutch drum contains a check valve, shake the drum to ensure the valve rattles. If the valve does not rattle, clean the orifice and try rattling it again, if this does not work, replace the drum.

h. Inspect the clutch drum bushing for wear and damage. Replace the bearing as necessary.

i. Remove the seals from the piston and drum and find the replacement seals in your rebuild kit.

j. Coat the new piston and drum seals with transmission fluid.

k. Install the clutch piston and clutch drum seals.

l. Install the piston in the drum and use a feeler gauge around the outer edge of the piston seal to ease the piston into position.

m. Install the return spring, spring retainer and the snapring.

n. Using a suitable tool, compress the spring and retainer until the snapring can be installed in its groove.

o. Install the clutch pack using the old pack as a reference.

p. Install the pressure plate, if applicable and the snap ring.

q. Check the clutch clearance. Refer to your factory information for the proper specification and procedure.

19. If your drum has a Belleville spring, proceed as follows:

a. Remove the Belleville spring retaining snapring.

b. Remove the Belleville spring and the piston.

c. Clean all parts thoroughly.

d. Inspect the clutch piston for damage and replace as necessary.

e. Inspect the clutch plate steel plates for damage and flatness. Replace as necessary.

f. If the piston has a check valve, shake the piston to ensure the valve rattles. If not , replace the piston.

g. Remove the seals from the piston and find the replacement seals in your rebuild kit.

h. Coat the new piston seals with transmission fluid.

i. Install the clutch piston seals.

j. Install the piston in the drum and use a feeler gauge around the outer edge of the piston seal to ease the piston into piston.

k. Install the Belleville spring and the snapring.

l. Install the clutch pack using the old pack as a reference.

m. Install the clutch pack retaining snap ring.

n. Check the clutch clearance. Refer to your factory information for the proper specification and procedure.

✻✻ WARNING

Always air check the clutch assemblies with the assembly facing down or away from you or anyone else.

20. Air check the clutch assemblies by applying reduced compressed air in short bursts to the hole in the drum or one of the holes on the shaft. Refer to your factory information for the proper air check technique and procedure.

21. Using your factory information, disassemble, clean, inspect and reassemble the valve body.

22. If equipped with a governor valve, clean the valve, check for proper valve operation and replace the valve seals.

23. Clean and inspect all accumulators. Replace the seals and any damaged parts.

24. Using your factory information, disassemble, clean, inspect and recondition the fluid pump.

25. Replace all seals and O-rings on the shafts.

26. Install the components to the transaxle assembly.

Transmission

REMOVAL & INSTALLATION

1. Disconnect the negative battery cable.

2. If equipped disconnect the throttle valve cable from the throttle assembly.

3. Raise and support the vehicle safely using jackstands.

4. Drain the transmission fluid.

5. Disconnect the shift linkage from the transmission.

6. Tag and disengage all hoses, electrical connections and hoses that would interfere with transmission removal.

7. Matchmark and remove the driveshaft(s).

8. Remove any components necessary for clearance.

9. Support the transmission assembly using a transmission jack.

10. Remove the transmission crossmember. Take care not to stretch or damage any cables or wiring when attempting to remove the crossmember.

11. Lower the transmission slightly for access, then remove the dipstick tube and seal. Cover or plug the opening in the transmission housing to prevent system contamination.

12. Disconnect the transmission fluid cooler lines. Plug or cap all openings to prevent system contamination or excessive fluid spillage.

13. Remove the transmission support braces. Be sure to tag or note the location of all support braces as they must be installed in their original positions.

14. Remove the torque converter housing cover.

15. Matchmark the flywheel-to-torque converter relationship, then remove the retaining bolts.

16. Support the engine using a jackstand and a block of wood.

17. Remove the transmission-to-engine retaining bolts. Note the positions of any brackets or clips and move them aside.

18. Using the transmission jack, slide the transmission straight back off the locating pins. Be careful not to drop the torque converter, then as soon as access is possible, install a torque converter retaining strap.

19. Carefully lower the transmission from the vehicle.

20. If your transmission was removed because of internal damage, flush the transmission fluid cooling system lines, fluid compartment (part of the radiator) or if equipped, an external cooler. Refer to your factory information for this procedure.

To install:

21. Make sure the torque converter is properly seated and a retaining strap is installed.

22. Using the transmission jack, carefully raise the transmission into position in the vehicle, then remove the converter retaining strap.

23. Slide the transmission straight onto the locating pins while aligning the marks on the flywheel and torque converter.

➡**The converter must be flush on the flywheel and rotate freely by hand.**

24. Install the transmission-to-engine retainers, taking care to properly reposition all brackets, clips and harness, as noted during removal. The retainers should not be tightened yet. Do NOT install the dipstick tube or transmission support brace screws at this time.

25. Install and finger-tighten the converter bolts, then tighten them to specification.

26. Remove the jackstand from the engine.

27. Install the converter housing cover.

28. Install the transmission support braces as noted during removal.

29. Uncap the openings, then connect the transmission cooler lines. Be careful not to twist or bend the lines.

30. Uncover the opening in the transmission housing and position a new seal, then install the dipstick tube.

31. Tighten the transmission-to-engine retainers and the dipstick tube retainer to specification.

32. Raise the rear of the transmission (taking care not to pinch or damage any cables, wires or components), then install the crossmember. Secure the transmission mount and any components that were removed for access.

33. Remove the transmission jack from the transmission, then reattach the exhaust system.

34. Align and install the driveshaft(s).

35. Attach any hoses, electrical connections and hoses that were removed.

36. Connect the shift linkage and adjust, as necessary.

37. Add the proper type and quantity of transmission fluid as follows:

    a. Add fluid to the transmission until you reach the **COLD** or **LOW** mark on the dipstick.

    b. Start the vehicle and let it reach normal operating temperature.

    c. Move the gear selector through all gears in the shift pattern.

    d. Check the transmission fluid level. Add fluid, as necessary, to obtain the correct level. Be careful not to overfill the transmission with fluid as this could cause internal damage to the assembly.

38. Remove the jackstands and carefully lower the vehicle.

39. If equipped, engage the throttle valve cable, then check and adjust, as necessary.

40. Connect the negative battery cable.

41. Road test the vehicle and check for proper operation.

GENERAL OVERHAUL

➡ **See Figures 109 thru 130**

The following is a general description of automatic transmission disassembly and assembly. After consulting a number of transmission specialty shops we chose a Ford AOD transmission for photographic purposes as this unit contains components which are common to most automatic transmissions, though the components in your transmission may be called different names, the function of the components are basically the same. The photographs used should give you a basic idea of what to expect when you disassemble your transmission.

When overhauling a transmission, the transmission should be completely dissembled and cleaned. All seals, friction materials and bushings should be replaced and all adjustments made to ensure proper transmission performance.

1. Remove the transmission from the vehicle.

2. Place the transmission on a suitable work bench or holding fixture.

3. Check the input shaft and output shaft end-play using a dial indicator and note the readings. Thrust washers in the transmission assembly control the end-play.

4. When disassembling a transmission assembly, note the locations of the bolts, washers, thrust washers, snaprings and seals. It is a good idea to make a template from Styrofoam or a similar material of your valve body so that the check balls and springs may be cleaned, stored and if being reused, replaced in their original position.

5. Tag all wires and hoses before unplugging or removing them so that they can be installed or attached in their original positions.

6. Prepare a large clean work surface on which to lay the components.

7. It is a good idea to clean and rebuild the components as they are removed, then set the components in the order they were removed to avoid confusion.

8. You should also draw or write down the component order of disassembly, this will make installation easier.

9. The transmission components should be cleaned in a suitable parts washer and air dried. Do not clean the parts with a shop rag as this could leave lint deposits that could cause internal component damage.

10. Remove the torque converter by pulling it from the converter housing.

11. Remove the oil pan and/or any side pans also remove the transmission fluid filter.

12. Remove the transmission fluid pump. If necessary use a suitable pulling device to remove the pump.

➡**The transmission fluid pump is usually removed from the case bell housing, but in some cases the pump may be located under a side cover, at the rear of the case or under the valve body.**

89448P01

Fig. 109 Your transmission may have an identification tag attached to the case. This tag contains information that may be helpful when ordering replacement parts.

Fig. 110 Unfasten the transmission pan retainers and separate the pan from the housing

Fig. 113 A slide hammer should be used to remove the transmission pump

Fig. 111 If equipped, remove the transmission filter retainers, then pull the filter from the valve body, be sure to remove any gaskets and O-rings associated with the filter

Fig. 114 Remove the pump from the housing

Fig. 112 Unfasten the valve body-to-transmission retainers (which can be of varying sizes and types) and remove the valve body assembly

Fig. 115 Grasp the turbine shaft and remove the intermediate clutch pack, intermediate one-way clutch. Reverse clutch and the forward clutch from the housing

13. Unfasten the valve body retainers and remove the valve body

14. Remove the accumulator piston assemblies and if equipped, governor assemblies from the transmission by removing their covers and withdrawing them.

15. If your transmission uses band(s), loosen the band adjusters and remove the band(s).

➡ **The components of the transmission are usually retained by a series of snaprings. The snaprings can be of varying sizes and installed in various locations. Some snaprings are located on the shafts, in the clutch drums or they may be on the transmission case. Never try to force any of the components if they are hard to remove.**

16. Remove the snaprings, thrust bearings, thrust washers, clutch assemblies, overrunning clutches and planetary gear sets from the transaxle/transmission case.

17. Remove all the oil rings, O-rings and seals.

18. Remove all of the transmission case seals.

To install:

19. Disassemble, clean and rebuild all the transmission components.

20. Thoroughly clean the transmission case.

21. Install new external seals on the transmission case.

Fig. 118 . . . and remove the servo assemblies

Fig. 116 Unfasten the 2–3 accumulator snapring and remove the assembly

Fig. 119 Remove the transmission band from the case

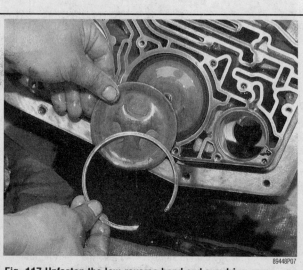

Fig. 117 Unfasten the low-reverse band and overdrive servo snaprings . . .

Fig. 120 Inspect the band friction surface for damage

Fig. 121 Remove the forward clutch hub and the needle bearing

Fig. 124 Unfasten the tailhousing retainers . . .

Fig. 122 Remove the forward sun gear and needle bearing

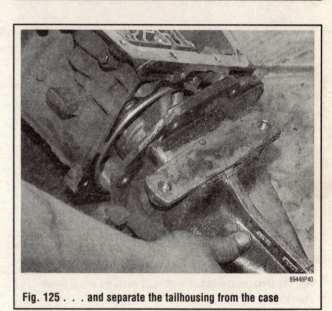

Fig. 125 . . . and separate the tailhousing from the case

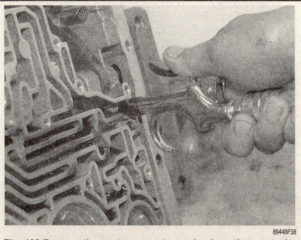

Fig. 123 Remove the center support snapring from the groove in the transmission case and remove the piston using compressed air

Fig. 126 Unfasten the governor assembly snapring and remove the governor assembly

Fig. 127 Use a magnet to remove the check ball from the output shaft

Fig. 128 Remove the output shaft, ring gear and direct clutch assembly

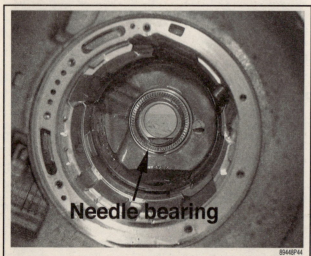

Fig. 129 Look for an output shaft needle bearing, if it is not on the shaft, it may still be in the transmission case

Fig. 130 Remove the output shaft needle bearing

22. By either consulting your own notes or using the correct factory information, install the clutch packs, overrun clutches, planetary gear sets, snaprings, seals, O-rings and gaskets in the reverse order of removal.

23. Install all the kickdown band(s), if equipped.

24. Install the accumulators and if equipped, the governor valve.

25. Air check the components using reduced compressed air and test plates. The test plates are designed by each manufacturer for a specific transmission and can be purchased at an auto parts store or your vehicle's dealership.

26. Install the valve body.

27. Using your factory information, make any adjustments required by the transmission manufacturer.

28. Install a new filter and the oil pan.

29. Install the transmission pump.

30. Check the input and output shaft end-play. Compare your readings to the factory specifications. If the readings are not within specifications you may have to install thicker or thinner thrust washers.

31. Install the torque converter making sure that it is properly seated.

32. Install the transaxle/transmission in the vehicle.

33. Add the proper type and quantity of transmission fluid as follows:

  a. Add fluid to the transmission until you reach the **COLD** or **LOW** mark on the dipstick.

  b. Start the vehicle and let it reach normal operating temperature.

  c. Move the gear selector through all gears in the shift pattern.

  d. Check the transmission fluid level. Add fluid, as necessary, to obtain the correct level. Be careful not to overfill the transmission with fluid as this could cause internal damage to the assembly.

34. Road test the vehicle and check for proper operation.

COMPONENT OVERHAUL

▶ **See Figures 131 thru 163**

The following is a general description of automatic transmission overhaul. After consulting a number of transmission specialty shops we chose a Ford AOD transmission for photographic purposes as this unit contains components which are common to most automatic transmissions, though the components in your transmission may be called different names, the function of the components are basically the same. The photographs used should give you a basic idea of what to expect when you perform a teardown and overhaul of your transmission assembly. This procedure is for a typical overhaul of transmission. It is intended to be used with factory materials by factory trained technicians. It is supplied here to provide the do-it-yourselfer with a general idea of the steps involved in overhauling a transmission.

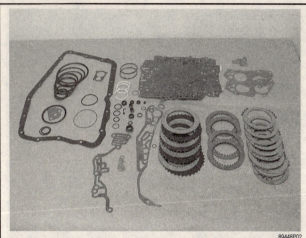

Fig. 131 A typical transmission overhaul kit should contain all the seals, gaskets and clutch packs necessary for transmission overhaul

1. Remove the components from the transmission assembly.
2. Throughly clean and dry all the components.
3. Once all components are cleaned and dried, they should be inspected for damage and if necessary, replaced.
4. Check that all bearings and thrust washers are damage free.
5. Check the planetary gear sets for damage such as worn gears, cracks and chipping.
6. Check that all oil passages are clean and free from any obstructions.
7. Soak the new friction discs in unused transmission fluid for about half an hour before installation.
8. Clean all the metal oil rings grooves.
9. Install the new oil rings, make sure they fit snugly and there is no movement.
10. Replace the O-rings with an O-ring of the same type and size. If the O-ring will not stay in place, coat it liberally with petroleum jelly, this will usually hold the O-ring in place.
11. Check the condition of the overrun clutches, theses clutches are also referred to sprag clutches and one way clutches.
12. Make sure the overrun clutches turn smoothly in one direction and do not turn in the other direction. Always follow the factory instructions when installing the overrun clutch.
13. Check the condition of the kick-down band(s) friction material and replace if cracked, worn or otherwise damaged.

Fig. 132 Remove and discard all O-rings and replace them with an O-ring of similar type and size

Fig. 133 If the O-ring will not stay in place, coat it liberally with petroleum jelly, this will usually hold the O-ring in place

14. Separate the clutch and hub assemblies. Note the location of each component as it is removed, especially the thrust washers.
15. Inspect the drums for cracks or damage and replace as necessary. Use a straightedge and check the outer drum surface for straightness where it may have been contacted by a band. If the drum is curved it may need to be replaced.
16. Remove the clutch assembly snapring and if applicable the clutch pressure plate.
17. Remove the clutch pack from the drum. Set the clutch pack aside in the order that they were removed, you can then use these for reference when installing the new clutch pack.
18. If the drum contains a piston return spring retainer and spring, proceed as follows:
    a. Using a suitable tool, compress the clutch piston return spring and release the return spring snapring.
    b. Remove the tool used to compress the clutch piston return spring.
    c. Remove the snapring, spring retainer, return spring and the piston.
    d. Clean all parts thoroughly.
    e. Inspect the clutch piston for damage and replace as necessary.
    f. Inspect the clutch plate steel plates for damage and flatness. Replace as necessary.
    g. If the clutch drum contains a check valve, shake the drum to ensure the valve rattles. If the valve does not rattle, clean the orifice and try rattling it again, if this does not work, replace the drum.
    h. Inspect the clutch drum bushing for wear and damage. Replace the bearing as necessary.
    i. Remove the seals from the piston and drum and find the replacement seals in your rebuild kit.
    j. Coat the new piston and drum seals with transmission fluid.
    k. Install the clutch piston and clutch drum seals.
    l. Install the piston in the drum and use a feeler gauge around the outer edge of the piston seal to ease the piston into position.
    m. Install the return spring, spring retainer and the snapring.
    n. Using a suitable tool, compress the spring and retainer until the snapring can be installed in its groove.
    o. Install the clutch pack using the old pack as a reference.
    p. Install the pressure plate, if applicable and the snap ring.
    q. Check the clutch clearance. Refer to your factory information for the proper specification and procedure.
19. If your drum has a Belleville spring, proceed as follows:
    a. Remove the Belleville spring retaining snapring.
    b. Remove the Belleville spring and the piston.
    c. Clean all parts thoroughly.
    d. Inspect the clutch piston for damage and replace as necessary.

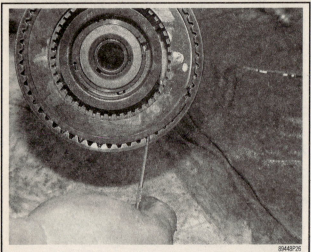

Fig. 134 If the drum contains a piston return spring retainer and spring, remove the first snapring . . .

Fig. 137 Remove the clutch piston return spring

Fig. 135 . . .then remove the clutch pack assembly from the drum

Fig. 138 Remove the piston assembly from the drum

Fig. 136 Using a suitable tool, compress the clutch piston return spring and release the return spring snapring

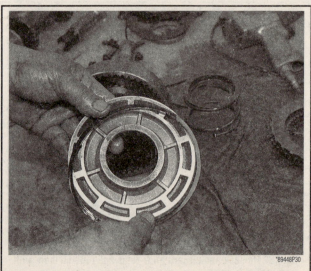

Fig. 139 Remove piston outside diameter seal . . .

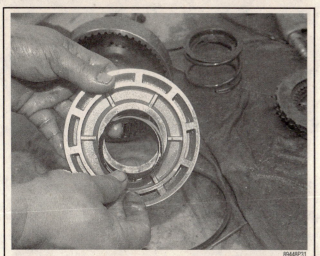

Fig. 140 . . . and outside diameter seal. Discard and replace the seals after thoroughly cleaning the piston assembly

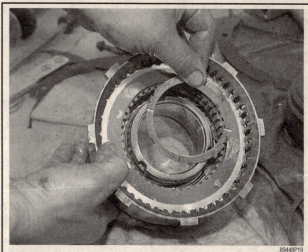

Fig. 143 If your clutch assembly is Belleville spring type, remove the clutch pack assembly snaprings . . .

Fig. 141 Install the piston in the drum and use a feeler gauge around the outer edge of the piston seal to ease the piston into position

Fig. 144 . . . and remove the clutch pack from the drum

Fig. 142 Remove the intermediate clutch pack from the intermediate one-way clutch

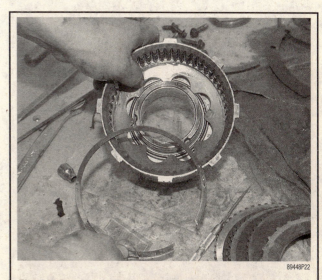

Fig. 145 Remove the Belleville spring retaining snapring

Fig. 146 Grasp the Belleville spring and remove it from the drum

Fig. 149 Remove the planetary center support assembly

Fig. 147 Remove the clutch piston and the piston inside diameter seal

Fig. 150 Remove the sprag clutch assembly, be careful not to drop any of the components

Fig. 148 Remove the outside diameter seal from the piston

Fig. 151 Separate the planetary fear assembly from the from the ring gear and park gear

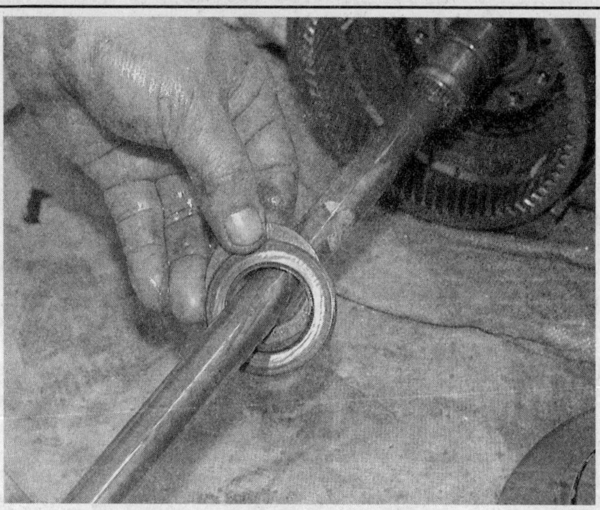

Fig. 152 Remove the direct clutch bearing

Fig. 155 Inspect the shaft for damage and replace the shaft seals

e. Inspect the clutch plate steel plates for damage and flatness. Replace as necessary.

f. If the piston has a check valve, shake the piston to ensure the valve rattles. If not , replace the piston.

g. Remove the seals from the piston and find the replacement seals in your rebuild kit.

h. Coat the new piston seals with transmission fluid.

i. Install the clutch piston seals.

j. Install the piston in the drum and use a feeler gauge around the outer edge of the piston seal to ease the piston into piston.

k. Install the Belleville spring and the snapring.

l. Install the clutch pack using the old pack as a reference.

m. Install the clutch pack retaining snap ring.

n. Check the clutch clearance. Refer to your factory information for the proper specification and procedure.

Fig. 153 Remove the intermediate shaft and inspect the splines for damage and wear

✱✱ WARNING

Always air check the clutch assemblies with the assembly facing down or away from you or anyone else.

20. Air check the clutch assemblies by applying reduced compressed air in short bursts to the hole in the drum, pump or transmission housing.

Fig. 154 Remove the output shaft-to-ring gear and park gear snap ring

Fig. 156 Some components can be air checked through passages in the pump housing or in passages in the transmission housing by applying reduced compressed air in short bursts

Fig. 157 Remove the separator plate from the valve body . . .

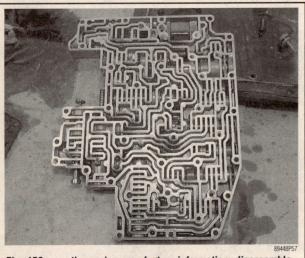

Fig. 158 . . . then using your factory information, disassemble, clean, inspect and reassemble the valve body

Fig. 159 Remove the lathe cut seal from the outside circumference of the pump

Fig. 160 Remove the gasket from the pump and clean the gasket mating surfaces

Fig. 161 Remove the seals from the pump shaft

Fig. 162 Unbolt and separate the pump assembly . . .

Fig. 163 . . . and inspect the pump gears for wear and damage

Refer to your factory information for the proper air check technique and procedure.

21. Using your factory information, disassemble, clean, inspect and reassemble the valve body.

22. If equipped with a governor valve, clean the valve, check for proper valve operation and replace the valve seals.

23. Clean and inspect all accumulators. Replace the seals and any damaged parts.

24. Using your factory information, disassemble, clean, inspect and recondition the fluid pump.

25. Replace all seals and O-rings on the shafts.

26. Install the components to the transmission assembly.

Transfer Case

On models equipped with four wheel drive systems, the transmission/transaxle may be removed with the transfer case still attached to the assembly, however in some cases it may be necessary to remove the transfer case before removing the transmission/transaxle . If the transfer case can be removed with the transmission/transaxle , remove the transmission/transaxle , unfasten all linkages and wires, unbolt the transfer case-to-transmission/transaxle bolts and separate the assemblies.

The following is a general idea of what to expect when removing a transfer case with the transmission still installed.

REMOVAL & INSTALLATION

1. Disconnect the negative battery cable.
2. Raise and support the vehicle safely using jackstands.
3. If equipped, remove the skid plate bolts, then remove the skid plate from under the transmission/transaxle transfer case assembly.
4. Remove the plug and drain the transfer case fluid.
5. Matchmark and remove the driveshafts from the transfer case.
6. Tag and unplug the vacuum lines and/or the electrical connectors, as equipped.
7. Tag and disconnect and/or remove any remaining linkages, brackets or crossmembers that would interfere with the transfer case removal.
8. Support the transfer case using a transmission jack, then remove the transfer case-to-transmission/transaxle retaining bolts
9. Slide the transfer case rearward and off the transmission/transaxle output shaft, then careful lower it from the vehicle.
10. Remove all traces of old gasket material from the mating surfaces.

To install:

11. Using the floor jack, carefully raise the transfer case into position on the transmission/transaxle . Position a new gasket, using sealer to hold it in position, then slide the transfer case onto the transmission/transaxle output shaft.

12. Install the transfer case-to-transmission/transaxle retaining bolts.

13. Connect and or install any linkages, brackets or crossmembers that were removed to facilitate the transfer case removal.

14. Remove the transmission jack from the transfer case.

15. Engage the vacuum lines and/or electrical connections, as necessary.

16. Align and install the driveshafts.

17. Properly refill the transfer case through the filler plug.

18. If equipped, install the skid plate.

19. If not done already, remove the jackstands and carefully lower the vehicle.

20. Connect the negative battery cable.

TRANSMISSION/TRANSAXLE EXPLODED AND SECTIONAL VIEWS

The following section shows various exploded or sectional views of automatic transmissions and transaxles. The transmissions and transaxles shown were the most commonly used models at the time this manual was published. Other models, which are available now, may not be shown.

In using these views, it must be remembered that not all transmissions and transaxles use the same components. Even when similar components are used, each manufacturer may not call them by the same name.

Most often, exploded views are used to understand the functioning of the transmission or transaxle. By seeing the interworking subassemblies laid out, a more complete understanding of the flow of power thorough the transmission or transaxle can be achieved. Other ways to use the views are when ordering parts or attempting to describe the components in a subassembly.

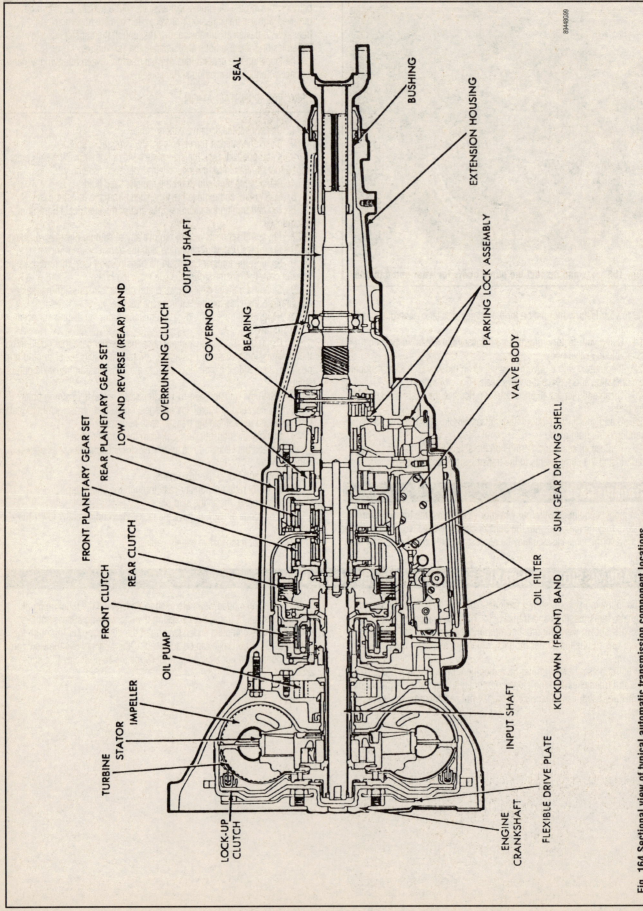

SEAL

BUSHING

EXTENSION HOUSING

OUTPUT SHAFT

PARKING LOCK ASSEMBLY

FRONT PLANETARY GEAR SET

REAR PLANETARY GEAR SET

LOW AND REVERSE (REAR) BAND

OVERRUNNING CLUTCH

GOVERNOR

BEARING

VALVE BODY

SUN GEAR DRIVING SHELL

FRONT CLUTCH

REAR CLUTCH

OIL FILTER

OIL PUMP

OIL (FRONT) BAND

KICKDOWN (FRONT) BAND

INPUT SHAFT

TURBINE

STATOR

IMPELLER

ENGINE CRANKSHAFT

FLEXIBLE DRIVE PLATE

LOCK-UP CLUTCH

89448G99

Fig. 164 Sectional view of typical automatic transmission component locations

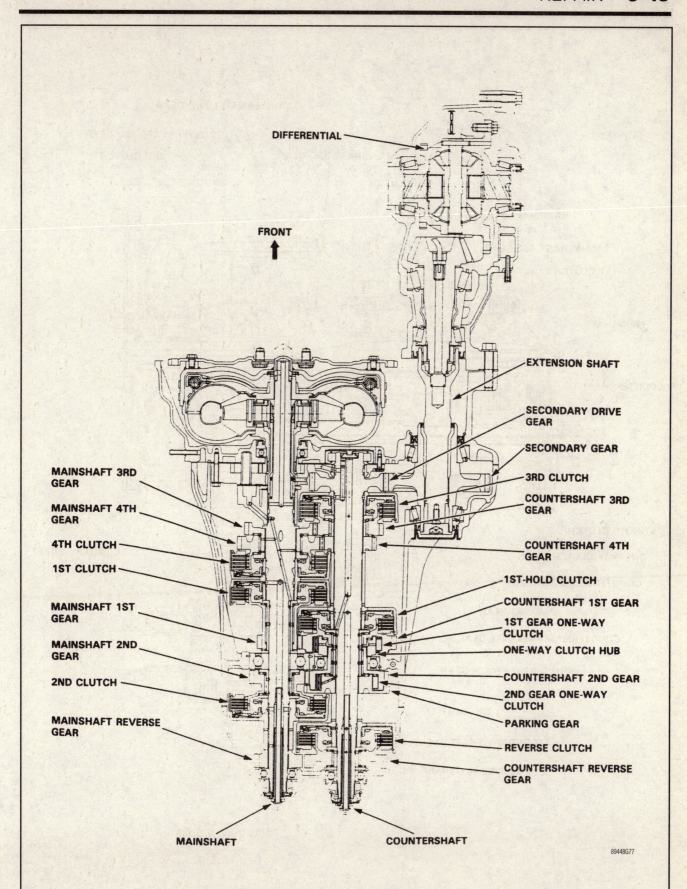

DIFFERENTIAL

FRONT

EXTENSION SHAFT

SECONDARY DRIVE GEAR

SECONDARY GEAR

MAINSHAFT 3RD GEAR

3RD CLUTCH

MAINSHAFT 4TH GEAR

COUNTERSHAFT 3RD GEAR

4TH CLUTCH

COUNTERSHAFT 4TH GEAR

1ST CLUTCH

1ST-HOLD CLUTCH

COUNTERSHAFT 1ST GEAR

MAINSHAFT 1ST GEAR

1ST GEAR ONE-WAY CLUTCH

MAINSHAFT 2ND GEAR

ONE-WAY CLUTCH HUB

COUNTERSHAFT 2ND GEAR

2ND CLUTCH

2ND GEAR ONE-WAY CLUTCH

PARKING GEAR

MAINSHAFT REVERSE GEAR

REVERSE CLUTCH

COUNTERSHAFT REVERSE GEAR

MAINSHAFT

COUNTERSHAFT

89448G77

Fig. 165 Sectional view of a Acura Legend transaxle assembly components

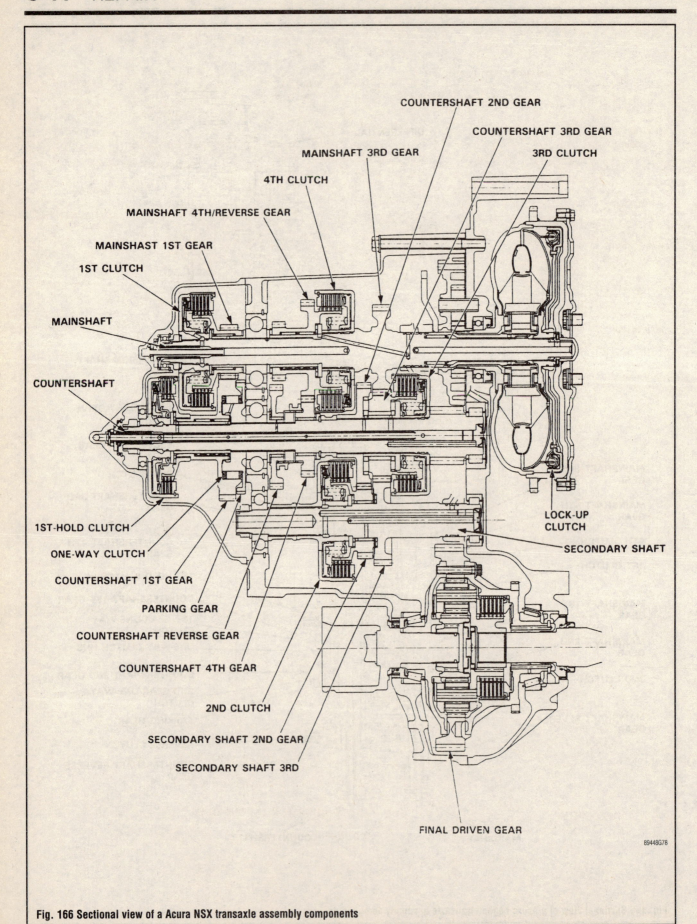

Fig. 166 Sectional view of a Acura NSX transaxle assembly components

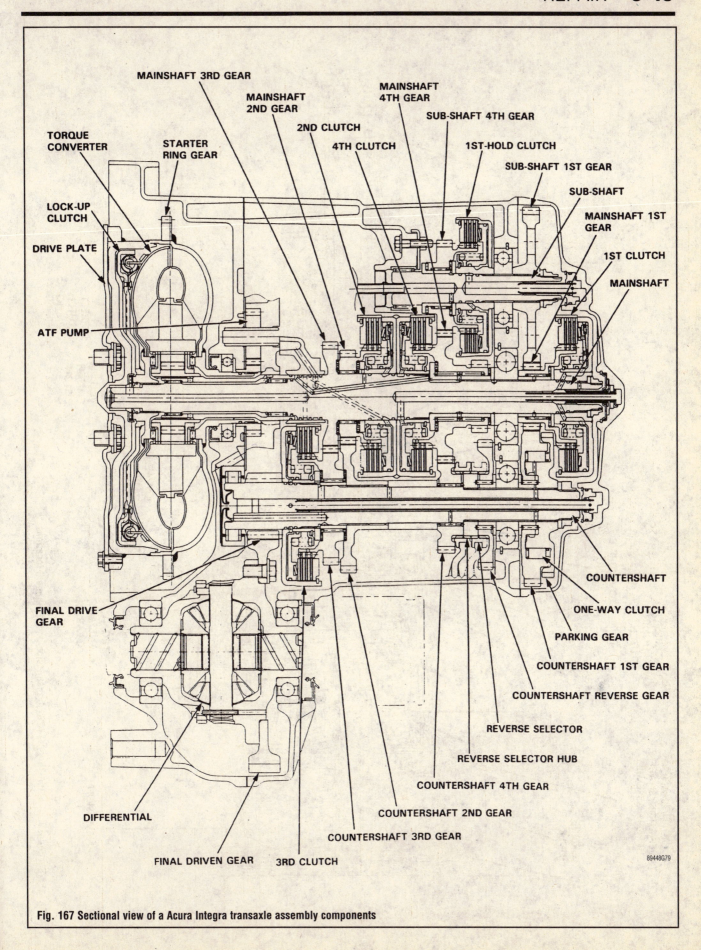

Fig. 167 Sectional view of a Acura Integra transaxle assembly components

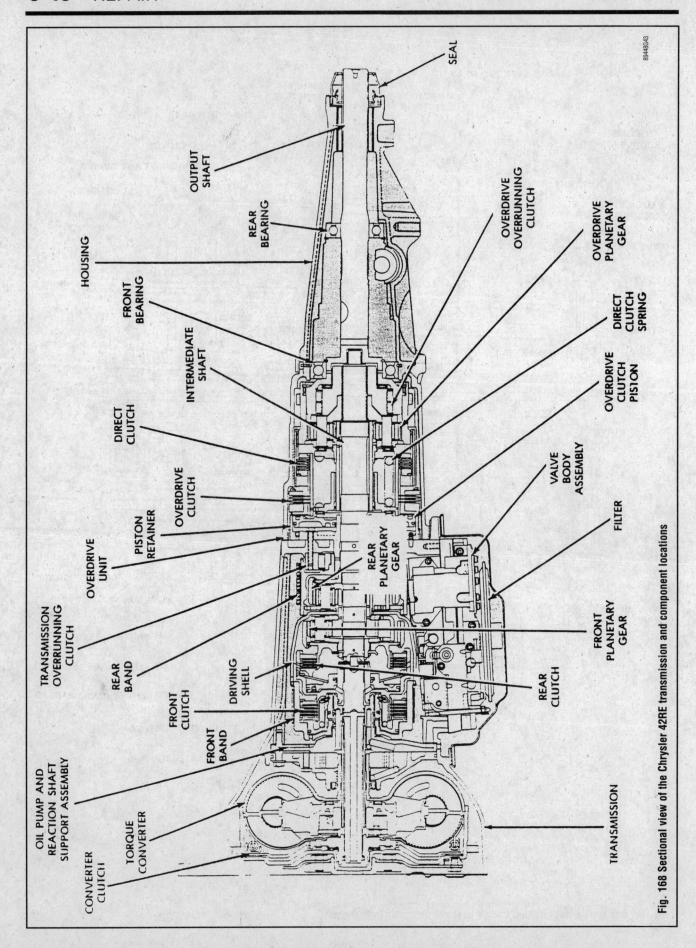

Fig. 168 Sectional view of the Chrysler 42RE transmission and component locations

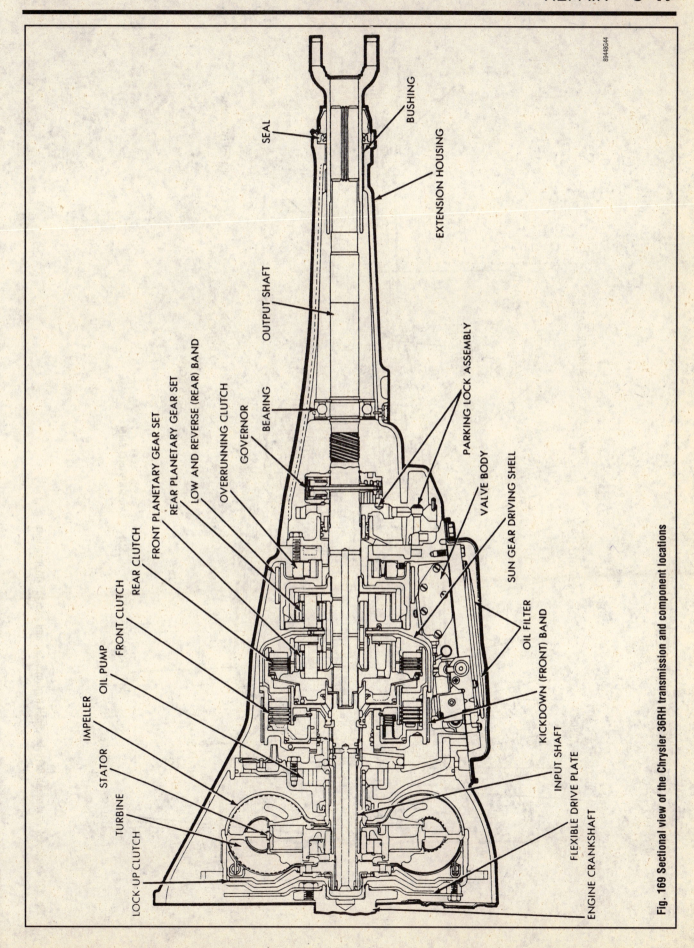

8948G44

Fig. 169 Sectional view of the Chrysler 36RH transmission and component locations

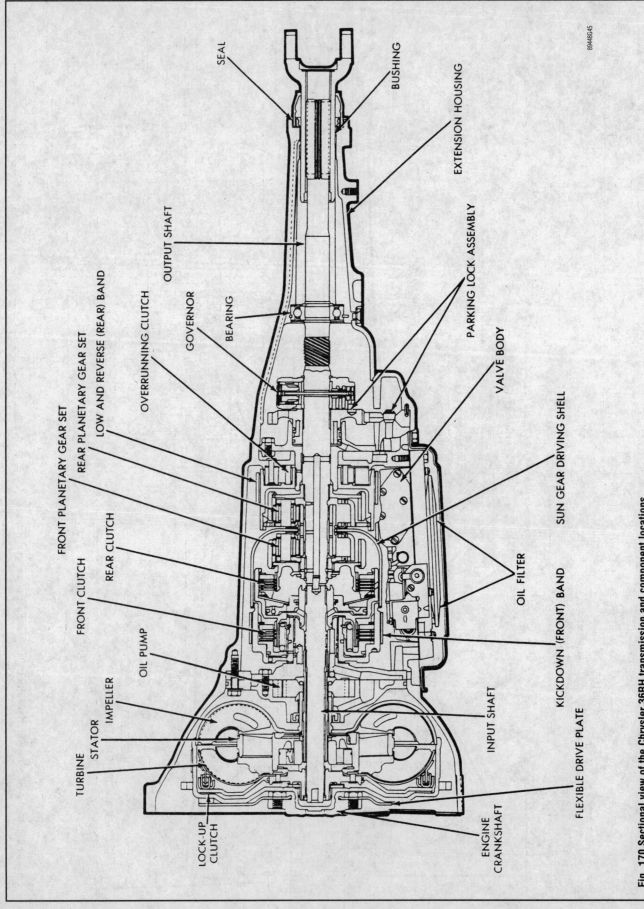

Fig. 170 Sectional view of the Chrysler 36RH transmission and component locations

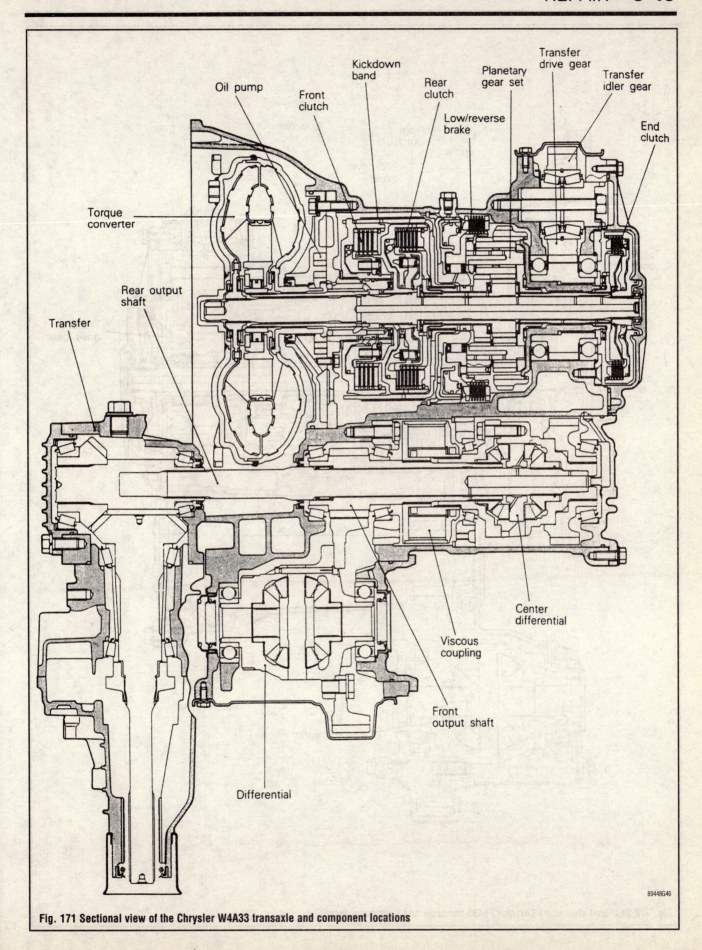

Fig. 171 Sectional view of the Chrysler W4A33 transaxle and component locations

89448G46

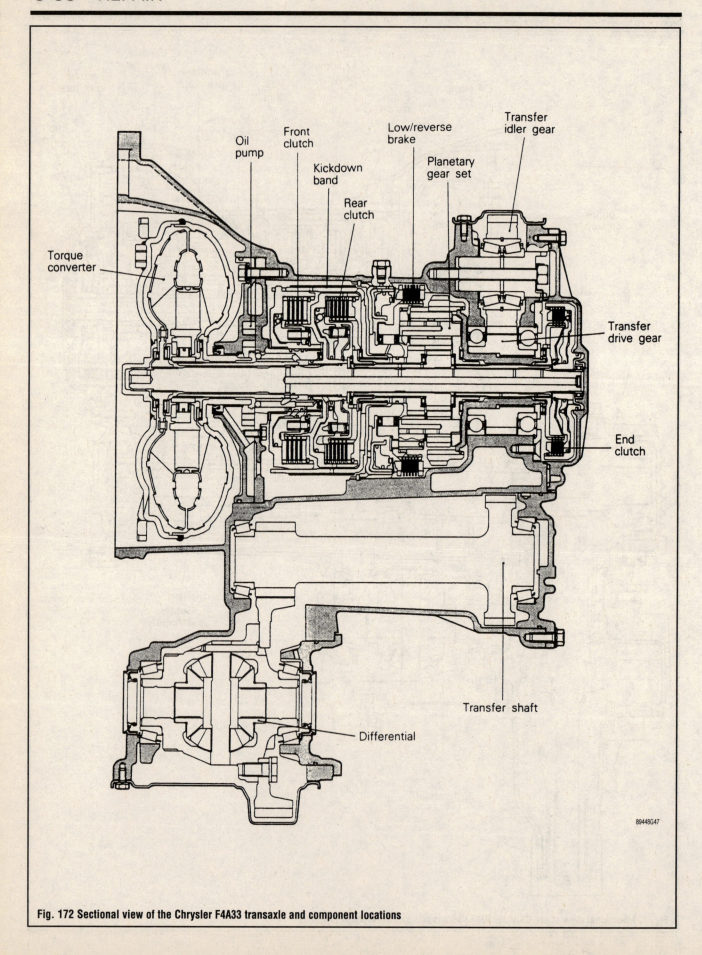

Fig. 172 Sectional view of the Chrysler F4A33 transaxle and component locations

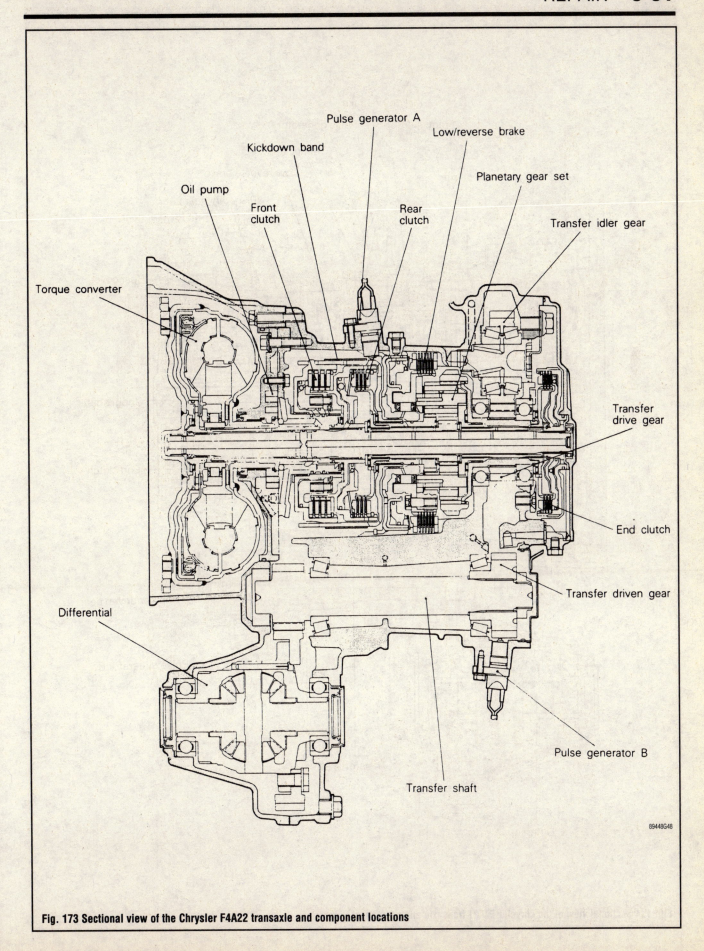

Fig. 173 Sectional view of the Chrysler F4A22 transaxle and component locations

89448G48

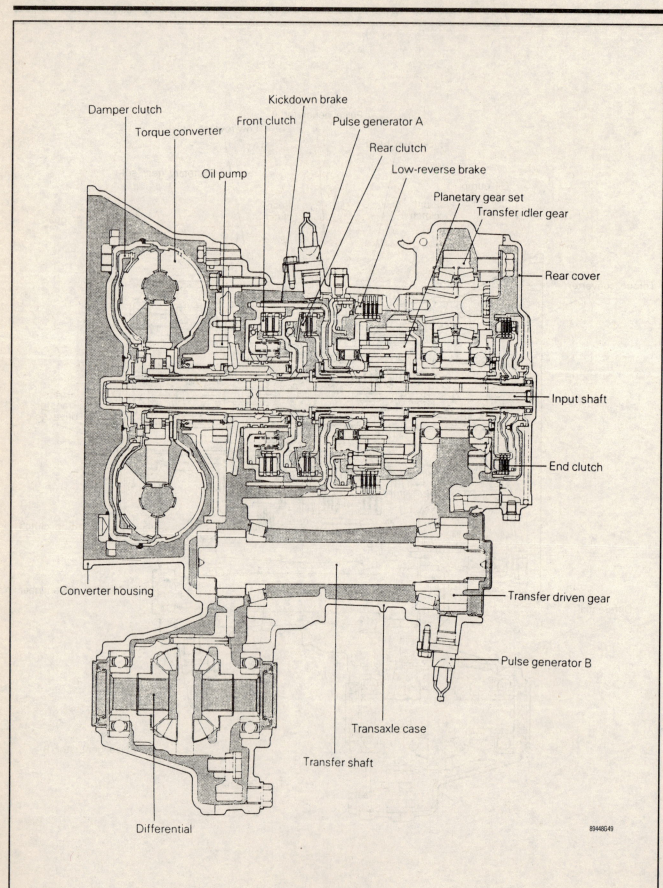

Damper clutch

Torque converter

Oil pump

Front clutch

Kickdown brake

Pulse generator A

Rear clutch

Low-reverse brake

Planetary gear set

Transfer idler gear

Rear cover

Input shaft

End clutch

Converter housing

Transfer driven gear

Pulse generator B

Transaxle case

Transfer shaft

Differential

89448G49

Fig. 174 Sectional view of the Chrysler F4A21 transaxle and component locations

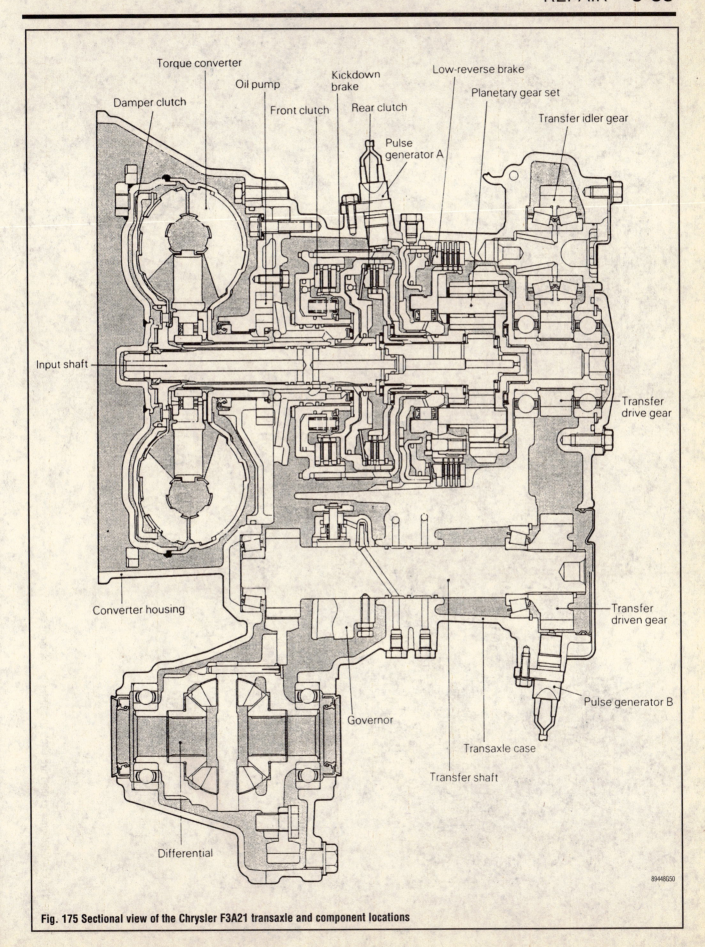

Damper clutch
Torque converter
Oil pump
Front clutch
Kickdown brake
Rear clutch
Pulse generator A
Low-reverse brake
Planetary gear set
Transfer idler gear
Input shaft
Transfer drive gear
Converter housing
Transfer driven gear
Governor
Pulse generator B
Differential
Transaxle case
Transfer shaft

89448G50

Fig. 175 Sectional view of the Chrysler F3A21 transaxle and component locations

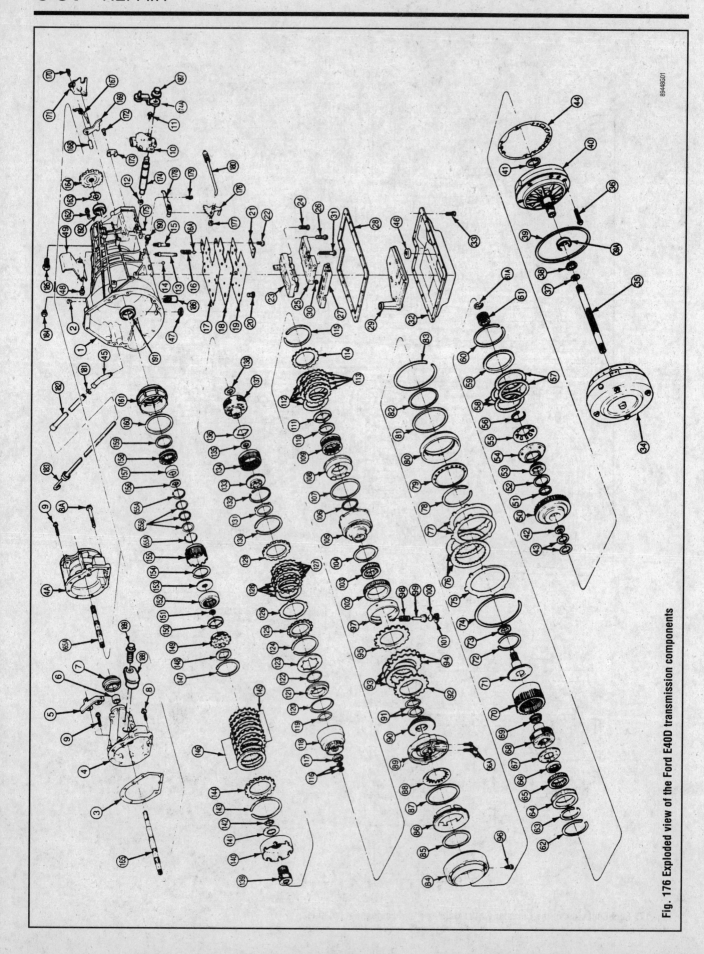

Fig. 176 Exploded view of the Ford E40D transmission components

Item No.	Description		Item No.	Description
1	Case Assembly		52	Seal — Outer
2	Vent Assembly		53	Piston Coast Clutch
3	Gasket — Extension Housing		54	Ring — Piston Apply
4	Extension Assembly (4x2)		55	Spring — Piston Return
#	— Extension & Bushing Assy		56	Ring — Retaining
#	— Extension		57	Plate — Coast Clutch External Spline (2 Pcs.)
4A	Extension Assembly (4x4)		58	Plate — Coast Clutch Internal Spline (2 Pcs.)
	Extension Assembly (Superduty)		59	Plate — Coast Clutch Pressure
5	Bracket — Wiring		60	Ring — Retaining (Selective Fit)
6	Bushing — Extension Housing (4x2)			Ring — Retaining
7	Seal — Extension Housing (4x2)			Ring — Retaining
8	Bolt — Extension (4x2 Bottom) (2 Pcs.) M10X1.5X35MM		61	Gear Assembly — Overdrive Sun
			61A	Ring — Retaining
8A	Bolt — Extension (Superduty & 4x4 Bottom) (2 Pcs.) M10X1.5X90MM		62	Ring — Retaining (Outer Race to Overdrive Ring Gear)
9	Bolt — Extension (Top) (7 Pcs.) M10X1.5X40MM		63	Ring — Retaining (Overdrive OWC to Outer Race)
10	Sensor — Manual Lever Position		*64	Race — Overdrive One Way Clutch Outer
11	Bolt Assembly (2 Pcs.) M6X1.0X30MM		*65	Clutch Assembly — Overdrive One Way
12	Seal — Manual Lever		*66	Race — Overdrive One Way Clutch Inner
13	Stud — Case to Solenoid Body (1 Pc.) M6X1.0X79MM		67	Washer — Thrust
			68	Planet Assembly — Overdrive
14	Ball — Rubber Check (9 Pcs.) (2 in Main Control)		#	— Carrier
			#	— Planet Gears (3 or 4 Pcs.)
	Ball — Steel Check (1 Pc.)		#	— Planet Shafts (3 or 4 Pcs.)
15	Stud — Case to Control Assembly (4 Pcs.) M6X1.0X61.25MM		#	— Thrust Washers (6 or 8 Pcs.)
			#	— Needle Bearings (60 or 80 Pcs.)
16	EPC Blow-off Spring		#	— Retaining Pins (3 or 4 Pcs.)
16A	EPC Blow-off Ball		#	— Needle Bearing Assembly
17	Gasket — Case to Separator Plate		69	Needle Bearing Assembly
18	Plate — Separator		70	Gear — Overdrive Ring
19	Gasket — Separator Plate to Control		71	Center Shaft
20	Screen — Solenoid		72	Ring — Retaining (Center Shaft to Overdrive Ring Gear)
#21	Plate — Separator Plate Reinforcing		73	Needle Bearing Assembly
22	Bolt (3 Pcs.) M6X1.0X16MM		74	Ring — Overdrive Retaining (Selective Fit)
23	Solenoid Body Assembly		75	Plate — Overdrive Clutch Pressure
24	Bolt — Torx Head (9 Pcs.) M6X1.0X40MM		76	Plate — Overdrive Clutch Internal Spline (2 Pcs.)
25	Main Control Body Assembly		77	Plate — Overdrive Clutch External Spline (2 Pcs.)
26	Bolt (18 Pcs.) M6X1.0X42.5MM			
27	Accumulator Body Assembly		78	Ring — Return Spring Retaining
28	Gasket — Oil Pan		79	Spring — Overdrive Return
*29	Filter and Seal Assembly (4x2)		80	Piston — Overdrive
*	Filter and Seal Assembly (4x4)		81	Seal — Overdrive Outer
30	Nut (5 Pcs.) M6X1.0		82	Seal — Overdrive Inner (Same as Intermediate Inner)
31	Bolt (7 Pcs.) M6X1.0X66MM			
32	Pan — Oil (4x2)		83	Ring — Int./O.D. Cylinder Retaining
	Pan — Oil (4x4)		84	Cylinder — Intermediate/Overdrive
33	Bolt — Oil Pan (20 Pcs.) M8X1.25X12MM		85	Seal — Intermediate Inner
34	Torque Converter Assembly		86	Piston — Intermediate
34A	— Plug — Converter Drain 1/8 in-27		87	Seal — Intermediate Outer
35	Shaft — Input		88	Spring — Intermediate Return
36	Bolt & Washer Assembly — Pump (9 Pcs.) M8X1.25X65MM		89	Support Assembly — Center
			90	Washer — Thrust
	— Washer — Replacement (9 Pcs.)		91	Seal — Direct Clutch Cast Iron (2 Pcs.)
37	Seal Ring — Teflon		92	Plate — Intermediate Clutch Apply
38	Seal — Converter Hub		93	Plate — Intermediate Clutch Internal Spline (2 or 3 Pcs.)
39	Seal — Square Cut O.D. Pump			
40	Pump Assembly		94	Plate — Intermediate Clutch External Spline (1 or 2 Pcs.)
41	Washer — Pump Thrust			
42	Needle Bearing Assembly		95	Plate — Intermediate Clutch Pressure
43	Seal Ring — Teflon (2 Pcs.)		96	Bolt — Cylinder Hydraulic Feed (1 Pc.) M10X1.5X24MM
44	Gasket — Pump			
45	Stub Tube		96A	Bolt — Center Support Hydraulic Feed (2 Pcs.) M12X1.75X31MM
46	Magnet — Pan			
47	Plug — Converter Access		97	Band Assembly
48	Bolt — Heat Shield (2 Pcs.)		98	Spring — Servo Return
49	Heat Shield — Solenoid Body Connector		99	Piston Assembly — Servo
50	Coast Clutch Cylinder Assembly			
51	Seal — Inner			

Not Serviced
* Serviced in Kits Only

89448G02

Fig. 177 Ford E40D transmission component keylist

Item No.	Description
100	Plate — Servo Cover
101	Ring — Servo Retaining
*102	Race — Intermediate One Way Clutch Outer
*103	Clutch Assembly — Intermediate One Way
104	Washer — Thrust (LG. Dia.)
105	Drum Assy. — Intermediate Brake
106	Seal — Inner
107	Seal — Outer
108	Piston Assembly
#	— Piston
#	— Check Ball (7/32 Inch Dia.)
#	— Ball Retainer
109	Spring — Piston Return
110	Ring — Spring Retaining
111	Washer — Thrust (Small Dia.)
112	Plate — Direct Clutch Internal Spline (3 or 4 pcs.)
113	Plate — Direct Clutch External Spline (3 or 4 pcs.)
114	Plate — Direct Clutch Pressure
115	Ring — Retaining (Selective Fit)
	Ring — Retaining
	Ring — Retaining
	Ring — Retaining
	Ring — Retaining
116	Seal Ring — Teflon® (2 pcs.)
117	Needle Bearing Assembly
118	Cylinder — Forward Clutch Assembly (3 or 4 plate)
119	Seal — Inner
120	Seal — Outer
121	Piston Assembly
#	— Piston
#	— Check Ball
#	— Ball Retainer
122	Ring — Piston Apply
123	Spring — Piston Return
124	Ring — Retaining (For Return Spring)
125	Plate — Forward Clutch Pressure
126	Spring — Cushion
127	Plate — Forward Clutch External Spline (3 or 4 pcs.)
128	Plate — Forward Clutch Internal Spline (3 or 4 pcs.)
129	Plate — Rear Clutch Pressure
130	Ring — Retaining (Selective Fit)
	Ring — Retaining
	Ring — Retaining
	Ring — Retaining
	Ring — Retaining
131	Washer — Plastic Thrust
132	Ring — Retaining
133	Hub — Forward
134	Gear — Forward Ring
135	Needle Bearing Assembly
136	Washer — Thrust
137	Planet Assembly — Forward
#	— Carrier
#	— Planet Gears (3 or 4 pcs.)
#	— Planet Gear Shafts (3 or 4 pcs.)
#	— Thrust Washers (6 or 8 pcs.)
#	— Needle Bearings (51 or 68 pcs.)
#	— Retaining Pins (3 or 4 pcs.)
138	Needle Bearing Assembly
139	Gear Assembly — Forward/Reverse Sun
140	Input Shell
141	Washer — Thrust
142	Ring — Retaining
143	Ring — Retaining

Item No.	Description
144	Plate — Reverse Clutch Pressure
145	Plate — Reverse Clutch External Spline (5 or 6 pcs.)
146	Plate — Reverse Clutch Internal Spline (5 or 6 pcs.)
147	Ring — Retaining
148	Washer — Thrust
149	Planet Assembly — Reverse
#	— Carrier
#	— Planet Gears (3 or 4 pcs.)
#	— Planet Gear Shafts (3 or 4 pcs.)
#	— Thrust Washers (6 or 8 pcs.)
#	— Needle Bearings (63 or 84 pcs.)
#	— Retaining Pins (3 or 4 pcs.)
150	Washer — Thrust
151	Ring — Retaining (for Output Shaft) (1-1/2 in dia.)
152	Gear — Reverse Ring
*153	Hub — Output Shaft
154	Ring — Retaining
155	Hub Assembly — Reverse Clutch
155A	Ring — 3-31/32 Retaining (2 — Att 7E194 IN 7E193 Assy.)
*155B	Spring Assy. — Overrunning Clutch
*155C	Roller — Overrunning Clutch (16 Req'd)
*155D	Bushing — Overrunning Clutch
156	Needle Bearing Assembly
*157	Race — Low/Reverse One Way Clutch Inner
158	Spring Assembly — Piston Return
159	Seal — Inner
160	Seal — Outer
161	Piston
162	Bolts (5 pcs.) 5/16 in-24 (One Way Clutch to Case)
163	Washer — Thrust
164	Parking Gear
165	Output Shaft Assembly (4x2)
165A	Output Shaft Assembly (4x4)
167	Spring — Parking Pawl Return
168	Pin — Parking Pawl
169	Parking Pawl
170	Bolt and Washer Assembly (2 pcs.)
171	Plate — Parking Rod Guide
172	Bolt (1 pc.)
173	Abutment — Parking Pawl Actuating
174	Shaft Manual Control Lever
174A	Lever Assy. — Man. Cont.
175	Pin — Manual Lever Retaining
176	Lever — Inner Detent
177	Nut — Inner Detent Lever M14x1.5 Hex
178	Spring Assembly — Manual Valve Detent
179	Bolt — Hex Flange Head M6x1.0x16.5mm
180	Rod Assembly — Parking Pawl Actuating
181	O-Ring Filler Tube
182	Tube Assy. — Oil Filler
183	Indicator Assy. — Oil Level
184	Connector Assembly — Oil Tube — Inlet (Front)
185	Valve Assembly — Oil Cooler Check — Outlet (Rear)
186	Filter Assembly, Accum. Regulator
187	Nut, M10x1.5 Hex
188	Plug Assembly, Ext. Hsg.
189	Screw and Washer Assy., 1/4-20x.62
190	Plug, Test Port, 1/8-27 Hex
191	Bushing, Case Front
192	Bushing, Case Rear

Not Serviced*
* Serviced in Kits Only

89448G03

Fig. 178 Ford E4OD transmission component keylist (continued)

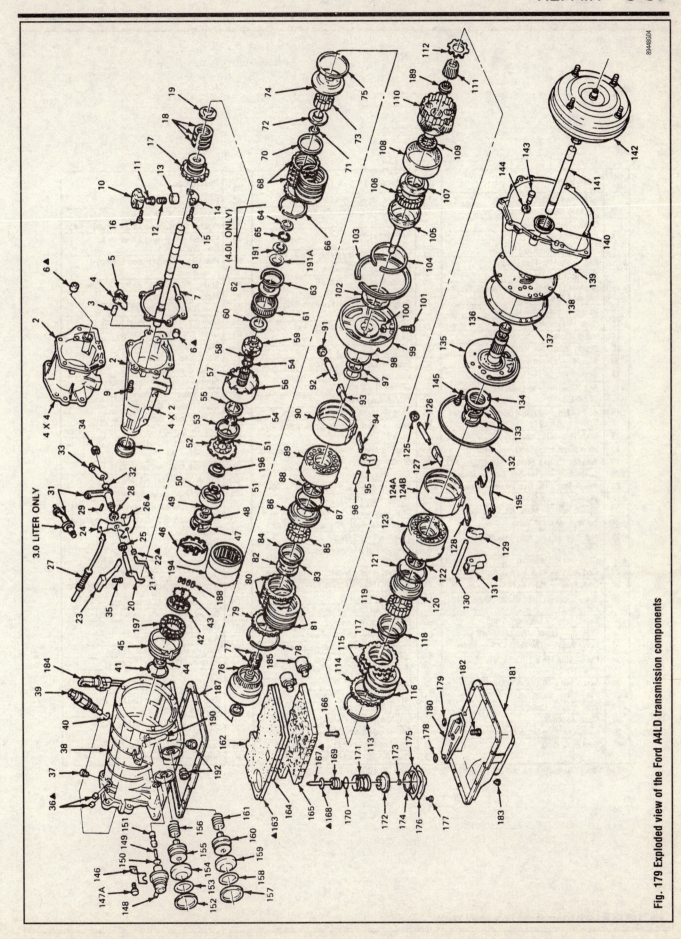

Fig. 179 Exploded view of the Ford A4LD transmission components

Ref. No.	Description	Ref. No.	Description	Ref. No.	Description
1	Seal Assy (Ext. Hsg.) Oil	41	Ring — Retaining 4.0L	80	Plate (Clutch) High
2	Housing (Extn.)	42	Retainer — Overrun Cl. Spring	81	Plate Assy. (Clutch) High
3	Shaft (Parking Pawl)	43	Spring — Overrun Clutch	82	Washer (Intm. Brake Drum Thrust) — #5
4	Pawl (Parking)			83	Ring 63 mm (High Cl. Pst. In Int. Brk. Drum)
5	Spring (Parking Pawl Return)	44	Washer (O.P. Shaft Hub Thrust #10)	
▲ 6	Cup — Parking Rod Guide	45	Band Assy. — Reverse	84	Ret. (Rev. Clutch Piston Spring) — 8 Tabs
7	Gasket (Extn. Hsg.)	46	Drum Assy. (Rev. Brake)	85	Spring (Rev. Clutch Piston) 20 Req'd
8	Shaft Assy (Output) 4.0L Vehicle, Shaft has no Lube Hole	47	Ring 87 mm Retain Forward Ring Gear to Hub	86	Piston (Rev. Clutch)
				87	Seal (Clutch Piston Oil)
* 9	Screw — Extension to Case	48	Hub — Output Shaft — 57 Ext — 34 Int Teeth	88	Seal (High Clutch Piston Inner)
10	Body — Gov. Valve			89	Drum Assy. (Interm. Brake)
11	Valve (Governor Primary)	49	Retaining Ring 25 x 1.2	90	Band Assy (Interm. Servo)
12	Spring (Governor Valve)	50	Gear — Output Shaft Ring	91	Net & Seal — Hex
13	Weight — (Governor Outer)	51	Washer — Planet Carrier Thrust — (2 Req'd) #8 and #9	92	Screw (Rev. Band Adj.)
14	Counterweight — Governor			93	Strut (Interm. Brk. Band Anchor)
15	Bolt (Gov. Body to Collector Body) — (2 Req'd)	52	Planet Assy (Rev.)	94	Strut (Interm. Brk. Band Apply)
		53	Ring (Planet to Drum) Except 4.0L	95	Lever (Interm. Band Servo)
16	Bolt — M6 x 20 (Gov. Body to Collector Body) — 2 Req'd	54	Ring — 39 mm (Input Shell to Sun Gr. Assy.) — (2 Req'd)	96	Shaft (Interm. Band Act. Lever)
				97	Seal Ring (High Clutch) 2 Req'd — Viton
17	Body (Gov. Oil Collector)	55	Washer (Input Shell Thrusts) Except 4.0L	
18	Ring (Gov. Hsg. Seal) — (3 Req'd)	56	Shell (Input)	98	Washer (Frt. Pump Input Thrust) — Sel. Fit — #4
19	Washer (Output Shaft Thrust Gr) — #11	57	Gear Assy (Sun)	
20	Lever Assy — Dwn/Shft Det. — Inner	58	Brg. Thrust — Sun Gear Race — RR	99	Support Assy — Center O/D
21	Pin — Man. Vlv. Det. Lever — Inner	59	Planet Assy (Fwd)	100	Nut & Cage Assy S/L Mtl. M6
▲22	Washer — Flat Steel	60	Brg. Assy. — Cl. Int. Drum Thrust — #7	101	Screw Cap — Hex 8.8 M6 x 15
23	Spring Assy — Man. Vlv. Detent	61	Gear (Ring Fwd) 72 Ext. 57 Int. Teeth	102	Washer (Center Support Thrust) — #3
24	Lever — Manual Valve	62	Hub (Fwd Ring Gear)	103	Ring Retaining (Retain 7G033 In Case)
25	Nut (Lev. to Lev. Assy Dwn/Shft Det. Inner)	63	Ring — Ret 87 mm (Fwd. Ring Gr. to Hub)	104	Ring — Ret. (110.1 mm x 1.6) Except 4.0L
▲26	Clip — Rod Retaining	64	Washer (Fwd. Cyl. Hub Thrust) — #6	
27	Rod Assy — Park Pawl Actu.	65	Needle Thrust Bearing #6	105	Shaft — Center Assy — O/D
28	Seal Assy — Main Control Lvr. Oil	66	Ret. Ring (Sel. Fit)	106	Clutch Assy — Overrun — O/D
29	Pin — Spring Roller (Retain Outer Man. Lvr. Assy.)	67	Plate (Fwd Clutch Pressure)	107	Washer — Over Clutch — O/D
		68	Plate (Clutch) Forward	108	Gear — O/D Ring
31	Lever Assy — Manual Control	69	Plate Assy — Clutch (Fwd)	109	Brg. Assy. — O/D Inner Race — #2
32	O Ring — Outer Man. Lvr. Shaft Oil	70	Spring — Forward Clutch Cushion	110	Carrier Assy — Plt. Gear — O/D
33	Lever Assy — Dwn/Shft Cntl — Outer	71	Ring 34 mm (Hub to Fwd. Ring Gear)	111	Gear Assy — Sun O/D
34	Nut — Hex M8 x 1 (Outer Man. Lvr. to Shaft)	72	Ret (Fwd. Cl. Piston Spring)	112	Adapter — O/D Clutch
		73	Spring — Fwd. Cl. Piston (15 Req'd)	113	Ring — Retaining (Sel. Fit)
35	Screw M6 x 30 (Valve Body to Case)	74	Piston Assy (Fwd. Clutch)	114	Plate — O/D Clutch Pressure
▲36	Pin — Rev. Band Anchor (2 Req'd)	75	Seal (Clutch Piston Oil)	115	Plate — O/D Clutch
37	Vent Assy — Case	76	Cyl. Assy. (Fwd. Clutch)	116	Plate Assy — O/D Clutch Int. Spline
38	Case Assembly	77	Seal (Fwd Clutch Cyl.) — 2 Req'd	117	Ring — Ret. 63 mm (O/D Cl. Pst. to O/D Brk. Drum)
39	Switch Assy — Gr. Shift Neutral	78	Ret Ring (Select Fit)	
40	Seal — O Ring	79	Plate (Clutch Press) Rev.	118	Retainer — O/D Cl. Pst. Spring — 8 Tabs
				119	Spring — O/D Cl. Piston (20 Req'd)
				120	Piston — O/D Clutch

*All light trucks transmissions use (4) E800152-S72 Screws and (2) E804137-S72 Studs.

89448G05

Fig. 180 Ford A4LD transmission component keylist

Ref. No.	Description	Ref. No.	Description	Ref. No.	Description
121	Seal — O/D Cl. Piston — Outer	146	Clamp — TV Control Diaphragm	174	Seal — Rev. Bnd. Servo Ret. Oil — Large
122	Seal — O/D Cl. Piston — Inner	147	Bolt M6 x 12 mm (Valve Clamp to Case)	175	Gasket — Rev. Servo Sep. Plate Cover
123	Drum Assy — O/D	147A	Stud-M6 × M6 × 12.0	176	Cover — Rev. Bnd. Servo Piston
124A	Band Assy — O/D			177	Bolt M6 x 20 (Rev. Servo to Vlv. Bdy.) 4 Req'd
124B	Band Assy — O/D	148	Diaphragm Assy — TV Control		
125	Nut & Seal — Hex	149	Rod — TV Control	178	O'Ring — Oil Screen Assy — Small
126	Screw — O/D Band Adj.	150	O'Ring — Throttle Valve	179	O'Ring — Oil Screen Assy — Large
127	Strut — O/D Brk. Drum Anchor	151	Valve — Throttle Control	180	Screen Assy — Oil Pan
128	Strut — O/D Brk. Drum Apply	152	Ring — Ret. 67 x 15 Intermediate	181	Oil Pan
		153	O-Ring — Servo Cover to Case — Interm.	182	Screw — M6 x 45 (Vlv. Bdy. to Case) 5 Req'd
129	Lever — O/D Band Servo	154	Cover & Seal Assy. Inter. Band Servo		
130	Shaft — O/D Band Adj. Lever	155	Piston & Rod Assy. — Intermediate	183	Screw — M8 x 14 (Oil Pan to Case) 18 Req'd
▲131	Bracket — O/D	156	Spring Interm. Band Servo Piston		
132	Seal (Front Oil Pump)	157	Ring — Ret. 67 x 15 — O/D	184	Connector — Conv. Cl. Override/3-4 Shift
133	Seal (Interm. Brk. Drum) — 2 Req'd	158	Seal — Servo Cover to Case — O/D	185	Solenoid Assy — Converter Clutch
134	Washer (Frt. Pump Input Thrust) Sel. Fit — #1	159	Cover & Seal Assy. — O/D Band Servo	186	Solenoid Assy — 3-4 Shift
		160	Piston & Rod Assy — O/D	187	Gasket — Oil Pan
135	Support & Gear Assy (Frt. Pump)	161	Spring — O/D Band Servo Piston	188	Roller Overrun Clutch
136	Seal (Front Pump Support)	162	Gasket — Cont. Vlv. Bdy. Sep.	189	Race-Sun Gear Thrust Brg. — Rear
137	Gasket — Oil Pump	▲163	Plate — Vlv. Bdy. Separating	190	Tube-Lube Oil Inlet — Short
138	Plate (Oil Pump Adaptor)	164	Gasket — Cont. Vlv. Bdy. Separating	191	Washer — Fwd. Ring Gear Thrust — #6
139	Hsg. Assy — Converter	165	Control Assy — Main	191A	Washer — Fwd. Clutch Thrust
140	Seal Assy (Frt. Oil Pump)	166	Screw M6 x 40 (Valve Body to Case) 19 Req'd	192	Connector Assy-Oil Tube — 2 Req'd
141	Input Shaft			193	Integral Thrust Washer
142	Converter Assembly	▲167	Rod — Rev. Band Servo Piston		
143	Screw M10 x 30 (Conv. Hsg. to Case) 8 Req'd	▲168	Ret. — Rev. Servo Cushion Spring	194	Drum Assy (Rev. Brake) 4.0L
		169	Spring — Rev. Servo Occum.	195	Strut-O/D Brk. Drum Apply (4.0L)
144	"O" Ring	170	Seal — Rev. Bnd. Servo Pst. Oil — Small	196	Output Shaft Sleeve (4.0L only)
145	Bolt — Flg. Hd. 8.8 x M8 x 35.0 (Pump Supt. Assy to Conv. as Assy) 5 Req'd	171	Spring — Rev. Servo Piston	197	Overrun Clutch — Sprag Type
		172	Piston & Rod Assy — Rev. Servo		
		▲173	Ring Retainer (Ret. Rod to Piston)		

89448G06

Fig. 181 Ford A4LD transmission component keylist (continued)

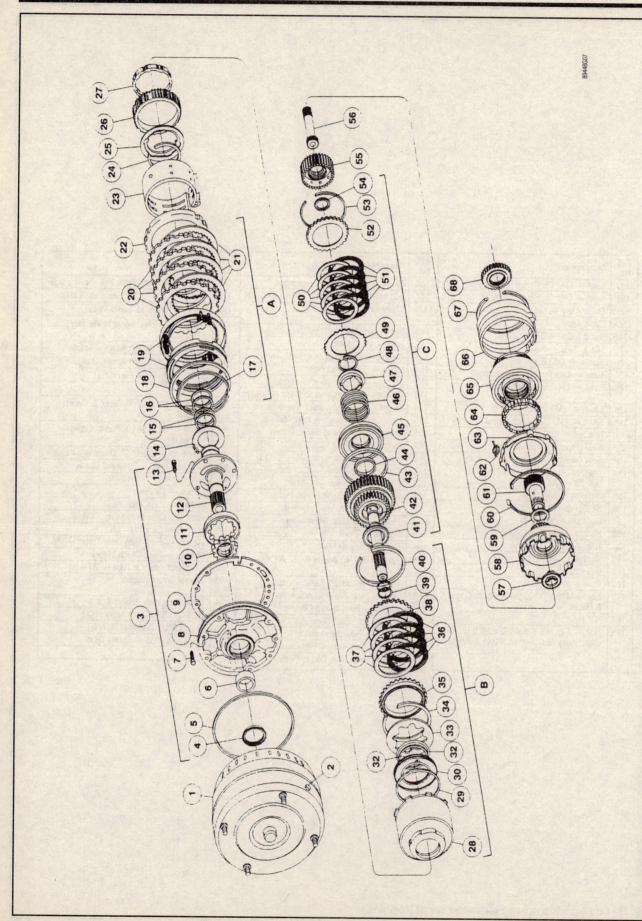

Fig. 182 Exploded view of the Ford 4R70W transmission components

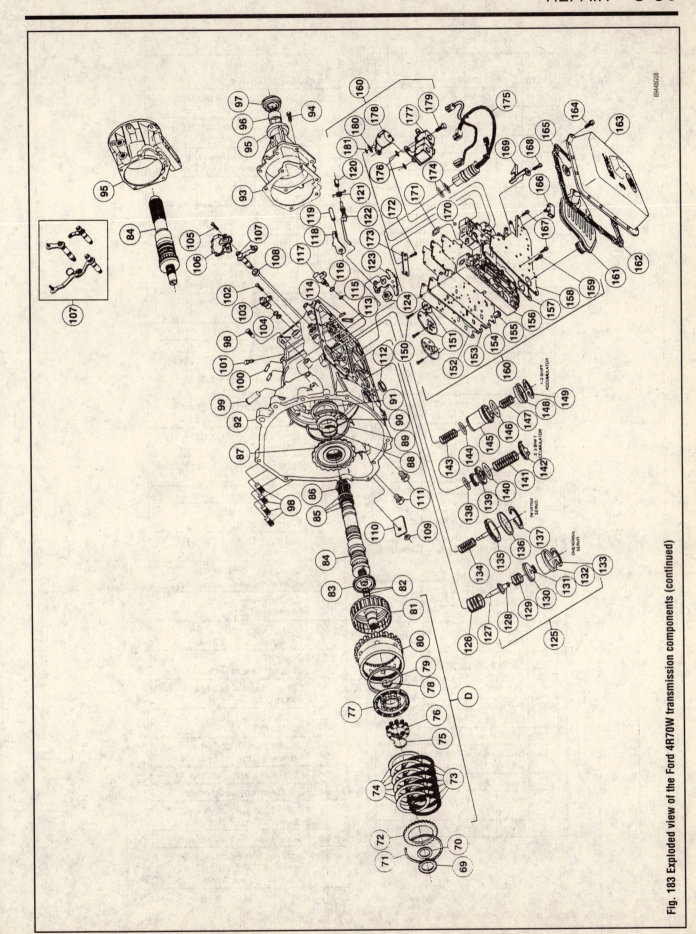

Fig. 183 Exploded view of the Ford 4R70W transmission components (continued)

Item	Description
1	Converter Assy
2	Plug - Converter Drain - 1/8-27 Dryseal
3	Pump Assy - Front
4	Seal Assy - Front Pump
5	Seal - Front Pump
6	Bushing - Front Pump
7	Bolt - M8-1.25 X 35 Hex Head
8	Body Assy - Front Pump
9	Gasket - Front Pump
10	Gear - Pump Inner Gerotor
11	Gear - Pump Outer Gerotor
12	Support Assy - Front Pump
13	Bolt - M8-1.25 X 25 Hex Flg Head
14	Washer - Front Pump Support Thrust - Select Fit No. 1
15	Seal - Reverse Clutch Cylinder (2 Req'd)
16	Seal - Forward Clutch Cylinder (2 Req'd)
17	Seal - Intermediate Clutch Piston - Inner
18	Seal - Intermediate Clutch Piston - Outer
19	Piston Kit - Intermediate Clutch
20	Plate - Intermediate Clutch External Spline (Select Fit) (Steel)
21	Plate Assy - Intermediate Clutch Internal Spline (Friction)
22	Plate - Intermediate Clutch Pressure
23	Band Assy - Overdrive
24	Ring - 3-21/64 Retains Type SU External
25	Retainer - Intermediate One-Way Clutch
26	Hub - Intermediate Clutch
27	Clutch Assy - Intermediate One-Way
28	Drum Assy - Reverse Clutch
29	Seal - Reverse Clutch Piston - Outer
30	Piston Assy - Reverse Clutch

Item	Description
31	Seal Reverse Clutch Piston - Inner
32	Ring - Reverse Clutch Piston Pressure
33	Spring - Reverse Clutch Piston Return
34	Spring - Reverse Clutch Piston Spring
35	Plate - Reverse Clutch Front Pressure
36	Plate - Reverse Clutch Internal Spline (Friction)
37	Plate - Reverse Clutch External Spline (Steel)
38	Plate - Reverse Clutch Rear Pressure
39	Seal - Input Shaft (2 Req'd)
40	Retainer - Reverse Clutch Pressure Plate - (Select Fit)
41	Bearing and Race Assy - Forward Clutch No.2
42	Cylinder and Input Shaft Assy - Forward Clutch
43	Seal - Forward Clutch Piston - Outer
44	Seal - Forward Clutch Piston - Inner
45	Piston - Forward Clutch
46	Spring - Forward Clutch Piston Return
47	Retainer Return Spring - Forward Clutch
48	Snap Ring - Retaining - 1-59/64
49	Spring - Rear Clutch Pressure Plate
50	Plate - Forward Clutch External Spline (Steel)
51	Plate - Forward Clutch Internal Spline (Friction)
52	Plate - Forward Clutch Pressure
53	Snap Ring - Retaining (Select Fit)
54	Bearing and Race Assy - Forward Clutch - Front No.3
55	Hub - Forward Clutch
56	Shaft - Intermediate Stub
57	Bearing and Race Assy - Forward Clutch Hub No.4
58	Gear Assy - Reverse Sun
59	Bearing and Race Assy - Forward Clutch Sun Gear No.5

89448G09

Fig. 184 Ford 4R70W transmission component keylist

Item	Description
60	Retaining Ring - Center Support - 7-7/92
61	Gear Assy - Forward Clutch Sun
62	Spring - Case to Planet Support
63	Support Assy - Planetary Gear
64	OWC Cage Spring and Roller Assy - Planetary
65	Planetary Assy
66	Band Assy - Reverse
67	Retaining Ring - 0.58 Thick (Locates Reverse Band During Assy)
68	Hub - Direct Clutch
69	Bearing and Race Assy - Direct Clutch Inner No.7
70	Support - Direct Clutch Inner Bearing
71	Retaining Ring - Direct Clutch Pressure Plate (Select Fit)
72	Plate - Direct Clutch Pressure
73	Plate - Direct Clutch Internal Spline (Friction)
74	Plate - Direct Clutch External Spline (Steel)
75	Retaining Ring - 1-19/32
76	Retainer and Spring Assy - Direct Clutch
77	Piston Assy - Direct Clutch
78	Seal - Direct Clutch Piston - Inner
79	Seal - Direct Clutch Piston - Outer
80	Gear - Output Shaft Ring
81	Cylinder Assy - Direct Clutch
82	Seal - Output Shaft Small - Direct Clutch (2 Req'd)
83	Bearing and Race Assy - Direct Clutch Outer No.8
84	Shaft Assy - Output
85	Seal - Output to Case Shaft Large (3 Req'd)
86	Seal - O-Ring (Piloted Output Shaft Only)
87	Hub - Output Shaft
88	Snap Ring - 1-13/16 Retaining
89	Snap Ring - Retaining

Item	Description
90	Bushing - Rear Case
91	Bearing and Race Assy - Case Rear No.9
92	Case Assy
93	Gasket - Extension
94	Bolt - M8-1.25 X 30
95	Extension Assy
96	Bushing - Extension Housing
97	Seal Assy - Extension Housing
98	Pipe Plug - 1/8-27 Dryseal Tapered (5 Req'd)
99	Pin - Overdrive Band Anchor
100	Pin - Reverse Band Anchor (Part of 7005)
101	Vent Assy - Case
102	Bolt - M6-1.0 X 14 Hex Flg Head (Attaches Output Shaft Speed Sensor to Case)
103	Sensor Assy - Transmission Output Shaft Speed
104	Seal - 14.0 X 1.78 O-Ring (2 Req'd)
105	Bolt and Washer Assy - M6-1.0 X 25MM
106	Sensor - Transmission Range
107	Lever Assy - Manual Control
108	Seal Assy - Manual Control Lever
109	Nut - 1/4 Spring
110	Tag - Identification (Part of 7005)
111	Connector Assy - Fluid Tube (2 Req'd)
112	Plug - Converter Housing Access
113	Screen Assy - Fluid
114	Pin - Manual Lever Shaft Retainer
115	Seal - 0.426 X 0.070 O-Ring
116	Seal - 14.0 X 1.78 O-Ring
117	Solenoid Valve - Transmission Pressure Control
118	Pawl - Parking Pawl
119	Shaft - Parking Pawl
120	Cup - Park Rod Guide
121	Spring - Parking Pawl Return

89448G10

Fig. 185 Ford 4R70W transmission component keylist (continued)

Item	Description
122	Rod Assy - Park Pawl Actuating
123	Lever Assy - Manual Valve Detent Lever
124	Nut - M14 X 1.5 Hex - Intermediate Detent Lever
125	Piston Assy - Overdrive Servo
126	Spring - Overdrive Servo Piston
127	Rod - Overdrive Servo Actuating
128	Washer - Backup Overdrive Servo
129	Spring - Overdrive Cushion Spring
130	Piston Assy - Overdrive Servo
131	Ring - Retaining
132	Sleeve Assy - Overdrive Servo
133	Ring - 2.85 Retaining Type TVP "H" Internal
134	Spring - Reverse Band Servo Piston
135	Piston Assy - Reverse Band Servo
136	Cover Assy - Reverse Band Servo Piston
137	Retaining Ring - Internal - 3-13/16
138	Seal - 2-3 Accumulator Piston - Upper
139	Piston - 2-3 Shift Accumulator
140	Seal - 2-3 Accumulator Piston - Lower
141	Spring - 2-3 Shift Accumulator Piston
142	Retainer - 2-3 Shift Accumulator Spring
143	Spring - 1-2 Shift Accumulator (Model Dependent)
144	Seal - 1-2 Shift Accumulator Piston
145	Piston - 1-2 Shift Accumulator
146	Seal - 1-2 Shift Accumulator Piston - Lower
147	Spring - 1-2 Shift Accumulator

Item	Description
148	Cover and Seal Assy - 1-2 Accumulator
149	Ring - 2-1/16 Retaining Type HU Internal
150	Bolt - M6-1.0 X 16 Hex Head
151	Plate - Valve Body Reinforcing
152	Gasket - Valve Body Separator - Upper
153	Plate - Control Valve Body Separator
154	Gasket - Valve Body Separating - Lower
155	Valve - Converter Drainback
156	Body Assy - Main Control
157	Gasket - Valve Body Cover Plate
158	Plate - Valve Body Cover
159	Bolt - M6-1.0 X 18 Hex Head
160	Control Assy - Main
161	Filter and Seal Assy - Fluid
162	Gasket - Transmission Pan
163	Pan - Transmission
164	Bolt - M8-1.25 X 18 Hex Flg Head
165	Magnet - Ceramic Case
166	Sensor - Transmission Fluid Temperature
167	Bolt - M8-1.25 X 46 Hex Shldr Pilot
168	Bolt - M6-1.0 X 52 Hex Flg Head
169	Retainer - Solenoid
170	Ball - 1/4 Diameter Coast Booster Valve Shuttle (8 Req'd)
171	Screen - Solenoid Pressure Supply
172	Bolt - M6-1.0 X 40 Hex Flg Head
173	Spring Assy - Manual Valve Detent
174	Seal - 0.864 X 0.070 O-Ring (2 Req'd)

89448G11

Fig. 186 Ford 4R70W transmission component keylist (continued)

Item	Description
175	Bulkhead Assy - Wiring Connector
176	Seal - 6.07 x 1.70 O-Ring (2 Req'd)
177	Solenoid Valve - Transmission Shift
178	Solenoid Valve - Transmission Torque Converter Clutch

Item	Description
179	Bolt - M6-1.0 X 16 Hex Head
180	Seal - 0.489 x 0.070 O-Ring
181	Seal - 0.176 x 0.070 O-Ring
A	Intermediate Clutch Assy
B	Reverse Clutch Assy
C	Forward Clutch Assy
D	Direct Clutch Assy

89448G12

Fig. 187 Ford 4R70W transmission component keylist (continued)

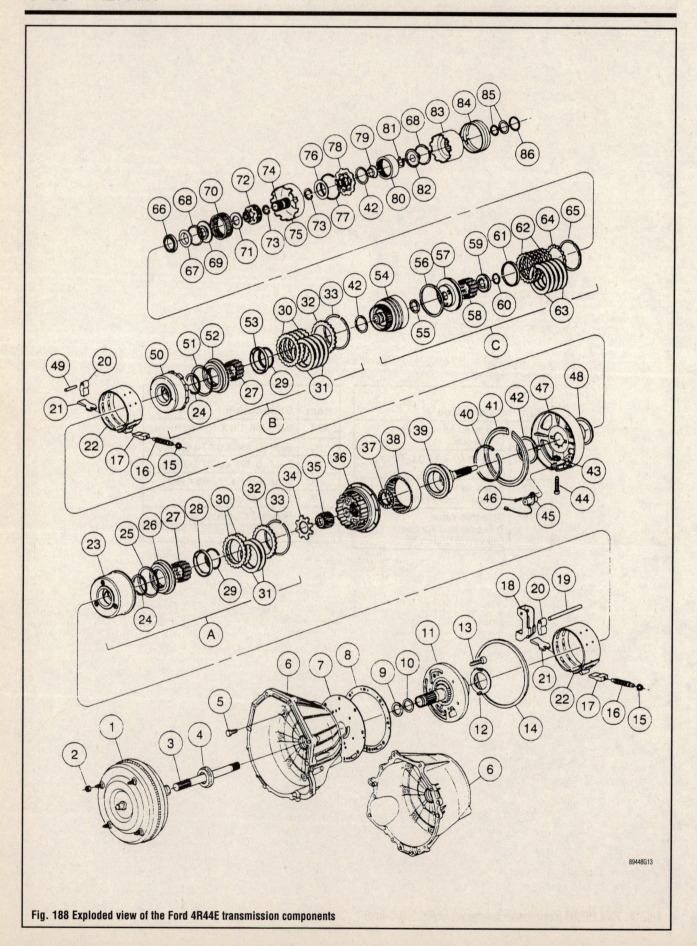

Fig. 188 Exploded view of the Ford 4R44E transmission components

89448G13

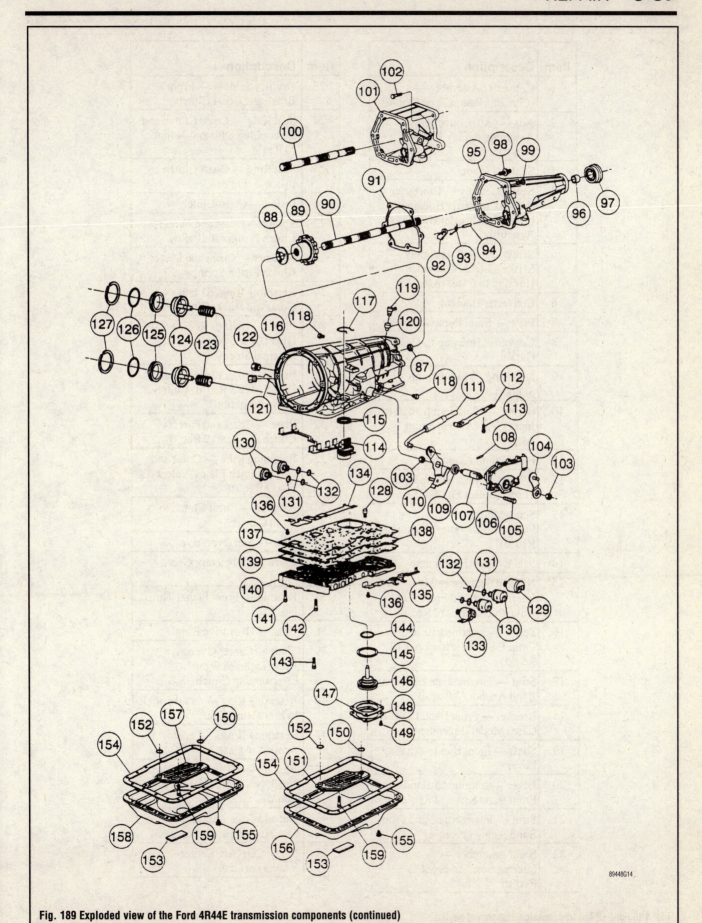

Fig. 189 Exploded view of the Ford 4R44E transmission components (continued)

89448G14

Item	Description
1	Converter Assembly — (Contains Piston Type Clutch)
2	Nut — (Att. Converter Assembly to Flex Plate) (4 Req'd)
3	Shaft — Input
4	Seal Assembly — Converter Hub to Converter Housing (Also in Converter Housing Assembly)
5	Screw and Seal Assembly — M10 x 33 (Att. Converter Housing to Case) (8 Req'd)
6	Converter Housing
7	Plate — Fluid Pump Adapter
8	Converter Housing-to-Case Gasket
9	Seal Ring — Fluid Pump Support
10	O-ring — Fluid Pump Shaft to Inner Gear (Also in Pump Assembly)
11	Fluid Pump Assembly
12	Washer — Fluid Pump Input Thrust (Select Fit) No. 1
13	Screw, Fluid Pump-to-Converter Housing (6 Req'd)
14	Seal Ring — Fluid Pump
15	Nut and Seal Assembly — Hex Intermediate and Front Band Adjustment/Lock (2 Req'd)
16	Screw — Intermediate and Front Band Adjuster/Lock (2 Req'd)
17	Strut — Intermediate and Front Band Anchor (2 Req'd)
18	Bracket — Front Band Lever to Case (Model Dependent)
19	Shaft — Front Band Actuating Lever
20	Lever — Intermediate and Front Band Servo (2 Req'd)
21	Strut — Intermediate and Front Band Apply (2 Req'd)
22	Band Assembly — Intermediate and Front (2 Req'd)

Item	Description
23	Drum Assembly — Front Brake and Coast Clutch
24	Seal Ring — Coast Clutch and Direct Clutch Piston — Inner (2 Req'd)
25	Seal Ring — Coast Clutch Piston — Outer
26	Piston — Coast Clutch
27	Spring — Coast and Direct Clutch Piston (40 Req'd)
28	Retainer — Coast and Direct Clutch Piston Springs
29	Retaining Ring, 63 mm — Coast and Direct Clutch Piston
30	Plate — Coast (2 Req'd) and Direct (4 or 5 Req'd) Clutch External Steel
31	Plate — Coast (2 Req'd) and Direct (4 or 5 Req'd) Clutch Internal Friction
32	Plate — Coast and Direct Clutch Pressure (2 Req'd)
33	Retaining Ring — Coast and Direct Clutch Plates (Select Fit) (2 Req'd)
34	Adapter — Coast Clutch to Front Carrier
35	Gear — Sun Overdrive
36	Carrier — Planetary Gear Front
37	Bearing — Front Planet Thrust No. 2
38	Gear — Overdrive Ring
39	Shaft — Center Overdrive Ring (Includes Front Overrunning Clutch
40	Retaining Ring — Center Shaft in Front Ring Gear
41	Retaining Ring — Center Support in Case
42	Bearing — Center Shaft Thrust No. 3, No. 5 and No. 9 (3 Req'd)
43	Nut and Cage Assembly — (Att. Center Support to Case)
44	Screw Cap (Att. Center Support to Case)

89448G15

Fig. 190 Ford 4R44E transmission component keylist

Item	Description
45	Sensor — Turbine Shaft Speed (TSS)
46	Screw — Turbine Shaft Speed Sensor
47	Support — Center
48	Bearing — Intermediate Brake Drum Thrust (Select Fit) No. 4
49	Shaft — Intermediate Band Actuating Lever
50	Drum — Intermediate Brake and Direct Clutch (Not Available Separately.)
51	Seal Ring — Direct Clutch Piston Inner
52	Piston — Direct Clutch
53	Retainer — Direct Clutch Piston Spring
54	Cylinder — Forward Clutch
55	Seal Ring — Forward Clutch Piston Inner
56	Seal Ring — Forward Clutch Piston Outer
57	Piston — Forward Clutch
58	Spring — Forward Clutch Piston (15 Req'd)
59	Retainer — Forward Clutch Piston Spring
60	Retaining Ring — Forward Clutch Piston and Spring in Forward Clutch Cylinder
61	Spring — Forward Clutch Cushion
62	Plate — Forward Clutch External Steel (6 Req'd)
63	Plate — Forward Clutch Internal Friction (6 Req'd)
64	Plate — Forward Clutch Pressure
65	Retaining Ring — Forward Clutch Plates in Forward Clutch Cylinder (Select Fit)
66	Bearing — Forward Ring Gear Hub Thrust No. 6A
67	Washer — Forward Clutch Thrust No. 6B
68	Retaining Ring — Forward and Output Shaft Ring Gears to Hubs (2 Req'd)
69	Hub — Forward Ring Gear

Item	Description
70	Gear — Forward Ring (72 External Teeth and 57 Internal Teeth)
71	Bearing — Forward Planet Thrust No. 7
72	Planetary — Forward (6 Pinion) (Not Available Separately)
73	Retaining Ring 39 mm — Input Shell to Sun Gear Assembly (2 Req'd)
74	Gear — Sun (Forward)
75	Shell — Input
76	Bearing — Low/Reverse Planet Carrier Thrust No. 8
77	Snap Ring
78	Planet — Low/Reverse (6 Pinion)
79	Sleeve — Output
80	Gear — Output Shaft Ring
81	Retaining Ring 25 x 2 mm — Output Shaft in Case
82	Hub — Output Shaft
83	Drum — Low/Reverse Brake (Includes Overrunning Clutch)
84	Band — Low/Reverse
85	Bearing Race (2 Pieces) — Output Shaft Hub Thrust No. 10A
86	Bearing — Output Shaft Hub Thrust No. 10B
87	Bearing — Output Shaft to Case (Part of Case Assembly)
88	Washer — Output Shaft Thrust No. 11
89	Gear — Transmission Parking
90	Shaft — Output (4x2)
91	Gasket — Extension Housing
92	Pawl — Parking
93	Spring — Parking Pawl Return
94	Shaft — Parking Pawl
95	Extension Housing (4x2)
96	Bushing — Extension Housing (4x2)
97	Seal — Extension Housing to Slip Yoke

Fig. 191 Ford 4R44E transmission component keylist (continued)

89448G16

Item	Description
98	Stud — Extension Housing
99	Screw — Extension Housing (5 Req'd)
100	Shaft — Output (4x4)
101	Extension Housing (4x4)
102	Screw — Extension Housing to Transfer Case (5 Req'd)
103	Nut — Att. Outer and Inner Manual Valves to Shaft (2 Req'd)
104	Lever — Manual Control Outer
105	Screw — Digital Transmission Range (TR) Sensor (2 Req'd)
106	Sensor — Digital Transmission Range (TR)
107	Shaft — Manual Valve Outer to Inner Lever
108	Pin — Spring (Retains Outer Manual Lever to Case)
109	Seal — Main Control Lever
110	Lever — Manual Valve Inner
111	Rod — Parking Pawl Actuating
112	Spring Assembly — Manual Valve Detent
113	Screw Detent Spring
114	Connector — Transmission Case (16-Pin With Wire Harness to 6 Solenoids)
115	O-Ring — Transmission Case (16-Pin) Connector
116	Case Assembly (Not Available Separately)
117	Spring — Transmission Case (16-Pin) Connector
118	Plug — Pipe Line and EPC Pressure (Part of Case Assembly) (2 Req'd) (Model Dependent)
119	Vent — (4x4)
120	Vent — (4x2)
121	Tube — Lube Fluid Inlet — Short
122	Connector — Fluid Tube (2 Req'd)
123	Spring — Intermediate Servo Piston (2 Req'd)

Item	Description
124	Piston and Rod — Intermediate and Front Servo (2 Req'd)
125	Cover and Seal — Intermediate and Front Servo (2 Req'd)
126	O-Ring — Intermediate and Front Servo (2 Req'd)
127	Retaining Ring — 67 x 1.5mm Intermediate and Front Servo (2 Req'd)
128	Screw — Att. Separating Plate to Main Control
129	Solenoid — Electronic Pressure Control (EPC)
130	Solenoid — Transmission Shift (SS) (4 Req'd)
131	O-Ring — Shift Solenoid Small (13x1.5) (4 Req'd)
132	O-Ring — Shift Solenoid Large (15x1.5) (4 Req'd)
133	Solenoid — Torque Converter Clutch (TCC)
134	Clamp — (SSA (1) and SSC (3)) Solenoids
135	Clamp — TCC/CCS/SSB (2) and EPC Solenoids (Model Dependent)
136	Screw — Clamp (2 Req'd)
137	Gasket — Control Valve Body to Case
138	Plate — Valve Body Separating (Not Available Separately)
139	Gasket — Control Valve Body Separating
140	Control Valve Body — Main (Not Available Separately)
141	Screw — Main Control Valve Body (4 Req'd)
142	Screw — Main Control Valve Body (16 Req'd)
143	Screw — Main Control Valve Body (3 Req'd)
144	Seal Ring — Low/Reverse Servo Piston Small
145	Seal Ring — Low/Reverse Servo Piston Large

89448G17

Fig. 192 Ford 4R44E transmission component keylist (continued)

Item	Description
146	Piston — Low/Reverse Band Servo (Select Fit)
147	Gasket — Low/Reverse Servo Cover
148	Cover — Low/Reverse Servo
149	Screw — Low/Reverse Servo Cover (4 Req'd)
150	O-Ring — Fluid Filter Small
151	Filter — Fluid Pan (4x2)
152	O-Ring — Fluid Filter Large
153	Magnet
154	Gasket — Fluid Pan

Item	Description
155	Screw — Transmission Fluid Pan (18 Req'd)
156	Pan — Transmission Fluid (4x2)
157	Filter — Transmission Fluid Pan (4x4)
158	Pan — Transmission Fluid (4x4)
159	Screw — Fluid Filter
A	Coast Clutch Assembly
B	Direct Clutch Assembly
C	Forward Clutch Assembly

89448G18

Fig. 193 Ford 4R44E transmission component keylist (continued)

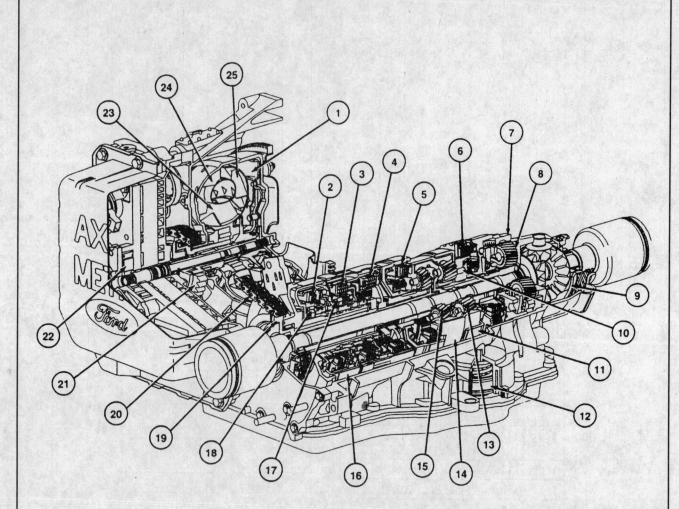

Item	Description
1	Torque Converter
2	Forward Clutch Assy
3	Direct Clutch Assy
4	Intermediate Clutch Assy
5	Reverse Clutch Assy
6	Low/Intermediate Clutch Assy
7	Final Drive Ring Gear
8	Final Drive Gearset
9	Differential Assy
10	Low/Intermediate One-Way Roller Clutch Assy
11	Coast Band
12	Coast Servo Piston

Item	Description
13	Rear Planetary Gearset
14	Rear Planetary Ring Gear
15	Front Planetary Gearset
16	Overdrive Band
17	Direct One-Way Roller Clutch Assy
18	Low One-Way Sprag Clutch
19	Driven Sprocket
20	Drive Chain
21	Drive Sprocket
22	Pump Assy
23	Reactor
24	Impeller
25	Turbine

89448G19

Fig. 194 Sectional view of the Ford AX4N transaxle and component locations

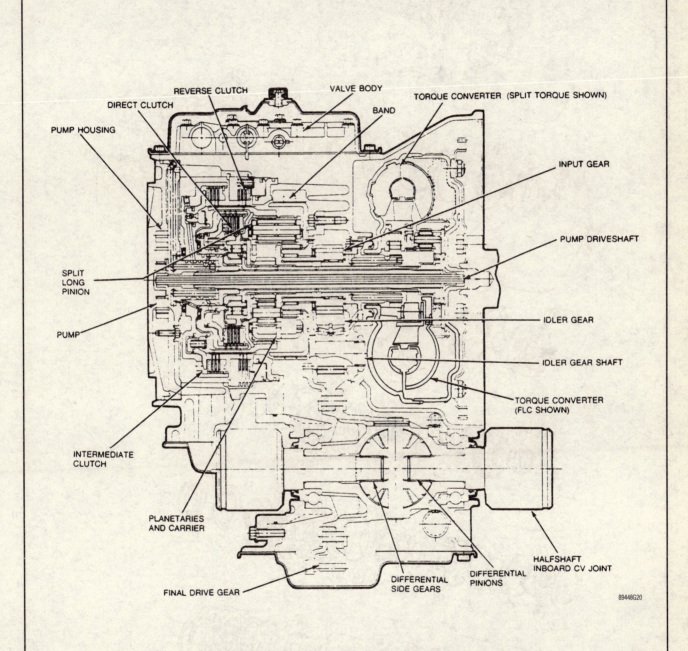

REVERSE CLUTCH

VALVE BODY

TORQUE CONVERTER (SPLIT TORQUE SHOWN)

DIRECT CLUTCH

BAND

PUMP HOUSING

INPUT GEAR

PUMP DRIVESHAFT

SPLIT
LONG
PINION

IDLER GEAR

PUMP

IDLER GEAR SHAFT

TORQUE CONVERTER
(FLC SHOWN)

INTERMEDIATE
CLUTCH

HALFSHAFT
INBOARD CV JOINT

PLANETARIES
AND CARRIER

DIFFERENTIAL
SIDE GEARS

DIFFERENTIAL
PINIONS

FINAL DRIVE GEAR

89448G20

Fig. 195 Sectional view of the Ford ATX transaxle component locations

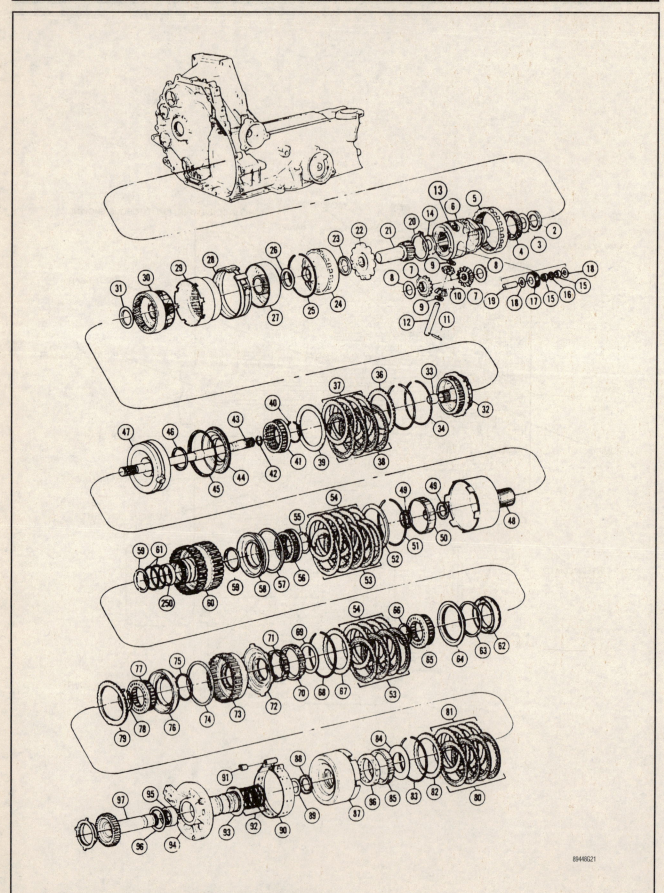

Fig. 196 Exploded view of the Ford AXOD-E transaxle components

89448G21

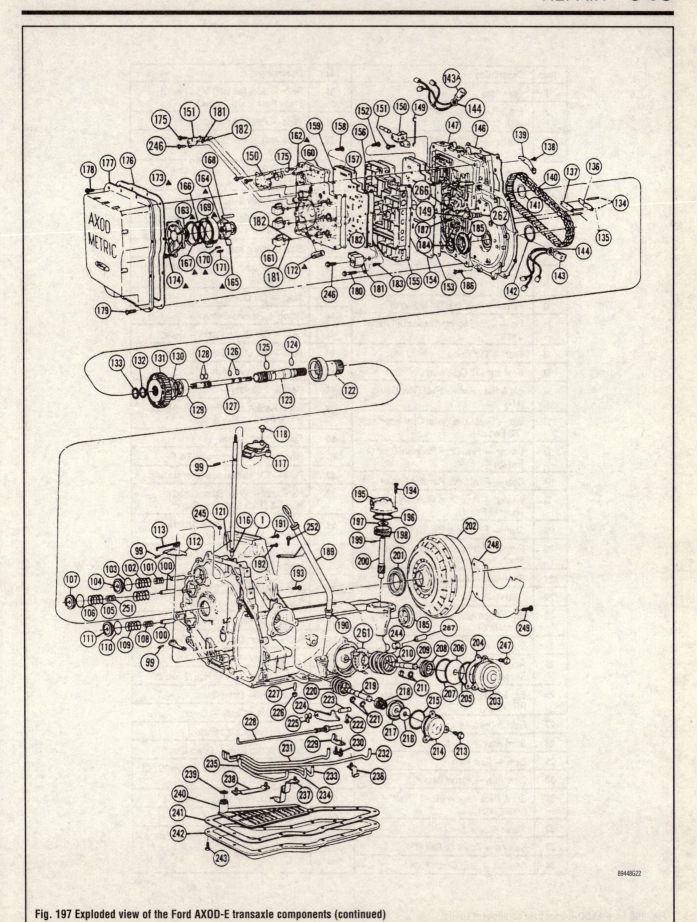

Fig. 197 Exploded view of the Ford AXOD-E transaxle components (continued)

89448G22

Item	Description	Item	Description
1	Case Assy	35	Ring — 153.9 Ret Int (Retain Rev Cl Press Plate to Cyl) — Sel Fit
2	Brg & Race Assy — Diff Carrier Thrust (#19)	36	Plate — Rev Clutch Pressure
3	Washer — Diff Carrier Thrust (#18) — Sel Fit	37	Plate Assy — Rev Cl Int Spline (Friction) (4 Req'd)
4	Gear — Governor Drive	38	Plate — Rev Cl Ext Spline (Steel) (4 Req'd)
5	Gear — Final Drive Ring	39	Spring — Rev Clutch Cushion
6	Case — Transaxle Diff Gear	40	Ring — 67.0 Ret Type Su Ext (Ret Rev Cl Spg & Ret to Cyl)
7	Gear — Diff Side (2 Req'd)	41	Supt & Spring Assy — Rev Clutch
8	Washer — Diff Side Gear Thrust (2 Req'd)	42	Ring — 27.0mm Ret Ext (Ret Diff Carrier Output Shaft)
9	Washer — RR Axle Diff Pinion Thrust (2 Req'd)	43	Shaft — Diff Output
10	Pinion — Rear Axle Diff (2 Req'd)	44	Piston — Reverse Clutch
11	Pin — Coiled Spring (Retains Diff Pinion Shaft)	45	Seal — Rev Clutch Piston — Outer
12	Shaft — Diff Pinion	46	Seal — Rev Clutch Piston — Inner
13	Gear and Diff Case Assy	47	Cylinder — Rev Clutch
14	Brg & Race Assy — Final Drive Carrier (#17)	48	Gear & Shell Assy — Frt Sun
15	Brg — Final Drive Planet Gear Needle (168 Req'd)	49	Brg & Race Assy — Frt Sun Gr Thrust (#10 & #11) — 2 Req'd
16	Spacer — Final Drive Planetary Gear (4 Req'd)	50	Hub — Interm Clutch
17	Gear — Final Drive Planet (4 Req'd)	51	Ring P Interm Clutch Plate (Sel Fit)
18	Washer — Final Drive Planetary Gear Thrust (8 Req'd)	52	Plate — Clutch Pressure (Intermediate)
19	Shaft — Final Drive Pinion (4 Req'd)	53	Plate Assy — Cl Int Spline (Used in Interm & Direct Clutch) as Req'd
20	Ring — 77.3 Ret Ext (Retain Pinion Shafts into Carrier)	54	Plate — Clutch Ext Spline (Used in Interm & Direct Clutch) as Req'd
21	Gear Assy — Final Drive Sun	55	Ring — 72.0 Ret Style Su Ext (Ret Interm Cl Spg & Ret to Cyl)
22	Gear — Parking	56	Supt & Spring Assy — Interm Clutch
23	Brg & Race Assy — Final Drive Gear Thrust (#16)	57	Seal — Interm Clutch — Outer
24	Support Assy — Planet Gear	58	Piston — Interm Clutch
25	Ring — 150.7 Ret Int (Used as Rear Support Ret Ring)	59	Seal — Interm/Dir Cl Inner (2 Req'd)
26	Brg & Race Assy — Sun Gear Thrust — RR (#15)	60	Cylinder Assy — Dir/Interm Clutch
27	Gear & Drum Assy — RR Sun	61	Seal — Interm & Dir Cl Hub (2 Req'd)
28	Band Assy — Low & Interm	62	Piston Assy — Direct Clutch
29	Gear — Rear Ring	63	Seal — Direct Clutch — Outer
30	Gear Assy — Planet Rear	64	Ring — Direct Clutch (Piston)
31	Brg & Race Assy — Planet Thrust — Center (#13)	65	Supt & Spring Assy — Direct Clutch
32	Planet Assy — Front	66	Ring — 77.0 Ret Style Su Ext (Ret Dir Cl Spg & Ret to Cyl)
33	Bearing — Frt Plt Gr Carrier	67	Plate — Clutch Pressure (Direct)
34	Retainer — Rear Clutch Plate	68	Ring — Dir Cl Plate (Sel Fit)
			Ring — 152.26 (Ret Dir Cl Press Plate to Cyl)

89448G23

Fig. 198 Ford AXOD-E transaxle component keylist

Item	Description	Item	Description
69	Washer — Dir Clutch Thrust (#7)	104	Piston — 1-2 Shift Accum
70	Race — Dir One-Way Cl — Outer	105	Spring — 3-4 Shift Accum
71	Clutch Assy — Direct One-Way	106	Seal — 3-4 Shift Accum — Piston
72	Race & Bshg Assy — Dir Owc — Inner	107	Piston — 3-4 Shift Accum
73	Cylinder Valve Assy — Fwd Clutch	108	Spring — Drive Shift Accum — Inner
74	Seal — Fwd Clutch — Outer	109	Spring Drive Shift Accum — Outer
75	Seal — Fwd Clutch — Inner	110	Seal — Drive Shift Accum — Piston
76	Piston — Forward Clutch	111	Piston — Drive Shift Accum
77	Supt & Spring Assy — Fwd Clutch	112	Lever Assy — Manual Detent
78	Ring — 85.0 Ret Type Su Ext (Ret Fwd Cl Spg & Ret to Cyl)	113	Rod — Man Control Valve Actu
79	Spring — Forward Clutch Wave	114	Shaft — Manual Control
80	Plate Assy — Fwd Cl Int Spline (Friction) as Req'd	115	Pin — Shaft Ret (Used as Man Lvr Shaft Ret Pin)
81	Plate — Fwd Cl Ext Spline (Steel) as Req'd	116	Seal Assy — Man Control Shaft
82	Plate — Fwd Cl Pressure	117	Sensor Assy — Main Lever Position
83	Ring — 152.26 Ret Int (Fwd) Sel Fit (Ret Fwd Cl Press Plt)	118	Bolt — M6 — 1.0 x 28 Hex Flg Hd (2-Neut Start Switch to Case)
84	Washer — Fwd Clutch Thrust (#6)	119	Bolt — M6 — 1.0 x 50 Hex Flg Hd (Att Chain Cover to Case)
85	Race — Low Owc — Outer	120	Plug — 1/8-27 Hex Hd Spd Fil (5 req'd — (3) in Chain Cover, (2) in Pump Body)
86	Clutch Assy — Low One-Way	121	Tag — Identification
87	Drum Assy — Overdrive	122	Support Assy — Stator
88	Brg & Race Assy — Dir Cl Hub (#9)	123	Shaft — Turbine
89	Washer — Driven Sprocket Supt Thrust — RR (#8) Sel Fit	124	Seal — O'Ring (Frt Turbine Shaft to Drive Sprkt)
90	Band Assy — Overdrive	125	Seal — Turbine Shaft — Rear
91	Retainer — O/D Band	126	Seal — Pump Shaft — Rear (2 Req'd)
92	Seal — Fwd Clutch Cyl (5 Req'd)	127	Shaft Assy — Oil Pump Drive
93	Washer — Support Thrust — Frt (#5) Sel Fit	128	Seal — Pump Shaft — Front
94	Support Assy — Driven Sprocket	129	Brg Assy — Drive Sprocket
95	Brg Assy — Driven Sprocket	130	Washer — Drive Sprocket Thrust (#2)
96	Washer — Driven Sprocket Thrust (#4)	131	Sprocket Assy — Drive
97	Sprocket Assy — Driven	132	Ring — 26.36 Ret Sty Su Ext (Ret Turb Shaft to Drive Sprkt)
98	Lever Assy — Manual Control	133	Seal — Turbine Shaft — Front (Metal)
99	Pin — 4mm x 28mm Spg Coiled Std (2-Used as Man Cntl Shft Pin)	134	Collar — Oil Level Thermo Retain
100	Shaft — Shift Accum Piston (3 Req'd)	135	Pin — 4mm x 22 Coiled (Locating By-Metal Element (3) Reg
101	Spring — 1-2 Shift Accum — Inner	136	Element — Oil Level Thermostatic
102	Spring — 1-2 Shift Accum — Outer		
103	Seal — 1-2 Shift Accum — Piston		

89448G24

Fig. 199 Ford AXOD-E transaxle component keylist (continued)

Item	Description	Item	Description
137	Plate — Oil Level Thermostat — Valve	170	Support — Oil Pump Bore Ring Radial Seal
138	Bolt — M6 x 1.0 14 Hex Flg Hd (Att Det Spring Assy to Chain Cover)	171	Seal — Oil Pump Bore Ring Radial
139	Spring Assy — Man Vlv Detent	172	Spring — Oil Pump Bore Ring
140	Chain Assy — Drive	173	Cover And Sleeve Assy — Oil Pump
141	Washer — Drive Sprocket Thrust (#1)	174	Bolt — M6 x 1.0 x 20 Hex Flg Plt (6 Req'd) Att Pump Cover to Pump Body
142	Washer — Chain Cover Thrust (#3)	175	Bolt — Hex Flg Hd (22-Att Pump Body & Main Contr to Chain Cover)
143, 143A	Bulkhead Assy — Wiring Conn	176	Gasket — Main Control Cover
144	Seal — 17.12 x 2.62 O'Ring (Wire Harness to Case)	177	Cover — Main Control
145	Vent Assy — Case	178	Bolt — M8-35.0 Hex Flg Hd (11-Att Chain Cover to Case)
146	Gasket — Chain Cover	179	Bolt — M8-1.25 x 25 Hex Flg Hd (12-Att Main Ctl Cvr to Chn Cvr.)
147	Cover Assy — Chain	180	Bolt — M6-1.00 x 40 Hex Flg Hd (3-Att Vlv Bdy to Chn Cvr and Sol Assy)
148	Conn Assy — 5/16 Tube x 1/4 Ex Pipe Plug (2 Req'd)		
149	Sensor Assy — Turbine Speed	181	Seal — 15.6 x 1.78 O-Ring
150	Solenoid Assy — Pressure Reg	182	Seal — 6.07 x 1.79 O-Ring (Bypass Solenoid Seal)
151	Solenoid Assy — By-Pass Clutch Control	183	Screen Assy — Bypass Clutch Solenoid
152	Screw — M6 x 1.0 x 14 Pan Hd (2-Att Vlv Bdy Sep Plt to Vlv Bdy)	184	Circle Clip — Output Shaft Retainer (Retains CV Joint)
153	Gasket — Control Assy	185	Seal Assy — Diff (2 Req'd)
154	Plate Assy — Valve Body Sep	186	Bolt — M8-1.25 x 45 Hex Flg Hd (2-Att Vlv Bdy to Chn Cvr and Sol Assy)
155	Gasket — Cntl Vlv Body Sep Plate		
156	Control Assy — Main	187	Bolt — M10-1.50 x 45 Hex (Att Chain Cover to Driven Support)
157	Gasket — Pump Assy		
158	Screw — M6 x 1.0 x 14 Pan Hd Torx T-30 (2 — Pump Sep Plate to Pump Body)	188	Indicator Assy — Oil Level
		189	Tube Assy — Oil Filler
159	Plate — Oil Pump Body Sep	190	Grommet — Oil Filler
160	Gasket — Oil Pump Body Sep Plate	191	Bolt — M10-1.50 x 45 Hex (Att Chain Cover to Driven Support)
161	Solenoid Assy — Switch Control (3 Req'd)	192	Bolt — M6 x 1.00 x 30 Hex Flg Hd (4-Att Case to Chain Cover)
162	Body Brg and Seal Assy — Oil Pump	193	Screw — M6-1.0 x 20 Pan Hd (6-Att Case to Stator Support)
163	Ring — Oil Pump Vane Support (2 Req'd)	194	Bolt — (2-Governor Cover to Case)
164	Rotor — Oil Pump	195	Cover — Governor
165	Vane — Oil Pump (7 Req'd)	196	Seal — 63.2 x 1.80 O'Ring (Used as Gov Cover Seal)
166	Seal — Oil Pump Bore Ring Side	197	Brg and Race Assy — Gov Thrust
167	Support — Oil Pump Bore Ring Side Seal	198	Gear — Speedo Drive (7TLH)
168	Pin — 8mm x 37.7 Straight Hrdn	199	Pin — 3.3 x 22 Spg Slot Hvy (Used as Speedo Gear Drive Pin)
169	Ring — Oil Pump Body		

89448G25

Fig. 200 Ford AXOD-E transaxle component keylist (continued)

Item	Description	Item	Description
200	Gear and Shaft Assy — Gov. Driven	234	Tube — Servo Rel. Oil Transfer
201	Seal Assy — Conv Imp Hub	235	Tube — Rev. Cl. Apply Oil Transfer
202	Converter Assy — 10-1/4	236	Brkt. Assy — Tube Support — Gov. Feed
203	Cover — Low/Interm Band Servo	237	Brkt. Assy — Tube Support — Rev. Clutch
204	Gasket — Low/Interm Band Servo	238	Brkt. Assy — Tube Support Main
205	Seal — Low/Interm Servo Piston Cover	239	Seal — (Used on Oil Filter)
206	Piston — Low/Interm Band Servo	240	Filter Assy — Oil
207	Seal — Low/Interm Band Servo Piston	241	Gasket — Oil Pan
208	Retainer and Spring Assy — Low/Interm Servo	242	Pan Oil
209	Rod — Low/Interm Servo Piston (Sel Fit)	243	Bolt — M8-1/25 x 14 Hex Flg. Hd. (17-Att Oil Pan to Case)
210	Spring — Low/Interm Servo Piston	244	Plug — 13.9mm Cup
211	Ring — 11mm Ret Type Rb Ext (2-Att L/I Servo Piston)	245	Nut — 1/4 Spring (Retain I.D. Tag)
212	Seal Assy — PR Lube Transfer Tube	246	Bolt — M6 x 1.0 x 14 Hex Flg. Hd. (Att Solenoid Assy to Viv. Body)
213	Bolt — (3-Att O/D Servo Cover to Case)	247	Bolt — (3-Att L/I Servo Cover to Case)
214	Cover — O/D Band Servo	248	Cover — Conv. Hsg. Lower
215	Seal — O/D Servo Cover	249	Bolt — (Att. Conv. Hsg. Cur. to Case)
216	Retainer — O/D Servo Piston	250	Bushing — Dir/Interm. Clutch Cylinder
217	Piston and Seal Assy — O/D Servo	251	Spring — 3-4 Shift Accum-Inner
218	Retainer and Cushion Spring Assy — O/D Servo	252	Bolt — M6 — 1.0 x 14 Hex Flg. Htd. (Att Filler Tube to Case)
219	Rod — O/D Servo Piston (Sel. Fit)	253	Bolt — M6 — 1.0 x 28 (2 Att — Oil Pump Assy to Main Control)
220	Spring — O/D Servo Return	254	Stud — M8 — 1.25 — 1.25 x 7.96 Hex Hd. Shoulder
221	Ring — Ret. Ext. Rod O/D Servo (2 Used on O/D Servo Rod)	255	Spring Fwd Clutch Wave (3.8L only)
222	Pin — Shaft Retainer (Used as Park Pawl Shaft Return Pin)	256	Ceramic Magnet Case
223	Shaft — Park Pawl	257	Spring 1-2 Shift Accum Center (3.8L only)
224	Pawl — Parking Brake	258	Wheel-Driven Sprocket Speed Senor
225	Spring — Park Pawl Return	259	Screen — Case Intermediate Circuit
226	Screw — M12 x 1.75mm Set Hd. Scket. (Rev. Cl. Assy Locator Blt.)	260	Sensor — Oil Temperature
227	Nut — M12 x 1.75 Hex (Rev. Cl. Assy Locator Bolt)	261	Retainer — L/I Servo Return Spring
228	Rod Assy — Park Pawl Actuating	262	Seal — 14.0 x 1.78 O-Ring 2-Req'D
229	Abutment — Park Pawl Actuating	263	Seal — 25.12 x 1.78 O-Ring
230	Bolt — M8-1.25 x 25 Hex Flg. Hd. (2-Att. Abutment Assy to Case)	264	Seal — 12.42 x 1.78 O-Ring
231	Tube — Rear Lube Oil Transfer	265	Ret — Clip Trans Cooler Tube
232	Tube — Gov. Feed Oil Transfer	266	Bolt — M6 — 1.0 x 20 Hex Flg Hd
233	Tube — Servo Apply Oil Transfer	267	Stud — M10 — 1.5 x 60.5

89448G26

Fig. 201 Ford AXOD-E transaxle component keylist (continued)

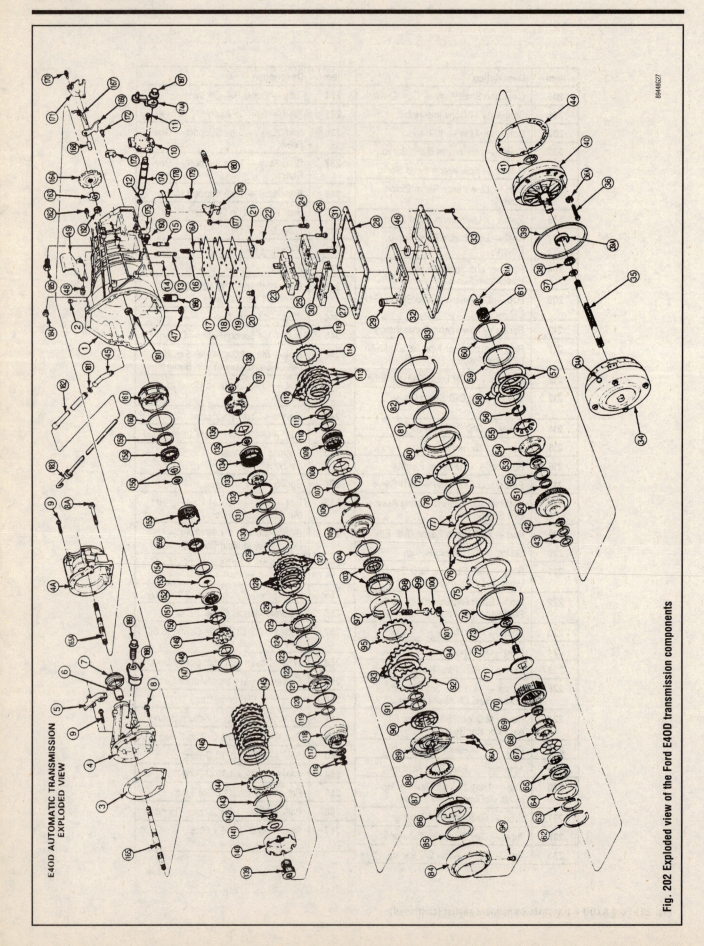

E4OD AUTOMATIC TRANSMISSION EXPLODED VIEW

Fig. 202 Exploded view of the Ford E4OD transmission components

Item	Description
1	Case
2	Vent
3	Extension Housing Gasket
4	Extension Housing (4x2)
4A	Extension Housing (4x4) and Super Duty
5	Wiring Bracket (Not Serviced)
6	Extension Housing Bushing (4x2)
7	Park / Neutral Position Switch (4x2)
8	Extension Bolt (4x2 Bottom) (2 Required) M10-1.5x35mm
8A	Extension Bolt (Super Duty and 4x4 Bottom) (2 Required) M10-1.5x90mm
9	Extension Bolt (Top) (7 Required) M10-1.5x40mm
10	Manual Lever Position Sensor
11	Bolt Assembly (2 Required) M6-1.0x30mm
12	Manual Control Lever Oil Seal
13	Stud M6-1.0x79mm
14	Rubber Check Ball (9 Required, 2 in Main Control)
15	Stud (4 Required) M6-1.0x61.25mm
16	Electronic Pressure Control (EPC) Blow-Off Ball
16A	EPC Blow-Off Ball
17	Valve Body Separator Plate Gasket
18	Valve Body Separator Plate
19	Valve Body Separating Plate-to-Case Gasket
20	Solenoid Screen Assembly
21	Separator Plate Reinforcing Plate (Not Serviced)
22	Bolt (3 Required) M6-1.0x16mm
23	Solenoid Valve Body Assembly — Transmission Control
24	Torx® Head Bolt (9 Required) M6-1.0x40mm
25	Main Control Valve Body
26	Bolt (18 Required) M6-1.0x42.5mm
27	Accumulator Body Assembly
28	Oil Pan Gasket
29	Filter and Seal Assembly — (Serviced in Kits Only)
30	Nut (5 Required) M6-1.0
31	Bolt (7 Required) M6-1.0x66mm
32	Transmission Oil Pan

Item	Description
33	Oil Pan Bolt (20 Pcs.) M8-1.25x12mm
34	Torque Converter
34A	Converter Drain Plug 1/8 In.-27
35	Input Shaft
36	Pump Bolt (9 Pcs.) M8-1.25x65mm
36A	Washer (9 Required)
37	Front Pump Support Seal
38	Front Oil Pump Seal
39	O-Ring
40	Front Oil Pump Assembly
41	Pump Thrust Washer
42	Overdrive Sun Gear Thrust Bearing
43	Coast Clutch Seal
44	Oil Pump Gasket
45	Lube Oil Inlet Short Tube
46	Oil Pan Magnet (Not Serviced Separately)
47	Converter Housing Access Plug
48	Heat Shield Bolt (2 Required)
49	Transmission Heat Shield
50	Coast Clutch Cylinder
51	Coast Clutch Seal — Inner
52	Coast Clutch Seal — Outer
53	Coast Clutch Piston
54	Piston Apply Ring
55	Coast Clutch Piston Spring — Disc
56	Coast Clutch Disc Spring Retaining Ring
57	Coast Clutch External Spline Plate — Steel
58	Coast Clutch Internal Spline Plate — Friction
59	Clutch Pressure Plate
60	Retaining Ring (Selective Fit)
61	Overdrive Sun Gear
61A	Retaining Ring
62	Retaining Ring (Outer Race to Overdrive Ring Gear)
63	Retaining Ring (Overdrive OWC to Outer Race)
65	Overdrive One-Way Clutch (Serviced in Kits Only)
67	Overdrive Overrunning Clutch Washer
68	Overdrive Planet Assembly
69	Overdrive Planet Thrust Bearing Assembly
70	Overdrive Ring Gear
71	Overdrive Center Shaft
72	Retaining Ring (Center Shaft to Overdrive Ring Gear)

89448G28

Fig. 203 Ford E4OD transmission component keylist

Item	Description
73	Overdrive Center Shaft Thrust Bearing Assembly
74	Clutch Pressure Plate Retainer Snap Ring
75	Clutch Pressure Plate
76	Overdrive Clutch Plate Internal Spline — Friction
77	Overdrive Clutch Plate External Spline — Steel
78	Overdrive Clutch Disc Spring Retaining Ring
79	Overdrive Clutch Piston Disc Spring
80	Overdrive Clutch Piston
81	Clutch Piston Seal — Outer
82	Clutch Piston Seal — Inner
83	Intermediate Cylinder Retaining Ring
84	Intermediate / Overdrive Clutch Cylinder
85	Intermediate Clutch Piston Inner Seal
86	Intermediate Clutch Piston
87	Intermediate Clutch Piston Outer Seal
88	Intermediate Clutch Piston Disc Spring
89	Center Support Assembly
90	Center Support Thrust Washer
91	Direct Clutch Cast Iron Seal
92	Clutch Pressure Plate
93	Intermediate Clutch Internal Spline Plate — Friction
94	Intermediate Clutch External Spline Plate — Steel
95	Clutch Pressure Plate — Rear
96	Bolt — Cylinder Hydraulic Feed (1 Required) M10-1.5x24mm
96A	Bolt — Center Support Hydraulic Feed (2 Required) M12-1.75x31mm
97	Intermediate Band Assembly
98	Servo Piston Spring
99	Intermediate Band Servo Piston
100	Rear Band Servo Retainer
101	Servo Piston Retaining Ring
103	One-Way Clutch (Serviced in Kits Only)
104	Intermediate One-Way Clutch Thrust Washer
105	Intermediate Brake Drum
106	Direct Clutch Piston Seal — Inner
107	Direct Clutch Piston Seal — Outer
108	Direct Clutch Piston

Item	Description
109	Direct Clutch Retainer and Spring Assembly
110	Direct Clutch Support Spring Retaining Ring
111	Intermediate Brake Drum Thrust Washer
112	Direct Clutch Internal Spline Plate — Friction
113	Direct Clutch External Spline Plate — Steel
114	Direct Clutch Pressure Plate
115	Clutch Plate Retaining Ring (Selective Fit)
116	Forward Clutch Cylinder Seal
117	Forward Clutch Needle Thrust Bearing
118	Forward Clutch Cylinder
119	Forward Clutch Piston Seal — Inner
120	Forward Clutch Piston Seal — Outer
121	Forward Clutch Piston
122	Forward Clutch Piston Spring Ring
123	Forward Clutch Piston Disc Spring
124	Forward Clutch Spring Ring
125	Forward Clutch Pressure Plate
126	Forward Clutch Pressure Spring
127	Forward Clutch External Spline Plate — Steel
128	Forward Clutch Internal Spline Plate — Friction
129	Forward Clutch Pressure Plate — Rear
130	Forward Clutch Pressure Retaining Ring (Selective Fit)
130	Retaining Ring
130	Retaining Ring
130	Forward Clutch Hub Thrust Washer
130	Retaining Ring
131	Forward Clutch Hub Thrust Washer
132	Forward Hub Retaining Ring
133	Forward Hub Ring Gear
134	Forward Ring Gear
135	Forward Clutch Thrust Bearing Assembly
136	Forward Planet Carrier Thrust Washer
137	Forward Planet Assembly
138	Forward Clutch Thrust Bearing Assembly
139	Forward / Reverse Sun Gear Assembly

89448G29

Fig. 204 Ford E4OD transmission component keylist (continued)

Item	Description
140	Input Shell
141	Input Shell Thrust Washer
142	Retaining Ring
143	Reverse Clutch Pressure Plate Retaining Ring
144	Reverse Clutch Pressure Plate
145	Reverse Clutch External Spline Plate — Steel
146	Reverse Clutch Internal Spline Plate — Friction
147	Reverse Planet Retaining Ring
148	Planet Carrier Thrust Washer
149	Reverse Planet
150	Planet Carrier Thrust Washer
151	Retaining Ring (for Output Shaft) (1-1/2 In. Dia.)
152	Output Shaft Ring Gear
153	Output Shaft Hub and Race (Serviced in Kits Only)
154	Retaining Ring
155	Reverse Clutch Hub Assembly
155B	Reverse One-Way Clutch (Serviced in Kits Only)
156	Output Shaft Hub Thrust Bearing
158	Reverse Clutch Retainer and Spring Assembly
159	Reverse Clutch Piston Inner Seal
160	Reverse Clutch Piston Outer Seal
161	Reverse Clutch Piston
162	Bolts (5 Required) 5/16 In.-24 (One Way Clutch to Case)
163	Output Shaft Thrust Washer — Rear
164	Overdrive Shaft Parking Gear

Item	Description
165	Output Shaft — (4x2)
165A	Output Shaft — (4x4)
167	Parking Pawl Return Spring
168	Parking Pawl Shaft
169	Parking Pawl
170	Bolt and Washer Assembly (2 Required) M8-1.25x23.8mm
171	Parking Rod Guide Plate
172	Bolt M8-1.25x25.9mm
173	Parking Pawl Actuating Abutment
174	Manual Control Lever Shaft
174A	Manual Control Lever
175	Manual Lever Shaft Retaining Pin
176	Manual Valve Detent Lever — Inner
177	Inner Detent Lever Nut M14-1.5 Hex
178	Manual Valve Detent Lever Spring
179	Bolt — Hex Flange Head M6-1.0x16.5mm
180	Parking Pawl Actuating Rod Assembly
181	Filler Tube O-Ring
182	Oil Filler Tube
183	Oil Level Indicator
184	Oil Tube Inlet Connector
185	Converter Drain Back Check Valve Assembly — Rear
186	Accum. Regulator Filter Assembly
187	Nut, M10-1.5 Hex
188	Extension Housing Plug Assembly
189	Screw and Washer Assembly, 1/4-20x.62
190	Test Port Hex Head Plug (2 Required) 1/8-27
191	Front Case Bushing
192	Rear Case Bushing

89448G30

Fig. 205 Ford E4OD transmission component keylist (continued)

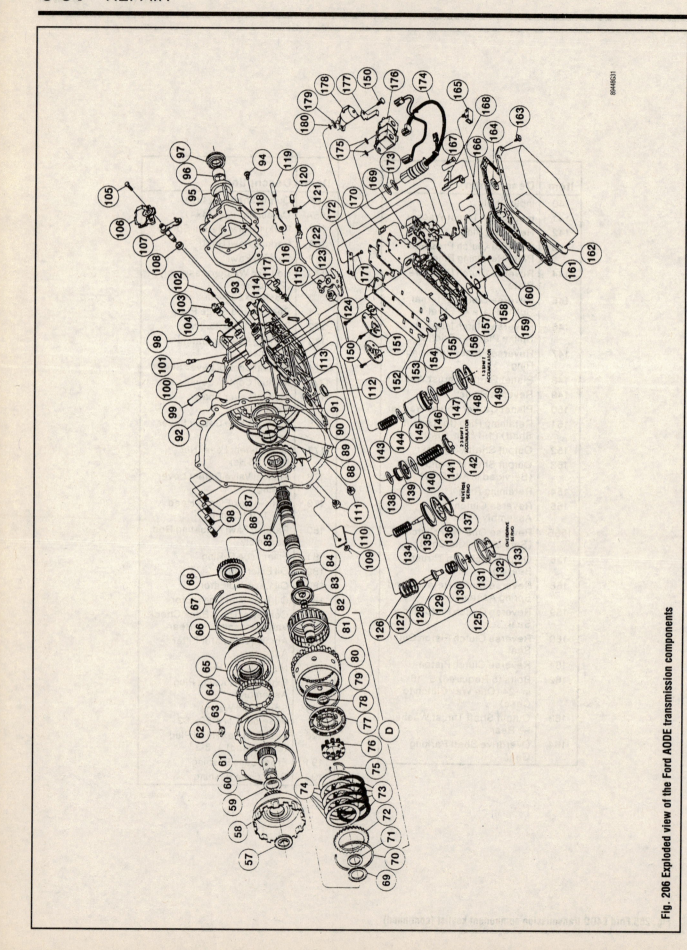

Fig. 206 Exploded view of the Ford AODE transmission components

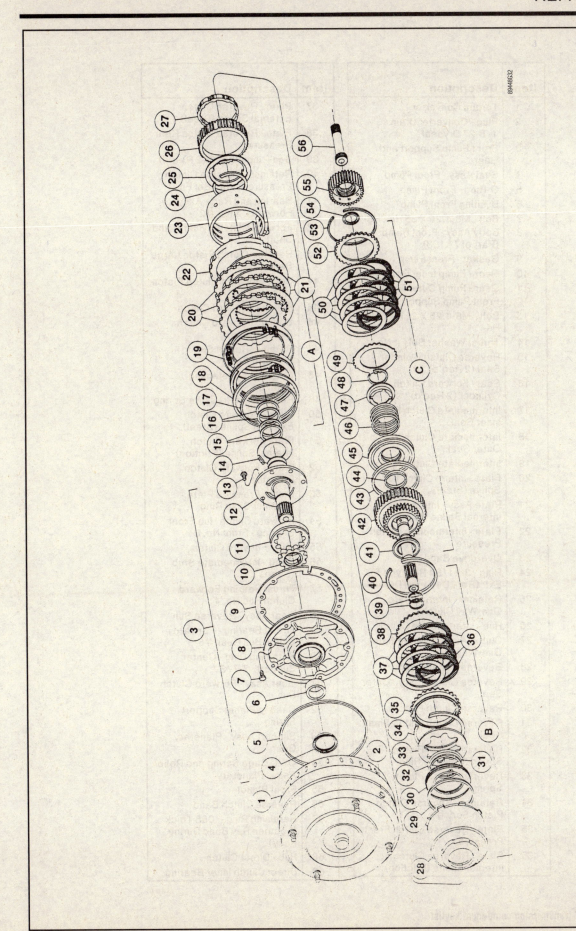

Fig. 207 Exploded view of the Ford AODE transmission components (continued)

Item	Description
1	Torque Converter
2	Plug - Converter Drain - 1/8-27 Dryseal
3	Front Pump Support and Gear
4	Seal Assy - Front Pump
5	O-Ring - Front Pump
6	Bushing Front Pump
7	Bolt - M8-1.25 X 35 Hex Hd
8	Body Assy - Front Pump (Part of 7A103)
9	Gasket - Front Pump
10	Gear - Pump Inner Gerotor
11	Gear - Pump Outer Gerotor
12	Front Pump Support
13	Bolt - M8-1.25 X 25 Hex Flg Hd
14	Thrust Washer Sel Fit No. 1
15	Reverse Clutch Cylinder Seal (2 Req'd)
16	Seal - Forward Clutch Cylinder (2 Req'd)
17	Intermediate Clutch Piston Inner Seal
18	Intermediate Clutch Piston Outer Seal
19	Intermediate Clutch Piston
20	Plate - Interm Clutch Ext Spline (Steel)
21	Plate Assy - Interm Clutch Internal Spline (Friction)
22	Plate - Intermediate Clutch Pressure
23	Overdrive Band
24	Ring - 3-21/64 Ret Type SU Ext. (Ret 7D191 to 7D044)
25	Retainer - Intermediate One-Way Clutch
26	Hub - Intermediate Clutch
27	Clutch Assy - Intermediate One-Way
28	Reverse Clutch Drum
29	Reverse Clutch Piston Outer Seal
30	Reverse Clutch Piston
31	Reverse Clutch Piston Small Seal
32	Reverse Clutch Piston Spring Pressure Ring
33	Reverse Clutch Piston Spring
34	Retainer - Reverse Clutch Piston Spring
35	Plate - Reverse Clutch Front Pressure
36	Plate - Reverse Clutch Internal Spline (Friction)

Item	Description
37	Plate - Reverse Clutch External Spline (Steel)
38	Plate - Reverse Clutch Rear Pressure
39	Seal - Input Shaft (2 Req'd)
40	Retainer - Reverse Clutch Pressure Plate - Sel Fit
41	Bearing and Race Assy - Forward Clutch #2
42	Forward Clutch Cylinder and Shaft
43	Reverse Clutch Piston Large Seal
44	Seal - Forward Clutch Piston - Inner
45	Piston - Forward Clutch
46	Spring - Forward Clutch Piston Return
47	Retainer Return Spring - Forward Clutch
48	Snap Ring - Retaining 1-59/64
49	Rear Clutch Pressure Spring
50	Plate - Forward Clutch External Spline (Steel)
51	Plate Forward Clutch Internal Spline (Friction)
52	Plate - Forward Clutch Pressure
53	Clutch Pressure Plate Retainer Snap Ring
54	Forward Clutch Hub Front Bearing - Front No. 3
55	Hub - Forward Clutch
56	Shaft - Intermediate Stub (Short)
57	Thrust Bearing Forward Clutch Hub No. 4
58	Gear Assy - Reverse Sun
59	Thrust Bearing - Forward Clutch Sun Gear #5
60	Retaining Ring - Center Support - 7-7/32
61	Gear Assy - Forward Clutch Sun
62	Case to Planet Support Spring
63	Support Assy - Planetary Gear
64	OWC Cage Spring and Roller Assy - Planetary
65	Front Planet
66	Reverse Clutch Band
67	Retaining Ring - .058 Thick (Locates Rev Band During Assy)
68	Hub - Direct Clutch
69	Direct Clutch Inner Bearing

89448G33

Fig. 208 Ford AODE transmission component keylist

Item	Description
70	Retaining Ring - Direct Clutch Press Plate -Select Fit
71	Support - Direct Clutch Inner Bearing Support
72	Plate - Direct Clutch Pressure
73	Plate - Direct Clutch Internal Spline (Friction)
74	Plate - Direct Clutch External Spline (Steel)
75	Retaining Ring - 1-19/32
76	Direct Clutch Support and Spring
77	Piston Assy - Direct Clutch
78	Direct Clutch Inner Seal
79	Reverse Clutch Piston Large Seal
80	Gear - Output Shaft Ring
81	Cylinder Assy - Direct Clutch
82	Seal - Output Shaft Small - Direct Clutch (2 Req'd)
83	Direct Clutch Hub Bearing and Race Outer No. 8
84	Output Shaft
85	Seal - Output to Case Shaft Large (3 Req'd)
86	Seal - O-Ring (Piloted Output Shaft Only)
87	output shaft hub
88	Snap Ring - 1-13/16 Retaining
89	N803175 - Retaining
90	Bushing - Rear Case
91	Case Rear Bearing No. 9
92	Case
93	Extension Housing Gasket
94	Bolt - M8-1.25 X 30
95	Extension Housing
96	Extension Housing Bushing
97	Seal Assy - Extension Housing (Booted)
98	Pipe Plug - 1/8-27 Dryseal Tapered (5 Req'd)
99	Overdrive Band Anchor Pin
100	Pin - Reverse Band Anchor
101	Vent
102	Bolt - M6-1.0 X 14 Hex Flg Hd
103	Turbine Shaft Speed (TSS) Sensor
104	Seal - 14.0 X 1.78 O-Ring (2 Req'd)

Item	Description
105	Bolt and Washer Assy - M6-1.0 X 25MM (2 Req'd Att 77F293 to 7005)
106	Transmission Range (TR) Sensor
107	Manual Control Lever
108	Seal Assy - Manual Control Lever
109	Nut - 1/4 Spring
110	Tag - Identification
111	Connector Assy - Fluid Tube (2 Req'd)
112	Converter Housing Access Plug
113	Screen Assy - Fluid
114	Manual Lever Shaft Retaining Pin
115	Seal - 0.426 X 0.070 O-Ring
116	Seal - 14.0 X 1.78 O-Ring
117	Solenoid Valve - Trans Pressure Control
118	Parking Pawl
119	Parking Pawl Shaft
120	Cup - Park Rod Guide
121	Parking Pawl Return Spring
122	Parking Lever Actuating Rod
123	Manual Valve Detent Lever
124	Nut - M14 X 1.5 Hex Interm Det Lever
125	Piston Assy - Overdrive Servo
126	Overdrive Servo Piston Return Spring
127	Rod - Overdrive Servo Actuating
128	Washer - Back-up Overdrive Servo Belleville
129	Spring - Belleville Overdrive Cushion Spring
130	Piston Assy - Overdrive Servo
131	Ring - Retaining
132	Sleeve Assy - Overdrive Servo
133	Ring - 2.85 Ret Type TVP "H" Int
134	Reverse Band Servo Spring
135	Reverse Band Servo Piston and Rod
136	Reverse Band Servo Cover
137	Retaining Ring - Internal - 3-13/16
138	Seal - 2-3 Accumulator Piston - Upper

89448G34

Fig. 209 Ford AODE transmission component keylist (continued)

Item	Description
139	Piston - 2-3 Shift Accumulator
140	Seal - 2-3 Accumulator Piston - Lower
141	Spring - 2-3 Shift Accumulator Piston
142	Retainer - 2-3 Shift Accumulator Spring
143	Spring - 1-2 Shift Accumulator (Model Dependant)
144	Seal - 1-2 Shift Accumulator Piston
145	Piston - 1-2 Shift Accumulator
146	Seal - 1-2 Accumulator Piston - Lower
147	Spring - 1-2 Shift Accumulator
148	Cover and Seal Assy - 1-2 Accumulator
149	Ring - 2-1/16 Retaining Type HU Internal
150	Bolt - M6-1.0 X 16 Hex Hd
151	Plate - Valve Body Reinforcing
152	Main Control to Case Gasket
153	Plate - Control Valve Body Separator
154	Gasket - Valve Body Separating - Lower
155	Valve - Converter Drain Back
156	Main Control Valve Body
157	Gasket - Valve Body Cover Plate
158	Plate - Valve Body Cover

Item	Description
159	Bolt - M6-1.0 X 18 Hex Hd
160	Filter and Seal Assy - Fluid
161	Gasket - Trans Pan
162	Pan - Trans
163	Bolt - M8-1.25 X 18 Hex Flg Hd
164	Magnet - Ceramic Case
165	Transmission Fluid Temperature (TFT) Sensor
166	Bolt - M8-1.25 x 46 Hex Shldr Pilot
167	Bolt - M6-1.0 X 52 Hex Flg Hd
168	Solenoid Retainer
169	Ball - 1/4 Dia. Coast Booster Valve Shuttle (8 Req'd)
170	Screen - Solenoid Pressure Supply
171	Bolt - M6-1.0 X 40 Hex Flg Hd
172	Manual Valve Detent Spring
173	Seal - Seal - 0.864 X 0.070 O-Ring (2 Req'd)
174	Wiring Connector Bulkhead
175	Seal - 6.07 X 1.70 O-Ring (2 Req'd)
176	Solenoid Valve - Transmission Shift
177	Shift Control Solenoid Bracket
178	Solenoid Valve - Torque Converter Clutch (TCC)
179	Seal - 0.489 X 0.070 O-Ring
180	Seal - 0.176 X 0.070 O-Ring
A	Intermediate Clutch Assy
B	Reverse Clutch Assy
C	Forward Clutch Assy
D	Direct Clutch Assy

89448G35

Fig. 210 Ford AODE transmission component keylist (continued)

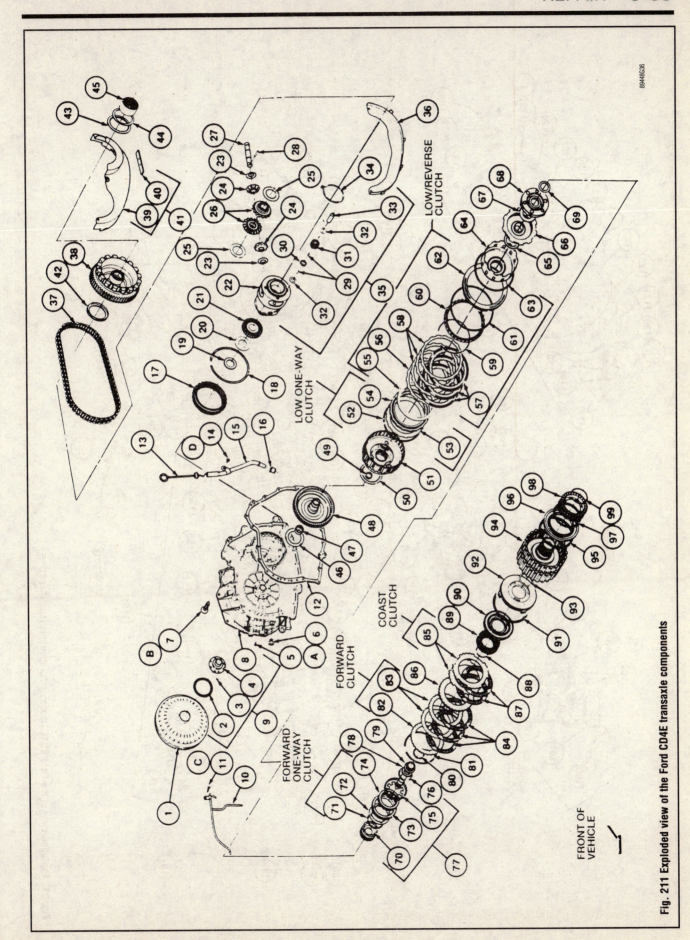

Fig. 211 Exploded view of the Ford CD4E transaxle components

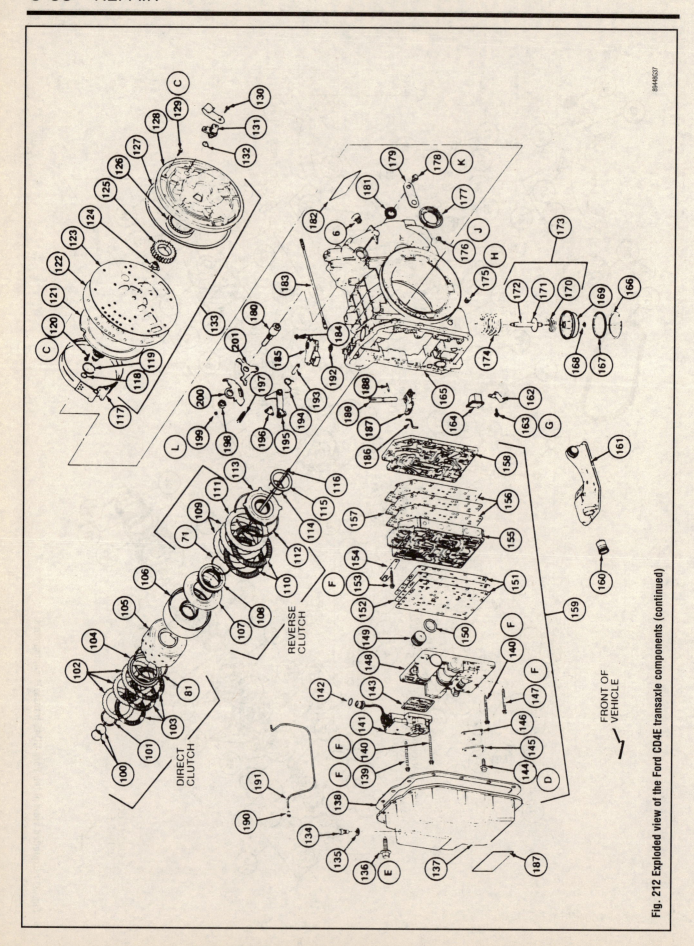

Fig. 212 Exploded view of the Ford CD4E transaxle components (continued)

Item	Description
1	Torque Converter
2	Converter Impeller Hub Seal
3	Stator Support Bolts
4	Stator Support (Part of 7005)
5	Pipe Plug - 1/8-27 Dry Seal
6	Oil Tube Connector
7	Converter Housing-to-Transaxle Case Bolts (20 Req'd)
8	Converter Housing
9	Converter Housing and Stator Support Assy
10	Differential Lube Tube
11	Differential Lube Tube Bolt
12	Transaxle Split Flange Gasket
13	Oil Level Indicator
14	Oil Filler Tube Bolt
15	Oil Filler Tube
16	Oil Filler Tube Grommet
17	Final Drive Ring Gear
18	Final Drive Ring Gear Retaining Ring
19	Differential Bearing, No. 15
20	Differential Bearing Shim, No. 14 (Selective Fit)
21	Speedometer Drive Gear
22	Differential Gear Case
23	Differential Pinion Thrust Washer
24	Differential Pinion Gear
25	Differential Side Gear Thrust Washer
26	Differential Side Gear
27	Differential Pinion Shaft
28	Pinion Shaft Roll Pin
29	Final Drive Planet Gear Needle Bearing (144 Req'd)
30	Final Drive Planet Gear Spacer (4 Req'd)
31	Final Drive Planet Gear (4 Req'd)
32	Final Drive Planet Gear Thrust Washer (8 Req'd)
33	Final Drive Pinion Shaft (Part of 7F465)
34	Final Drive Pinion Shaft Retaining Ring
35	Final Drive Planet and Carrier
36	Front Chain Cover
37	Drive Chain Assy

Item	Description
38	Drive Sprocket Assy
39	Rear Chain Cover
40	Magnet - Case
41	Chain Cover Pan Assy
42	No. 13 Driven Sprocket Thrust Bearing Assy
43	No. 12 Driven Sprocket Bearing Assy
44	No. 11 Driven Sprocket Shim (Selective Fit)
45	No. 18 Driven Sprocket Bearing Assy
46	Drive Sprocket Thrust Washer, No. 10 (Selective Fit)
47	No. 17 Stator Support Bearing Assy
48	Reverse/Overdrive Ring Gear Assy
49	No. 16 Reverse/Overdrive Ring Gear Bearing Assy
50	No. 9 Reverse/Overdrive Ring Gear Thrust Bearing Assy
51	Reverse/Overdrive Carrier Assy (with Captured No. 8 Thrust Bearing)
52	Low One-Way Clutch Retaining Ring
53	Low One-Way Clutch Thrust Plate
54	Low One-Way Clutch Assy
55	Low/Reverse Clutch Wave Spring
56	Low/Reverse Clutch Pressure Plate
57	Low/Reverse Clutch Internal Spline Clutch Plates (Friction) (3 Req'd)
58	Low/Reverse Clutch External Spline Clutch Plates (Steel) (3 Req'd)
59	Low/Reverse Clutch Return Spring Retaining Ring
60	Low/Reverse Clutch Return Spring Assy
61	Low/Reverse Clutch Piston Outer Seal
62	Low/Reverse Clutch Piston
63	Low/Reverse Clutch Piston Inner Seal
64	Reverse/Overdrive Sun Gear and Shell Assy
65	No. 7 Reverse/Overdrive Sun Gear Thrust Bearing Assy
66	Low-Intermediate Ring Gear Assy

89448G38

Fig. 213 Ford CD4E transaxle component keylist

Item	Description
67	No. 6 Low-Intermediate Carrier Thrust Bearing Assy
68	Low-Intermediate Carrier Assy
69	No. 5 Low-Intermediate Sun Gear Thrust Bearing Assy
70	Low-Intermediate Sun Gear Assy
71	Retaining Ring (2 Req'd)
72	Forward One-Way Clutch Retainer
73	Forward One-Way Clutch Outer Race
74	Forward One-Way Clutch Sprag Assy
75	Coast Clutch Hub
76	Coast Clutch Hub Retaining Ring
77	Forward One-Way Clutch and Low-Intermediate Sun Gear Assy
78	No. 4 Turbine Shaft Thrust Bearing Assy
79	Turbine Shaft Assy
80	Forward / Coast / Direct Clutch Cylinder Hub Seal
81	Clutch Pressure Plate Retaining Ring (2 Req'd)
82	Forward Clutch Pressure Plate
83	Forward Clutch External Spline Clutch Plates (Steel) (2 Req'd)
84	Forward Clutch Internal Spline Clutch Plates (Friction) (3 Req'd)
85	Coast Clutch Pressure Plate
86	Coast Clutch External Spline Clutch Plates (Steel) (2 Req'd)
87	Coast Clutch Internal Spline Clutch Plates (Friction) (2 Req'd)
88	Forward Clutch Spring Retaining Ring
89	Forward / Coast Return Clutch and Spring Assy
90	Coast Clutch Piston and Seal Assy
91	Forward Clutch Piston Outer Lip Seal
92	Forward Clutch Piston Assy
93	Forward Clutch Piston Inner Lip Seal
94	Forward / Coast / Direct (F / C / D) Clutch Cylinder Assy
95	Direct Clutch Piston Outer Lip Seal
96	Direct Clutch Piston Assy

Item	Description
97	Direct Clutch Piston Inner Lip Seal
98	Direct Clutch Return Spring Assy
99	Direct Clutch Spring Retaining Ring
100	Reverse Clutch Cylinder Seals (2 Req'd)
101	No. 2 Direct Clutch Thrust Washer
102	Direct Clutch External Spline Clutch Plates (Steel) (3 Req'd)
103	Direct Clutch Internal Spline Clutch Plates (Friction) (3 Req'd)
104	Direct Clutch Pressure Plate
105	Direct Clutch Shell Assy
106	Reverse Clutch Drum Assy
107	Reverse Clutch Piston Assy
108	Reverse Clutch Return Spring Assy
109	Reverse Clutch External Spline Clutch Plates (Steel) (2 Req'd)
110	Reverse Clutch Internal Spline Clutch Plates (Friction) (2 Req'd)
111	Reverse Clutch Pressure Plate
112	Reverse Clutch Pressure Plate Retainer Snap Ring (Selective Fit)
113	Reverse Clutch Hub
114	Reverse Clutch Hub Retaining Ring
115	No. 1 Pump Support Thrust Bearing Assy
116	Oil Pump Drive Shaft
117	Intermediate and Overdrive Band Assy
118	Pump Support Seal Ring
119	Pump Support Seal Rings (6 Req'd)
120	Pump Support Bolts (6 Req'd)
121	Pump Support Assy
122	Pump Body Separator Plate Gasket
123	Oil Pump Body Separator Plate
124	Oil Pump Drive Gear Insert
125	Oil Pump Drive Gear Assy
126	Oil Pump Driven Gear
127	Oil Pump Seal
128	Oil Pump Body
129	Oil Pump Assembly Bolts (9 Req'd)

89448G39

Fig. 214 Ford CD4E transaxle component keylist (continued)

Item	Description
130	Turbine Shaft Speed (TSS) Sensor Bolt
131	Turbine Shaft Speed (TSS) Sensor
132	Turbine Shaft Speed (TSS) Sensor O-Ring
133	Oil Pump Assembly
134	Vent Assy - Main Control Cover
135	Grommet - Main Control Cover
136	Main Control Cover Bolts (14 Req'd)
137	Main Control Cover Assy
138	Main Control Cover Gasket
139	Solenoid Valve Body Bolts (3 Req'd)
140	Main Control Bolts (11 Req'd)
141	Solenoid Valve Body
142	O-Ring - Solenoid Valve Body
143	Solenoid Valve Body Gasket
144	Pressure Tap Plate Bolts (3 Req'd)
145	Pressure Tap Plate
146	Pressure Tap Plate Gasket
147	Accumulator Body-to-Transfer Plate Bolts (3 Req'd)
148	Accumulator Body Assy
149	Intermediate and Overdrive Accumulator Valve Plug
150	Intermediate and Overdrive Accumulator Plug Seal
151	Accumulator Body Separator Gaskets (2 Req'd)
152	Accumulator Body Separator Plate
153	Control Valve Body-to-Transfer Plate Bolts (2 Req'd)
154	Manual Valve Detent Spring Assy
155	Control Valve Body Assy
156	Valve Body Separator Plate Gaskets (2 Req'd)
157	Valve Body Separating Plate
158	Transfer Plate
159	Main Control
160	Oil Filter Recirculating Regulator Exhaust Seal
161	Oil Filter and Seal Assy
162	Thermostatic Oil Level Control Valve Bracket

Item	Description
163	Thermostatic Oil Level Control Valve Bracket Bolt
164	Thermostatic Oil Level Control Valve
165	Transaxle Case Assy
166	Servo Cover Retaining Ring
167	Servo Cover Assy
168	Servo to Apply Rod Retaining Ring
169	Intermediate and Overdrive Servo Piston Assy
170	Intermediate and Overdrive Servo Cushion Spring
171	Servo Cushion Spring Backup Washer
172	Intermediate and Overdrive Servo Apply Rod
173	Intermediate and Overdrive Servo Piston and Rod Assy
174	Intermediate and Overdrive Servo Return Spring Assy
175	Line Pressure Port Plug
176	Transaxle Drain Plug
177	Differential Oil Seal Assy (2 Req'd)
178	Manual Control Lever Bolt
179	Manual Control Lever Assy
180	Manual Control Lever Shaft
181	Manual Control Lever Seal
182	Transaxle Identification Tag (2 Req'd)
183	Manual Valve Detent Lever Actuating Rod Assy
184	Transmission Range (TR) Sensor Bolts (2 Req'd)
185	Transmission Range (TR) Sensor
186	Manual Valve Actuator Rod (Z-Link)
187	Manual Valve Detent Lever Assy
188	Retaining Pin (2 Req'd)
189	Manual Valve Detent Lever Shaft
190	Final Drive Lube Tube Seal
191	Final Drive Lube Tube
192	Manual Control Seal Assy
193	Parking Pawl Shaft
194	Parking Pawl Return Spring
195	Parking Brake Pawl Assy
196	Parking Pawl Shaft Retainer and Bolt Assy
197	Park Pawl Ratcheting Spring
198	Park Lever Spacer

89448G40

Fig. 215 Ford CD4E transaxle component keylist (continued)

Item	Description
199	Manual Control Lever Shaft Nut
200	Parking Pawl Actuating Cam
201	Parking Cam Actuator Lever Assy
A	Tighten to 36-44 N·m (27-32 Lb-Ft)
B	Tighten to 20-25 N·m (15-18 Lb-Ft)
C	Tighten to 12-14 N·m (104-127 Lb-In)
D	Tighten to 7-9 N·m (62-80 Lb-In)

Item	Description
E	Tighten to 14-18 N·m (127-138 Lb-In)
F	Tighten to 9-11 N·m (80-97 Lb-In)
G	Tighten to 11-13 N·m (96-117 Lb-In)
H	Tighten to 22-26 N·m (16-19 Lb-Ft)
J	Tighten to 15-35 N·m (27-32 Lb-Ft)
K	Tighten to 24-30 N·m (18-22 Lb-Ft)
L	Tighten to 67-81 N·m (49-60 Lb-Ft)

89448G41

Fig. 216 Ford CD4E transaxle component keylist (continued)

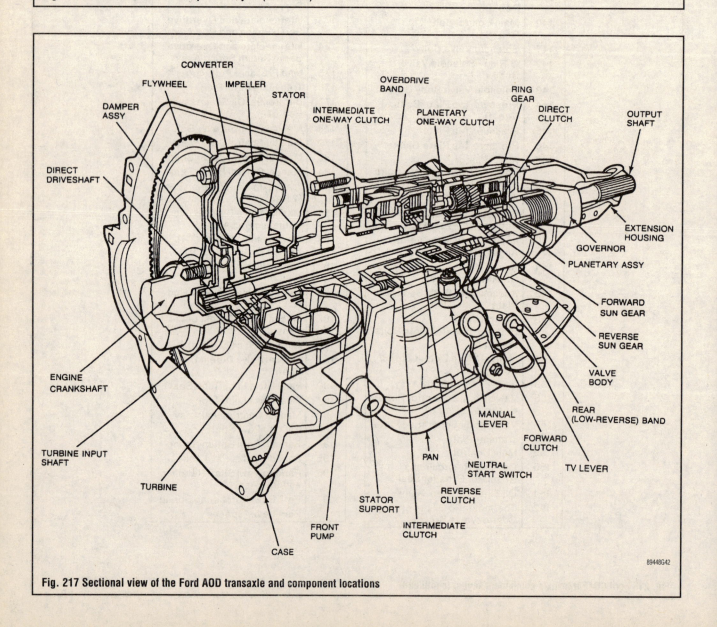

89448G42

Fig. 217 Sectional view of the Ford AOD transaxle and component locations

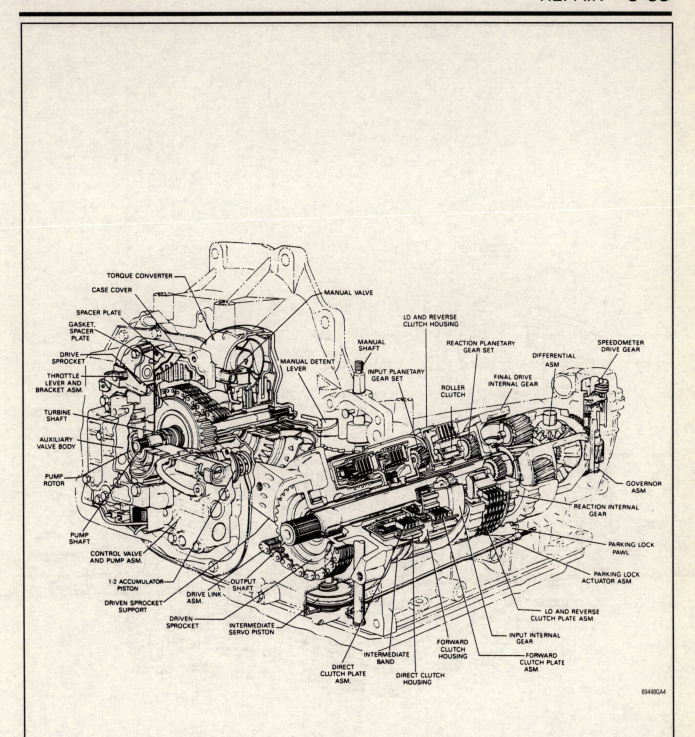

Fig. 218 Sectional view of the General Motors 3T40 transaxle components

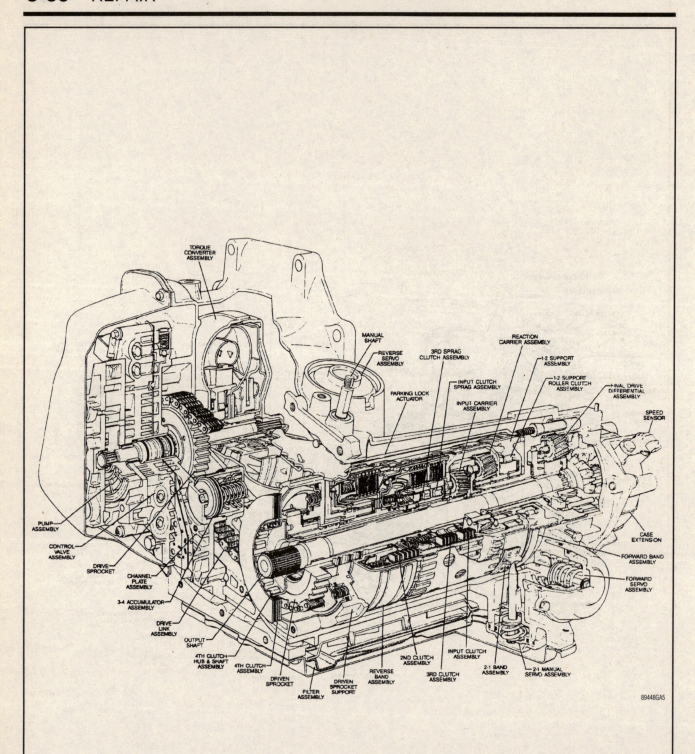

Fig. 219 Sectional view of the General Motors 4T60-E transaxle components

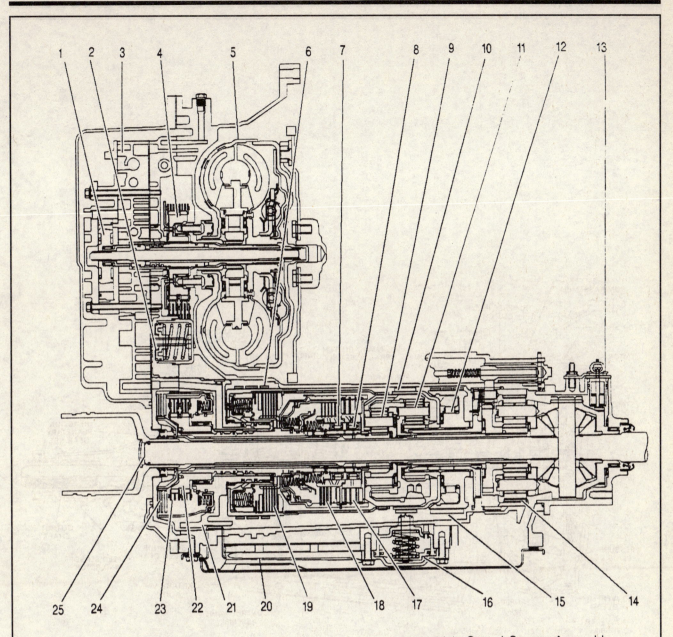

Legend

(1) Oil Pump Assembly
(2) Case Cover Assembly
(3) Control Valve Body Assembly
(4) Drive Sprocket
(5) Torque Converter Clutch Assembly
(6) Reverse Band Assembly
(7) Third Clutch Sprag Assembly
(8) Input Clutch Sprag Assembly
(9) Input Carrier Assembly
(10) 2–1 Manual Band Assembly
(11) Reaction Carrier Assembly
(12) 1–2 Roller Clutch Assembly
(13) Vehicle Speed Sensor Assembly
(14) Final Drive/Differential Carrier Assembly
(15) Forward Band Assembly
(16) 2–1 Manual Band Servo Assembly
(17) Input Clutch Assembly
(18) Third Clutch Assembly
(19) Second Clutch Assembly
(20) Oil Filter Assembly
(21) Driven Sprocket Support Assembly
(22) Driven Sprocket
(23) Drive Link Assembly
(24) Fourth Clutch Assembly
(25) Output Shaft

89448GA6

Fig. 220 Sectional view of the General Motors 4T65-E transaxle components

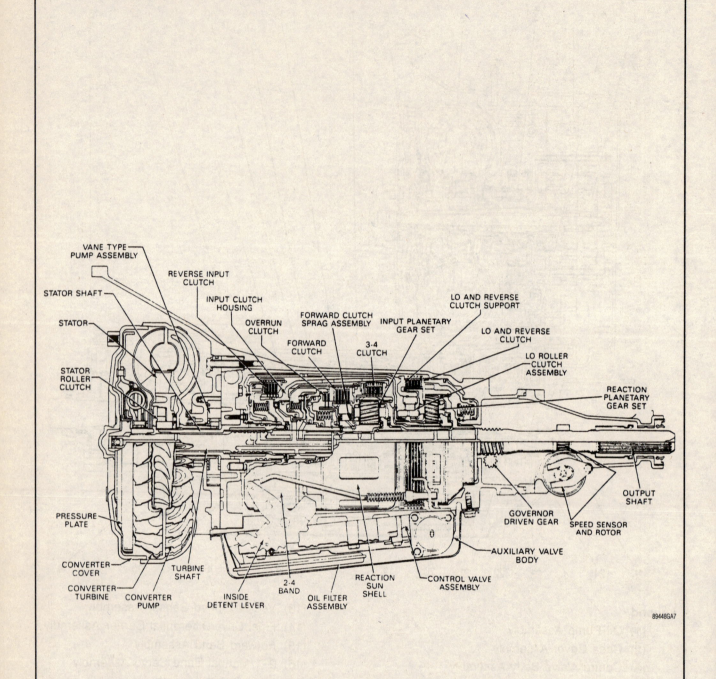

Fig. 221 Sectional view of the General Motors 4L60 (700-R4) transmission components

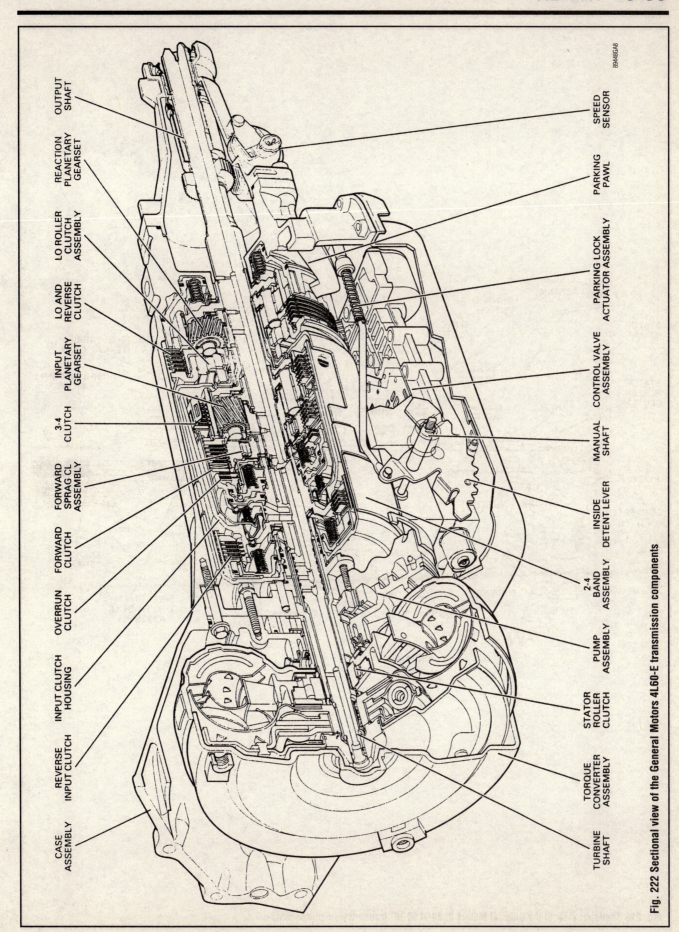

89446GA8

OUTPUT SHAFT

REACTION PLANETARY GEARSET

LO ROLLER CLUTCH ASSEMBLY

LO AND REVERSE CLUTCH

INPUT PLANETARY GEARSET

3-4 CLUTCH

FORWARD SPRAG CL ASSEMBLY

FORWARD CLUTCH

OVERRUN CLUTCH

INPUT CLUTCH HOUSING

REVERSE INPUT CLUTCH

CASE ASSEMBLY

SPEED SENSOR

PARKING PAWL

PARKING LOCK ACTUATOR ASSEMBLY

CONTROL VALVE ASSEMBLY

MANUAL SHAFT

INSIDE DETENT LEVER

2-4 BAND ASSEMBLY

PUMP ASSEMBLY

STATOR ROLLER CLUTCH

TORQUE CONVERTER ASSEMBLY

TURBINE SHAFT

Fig. 222 Sectional view of the General Motors 4L60-E transmission components

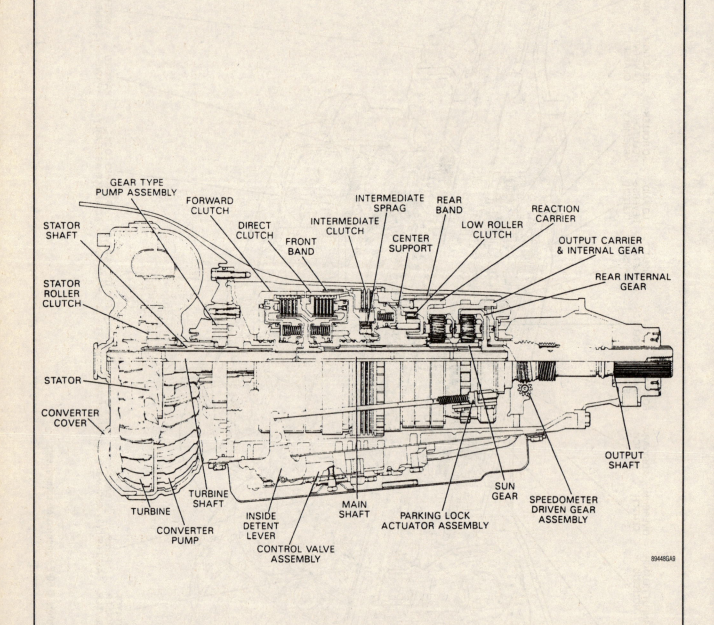

Fig. 223 Sectional view of the General Motors 3L80/3L80-HD transmission components

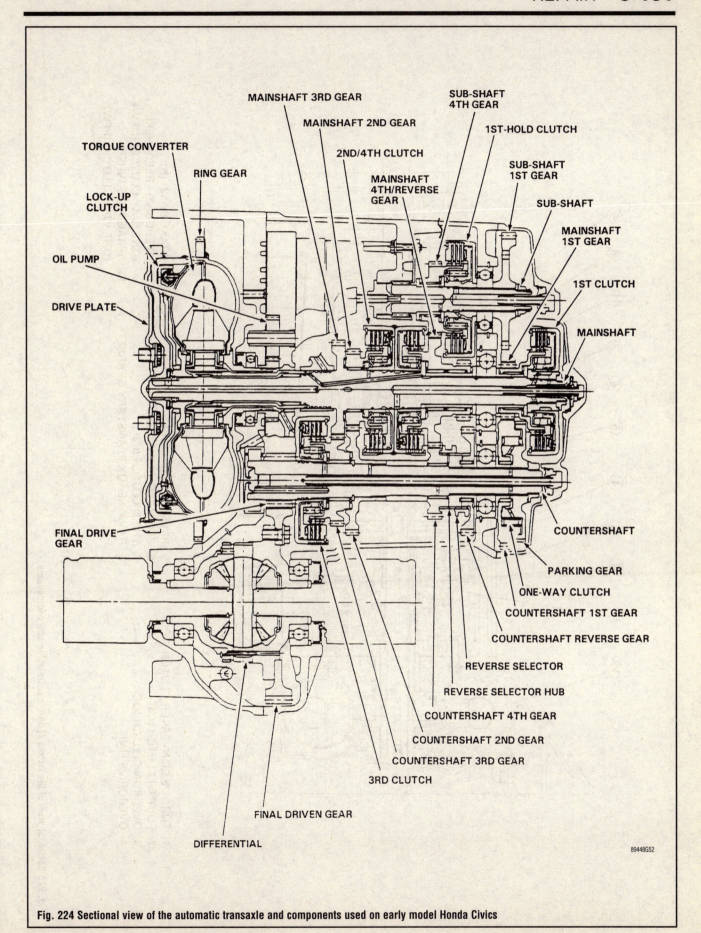

Fig. 224 Sectional view of the automatic transaxle and components used on early model Honda Civics

89448G52

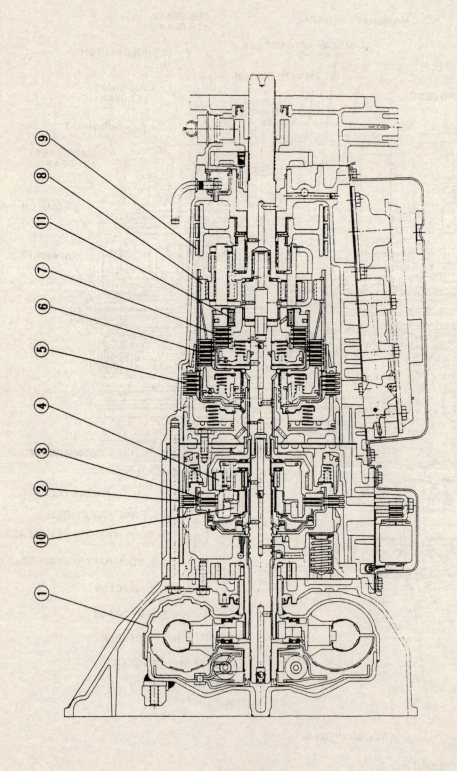

1. TORQUE CONVERTER-CLUTCH (TCC)
2. FOURTH CLUTCH (C4)
3. OVERRUN CLUTCH (OC)
4. OVERDRIVE UNIT

5. REVERSE CLUTCH (RC)
6. SECOND CLUTCH (C2)
7. THIRD CLUTCH (C3)
8. RAVIGNEAUX PLANETARY GEARSET

9. BRAKE BAND (B)
10. OVERDRIVE FREE WHEEL
    (ONE WAY CLUTCH) (OFW)
11. SPRAG FREE WHEEL
    (ONE WAY CLUTCH) (PFW)

Fig. 225 Sectional view of the Honda 4L30-E transmission and components

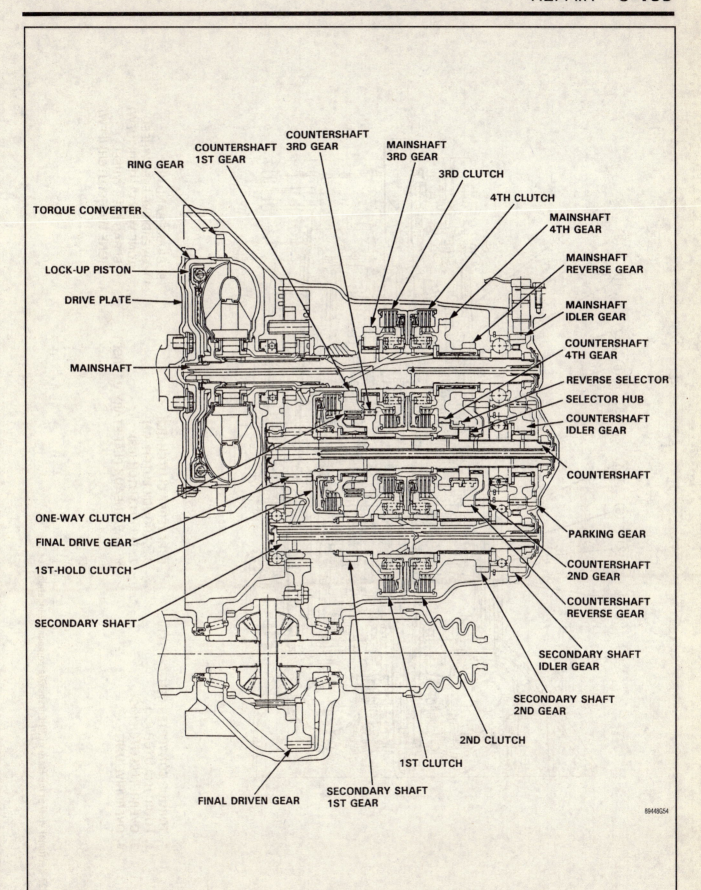

Fig. 226 Sectional view of the automatic transaxle and components used on a 1997 Honda Accord

89448G54

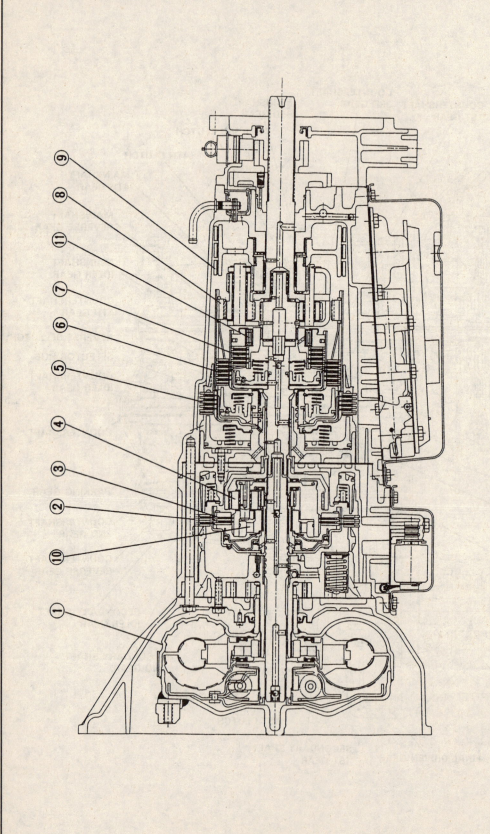

1. TORQUE CONVERTER-CLUTCH (TCC)
2. FOURTH CLUTCH (C4)
3. OVERRUN CLUTCH (OC)
4. OVERDRIVE UNIT

5. REVERSE CLUTCH (RC)
6. SECOND CLUTCH (C2)
7. THIRD CLUTCH (C3)
8. RAVIGNEAUX PLANETARY GEARSET

9. BRAKE BAND (B)
10. OVERDRIVE FREE WHEEL
    (ONE WAY CLUTCH) (OFW)
11. SPRAG FREE WHEEL
    (ONE WAY CLUTCH) (PFW)

Fig. 227 Sectional view of the Isuzu 4L30-E transaxle component locations

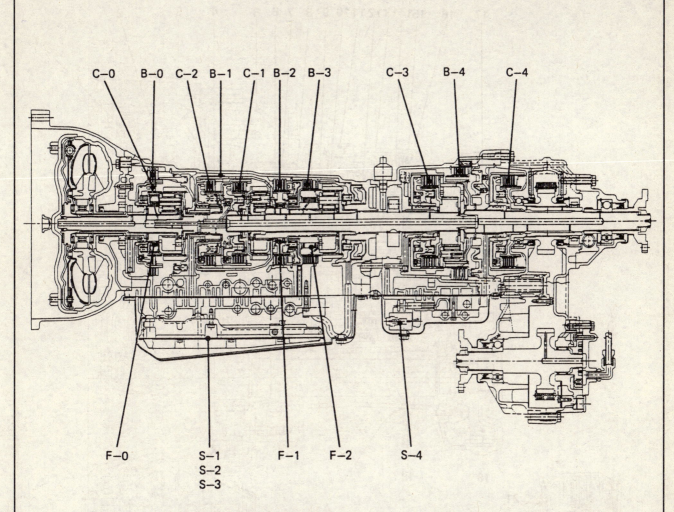

C-0 : Overdrive (OD) direct clutch
C-1 : Forward clutch
C-2 : Direct clutch
C-3 : Direct clutch (Transfer)
C-4 : Front drive clutch (Transfer)
B-0 : OD brake
B-1 : Second coast brake
B-2 : Second brake
B-3 : First and reverse brake
B-4 : Low speed brake
F-0 : OD oneway clutch
F-1 : Oneway clutch (No. 1)
F-2 : Oneway clutch (No. 2)
S-1 : Solenoid No. 1
S-2 : Solenoid No. 2
S-3 : Solenoid No. 3
S-4 : Solenoid No. 4

89448G81

Fig. 228 Sectional view of the Isuzu AW30-80LE transmission component locations

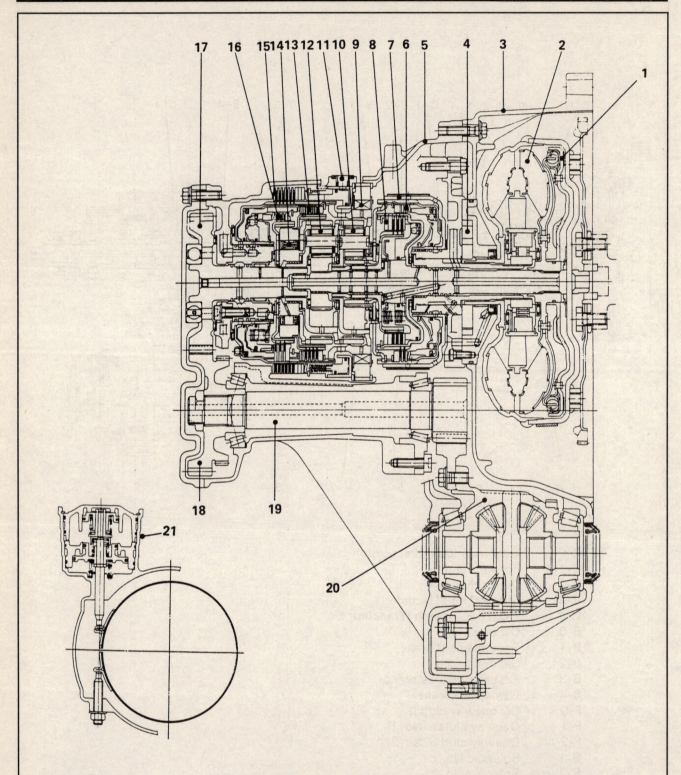

1. Lock-up piston
2. Torque converter
3. Converter housing
4. Oil pump
5. Transaxle case
6. Reverse clutch
7. Brake band
8. High clutch
9. Low one-way clutch
10. Front planetary carrier
(11. Control valve-in)
12. Rear planetary carrier
13. Forward clutch
14. Forward one-way clutch
15. Low & reverse brake
16. Overrun clutch
17. Output gear
18. Idle gear
19. Reduction gear
20. Final gear & differential gear
21. Band servo

89448G82

Fig. 229 Sectional view of the Isuzu JF403E transaxle component locations

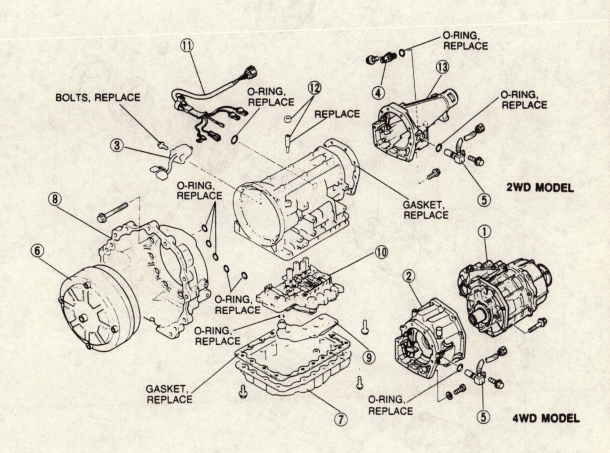

BOLTS, REPLACE

O-RING, REPLACE

O-RING, REPLACE

REPLACE

O-RING, REPLACE

O-RING, REPLACE

2WD MODEL

GASKET, REPLACE

O-RING, REPLACE

O-RING, REPLACE

GASKET, REPLACE

O-RING, REPLACE

4WD MODEL

1. Transfer case (4WD)
2. Adapter case (4WD)
3. Inhibitor switch
4. Speedometer driven gear
5. Speed sensor 1

6. Torque converter
7. Oil pan
8. Converter housing
9. Oil strainer

10. Control valve body
11. Solenoid valve connectors
12. Anchor end bolt and nut
13. Extension housing

89448G94

Fig. 230 Exploded view of a typical Mazda transmission components

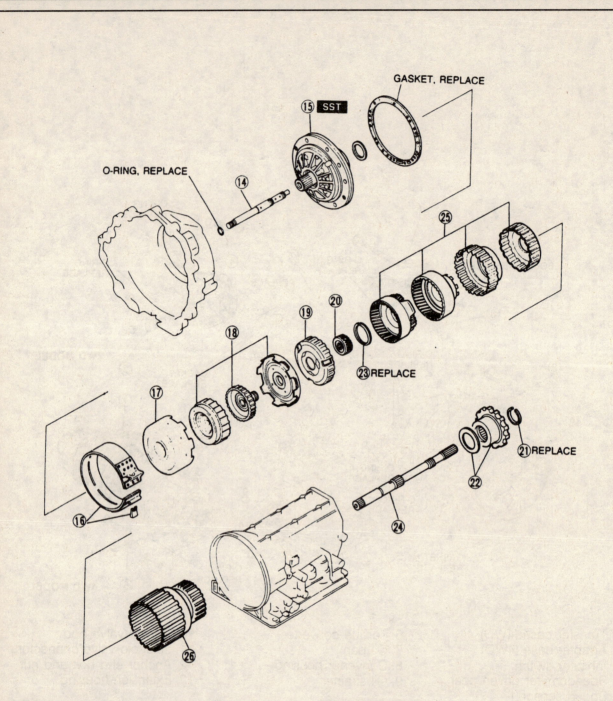

14. Input shaft
15. Oil pump
16. Brake band and strut
17. Reverse clutch
18. High clutch and front sun gear
19. Front planetary carrier
20. Rear sun gear

21. Snap ring
22. Parking gear and bearing
23. Snap ring
24. Output shaft
25. Front internal gear, rear internal gear, forward clutch hub, overrunning clutch hub
26. Forward clutch drum (forward clutch, overrunning clutch, low one-way clutch)

89448G95

Fig. 231 Exploded view of a typical Mazda transmission components (continued)

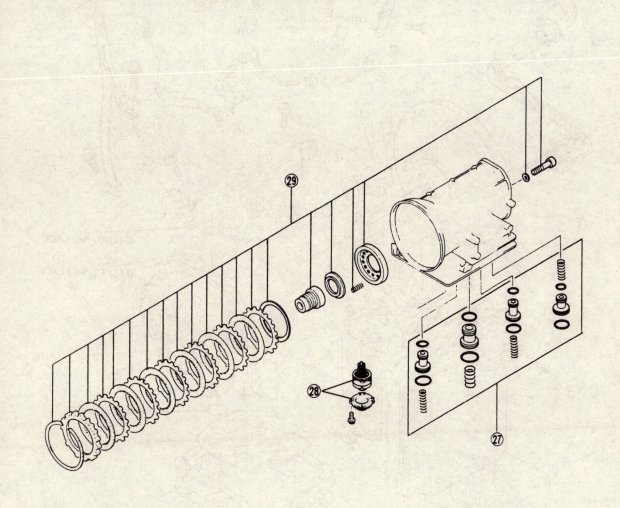

27. Accumulator spring and piston

28. Band servo

29. Low and reverse brake piston and spring

89448G96

Fig. 232 Exploded view of a typical Mazda transmission components (continued)

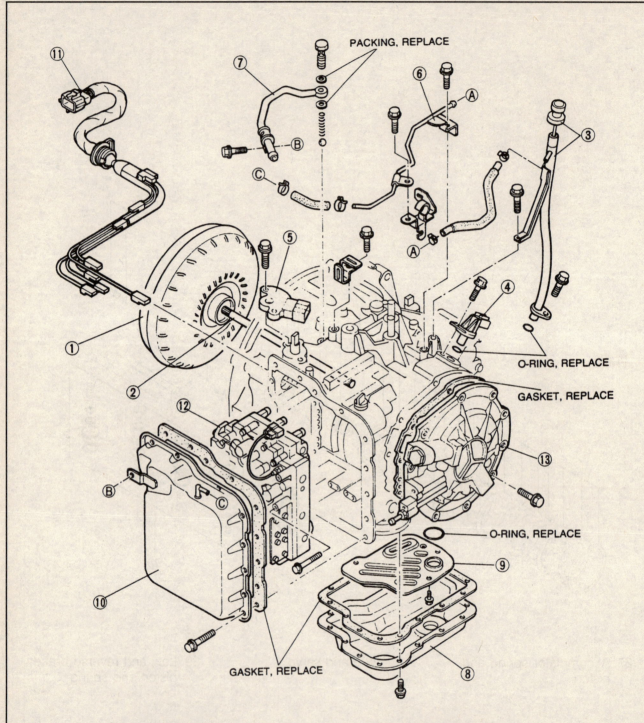

PACKING, REPLACE

O-RING, REPLACE

GASKET, REPLACE

O-RING, REPLACE

GASKET, REPLACE

1. Torque converter
2. Oil pump shaft
3. ATF dipstick and oil filler tube
4. Vehicle speed pulse generator
5. Park/Neutral switch
6. Breather hose
7. Oil pipe
8. Oil pan

9. Oil strainer
10. Control valve body cover
11. Coupler assembly
12. Control valve body
13. Oil pump

89448G97

Fig. 233 Exploded view of a typical Mazda transaxle components

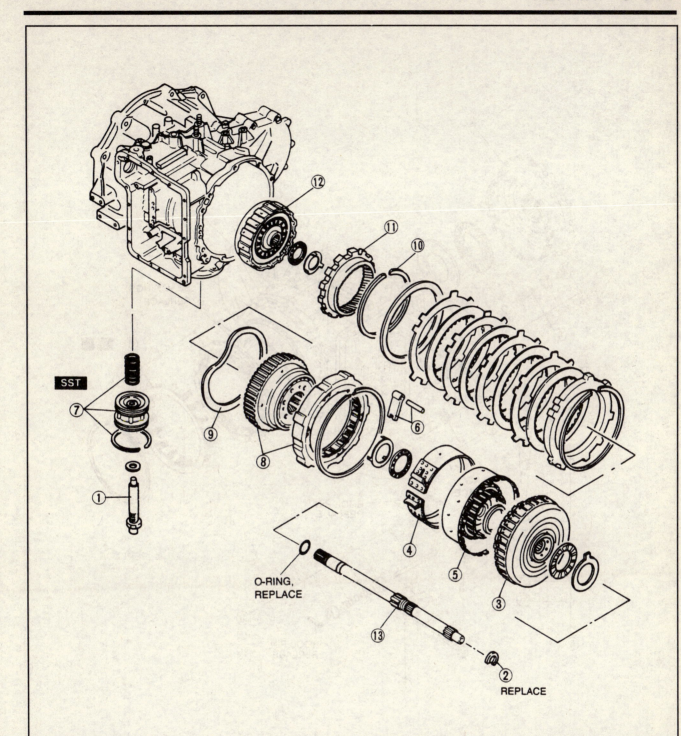

SST

O-RING,
REPLACE

REPLACE

1. Piston stem
2. Snap ring
3. Clutch assembly
4. 2-4 brake band
5. Small sun gear and one-way clutch 1
6. Anchor strut and shaft
7. Band servo

8. One-way clutch 2 and carrier hub assembly
9. Friction plate
10. Snap ring
11. Internal gear
12. 3-4 clutch
13. Turbine shaft

89448G98

Fig. 234 Exploded view of a typical Mazda transaxle components (continued)

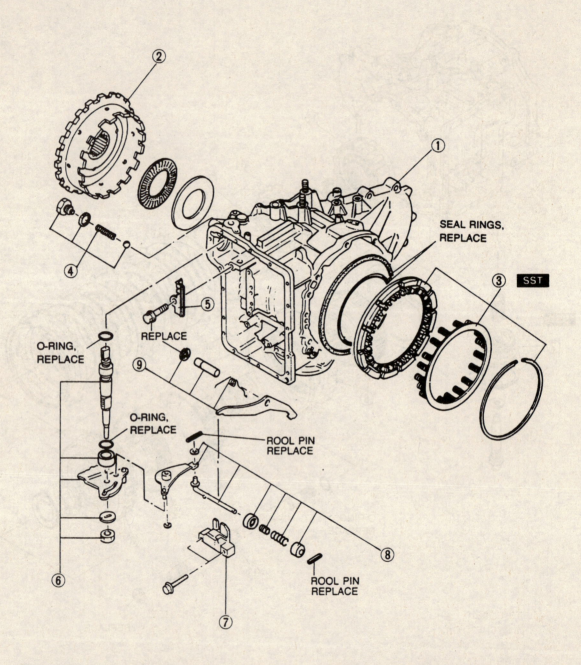

SEAL RINGS, REPLACE

SST

O-RING, REPLACE

REPLACE

O-RING, REPLACE

ROOL PIN REPLACE

ROOL PIN REPLACE

1. Transaxle case
2. Output shell
3. Low and reverse brake
4. Plug, packing, spring, and detent ball
5. Bracket

6. Manual shaft and manual plate
7. Actuator support
8. Parking assist lever
9. Parking pawl

89448GA1

Fig. 235 Exploded view of a typical Mazda transaxle components (continued)

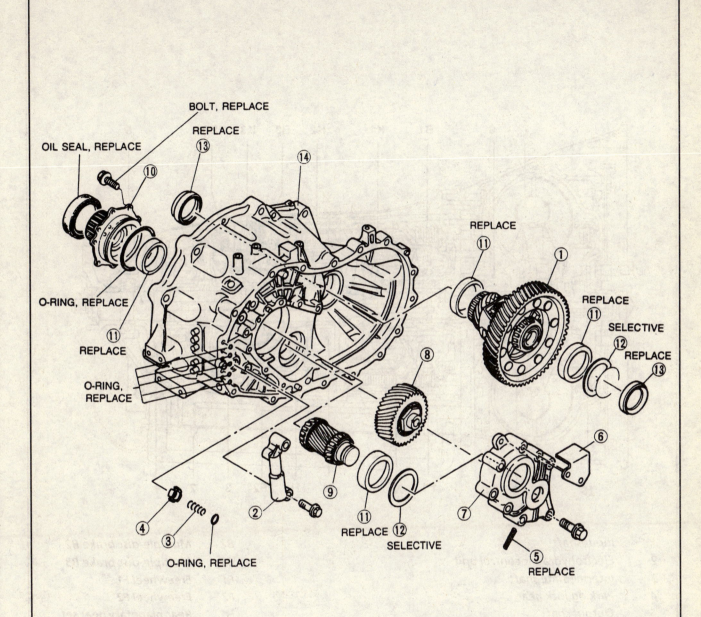

1. Differential
2. 2-3 accumulator
3. Orifice check valve spring
4. Orifice check valve
5. Roll pin
6. Baffle plate
7. Bearing housing

8. Idler gear
9. Output gear
10. Bearing cover assembly
11. Bearing races
12. Adjustment shim (bearing preload)
13. Oil seal
14. Converter housing

89448GA2

Fig. 236 Exploded view of a typical Mazda transaxle components (continued)

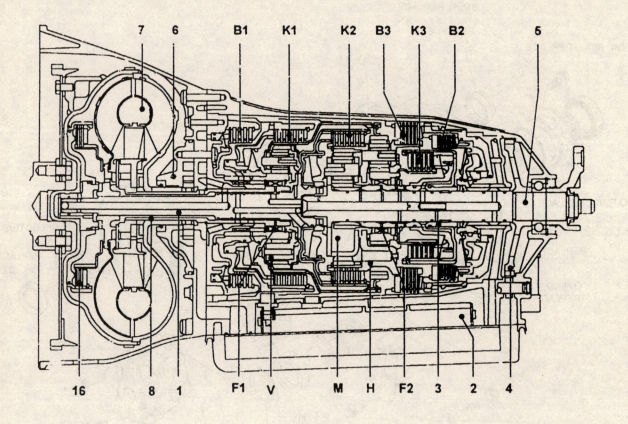

1	Input shaft	B2	Multiple-disc brake B2
2	Electrohydraulic control unit	B3	Multiple-disc brake B3
3	Intermediate shaft	F1	Freewheel F1
4	Parking lock gear	F2	Freewheel F2
5	Output shaft	H	Rear planetary gear set
6	Oil pump	K1	Multiple-disc clutch K1
7	Torque converter	K2	Multiple-disc clutch K2
8	Stator shaft	K3	Multiple-disc clutch K3
9	Torque converter lock-up clutch (KÜB)	M	Center planetary gear set
B1	Multiple-disc brake B1	V	Front planetary gear set

Fig. 237 Sectional view of the Mercedes Benz 722.6 transmission components

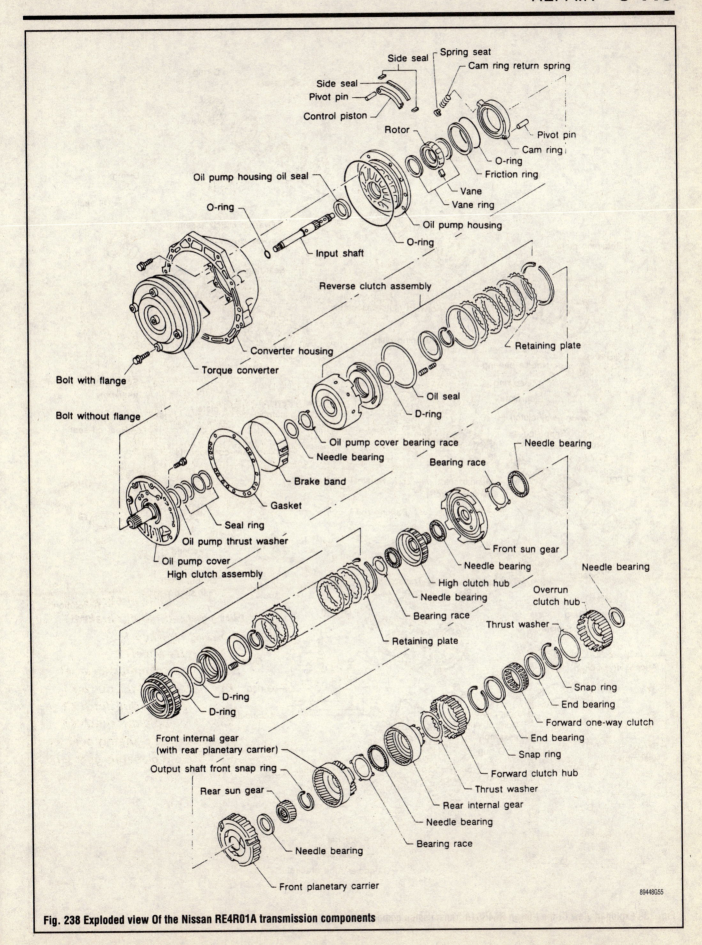

Fig. 238 Exploded view Of the Nissan RE4R01A transmission components

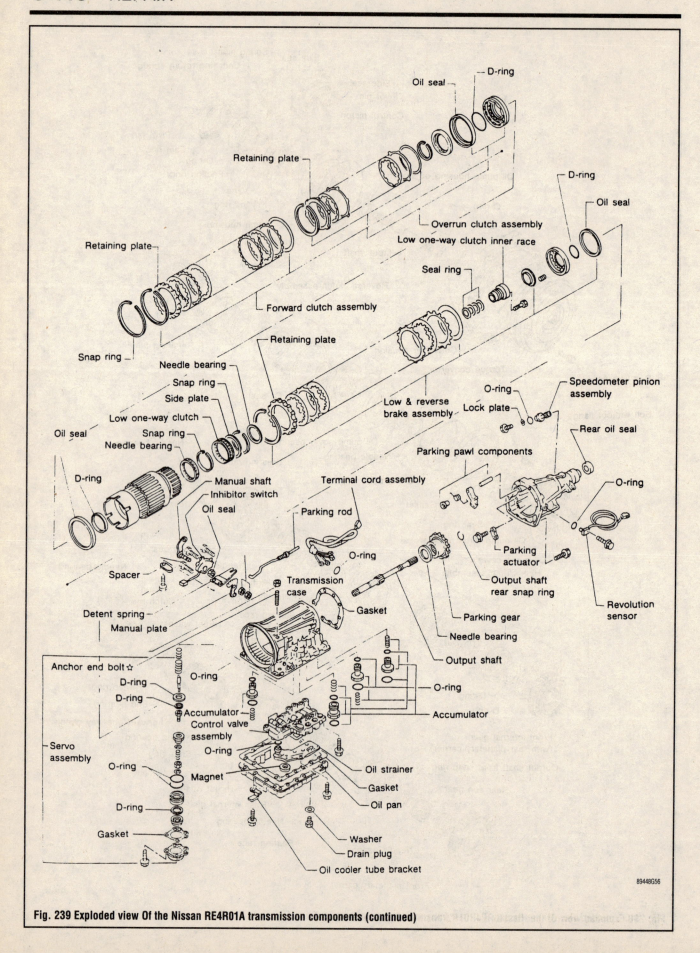

Fig. 239 Exploded view Of the Nissan RE4R01A transmission components (continued)

89448G56

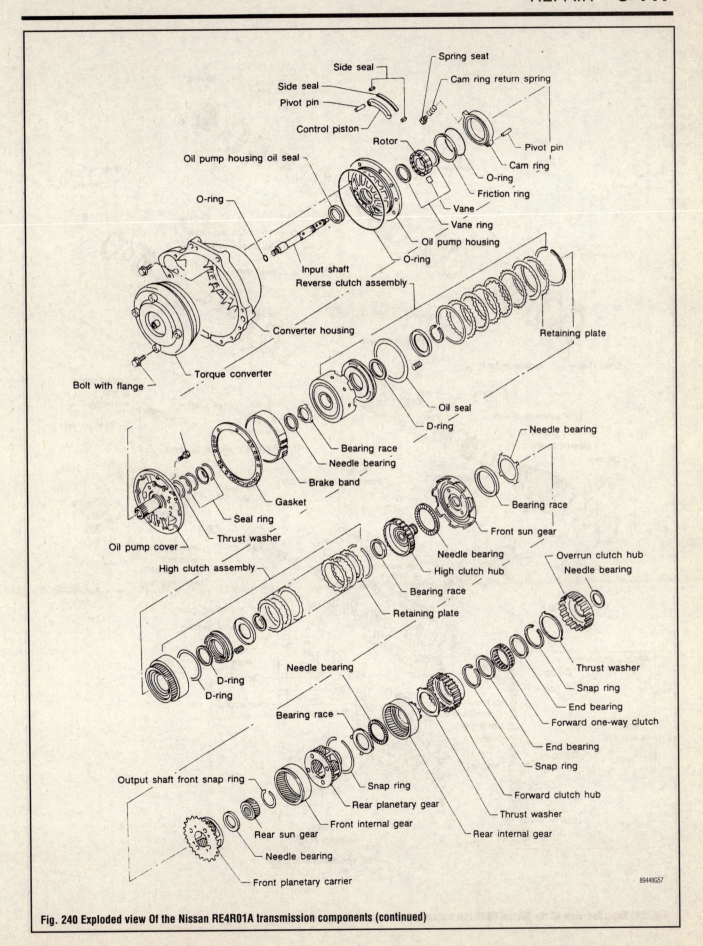

Fig. 240 Exploded view Of the Nissan RE4R01A transmission components (continued)

89448G57

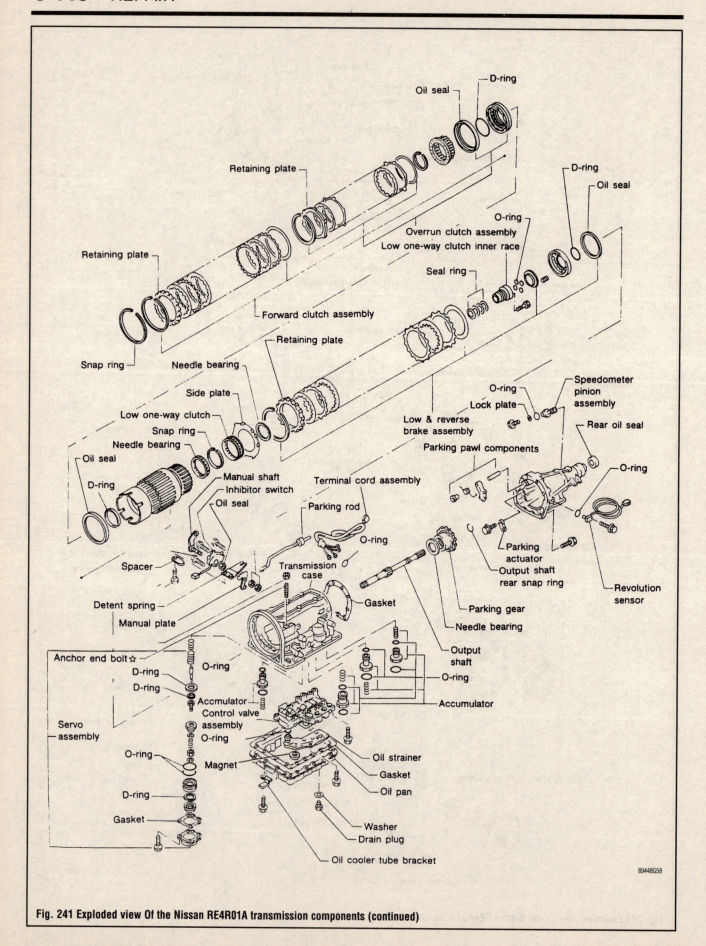

Fig. 241 Exploded view Of the Nissan RE4R01A transmission components (continued)

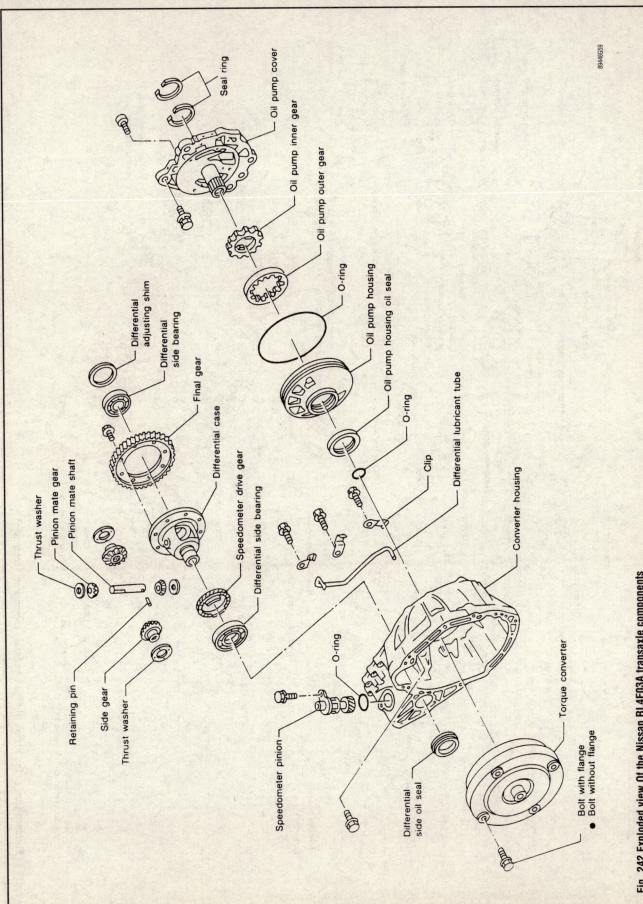

Fig. 242 Exploded view Of the Nissan RL4F03A transaxle components

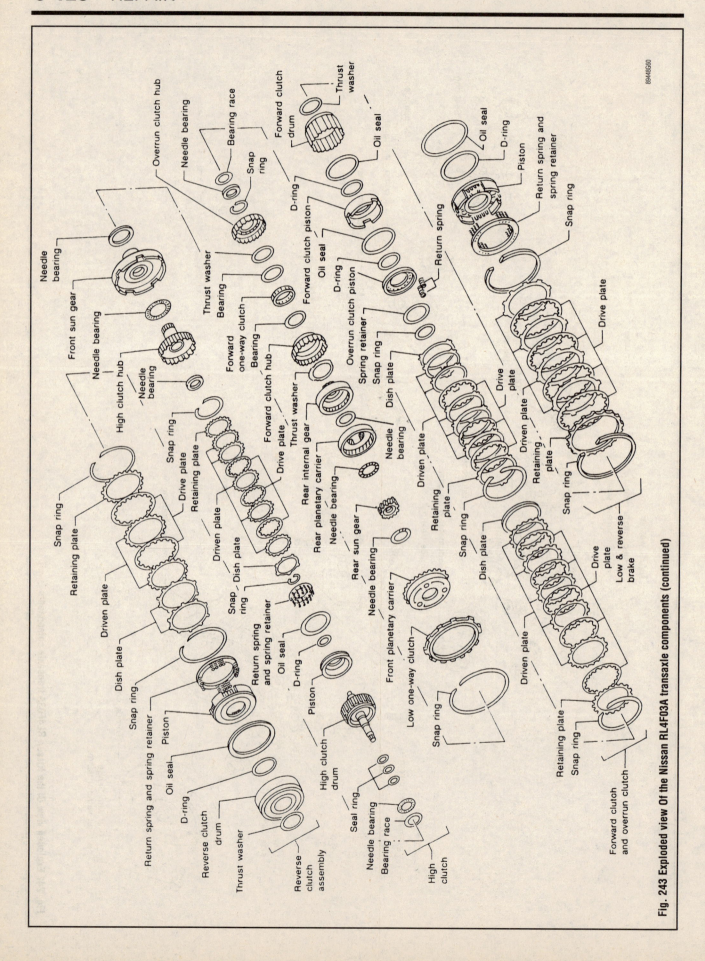

Fig. 243 Exploded view Of the Nissan RL4F03A transaxle components (continued)

89448661

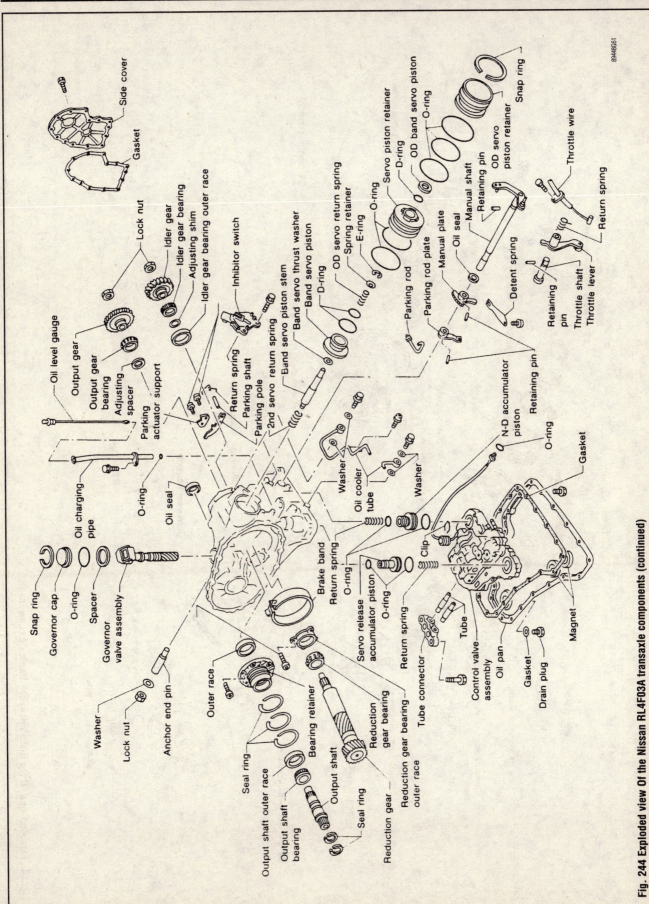

Fig. 244 Exploded view Of the Nissan RL4F03A transaxle components (continued)

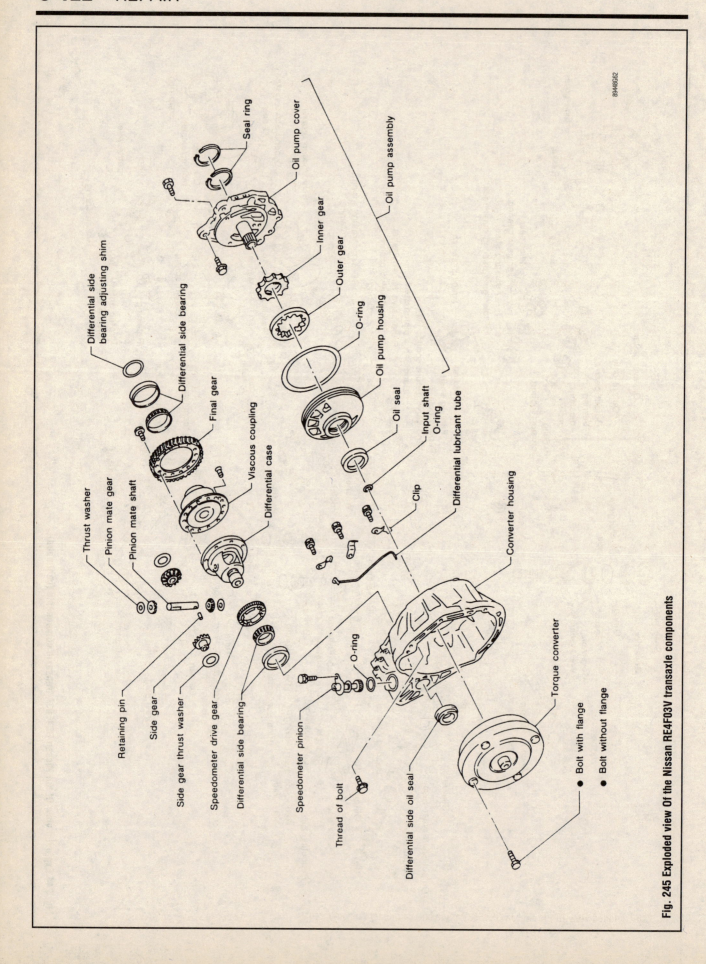

Fig. 245 Exploded view Of the Nissan RE4F03V transaxle components

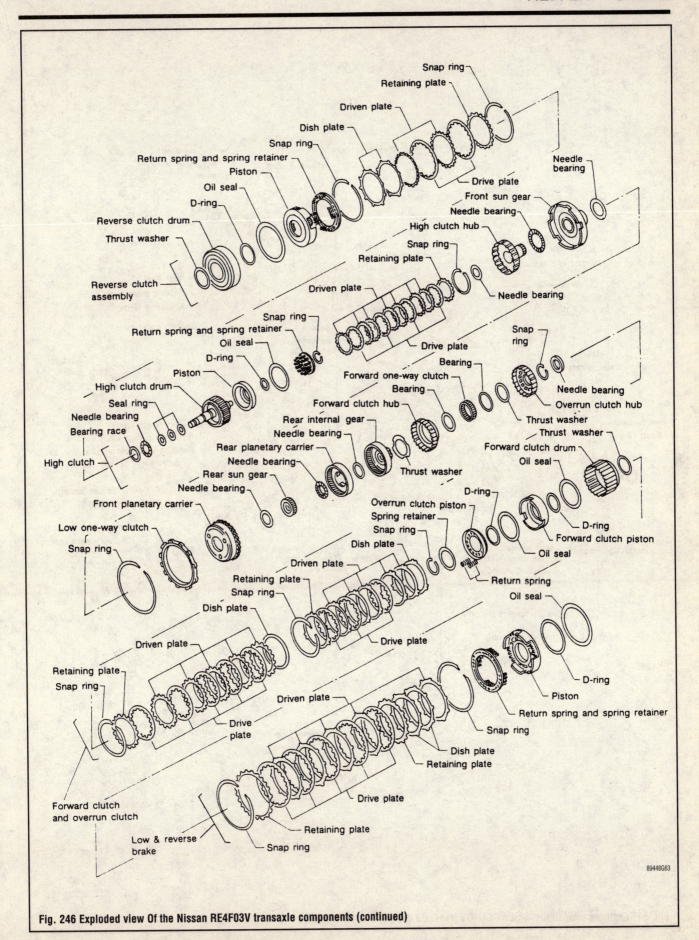

Fig. 246 Exploded view Of the Nissan RE4F03V transaxle components (continued)

89448G63

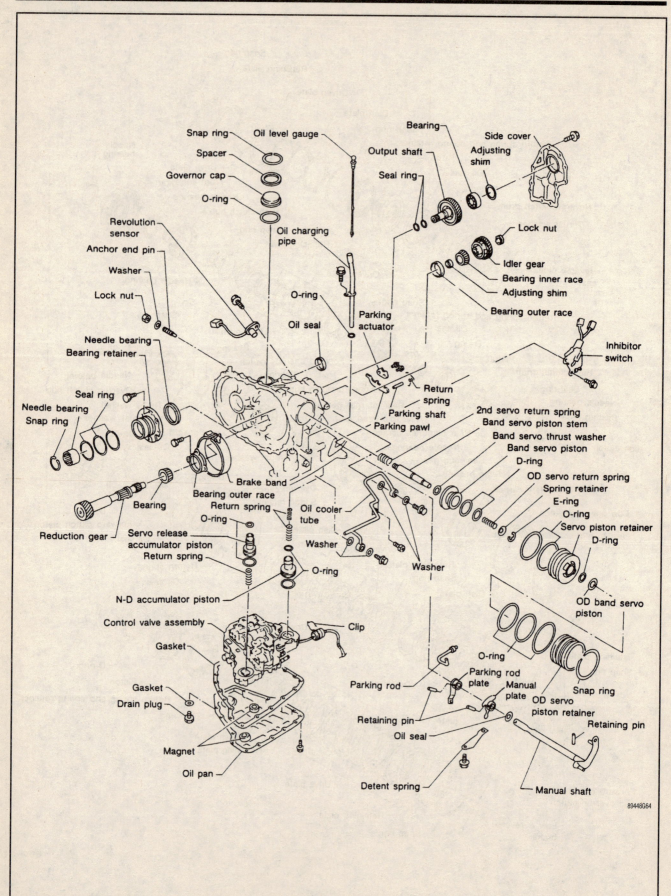

Fig. 247 Exploded view Of the Nissan RE4F03V transaxle components (continued)

89448G64

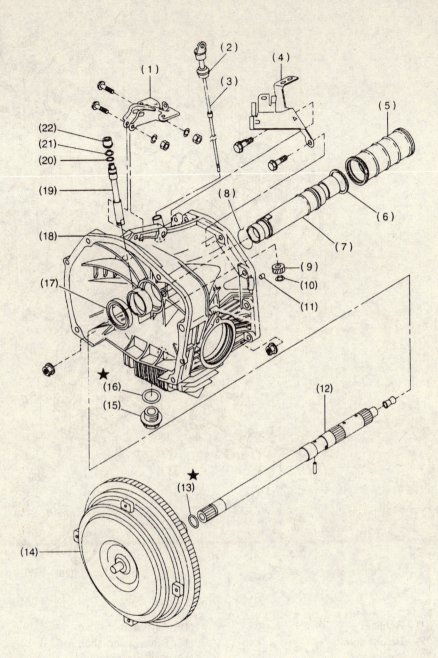

(1) Pitching stopper bracket
(2) O-ring
(3) Oil level gauge
(4) Stay
(5) Seal pipe
(6) Seal ring
(7) Oil pump shaft
(8) Clip
(9) Speedometer driven gear
(10) Snap ring
(11) Oil drain pipe
(12) Input shaft
(13) O-ring
(14) Torque converter clutch
(15) Drain plug
(16) Gasket
(17) Oil seal
(18) Torque converter clutch case
(19) Speedometer shaft
(20) Washer
(21) Snap ring
(22) Oil seal

89448G83

Fig. 248 Exploded view of a typical Subaru transmission components

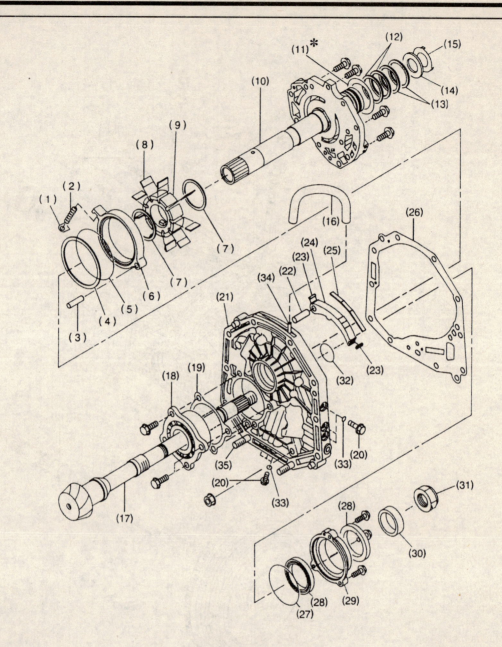

(1) Retainer
(2) Return spring
(3) Pin
(4) Friction ring
(5) O-ring
(6) Cam ring
(7) Vane ring
(8) Vane
(9) Rotor
(10) Oil pump cover
(11) Thrust washer
(12) Seal ring (R)
(13) Seal ring (H)
(14) Thrust needle bearing
(15) Thrust washer

(16) Air breather hose
(17) Drive pinion shaft
(18) Roller bearing
(19) Shim
(20) Test plug
(21) Oil pump housing
(22) Pin
(23) Side seal
(24) Control piston
(25) Plane seal
(26) Gasket
(27) O-ring
(28) Oil seal
(29) Oil seal retainer
(30) Drive pinion collar

(31) Lock nut
(32) O-ring
(33) O-ring
(34) Nipple
(35) Stud bolt

Fig. 249 Exploded view of a typical Subaru transmission components (continued)

89448G84

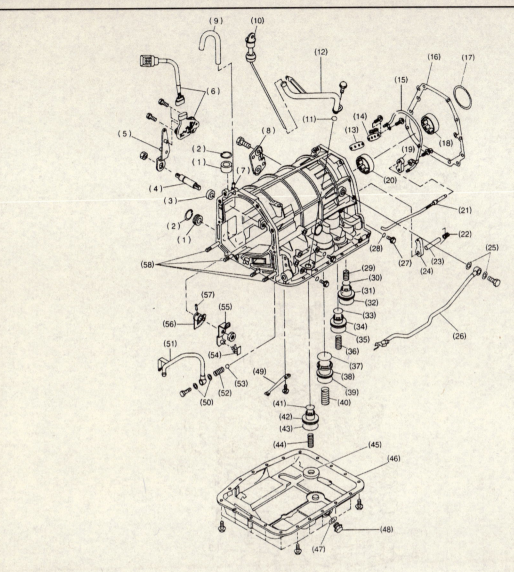

(1) Plug
(2) Snap ring
(3) Oil seal
(4) Manual shaft
(5) Range select lever
(6) Inhibitor switch ASSY
(7) Nipple
(8) Plate ASSY
(9) Air breather hose
(10) Oil level gauge
(11) O-ring
(12) Oil charger pipe
(13) Gasket
(14) Relief valve
(15) Pipe
(16) Gasket
(17) Shim
(18) Roller bearing
(19) Parking support
(20) Ball bearing

(21) Parking rod
(22) Return spring
(23) Shaft
(24) Parking pawl
(25) Gasket
(26) Inlet pipe
(27) Test plug
(28) O-ring
(29) Spring
(30) O-ring
(31) Accumulator piston (N-D)
(32) O-ring
(33) O-ring
(34) Accumulator piston (2-3)
(35) O-ring
(36) Spring
(37) O-ring
(38) Accumulator piston (1-2)
(39) O-ring
(40) Spring

(41) O-ring
(42) Accumulator piston (3-4)
(43) O-ring
(44) Spring
(45) Magnet
(46) Oil pan
(47) Gasket
(48) Drain plug
(49) Detention spring
(50) Gasket
(51) Outlet pipe
(52) Spring
(53) Ball
(54) Stopper
(55) Manual lever
(56) Manual plate
(57) Spring pin
(58) Stud bolt

Fig. 250 Exploded view of a typical Subaru transmission components (continued)

89448G85

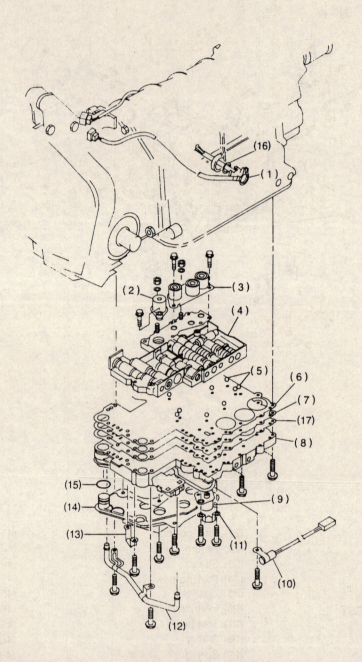

(1) Transmission harness	(7) Lower separator gasket	(13) Bracket
(2) Duty solenoid A (Line pressure)	(8) Lower valve body	(14) Oil strainer
(3) Shift solenoid ASSY	(9) Duty solenoid B (Lock-up)	(15) O-ring
(4) Upper valve body	(10) ATF temperature sensor	(16) O-ring
(5) Ball	(11) Bracket	(17) Separator plate
(6) Upper separator gasket	(12) Pipe	

89448G87

Fig. 251 Exploded view of a typical Subaru transmission components (continued)

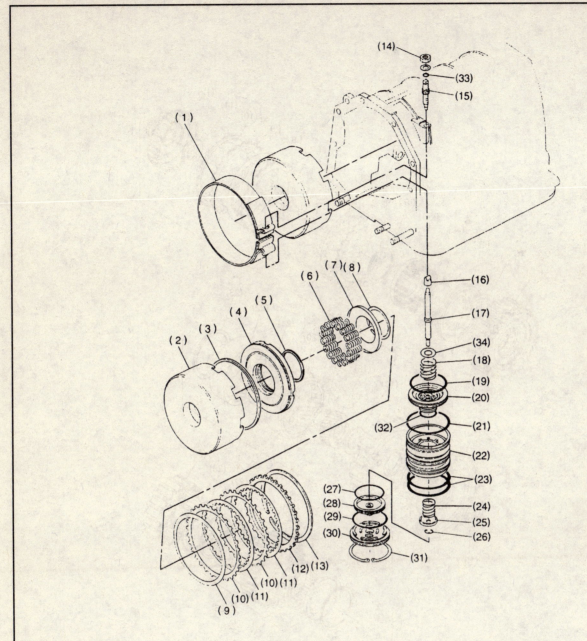

(1) Brake band
(2) Reverse clutch drum
(3) Lip seal
(4) Piston
(5) Lathe cut seal ring
(6) Spring
(7) Spring retainer
(8) Snap ring
(9) Dish plate
(10) Driven plate
(11) Drive plate
(12) Retaining plate
(13) Snap ring

(14) Lock nut
(15) Brake band adjusting screw
(16) Strut
(17) Band servo piston stem
(18) Spring
(19) Lathe cut seal ring
(20) Band servo piston (1-2)
(21) O-ring
(22) Retainer
(23) O-ring
(24) Spring
(25) Retainer
(26) Circlip

(27) Lathe cut seal ring
(28) Band servo piston (3-4)
(29) O-ring
(30) O.D. servo retainer
(31) Snap ring
(32) Lathe cut seal ring
(33) O-ring
(34) Washer

89448G88

Fig. 252 Exploded view of a typical Subaru transmission components (continued)

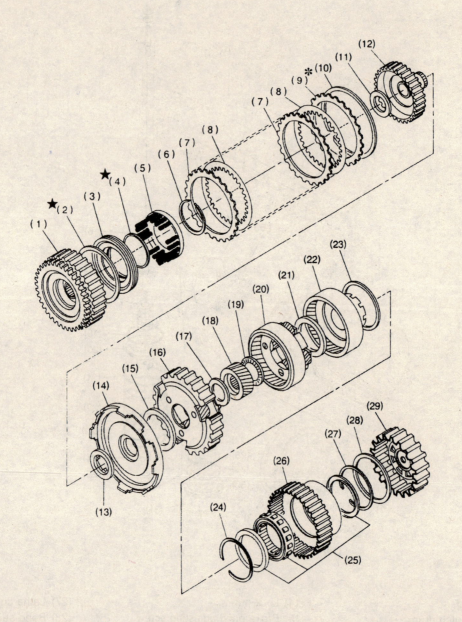

(1)	High clutch drum	(10)	Snap ring	(19)	Thrust needle bearing
(2)	Lathe cut seal ring	(11)	Thrust needle bearing	(20)	Rear planetary carrier
(3)	Piston	(12)	High clutch hub	(21)	Thrust needle bearing
(4)	Lathe cut seal ring	(13)	Thrust needle bearing	(22)	Rear internal gear
(5)	Spring retainer	(14)	Front sun gear	(23)	Thrust washer
(6)	Snap ring	(15)	Thrust needle bearing	(24)	Snap ring
(7)	Driven plate	(16)	Front planetary carrier	(25)	One-way clutch (3-4)
(8)	Drive plate	(17)	Thrust needle bearing	(26)	One-way clutch outer race (3-4)
(9)	Retaining plate	(18)	Rear sun gear	(27)	Overrunning clutch hub

89448G89

Fig. 253 Exploded view of a typical Subaru transmission components (continued)

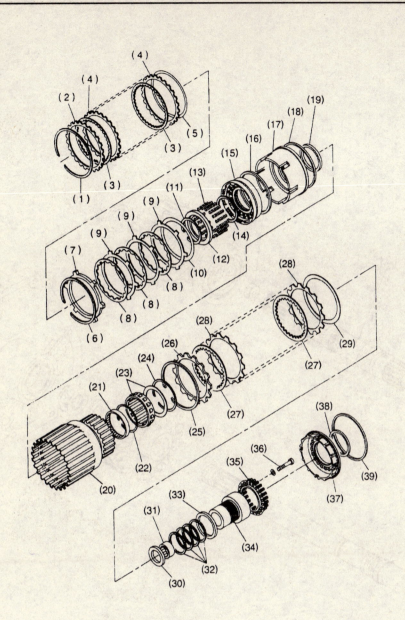

(1) Snap ring	(15) Overrunning piston	(29) Dish plate
(2) Retaining plate	(16) Lathe cut seal ring	(30) Thrust needle bearing
(3) Drive plate	(17) Forward piston	(31) Needle bearing
(4) Driven plate	(18) Lip seal	(32) Seal ring
(5) Dish plate	(19) Lathe cut seal ring	(33) Thrust washer
(6) Snap ring	(20) Forward clutch drum	(34) One-way clutch inner race (1-2)
(7) Retaining plate	(21) Needle bearing	(35) Spring retainer
(8) Drive plate	(22) Snap ring	(36) Socket bolt
(9) Driven plate	(23) One-way clutch (1-2)	(37) Low & reverse piston
(10) Dish plate	(24) Snap ring	(38) Lathe cut seal ring
(11) Snap ring	(25) Snap ring	(39) Lathe cut seal ring
(12) Spring retainer	(26) Retaining plate	
(13) Spring	(27) Drive plate	
(14) Lathe cut seal ring	(28) Driven plate	

89448G90

Fig. 254 Exploded view of a typical Subaru transmission components (continued)

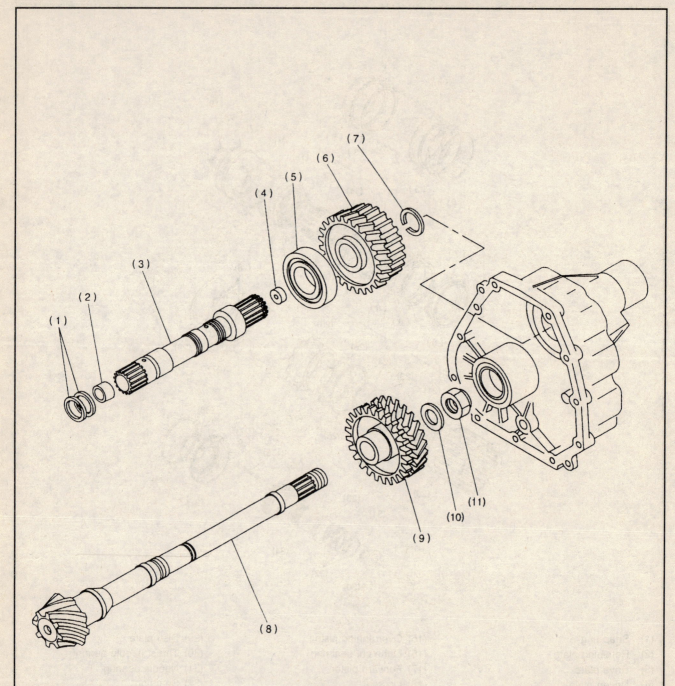

(1) Seal ring
(2) Bushing
(3) Reduction drive shaft
(4) Plug
(5) Ball bearing

(6) Reduction drive gear
(7) Snap ring
(8) Drive pinion shaft
(9) Reduction driven gear
(10) Washer

(11) Lock nut

89448G91

Fig. 255 Exploded view of a typical Subaru transmission components (continued)

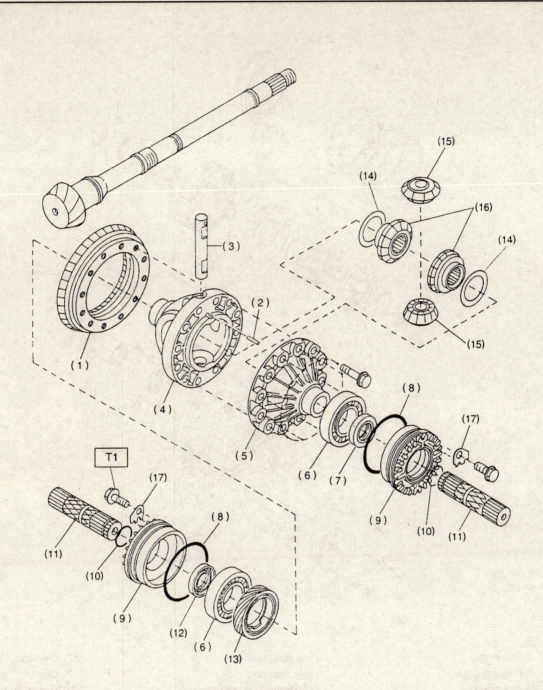

(1) Crown gear
(2) Straight pin
(3) Pinion shaft
(4) Differential case (RH)
(5) Differential case (LH)
(6) Taper roller bearing
(7) Oil seal (LH)
(8) O-ring

(9) Differential side retainer
(10) Circlip
(11) Axle shaft
(12) Oil seal (RH)
(13) Speedometer drive gear
(14) Washer
(15) Differential bevel pinion
(16) Differential bevel gear

(17) Lock plate

89448G92

Fig. 256 Exploded view of a typical Subaru transmission components (continued)

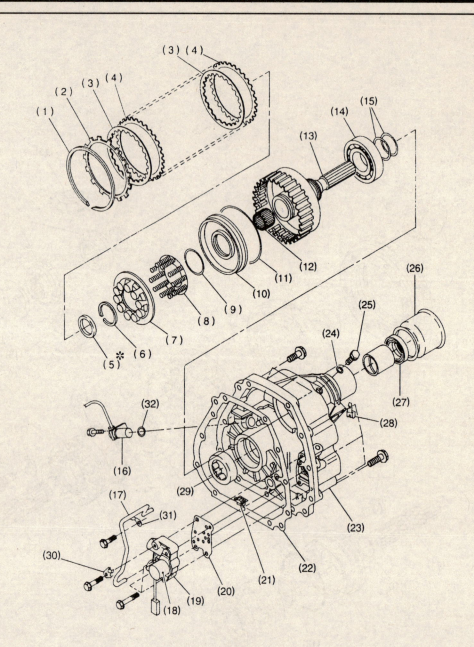

(1)	Snap ring	(14)	Ball bearing	(27)	Oil seal
(2)	Pressure plate	(15)	Seal ring	(28)	Clip
(3)	Drive plate	(16)	Vehicle speed sensor 1	(29)	Roller bearing
(4)	Driven plate	(17)	Transfer clutch pipe	(30)	Clip
(5)	Thrust needle bearing	(18)	Duty solenoid C (Transfer clutch)	(31)	Stay
(6)	Snap ring	(19)	Transfer valve body	(32)	O-ring
(7)	Seal transfer piston	(20)	Transfer valve plate		
(8)	Spring retainer	(21)	Filter		
(9)	Lathe cut seal ring	(22)	Gasket		
(10)	Transfer clutch piston	(23)	Extension case		
(11)	Lathe cut seal ring	(24)	O-ring		
(12)	Needle bearing	(25)	Test plug		
(13)	Rear drive shaft	(26)	Dust seal		

89448G93

Fig. 257 Exploded view of a typical Subaru transmission components (continued)

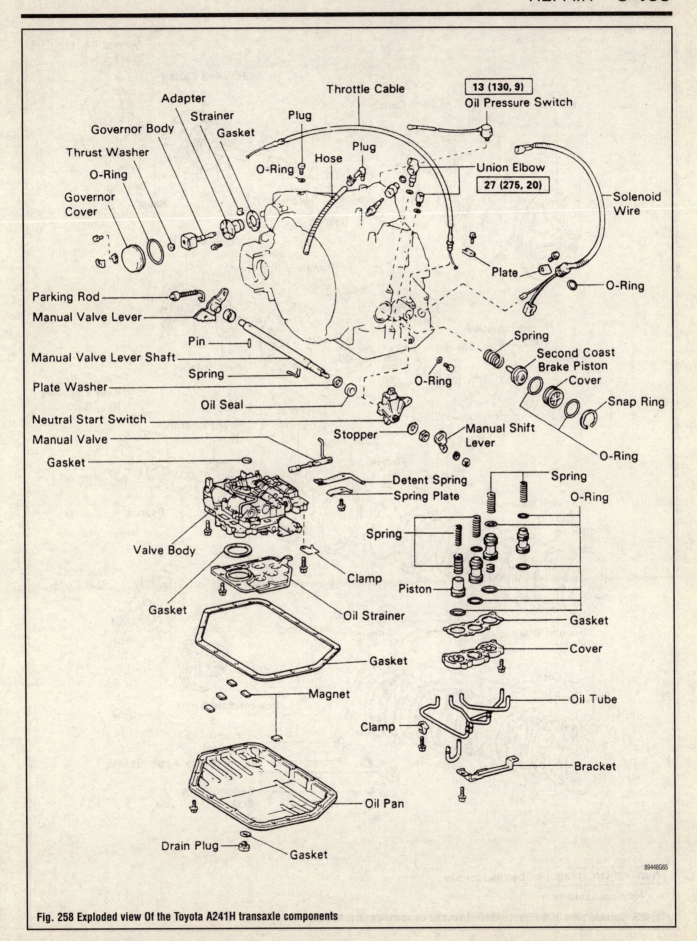

Fig. 258 Exploded view Of the Toyota A241H transaxle components

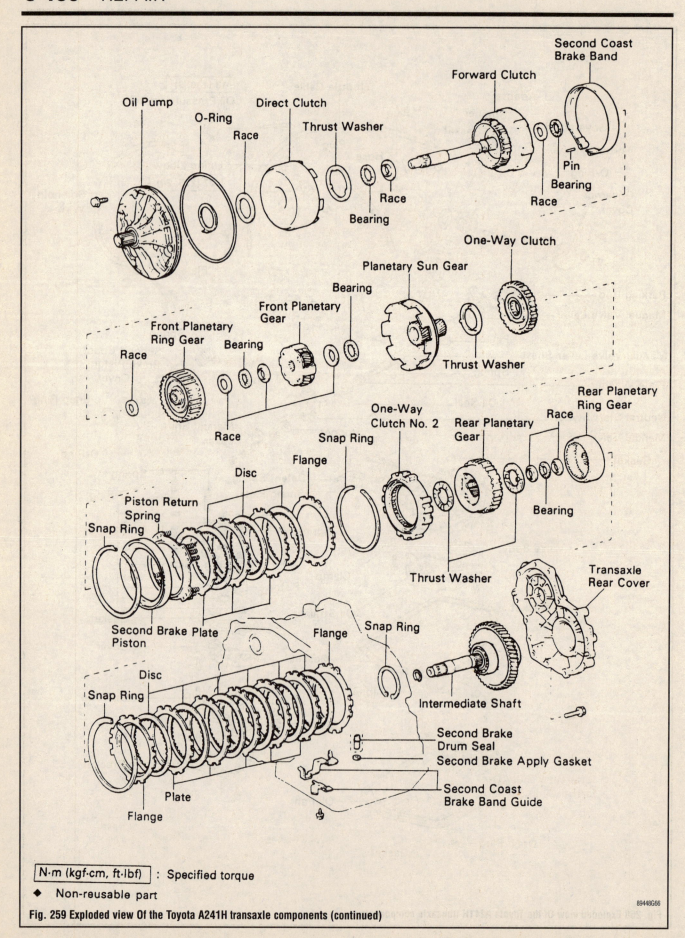

N·m (kgf·cm, ft·lbf) : Specified torque

◆ Non-reusable part

Fig. 259 Exploded view Of the Toyota A241H transaxle components (continued)

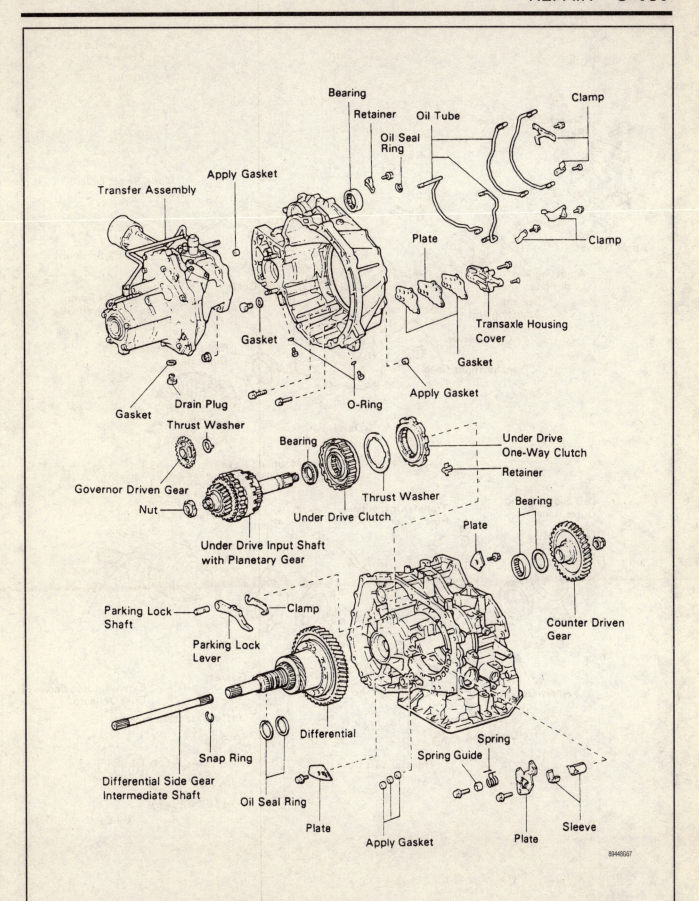

Fig. 260 Exploded view Of the Toyota A241H transaxle components (continued)

89448G67

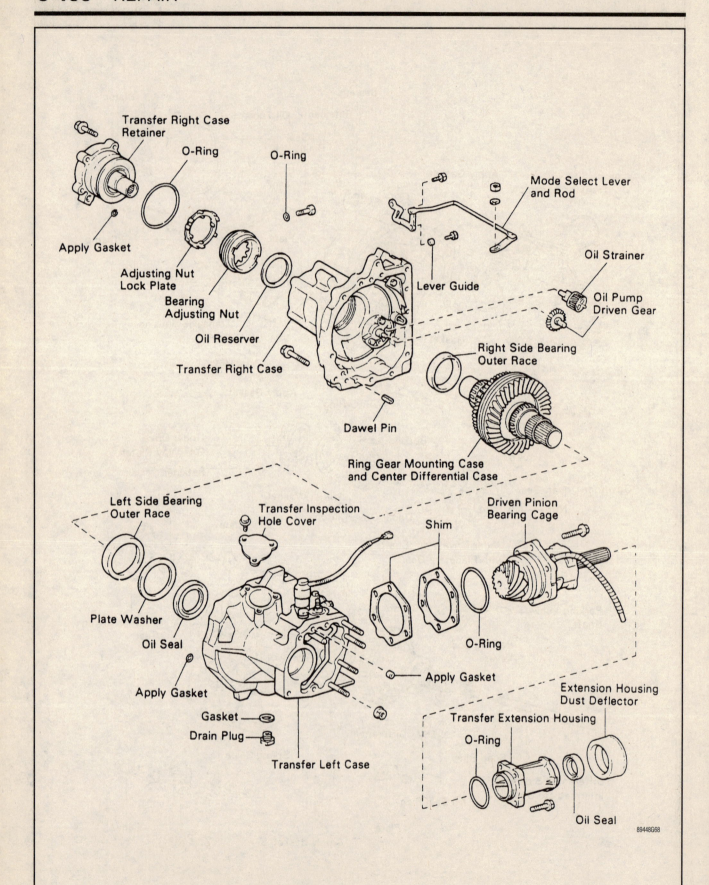

Transfer Right Case Retainer

O-Ring

O-Ring

Apply Gasket

Adjusting Nut Lock Plate

Bearing Adjusting Nut

Oil Reserver

Transfer Right Case

Dawel Pin

Mode Select Lever and Rod

Lever Guide

Oil Strainer

Oil Pump Driven Gear

Right Side Bearing Outer Race

Ring Gear Mounting Case and Center Differential Case

Left Side Bearing Outer Race

Transfer Inspection Hole Cover

Shim

Driven Pinion Bearing Cage

Plate Washer

Oil Seal

Apply Gasket

Gasket

Drain Plug

Transfer Left Case

O-Ring

Apply Gasket

Extension Housing Dust Deflector

Transfer Extension Housing

O-Ring

Oil Seal

89448G68

Fig. 261 Exploded view Of the Toyota A241H transaxle components (continued)

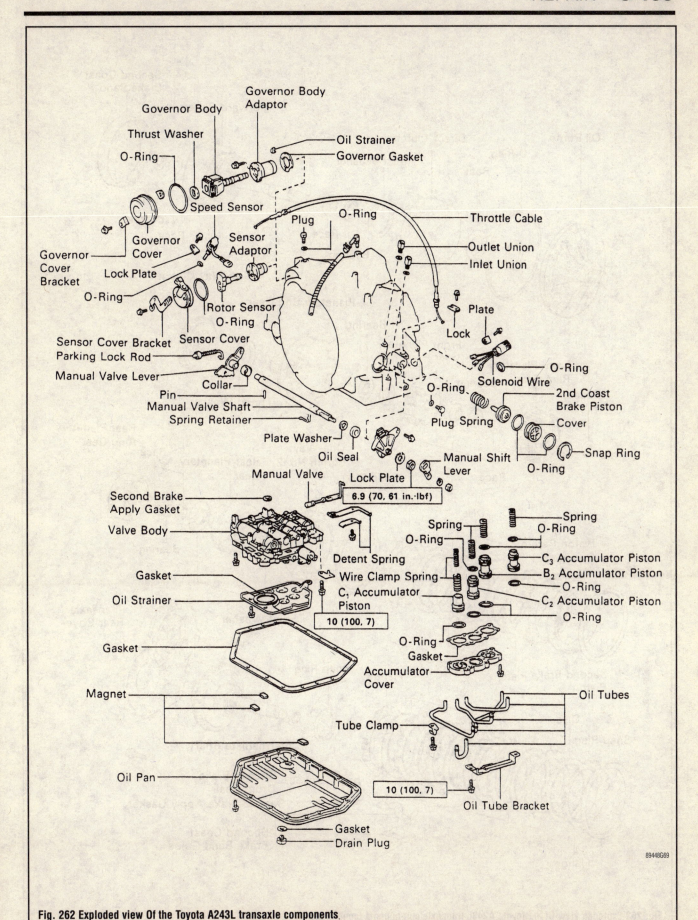

Fig. 262 Exploded view Of the Toyota A243L transaxle components

89448G69

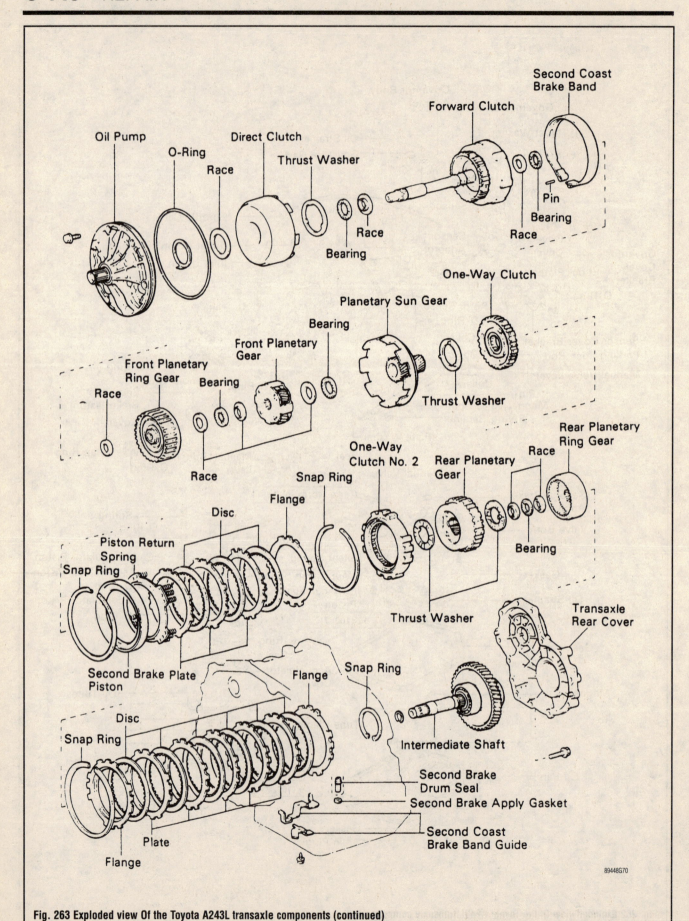

Fig. 263 Exploded view Of the Toyota A243L transaxle components (continued)

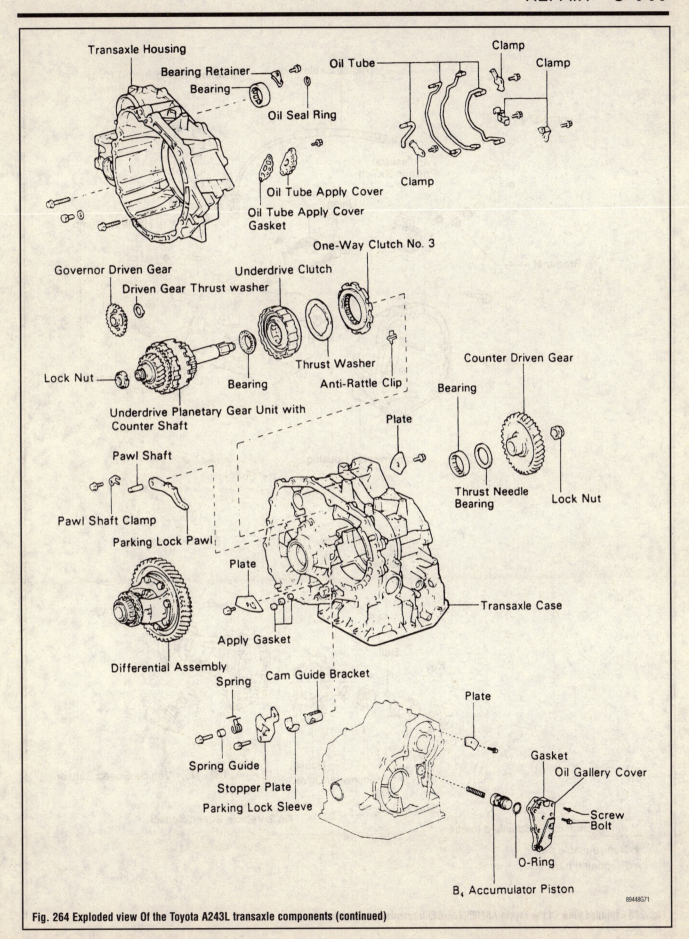

Fig. 264 Exploded view Of the Toyota A243L transaxle components (continued)

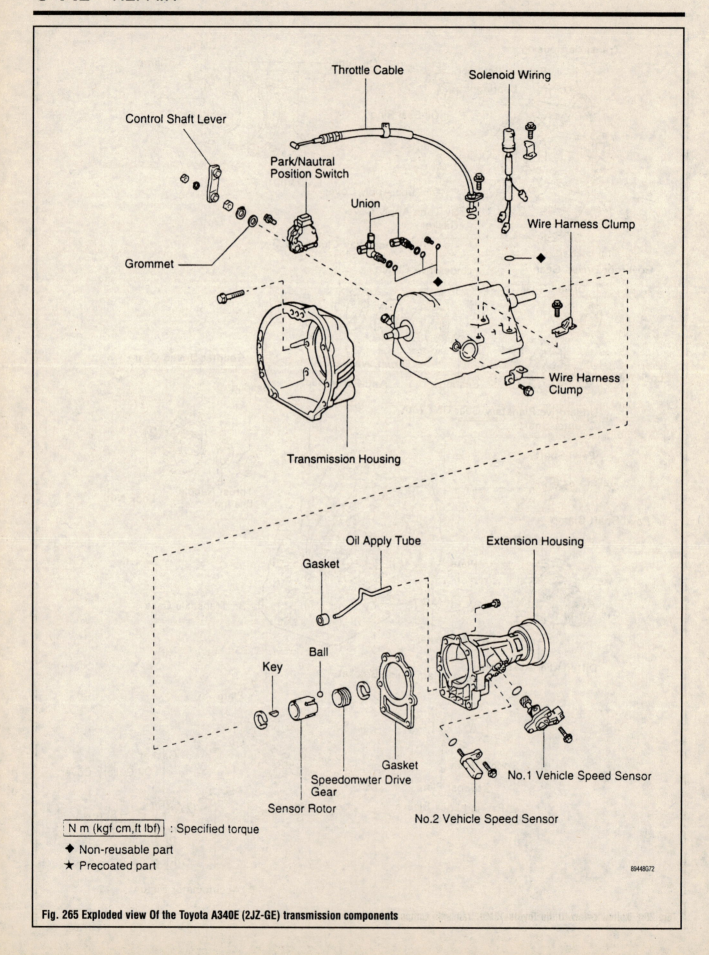

Fig. 265 Exploded view Of the Toyota A340E (2JZ-GE) transmission components

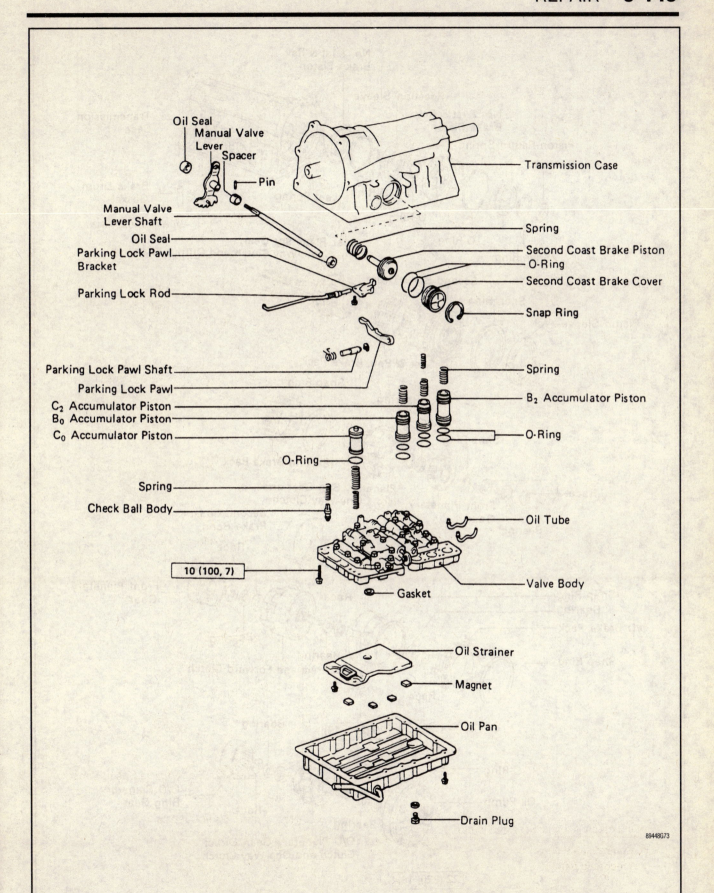

Fig. 266 Exploded view Of the Toyota A340E (2JZ-GE) transmission components (continued)

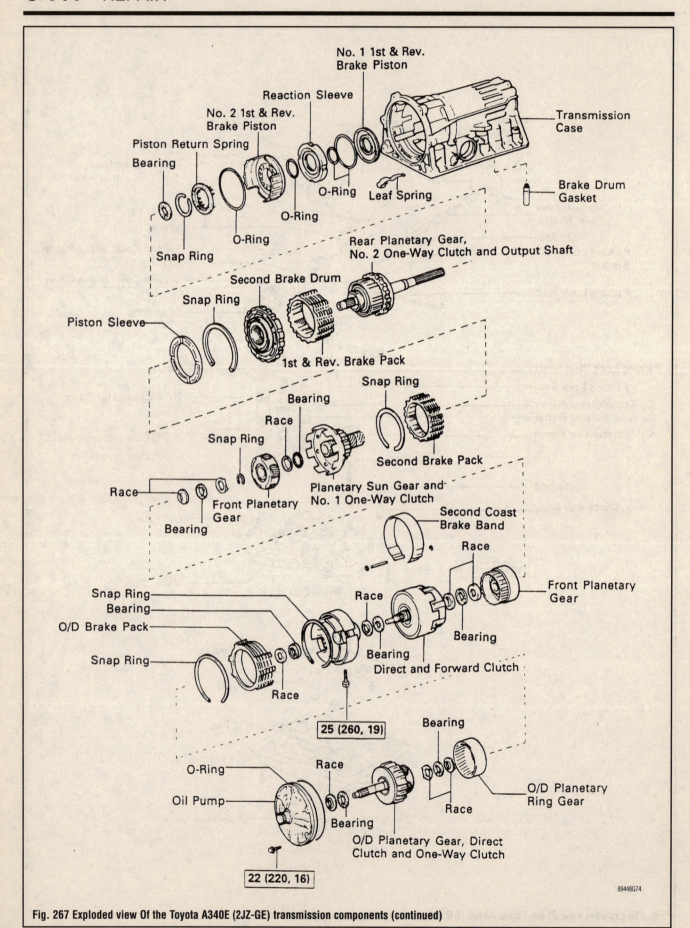

Fig. 267 Exploded view Of the Toyota A340E (2JZ-GE) transmission components (continued)

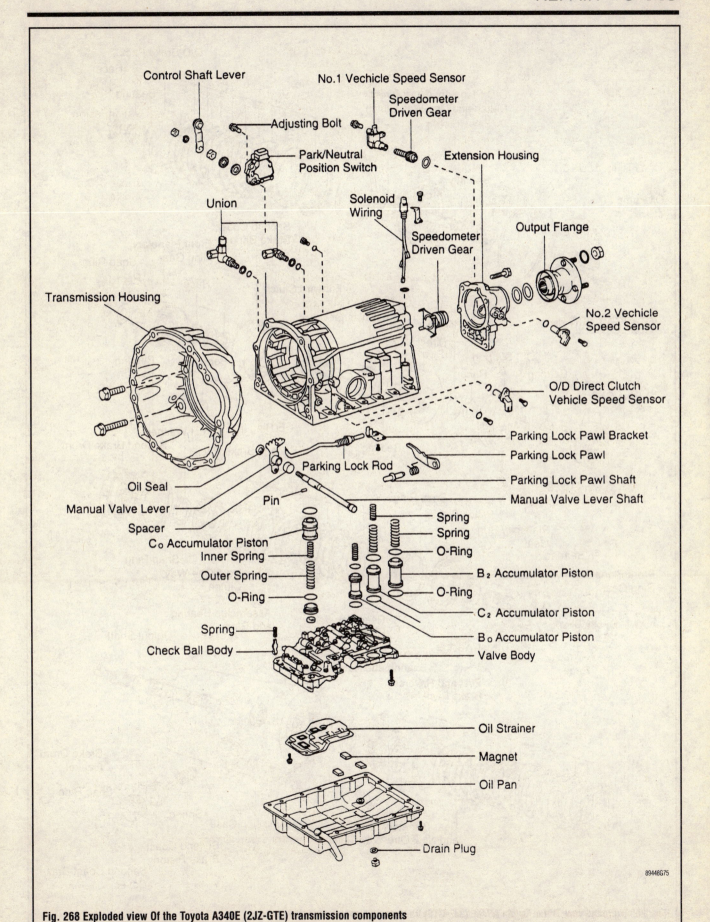

Fig. 268 Exploded view Of the Toyota A340E (2JZ-GTE) transmission components

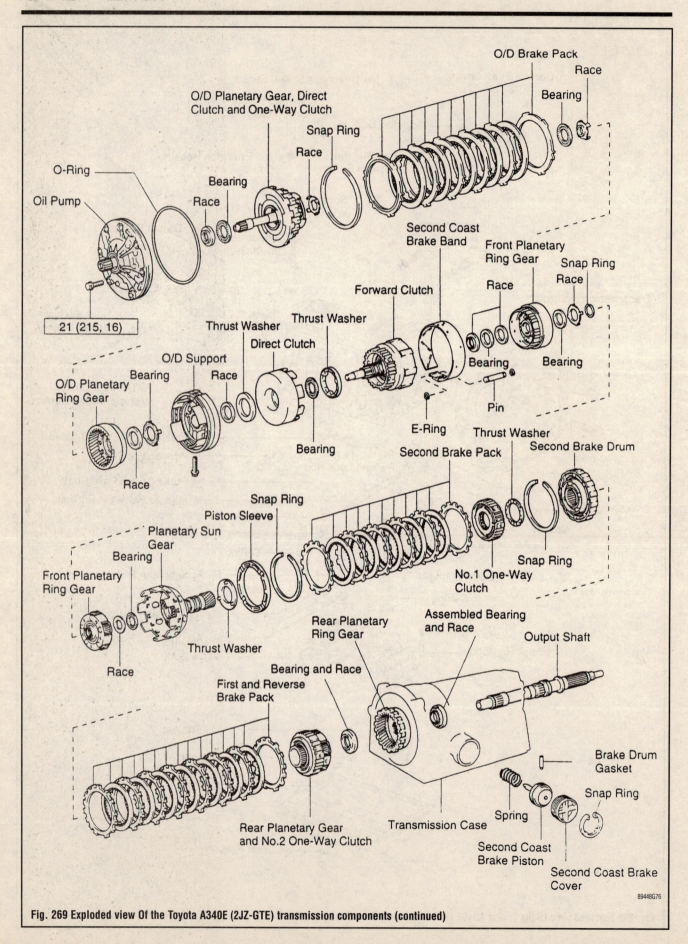

Fig. 269 Exploded view Of the Toyota A340E (2JZ-GTE) transmission components (continued)

9

GLOSSARY

GLOSSARY

Understanding your mechanic is as important as understanding your car. Just about everyone drives a car, but many drivers have difficulty understanding automotive terminology. Talking the language of cars makes it easier to effectively communicate with professional mechanics. It isn't necessary (or recommended) that you diagnose the problem for him, but it will save him time, and you money, if you can accurately describe what is happening. It will also help you to know why your car does what it is doing, and what repairs were made.

ABS: Anti-lock braking system. An electro-mechanical braking system which is designed to minimize or prevent wheel lock-up during braking.

ABSOLUTE PRESSURE: Atmospheric (barometric) pressure plus the pressure gauge reading.

ACCELERATOR PUMP: A small pump located in the carburetor that feeds fuel into the air/fuel mixture during acceleration.

ACCUMULATOR: A device that controls shift quality by cushioning the shock of hydraulic oil pressure being applied to a clutch or band.

ACTUATING MECHANISM: The mechanical output devices of a hydraulic system, for example, clutch pistons and band servos.

ACTUATOR: The output component of a hydraulic or electronic system.

ADVANCE: Setting the ignition timing so that spark occurs earlier before the piston reaches top dead center (TDC).

ADAPTIVE MEMORY (ADAPTIVE STRATEGY): The learning ability of the TCM or PCM to redefine its decision-making process to provide optimum shift quality.

AFTER TOP DEAD CENTER (ATDC): The point after the piston reaches the top of its travel on the compression stroke.

AIR BAG: Device on the inside of the car designed to inflate on impact of crash, protecting the occupants of the car.

AIR CHARGE TEMPERATURE (ACT) SENSOR: The temperature of the airflow into the engine is measured by an ACT sensor, usually located in the lower intake manifold or air cleaner. ALDL (assembly line diagnostic link): Electrical connector for scanning ECM/PCM/TCM input and output devices.

AIR CLEANER: An assembly consisting of a housing, filter and any connecting ductwork. The filter element is made up of a porous paper, sometimes with a wire mesh screening, and is designed to prevent airborne particles from entering the engine through the carburetor or throttle body.

AIR INJECTION: One method of reducing harmful exhaust emissions by injecting air into each of the exhaust ports of an engine. The fresh air entering the hot exhaust manifold causes any remaining fuel to be burned before it can exit the tailpipe.

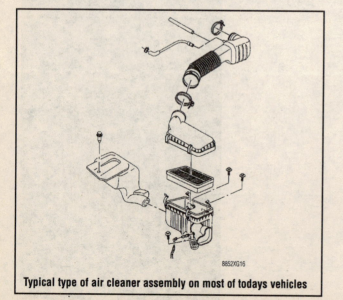

8852XG16

Typical type of air cleaner assembly on most of todays vehicles

AIR PUMP: An emission control device that supplies fresh air to the exhaust manifold to aid in more completely burning exhaust gases.

AIR/FUEL RATIO: The ratio of air-to-gasoline by weight in the fuel mixture drawn into the engine.

ALIGNMENT RACK: A special drive-on vehicle lift apparatus/measuring device used to adjust a vehicle's toe, caster and camber angles.

ALL WHEEL DRIVE: Term used to describe a full time four wheel drive system or any other vehicle drive system that continuously delivers power to all four wheels. This system is found primarily on station wagon vehicles and SUVs not utilized for significant off road use.

ALTERNATING CURRENT (AC): Electric current that flows first in one direction, then in the opposite direction, continually reversing flow.

ALTERNATOR: A device which produces AC (alternating current) which is converted to DC (direct current) to charge the car battery.

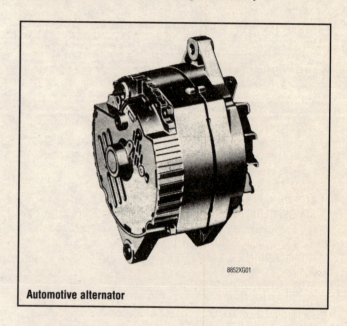

8852XG01

Automotive alternator

AMMETER: An instrument, calibrated in amperes, used to measure the flow of an electrical current in a circuit. Ammeters are always connected in series with the circuit being tested.

AMPERAGE: The total amount of current (amperes) flowing in a circuit.

AMPLIFIER: A device used in an electrical circuit to increase the voltage of an output signal.

AMP/HR. RATING (BATTERY): Measurement of the ability of a battery to deliver a stated amount of current for a stated period of time. The higher the amp/hr. rating, the better the battery.

AMPERE: The rate of flow of electrical current present when one volt of electrical pressure is applied against one ohm of electrical resistance.

ANALOG COMPUTER: Any microprocessor that uses similar (analogous) electrical signals to make its calculations.

ANODIZED: A special coating applied to the surface of aluminum valves for extended service life.

ANTIFREEZE: A substance (ethylene or propylene glycol) added to the coolant to prevent freezing in cold weather.

ANTI-FOAM AGENTS: Minimize fluid foaming from the whipping action encountered in the converter and planetary action.

ANTI-LOCK BRAKING SYSTEM: A supplementary system to the base hydraulic system that prevents sustained lock-up of the wheels during braking as well as automatically controlling wheel slip.

ANTI-ROLL BAR: See stabilizer bar.

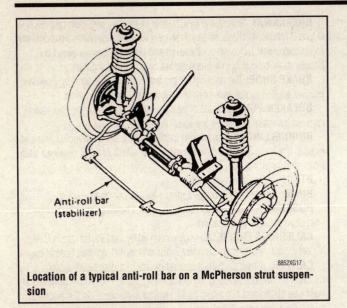

Location of a typical anti-roll bar on a McPherson strut suspension

ANTI-WEAR AGENTS: Zinc agents that control wear on the gears, bushings, and thrust washers.

ARC: A flow of electricity through the air between two electrodes or contact points that produces a spark.

ARMATURE: A laminated, soft iron core wrapped by a wire that converts electrical energy to mechanical energy as in a motor or relay. When rotated in a magnetic field, it changes mechanical energy into electrical energy as in a generator.

ATDC: After Top Dead Center.

ATF: Automatic transmission fluid.

ATMOSPHERIC PRESSURE: The pressure on the Earth's surface caused by the weight of the air in the atmosphere. At sea level, this pressure is 14.7 psi at 32°F (101 kPa at 0°C).

ATOMIZATION: The breaking down of a liquid into a fine mist that can be suspended in air.

AUXILIARY ADD-ON COOLER: A supplemental transmission fluid cooling device that is installed in series with the heat exchanger (cooler), located inside the radiator, to provide additional support to cool the hot fluid leaving the torque converter.

AUXILIARY PRESSURE: An added fluid pressure that is introduced into a regulator or balanced valve system to control valve movement. The auxiliary pressure itself can be either a fixed or a variable value. (See balanced valve; regulator valve.)

AWD: All wheel drive.

AXIAL FORCE: A side or end thrust force acting in or along the same plane as the power flow.

AXIAL PLAY: Movement parallel to a shaft or bearing bore.

AXLE CAPACITY: The maximum load-carrying capacity of the axle itself, as specified by the manufacturer. This is usually a higher number than the GAWR.

AXLE RATIO: This is a number (3.07:1, 4.56:1, for example) expressing the ratio between driveshaft revolutions and wheel revolutions. A low numerical ratio allows the engine to work easier because it doesn't have to turn as fast. A high numerical ratio means that the engine has to turn more rpm's to move the wheels through the same number of turns.

BACKFIRE: The sudden combustion of gases in the intake or exhaust system that results in a loud explosion.

BACKLASH: The clearance or play between two parts, such as meshed gears.

BACKPRESSURE: Restrictions in the exhaust system that slow the exit of exhaust gases from the combustion chamber.

BAKELITE®: A heat resistant, plastic insulator material commonly used in printed circuit boards and transistorized components.

BALANCED VALVE: A valve that is positioned by opposing auxiliary hydraulic pressures and/or spring force. Examples include mainline regulator, throttle, and governor valves. (See regulator valve.)

BAND: A flexible ring of steel with an inner lining of friction material. When tightened around the outside of a drum, a planetary member is held stationary to the transmission/transaxlecase.

BALL BEARING: A bearing made up of hardened inner and outer races between which hardened steel balls roll.

BALL JOINT: A ball and matching socket connecting suspension components (steering knuckle to lower control arms). It permits rotating movement in any direction between the components that are joined.

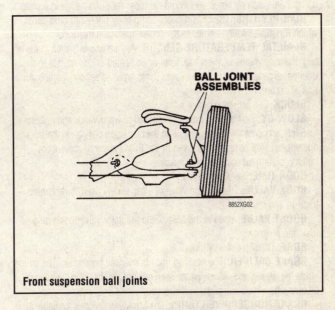

Front suspension ball joints

BARO (BAROMETRIC PRESSURE SENSOR): Measures the change in the intake manifold pressure caused by changes in altitude.

BAROMETRIC MANIFOLD ABSOLUTE PRESSURE (BMAP) SENSOR: Operates similarly to a conventional MAP sensor; reads intake manifold pressure and is also responsible for determining altitude and barometric pressure prior to engine operation.

BAROMETRIC PRESSURE: (See atmospheric pressure.)

BALLAST RESISTOR: A resistor in the primary ignition circuit that lowers voltage after the engine is started to reduce wear on ignition components.

BATTERY: A direct current electrical storage unit, consisting of the

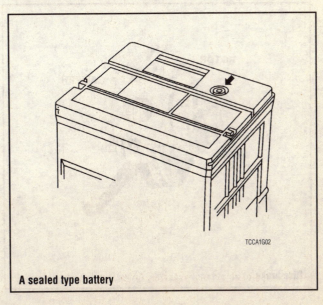

A sealed type battery

basic active materials of lead and sulphuric acid, which converts chemical energy into electrical energy. Used to provide current for the operation of the starter as well as other equipment, such as the radio, lighting, etc.

BEAD: The portion of a tire that holds it on the rim.

BEARING: A friction reducing, supportive device usually located between a stationary part and a moving part.

BEFORE TOP DEAD CENTER (BTDC): The point just before the piston reaches the top of its travel on the compression stroke.

BELTED TIRE: Tire construction similar to bias-ply tires, but using two or more layers of reinforced belts between body plies and the tread.

BEZEL: Piece of metal surrounding radio, headlights, gauges or similar components; sometimes used to hold the glass face of a gauge in the dash.

BIAS-PLY TIRE: Tire construction, using body ply reinforcing cords which run at alternating angles to the center line of the tread.

BI-METAL TEMPERATURE SENSOR: Any sensor or switch made of two dissimilar types of metal that bend when heated or cooled due to the different expansion rates of the alloys. These types of sensors usually function as an on/off switch.

BLOCK: See Engine Block.

BLOW-BY: Combustion gases, composed of water vapor and unburned fuel, that leak past the piston rings into the crankcase during normal engine operation. These gases are removed by the PCV system to prevent the buildup of harmful acids in the crankcase.

BOOK TIME: See Labor Time.

BOOK VALUE: The average value of a car, widely used to determine trade-in and resale value.

BOOST VALVE: Used at the base of the regulator valve to increase mainline pressure.

BORE: Diameter of a cylinder.

BRAKE CALIPER: The housing that fits over the brake disc. The caliper holds the brake pads, which are pressed against the discs by the caliper pistons when the brake pedal is depressed.

BRAKE HORSEPOWER (BHP): The actual horsepower available at the engine flywheel as measured by a dynamometer.

BRAKE FADE: Loss of braking power, usually caused by excessive heat after repeated brake applications.

BRAKE HORSEPOWER: Usable horsepower of an engine measured at the crankshaft.

BRAKE PAD: A brake shoe and lining assembly used with disc brakes.

BRAKE PROPORTIONING VALVE: A valve on the master cylinder which restricts hydraulic brake pressure to the wheels to a specified amount, preventing wheel lock-up.

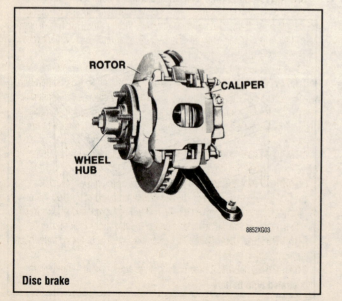

Disc brake

BREAKAWAY: Often used by Chrysler to identify first-gear operation in D and 2 ranges. In these ranges, first-gear operation depends on a one-way roller clutch that holds on acceleration and releases (breaks away) on deceleration, resulting in a freewheeling coastdown condition.

BRAKE SHOE: The backing for the brake lining. The term is, however, usually applied to the assembly of the brake backing and lining.

BREAKER POINTS: A set of points inside the distributor, operated by a cam, which make and break the ignition circuit.

BRINNELLING: A wear pattern identified by a series of indentations at regular intervals. This condition is caused by a lack of lube, overload situations, and/or vibrations.

BTDC: Before Top Dead Center.

BUMP: Sudden and forceful apply of a clutch or band.

BUSHING: A liner, usually removable, for a bearing; an anti-friction liner used in place of a bearing.

CALIFORNIA ENGINE: An engine certified by the EPA for use in California only; conforms to more stringent emission regulations than Federal engine.

CALIPER: A hydraulically activated device in a disc brake system, which is mounted straddling the brake rotor (disc). The caliper contains at least one piston and two brake pads. Hydraulic pressure on the piston(s) forces the pads against the rotor.

CAMBER: One of the factors of wheel alignment. Viewed from the front of the car, it is the inward or outward tilt of the wheel. The top of the tire will lean outward (positive camber) or inward (negative camber).

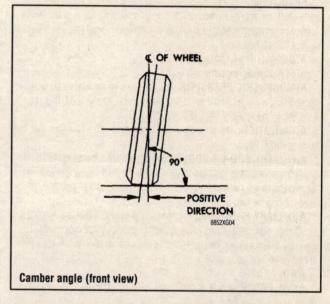

Camber angle (front view)

CAMSHAFT: A shaft in the engine on which are the lobes (cams) which operate the valves. The camshaft is driven by the crankshaft, via a belt, chain or gears, at one half the crankshaft speed.

CANCER: Rust on a car body.

CAPACITOR: A device which stores an electrical charge.

CAPACITY: The quantity of electricity that can be delivered from a unit, as from a battery in ampere-hours, or output, as from a generator.

CARBON MONOXIDE (CO): A colorless, odorless gas given off as a normal byproduct of combustion. It is poisonous and extremely dangerous in confined areas, building up slowly to toxic levels without warning if adequate ventilation is not available.

CARBURETOR: A device, usually mounted on the intake manifold of an engine, which mixes the air and fuel in the proper proportion to allow even combustion.

CASTER: The forward or rearward tilt of an imaginary line drawn through the upper ball joint and the center of the wheel. Viewed from the

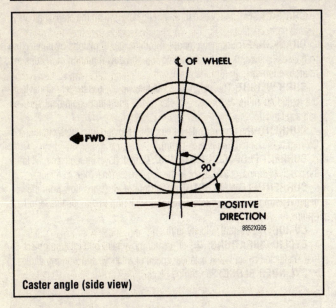

Caster angle (side view)

sides, positive caster (forward tilt) lends directional stability, while negative caster (rearward tilt) produces instability.

CATALYTIC CONVERTER: A device installed in the exhaust system, like a muffler, that converts harmful byproducts of combustion into carbon dioxide and water vapor by means of a heat-producing chemical reaction.

CENTRIFUGAL ADVANCE: A mechanical method of advancing the spark timing by using flyweights in the distributor that react to centrifugal force generated by the distributor shaft rotation.

CENTRIFUGAL FORCE: The outward pull of a revolving object, away from the center of revolution. Centrifugal force increases with the speed of rotation.

CETANE RATING: A measure of the ignition value of diesel fuel. The higher the cetane rating, the better the fuel. Diesel fuel cetane rating is roughly comparable to gasoline octane rating.

CHECK VALVE: Any one-way valve installed to permit the flow of air, fuel or vacuum in one direction only.

CHOKE: The valve/plate that restricts the amount of air entering an engine on the induction stroke, thereby enriching the air:fuel ratio.

CHUGGLE: Bucking or jerking condition that may be engine related and may be most noticeable when converter clutch is engaged; similar to the feel of towing a trailer.

CIRCLIP: A split steel snapring that fits into a groove to hold various parts in place.

CIRCUIT BREAKER: A switch which protects an electrical circuit from overload by opening the circuit when the current flow exceeds a pre-determined level. Some circuit breakers must be reset manually, while most reset automatically.

CIRCUIT: Any unbroken path through which an electrical current can flow. Also used to describe fuel flow in some instances.

CIRCUIT, BYPASS: Another circuit in parallel with the major circuit through which power is diverted.

CIRCUIT, CLOSED: An electrical circuit in which there is no interruption of current flow.

CIRCUIT, GROUND: The non-insulated portion of a complete circuit used as a common potential point. In automotive circuits, the ground is composed of metal parts, such as the engine, body sheet metal, and frame and is usually a negative potential.

CIRCUIT, HOT: That portion of a circuit not at ground potential. The hot circuit is usually insulated and is connected to the positive side of the battery.

CIRCUIT, OPEN: A break or lack of contact in an electrical circuit, either intentional (switch) or unintentional (bad connection or broken wire).

CIRCUIT, PARALLEL: A circuit having two or more paths for current flow with common positive and negative tie points. The same voltage is applied to each load device or parallel branch.

CIRCUIT, SERIES: An electrical system in which separate parts are connected end to end, using one wire, to form a single path for current to flow.

CIRCUIT, SHORT: A circuit that is accidentally completed in an electrical path for which it was not intended.

CLAMPING (ISOLATION) DIODES: Diodes positioned in a circuit to prevent self-induction from damaging electronic components.

CLEARCOAT: A transparent layer which, when sprayed over a vehicle's paint job, adds gloss and depth as well as an additional protective coating to the finish.

CLUTCH: Part of the power train used to connect/disconnect power to the rear wheels.

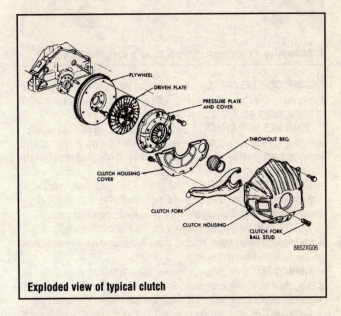

Exploded view of typical clutch

CLUTCH, FLUID: The same as a fluid coupling. A fluid clutch or coupling performs the same function as a friction clutch by utilizing fluid friction and inertia as opposed to solid friction used by a friction clutch. (See fluid coupling.)

CLUTCH, FRICTION: A coupling device that provides a means of smooth and positive engagement and disengagement of engine torque to the vehicle powertrain. Transmission of power through the clutch is accomplished by bringing one or more rotating drive members into contact with complementing driven members.

COAST: Vehicle deceleration caused by engine braking conditions.

COEFFICIENT OF FRICTION: The amount of surface tension between two contacting surfaces; identified by a scientifically calculated number.

COIL: Part of the ignition system that boosts the relatively low voltage supplied by the car's electrical system to the high voltage required to fire the spark plugs.

COMBINATION MANIFOLD: An assembly which includes both the intake and exhaust manifolds in one casting.

COMBINATION VALVE: A device used in some fuel systems that routes fuel vapors to a charcoal storage canister instead of venting them into the atmosphere. The valve relieves fuel tank pressure and allows fresh air into the tank as the fuel level drops to prevent a vapor lock situation.

COMBUSTION CHAMBER: The part of the engine in the cylinder head where combustion takes place.

COMPOUND GEAR: A gear consisting of two or more simple gears with a common shaft.

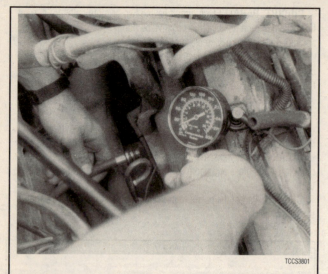

TCCS3801

Performing a compression check using a compression gauge

COMPOUND PLANETARY: A gearset that has more than the three elements found in a simple gearset and is constructed by combining members of two planetary gearsets to create additional gear ratio possibilities.

COMPRESSION CHECK: A test involving removing each spark plug and inserting a gauge. When the engine is cranked, the gauge will record a pressure reading in the individual cylinder. General operating condition can be determined from a compression check.

COMPRESSION RATIO: The ratio of the volume between the piston and cylinder head when the piston is at the bottom of its stroke (bottom dead center) and when the piston is at the top of its stroke (top dead center).

COMPUTER: An electronic control module that correlates input data according to prearranged engineered instructions; used for the management of an actuator system or systems.

CONDENSER: 1. An electrical device which acts to store an electrical charge, preventing voltage surges. 2. A radiator-like device in the air conditioning system in which refrigerant gas condenses into a liquid, giving off heat.

CONDUCTOR: Any material through which an electrical current can be transmitted easily.

CONNECTING ROD: The connecting link between the crankshaft and piston.

CONSTANT VELOCITY JOINT: Type of universal joint in a halfshaft assembly in which the output shaft turns at a constant angular velocity without variation, provided that the speed of the input shaft is constant.

CONTINUITY: Continuous or complete circuit. Can be checked with an ohmmeter.

CONTROL ARM: The upper or lower suspension components which are mounted on the frame and support the ball joints and steering knuckles.

CONVENTIONAL IGNITION: Ignition system which uses breaker points.

CONVERTER: (See torque converter.)

CONVERTER LOCKUP: The switching from hydrodynamic to direct mechanical drive, usually through the application of a friction element called the converter clutch.

COOLANT: Mixture of water and anti-freeze circulated through the engine to carry off heat produced by the engine.

CORROSION INHIBITOR: An inhibitor in ATF that prevents corrosion of bushings, thrust washers, and oil cooler brazed joints.

COUNTERSHAFT: An intermediate shaft which is rotated by a mainshaft and transmits, in turn, that rotation to a working part.

COUPLING PHASE: Occurs when the torque converter is operating at its greatest hydraulic efficiency. The speed differential between the impeller and the turbine is at its minimum. At this point, the stator freewheels, and there is no torque multiplication.

CRANKCASE: The lower part of an engine in which the crankshaft and related parts operate.

CRANKSHAFT: Engine component (connected to pistons by connecting rods) which converts the reciprocating (up and down) motion of pistons to rotary motion used to turn the driveshaft.

CURB WEIGHT: The weight of a vehicle without passengers or payload, but including all fluids (oil, gas, coolant, etc.) and other equipment specified as standard.

CURRENT: The flow (or rate) of electrons moving through a circuit. Current is measured in amperes (amp).

CURRENT FLOW CONVENTIONAL: Current flows through a circuit from the positive terminal of the source to the negative terminal (plus to minus).

CURRENT FLOW, ELECTRON: Current or electrons flow from the negative terminal of the source, through the circuit, to the positive terminal (minus to plus).

CV-JOINT: Constant velocity joint.

CYCLIC VIBRATIONS: The off-center movement of a rotating object that is affected by its initial balance, speed of rotation, and working angles.

CYLINDER BLOCK: See engine block.

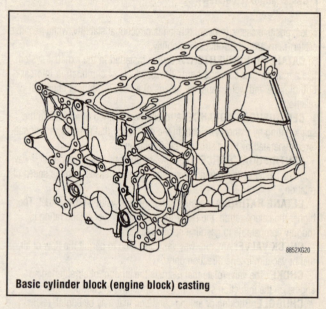

8852XG20

Basic cylinder block (engine block) casting

CYLINDER HEAD: The detachable portion of the engine, usually fastened to the top of the cylinder block and containing all or most of the combustion chambers. On overhead valve engines, it contains the valves and their operating parts. On overhead cam engines, it contains the camshaft as well.

CYLINDER: In an engine, the round hole in the engine block in which the piston(s) ride.

DATA LINK CONNECTOR (DLC): Current acronym/term applied to the federally mandated, diagnostic junction connector that is used to monitor ECM/PC/TCM inputs, processing strategies, and outputs including diagnostic trouble codes (DTCs).

DEAD CENTER: The extreme top or bottom of the piston stroke.

DECELERATION BUMP: When referring to a torque converter clutch in the applied position, a sudden release of the accelerator pedal causes a forceful reversal of power through the drivetrain (engine braking), just prior to the apply plate actually being released.

DELAYED (LATE OR EXTENDED): Condition where shift is expected but does not occur for a period of time, for example, where clutch or band engagement does not occur as quickly as expected during part throttle or wide open throttle apply of accelerator or when manually downshifting to a lower range.

DETENT: A spring-loaded plunger, pin, ball, or pawl used as a holding device on a ratchet wheel or shaft. In automatic transmissions, a detent mechanism is used for locking the manual valve in place.

DETENT DOWNSHIFT: (See kickdown.)

DETERGENT: An additive in engine oil to improve its operating characteristics.

DETONATION: An unwanted explosion of the air/fuel mixture in the combustion chamber caused by excess heat and compression, advanced timing, or an overly lean mixture. Also referred to as "ping".

DEXRON®: A brand of automatic transmission fluid.

DIAGNOSTIC TROUBLE CODES (DTCS): A digital display from the control module memory that identifies the input, processor, or output device circuit that is related to the powertrain emission/driveability malfunction detected. Diagnostic trouble codes can be read by the MIL to flash any codes or by using a handheld scanner.

DIAPHRAGM: A thin, flexible wall separating two cavities, such as in a vacuum advance unit.

DIESELING: The engine continues to run after the car is shut off; caused by fuel continuing to be burned in the combustion chamber.

DIFFERENTIAL: A geared assembly which allows the transmission of motion between drive axles, giving one axle the ability to rotate faster than the other, as in cornering.

DIFFERENTIAL AREAS: When opposing faces of a spool valve are acted upon by the same pressure but their areas differ in size, the face with the larger area produces the differential force and valve movement. (See spool valve.)

DIFFERENTIAL FORCE: (See differential areas.) digital readout: A display of numbers or a combination of numbers and letters.

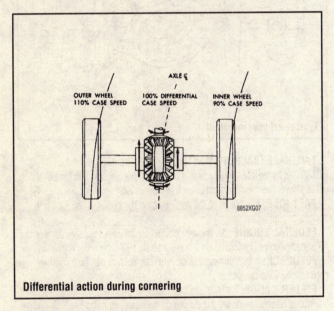

Differential action during cornering

DIGITAL VOLT OHMMETER: An electronic diagnostic tool used to measure voltage, ohms and amps as well as several other functions, with the readings displayed on a digital screen in tenths, hundredths and thousandths.

DIODE: An electrical device that will allow current to flow in one direction only.

DIRECT CURRENT (DC): Electrical current that flows in one direction only.

DIRECT DRIVE: The gear ratio is 1:1, with no change occurring in the torque and speed input/output relationship.

DISC BRAKE: A hydraulic braking assembly consisting of a brake disc, or rotor, mounted on an axleshaft, and a caliper assembly containing, usually two brake pads which are activated by hydraulic pressure. The pads are forced against the sides of the disc, creating friction which slows the vehicle.

DISPERSANTS: Suspend dirt and prevent sludge buildup. double bump (double feel): Two sudden and forceful applies of a clutch or band.

DISPLACEMENT: The total volume of air that is displaced by all pistons as the engine turns through one complete revolution.

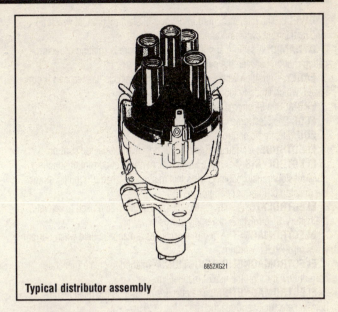

Typical distributor assembly

DISTRIBUTOR: A mechanically driven device on an engine which is responsible for electrically firing the spark plug at a pre-determined point of the piston stroke.

DOHC: Double overhead camshaft.

DOUBLE OVERHEAD CAMSHAFT: The engine utilizes two camshafts mounted in one cylinder head. One camshaft operates the exhaust valves, while the other operates the intake valves.

DOWEL PIN: A pin, inserted in mating holes in two different parts allowing those parts to maintain a fixed relationship.

DRIVELINE: The drive connection between the transmission and the drive wheels.

DRIVE TRAIN: The components that transmit the flow of power from the engine to the wheels. The components include the clutch, transmission, driveshafts (or axle shafts in front wheel drive), U-joints and differential.

DRUM BRAKE: A braking system which consists of two brake shoes and one or two wheel cylinders, mounted on a fixed backing plate, and a brake drum, mounted on an axle, which revolves around the assembly.

DRY CHARGED BATTERY: Battery to which electrolyte is added when the battery is placed in service.

DVOM: Digital volt ohmmeter

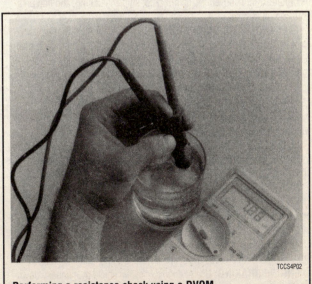

Performing a resistance check using a DVOM

DWELL: The rate, measured in degrees of shaft rotation, at which an electrical circuit cycles on and off.

DYNAMIC: A sealing application in which there is rotating or reciprocating motion between the parts.

EARLY: Condition where shift occurs before vehicle has reached proper speed, which tends to labor engine after upshift.

EBCM: See Electronic Control Unit (ECU).

ECM: See Electronic Control Unit (ECU).

ECU: Electronic control unit.

ELECTRODE: Conductor (positive or negative) of electric current.

ELECTROLYSIS: A surface etching or bonding of current conducting transmission/transaxle components that may occur when grounding straps are missing or in poor condition.

ELECTROLYTE: A solution of water and sulfuric acid used to activate the battery. Electrolyte is extremely corrosive.

ELECTROMAGNET: A coil that produces a magnetic field when current flows through its windings.

ELECTROMAGNETIC INDUCTION: A method to create (generate) current flow through the use of magnetism.

ELECTROMAGNETISM: The effects surrounding the relationship between electricity and magnetism.

ELECTROMOTIVE FORCE (EMF): The force or pressure (voltage) that causes current movement in an electrical circuit.

ELECTRONIC CONTROL UNIT: A digital computer that controls engine (and sometimes transmission, brake or other vehicle system) functions based on data received from various sensors. Examples used by some manufacturers include Electronic Brake Control Module (EBCM), Engine Control Module (ECM), Powertrain Control Module (PCM) or Vehicle Control Module (VCM).

ELECTRONIC IGNITION: A system in which the timing and firing of the spark plugs is controlled by an electronic control unit, usually called a module. These systems have no points or condenser.

ELECTRONIC PRESSURE CONTROL (EPC) SOLENOID: A specially designed solenoid containing a spool valve and spring assembly to control fluid mainline pressure. A variable current flow, controlled by the ECM/PCM, varies the internal force of the solenoid on the spool valve and resulting mainline pressure. (See variable force solenoid.)

ELECTRONICS: Miniaturized electrical circuits utilizing semiconductors, solid-state devices, and printed circuits. Electronic circuits utilize small amounts of power.

ELECTRONIFICATION: The application of electronic circuitry to a mechanical device. Regarding automatic transmissions, electrification is incorporated into converter clutch lockup, shift scheduling, and line pressure control systems.

ELECTROSTATIC DISCHARGE (ESD): An unwanted, high-voltage electrical current released by an individual who has taken on a static charge of electricity. Electronic components can be easily damaged by ESD.

ELEMENT: A device within a hydrodynamic drive unit designed with a set of blades to direct fluid flow.

ENAMEL: Type of paint that dries to a smooth, glossy finish.

END BUMP (END FEEL OR SLIP BUMP): Firmer feel at end of shift when compared with feel at start of shift.

END-PLAY: The clearance/gap between two components that allows for expansion of the parts as they warm up, to prevent binding and to allow space for lubrication.

ENERGY: The ability or capacity to do work.

ENGINE: The primary motor or power apparatus of a vehicle, which converts liquid or gas fuel into mechanical energy.

ENGINE BLOCK: The basic engine casting containing the cylinders, the crankshaft main bearings, as well as machined surfaces for the mounting of other components such as the cylinder head, oil pan, transmission, etc.

ENGINE BRAKING: Use of engine to slow vehicle by manually downshifting during zero-throttle coast down.

ENGINE CONTROL MODULE (ECM): Manages the engine and incorporates output control over the torque converter clutch solenoid. (Note: Current designation for the ECM in late model vehicles is PCM.)

ENGINE COOLANT TEMPERATURE (ECT) SENSOR: Prevents converter clutch engagement with a cold engine; also used for shift timing and shift quality.

EP LUBRICANT: EP (extreme pressure) lubricants are specially formulated for use with gears involving heavy loads (transmissions, differentials, etc.).

ETHYL: A substance added to gasoline to improve its resistance to knock, by slowing down the rate of combustion.

ETHYLENE GLYCOL: The base substance of antifreeze.

EXHAUST MANIFOLD: A set of cast passages or pipes which conduct exhaust gases from the engine.

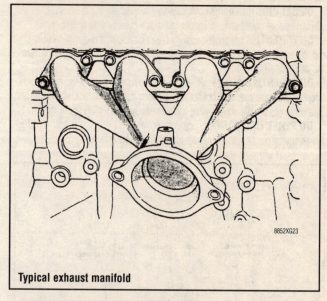

8852XG23

Typical exhaust manifold

FAIL-SAFE (BACKUP) CONTROL: A substitute value used by the PCM/TCM to replace a faulty signal from an input sensor. The temporary value allows the vehicle to continue to be operated.

FAST IDLE: The speed of the engine when the choke is on. Fast idle speeds engine warm-up.

FEDERAL ENGINE: An engine certified by the EPA for use in any of the 49 states (except California).

FEEDBACK: A circuit malfunction whereby current can find another path to feed load devices.

FEELER GAUGE: A blade, usually metal, of precisely predetermined thickness, used to measure the clearance between two parts.

FILAMENT: The part of a bulb that glows; the filament creates high resistance to current flow and actually glows from the resulting heat.

FINAL DRIVE: An essential part of the axle drive assembly where final gear reduction takes place in the powertrain. In RWD applications and north-south FWD applications, it must also change the power flow direction to the axle shaft by ninety degrees. (Also see axle ratio).

FIRING ORDER: The order in which combustion occurs in the cylinders of an engine. Also the order in which spark is distributed to the plugs by the distributor.

FIRM: A noticeable quick apply of a clutch or band that is considered normal with medium to heavy throttle shift; should not be confused with harsh or rough.

FLAME FRONT: The term used to describe certain aspects of the fuel explosion in the cylinders. The flame front should move in a controlled pattern across the cylinder, rather than simply exploding immediately.

FLARE (SLIPPING): A quick increase in engine rpm accompanied by momentary loss of torque; generally occurs during shift.

FLAT ENGINE: Engine design in which the pistons are horizontally opposed. Porsche, Subaru and some old VWs are common examples of flat engines.

FLAT RATE: A dealership term referring to the amount of money paid to a technician for a repair or diagnostic service based on that particular service versus dealership's labor time (NOT based on the actual time the technician spent on the job).

FLAT SPOT: A point during acceleration when the engine seems to lose power for an instant.

FLOODING: The presence of too much fuel in the intake manifold and combustion chamber which prevents the air/fuel mixture from firing, thereby causing a no-start situation.

FLUID: A fluid can be either liquid or gas. In hydraulics, a liquid is used for transmitting force or motion.

FLUID COUPLING: The simplest form of hydrodynamic drive, the fluid coupling consists of two look-alike members with straight radial varies referred to as the impeller (pump) and the turbine. input torque is always equal to the output torque.

FLUID DRIVE: Either a fluid coupling or a fluid torque converter. (See hydrodynamic drive units.)

FLUID TORQUE CONVERTER: A hydrodynamic drive that has the ability to act both as a torque multiplier and fluid coupling. (See hydrodynamic drive units; torque converter.)

FLUID VISCOSITY: The resistance of a liquid to flow. A cold fluid (oil) has greater viscosity and flows more slowly than a hot fluid (oil).

FLYWHEEL: A heavy disc of metal attached to the rear of the crankshaft. It smoothes the firing impulses of the engine and keeps the crankshaft turning during periods when no firing takes place. The starter also engages the flywheel to start the engine.

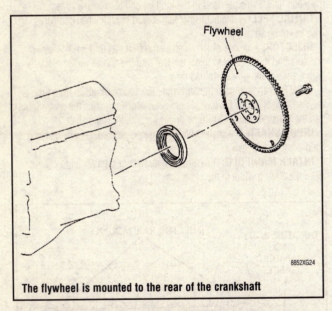

The flywheel is mounted to the rear of the crankshaft

FOOT POUND (ft. lbs. or sometimes, ft. lb.): The amount of energy or work needed to raise an item weighing one pound, a distance of one foot.

FREEZE PLUG: A plug in the engine block which will be pushed out if the coolant freezes. Sometimes called expansion plugs, they protect the block from cracking should the coolant freeze.

FRICTION: The resistance that occurs between contacting surfaces. This relationship is expressed by a ratio called the coefficient of friction (CL).

FRICTION, COEFFICIENT OF: The amount of surface tension between two contacting surfaces; expressed by a scientifically calculated number.

FRONT END ALIGNMENT: A service to set caster, camber and toe-in to the correct specifications. This will ensure that the car steers and handles properly and that the tires wear properly.

FRICTION MODIFIER: Changes the coefficient of friction of the fluid between the mating steel and composition clutch/band surfaces during the engagement process and allows for a certain amount of intentional slipping for a good "shift-feel." full throttle detent downshift: A quick apply of accelerator pedal to its full travel, forcing a downshift.

FRONTAL AREA: The total frontal area of a vehicle exposed to air flow.

FUEL FILTER: A component of the fuel system containing a porous paper element used to prevent any impurities from entering the engine through the fuel system. It usually takes the form of a canister-like housing, mounted in-line with the fuel hose, located anywhere on a vehicle between the fuel tank and engine.

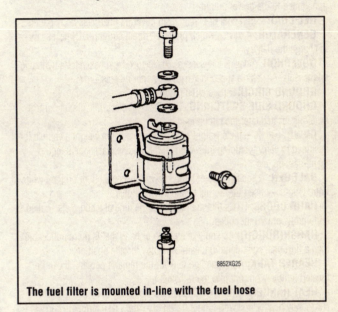

The fuel filter is mounted in-line with the fuel hose

FUEL INJECTION: A system replacing the carburetor that sprays fuel into the cylinder through nozzles. The amount of fuel can be more precisely controlled with fuel injection.

FULL FLOATING AXLE: An axle in which the axle housing extends through the wheel giving bearing support on the outside of the housing. The front axle of a four-wheel drive vehicle is usually a full floating axle, as are the rear axles of many larger (1 ton and over) pick-ups and vans.

FULL-TIME FOUR-WHEEL DRIVE: A four-wheel drive system that continuously delivers power to all four wheels. A differential between the front and rear driveshafts permits variations in axle speeds to control gear wind-up without damage.

FUSE: A protective device in a circuit which prevents circuit overload by breaking the circuit when a specific amperage is present. The device is constructed around a strip or wire of a lower amperage rating than the circuit it is designed to protect. When an amperage higher than that stamped on the fuse is present in the circuit, the strip or wire melts, opening the circuit.

FUSIBLE LINK: A piece of wire in a wiring harness that performs the same job as a fuse. If overloaded, the fusible link will melt and interrupt the circuit.

FWD: Front wheel drive.

GAWR: (Gross axle weight rating) the total maximum weight an axle is designed to carry.

GCW: (Gross combined weight) total combined weight of a tow vehicle and trailer.

GARAGE SHIFT: initial engagement feel of transmission, neutral to reverse or neutral to a forward drive.

GARAGE SHIFT FEEL: A quick check of the engagement quality and responsiveness of reverse and forward gears. This test is done with the vehicle stationary.

GEAR: A toothed mechanical device that acts as a rotating lever to transmit power or turning effort from one shaft to another. (See gear ratio.)

GEAR RATIO: A ratio expressing the number of turns a smaller gear will make to turn a larger gear through one revolution. The ratio is found by dividing the number of teeth on the smaller gear into the number of teeth on the larger gear.

GEARBOX: Transmission

GEAR REDUCTION: Torque is multiplied and speed decreased by the factor of the gear ratio. For example, a 3:1 gear ratio changes an input torque of 180 ft. lbs. and an input speed of 2700 rpm to 540 Ft. lbs. and 900 rpm, respectively. (No account is taken of frictional losses, which are always present.)

GEARTRAIN: A succession of intermeshing gears that form an assembly and provide for one or more torque changes as the power input is transmitted to the power output.

GEL COAT: A thin coat of plastic resin covering fiberglass body panels.

GENERATOR: A device which produces direct current (DC) necessary to charge the battery.

GOVERNOR: A device that senses vehicle speed and generates a hydraulic oil pressure. As vehicle speed increases, governor oil pressure rises.

GROUND CIRCUIT: (See circuit, ground.)

GROUND SIDE SWITCHING: The electrical/electronic circuit control switch is located after the circuit load.

GVWR: (Gross vehicle weight rating) total maximum weight a vehicle is designed to carry including the weight of the vehicle, passengers, equipment, gas, oil, etc.

HALOGEN: A special type of lamp known for its quality of brilliant white light. Originally used for fog lights and driving lights.

HARD CODES: DTCs that are present at the time of testing; also called continuous or current codes.

HARSH(ROUGH): An apply of a clutch or band that is more noticeable than a firm one; considered undesirable at any throttle position.

HEADER TANK: An expansion tank for the radiator coolant. It can be located remotely or built into the radiator.

HEAT RANGE: A term used to describe the ability of a spark plug to carry away heat. Plugs with longer nosed insulators take longer to carry heat off effectively.

HEAT RISER: A flapper in the exhaust manifold that is closed when the engine is cold, causing hot exhaust gases to heat the intake manifold providing better cold engine operation. A thermostatic spring opens the flapper when the engine warms up.

HEAVY THROTTLE: Approximately three-fourths of accelerator pedal travel.

HEMI: A name given an engine using hemispherical combustion chambers.

HERTZ (HZ): The international unit of frequency equal to one cycle per second (10,000 Hertz equals 10,000 cycles per second).

HIGH-IMPEDANCE DVOM (DIGITAL VOLT-OHMMETER): This styled device provides a built-in resistance value and is capable of limiting circuit current flow to safe milliamp levels.

HIGH RESISTANCE: Often refers to a circuit where there is an excessive amount of opposition to normal current flow.

HORSEPOWER: A measurement of the amount of work; one horsepower is the amount of work necessary to lift 33,000 lbs. one foot in one minute. Brake horsepower (bhp) is the horsepower delivered by an engine on a dynamometer. Net horsepower is the power remaining (measured at the flywheel of the engine) that can be used to turn the wheels after power is consumed through friction and running the engine accessories (water pump, alternator, air pump, fan etc.)

HOT CIRCUIT: (See circuit, hot; hot lead.) hot lead: A wire or conductor in the power side of the circuit. (See circuit, hot.)

HOT SIDE SWITCHING: The electrical/electronic circuit control switch is located before the circuit load.

HUB: The center part of a wheel or gear.

HUNTING (BUSYNESS): Repeating quick series of upshifts and downshifts that causes noticeable change in engine rpm, for example, as in a 4-3-4 shift pattern.

HYDRAULICS: The use of liquid under pressure to transfer force of motion.

HYDROCARBON (HC): Any chemical compound made up of hydrogen and carbon. A major pollutant formed by the engine as a by-product of combustion.

HYDRODYNAMIC DRIVE UNITS: Devices that transmit power solely by the action of a kinetic fluid flow in a closed recirculating path. An impeller energizes the fluid and discharges the high-speed jet stream into the turbine for power output.

HYDROMETER: An instrument used to measure the specific gravity of a solution.

HYDROPLANING: A phenomenon of driving when water builds up under the tire tread, causing it to lose contact with the road. Slowing down will usually restore normal tire contact with the road.

HYPOID GEARSET: The drive pinion gear may be placed below or above the centerline of the driven gear; often used as a final drive gearset.

IDLE MIXTURE: The mixture of air and fuel (usually about 14:1) being fed to the cylinders. The idle mixture screw(s) are sometimes adjusted as part of a tune-up.

IDLER ARM: Component of the steering linkage which is a geometric duplicate of the steering gear arm. It supports the right side of the center steering link.

IMPELLER: Often called a pump, the impeller is the power input (drive) member of a hydrodynamic drive. As part of the torque converter cover, it acts as a centrifugal pump and puts the fluid in motion.

INCH POUND (inch lbs.; sometimes in. lb. or in. lbs.): One twelfth of a foot pound.

INDUCTANCE: The force that produces voltage when a conductor is passed through a magnetic field.

INDUCTION: A means of transferring electrical energy in the form of a magnetic field. Principle used in the ignition coil to increase voltage.

INITIAL FEEL: A distinct firmer feel at start of shift when compared with feel at finish of shift.

INJECTOR: A device which receives metered fuel under relatively low pressure and is activated to inject the fuel into the engine under relatively high pressure at a predetermined time.

INPUT: In an automatic transmission, the source of power from the engine is absorbed by the torque converter, which provides the power input into the transmission. The turbine drives the input(turbine)shaft.

INPUT SHAFT: The shaft to which torque is applied, usually carrying the driving gear or gears.

INTAKE MANIFOLD: A casting of passages or pipes used to conduct air or a fuel/air mixture to the cylinders.

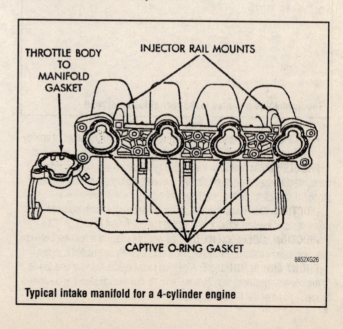

THROTTLE BODY TO MANIFOLD GASKET

INJECTOR RAIL MOUNTS

CAPTIVE O-RING GASKET

8852XG26

Typical intake manifold for a 4-cylinder engine

INTERNAL GEAR: The ring-like outer gear of a planetary gearset with the gear teeth cut on the inside of the ring to provide a mesh with the planet pinions.

ISOLATION (CLAMPING) DIODES: Diodes positioned in a circuit to prevent self-induction from damaging electronic components.

IX ROTARY GEAR PUMP: Contains two rotating members, one shaped with internal gear teeth and the other with external gear teeth. As the gears separate, the fluid fills the gaps between gear teeth, is pulled across a crescent-shaped divider, and then is forced to flow through the outlet as the gears mesh.

IX ROTARY LOBE PUMP: Sometimes referred to as a gerotor type pump. Two rotating members, one shaped with internal lobes and the other with external lobes, separate and then mesh to cause fluid to flow.

JOURNAL: The bearing surface within which a shaft operates.

JUMPER CABLES: Two heavy duty wires with large alligator clips used to provide power from a charged battery to a discharged battery mounted in a vehicle.

JUMPSTART: Utilizing the sufficiently charged battery of one vehicle to start the engine of another vehicle with a discharged battery by the use of jumper cables.

KEY: A small block usually fitted in a notch between a shaft and a hub to prevent slippage of the two parts.

KICKDOWN: Detent downshift system; either linkage, cable, or electrically controlled.

KILO: A prefix used in the metric system to indicate one thousand.

KNOCK: Noise which results from the spontaneous ignition of a portion of the air-fuel mixture in the engine cylinder caused by overly advanced ignition timing or use of incorrectly low octane fuel for that engine.

KNOCK SENSOR: An input device that responds to spark knock, caused by over advanced ignition timing.

LABOR TIME: A specific amount of time required to perform a certain repair or diagnostic service as defined by a vehicle or after-market manufacturer.

LACQUER: A quick-drying automotive paint.

LATE: Shift that occurs when engine is at higher than normal rpm for given amount of throttle.

LIGHT-EMITTING DIODE (LED): A semiconductor diode that emits light as electrical current flows through it; used in some electronic display devices to emit a red or other color light.

LIGHT THROTTLE: Approximately one-fourth of accelerator pedal travel.

LIMITED SLIP: A type of differential which transfers driving force to the wheel with the best traction.

LIMP-IN MODE: Electrical shutdown of the transmission/ transaxle output solenoids, allowing only forward and reverse gears that are hydraulically energized by the manual valve. This permits the vehicle to be driven to a service facility for repair.

LIP SEAL: Molded synthetic rubber seal designed with an outer sealing edge (lip) that points into the fluid containing area to be sealed. This type of seal is used where rotational and axial forces are present.

LITHIUM-BASE GREASE: Chassis and wheel bearing grease using lithium as a base. Not compatible with sodium-base grease.

LOAD DEVICE: A circuit's resistance that converts the electrical energy into light, sound, heat, or mechanical movement.

LOAD RANGE: Indicates the number of plies at which a tire is rated. Load range B equals four-ply rating; C equals six-ply rating; and, D equals an eight-ply rating.

LOAD TORQUE: The amount of output torque needed from the transmission/transaxle to overcome the vehicle load.

LOCKING HUBS: Accessories used on part-time four-wheel drive systems that allow the front wheels to be disengaged from the drive train when four-wheel drive is not being used. When four-wheel drive is desired, the hubs are engaged, locking the wheels to the drive train.

LOCKUP CONVERTER: A torque converter that operates hydraulically and mechanically. When an internal apply plate (lockup plate) clamps to the torque converter cover, hydraulic slippage is eliminated.

LOCK RING: See Circlip or Snapring

MAGNET: Any body with the property of attracting iron or steel.

MAGNETIC FIELD: The area surrounding the poles of a magnet that is affected by its attraction or repulsion forces.

MAIN LINE PRESSURE: Often called control pressure or line pressure, it refers to the pressure of the oil leaving the pump and is controlled by the pressure regulator valve.

MALFUNCTION INDICATOR LAMP (MIL): Previously known as a check engine light, the dash-mounted MIL illuminates and signals the driver that an emission or driveability problem with the powertrain has been detected by the ECM/PCM. When this occurs, at least one diagnostic trouble code (DTC) has been stored into the control module memory.

MANIFOLD ABSOLUTE PRESSURE (MAP) SENSOR: Reads the amount of air pressure (vacuum) in the engine's intake manifold system; its signal is used to analyze engine load conditions.

MANIFOLD VACUUM: Low pressure in an engine intake manifold formed just below the throttle plates. Manifold vacuum is highest at idle and drops under acceleration.

MANIFOLD: A casting of passages or set of pipes which connect the cylinders to an inlet or outlet source.

MANUAL LEVER POSITION SWITCH (MLPS): A mechanical switching unit that is typically mounted externally to the transmission/ transaxle to inform the PCM/ECM which gear range the driver has selected.

MANUAL VALVE: Located inside the transmission/transaxle, it is directly connected to the driver's shift lever. The position of the manual valve determines which hydraulic circuits will be charged with oil pressure and the operating mode of the transmission.

MANUAL VALVE LEVER POSITION SENSOR (MVLPS): The input from this device tells the TCM what gear range was selected.

MASS AIR FLOW (MAF) SENSOR: Measures the airflow into the engine.

MASTER CYLINDER: The primary fluid pressurizing device in a hydraulic system. In automotive use, it is found in brake and hydraulic clutch systems and is pedal activated, either directly or, in a power brake system, through the power booster.

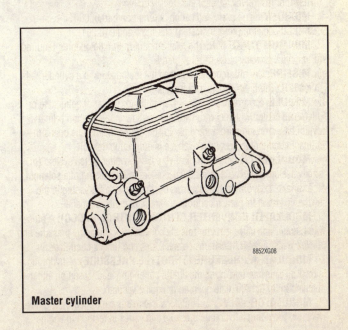

8852XG08

Master cylinder

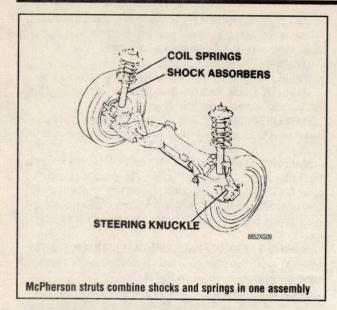

McPherson struts combine shocks and springs in one assembly

McPHERSON STRUT: A suspension component combining a shock absorber and spring in one unit.

MEDIUM THROTTLE: Approximately one-half of accelerator pedal travel.

MEGA: A metric prefix indicating one million.

MEMBER: An independent component of a hydrodynamic unit such as an impeller, a stator, or a turbine. It may have one or more elements.

MERCON: A fluid developed by Ford Motor Company in 1988. It contains a friction modifier and closely resembles operating characteristics of Dexron.

METAL SEALING RINGS: Made from cast iron or aluminum, their primary application is with dynamic components involving pressure sealing circuits of rotating members. These rings are designed with either butt or hook lock end joints.

METER (ANALOG): A linear-style meter representing data as lengths; a needle-style instrument interfacing with logical numerical increments. This style of electrical meter uses relatively low impedance internal resistance and cannot be used for testing electronic circuitry.

METER (DIGITAL): Uses numbers as a direct readout to show values. Most meters of this style use high impedance internal resistance and must be used for testing low current electronic circuitry.

MICRO: A metric prefix indicating one-millionth (0.000001).

MILLI: A metric prefix indicating one-thousandth (0.001).

MINIMUM THROTTLE: The least amount of throttle opening required for upshift; normally close to zero throttle.

MISFIRE: Condition occurring when the fuel mixture in a cylinder fails to ignite, causing the engine to run roughly.

MODULE: Electronic control unit, amplifier or igniter of solid state or integrated design which controls the current flow in the ignition primary circuit based on input from the pick-up coil. When the module opens the primary circuit, high secondary voltage is induced in the coil.

MODULATED: In an electronic-hydraulic converter clutch system (or shift valve system), the term modulated refers to the pulsing of a solenoid, at a variable rate. This action controls the buildup of oil pressure in the hydraulic circuit to allow a controlled amount of clutch slippage.

MODULATED CONVERTER CLUTCH CONTROL (MCCC): A pulse width duty cycle valve that controls the converter lockup apply pressure and maximizes smoother transitions between lock and unlock conditions.

MODULATOR PRESSURE (THROTTLE PRESSURE): A hydraulic signal oil pressure relating to the amount of engine load, based on either the amount of throttle plate opening or engine vacuum.

MODULATOR VALVE: A regulator valve that is controlled by engine vacuum, providing a hydraulic pressure that varies in relation to engine

torque. The hydraulic torque signal functions to delay the shift pattern and provide a line pressure boost. (See throttle valve.)

MOTOR: An electromagnetic device used to convert electrical energy into mechanical energy.

MULTIPLE-DISC CLUTCH: A grouping of steel and friction lined plates that, when compressed together by hydraulic pressure acting upon a piston, lock or unlock a planetary member.

MULTI-WEIGHT: Type of oil that provides adequate lubrication at both high and low temperatures.

needed to move one amp through a resistance of one ohm.

MUSHY: Same as soft; slow and drawn out clutch apply with very little shift feel.

MUTUAL INDUCTION: The generation of current from one wire circuit to another by movement of the magnetic field surrounding a current-carrying circuit as its ampere flow increases or decreases.

NEEDLE BEARING: A bearing which consists of a number (usually a large number) of long, thin rollers.

NITROGEN OXIDE (NOx): One of the three basic pollutants found in the exhaust emission of an internal combustion engine. The amount of NOx usually varies in an inverse proportion to the amount of HC and CO.

NONPOSITIVE SEALING: A sealing method that allows some minor leakage, which normally assists in lubrication.

O2 SENSOR: Located in the engine's exhaust system, it is an input device to the ECM/PCM for managing the fuel delivery and ignition system. A scanner can be used to observe the fluctuating voltage readings produced by an O2 sensor as the oxygen content of the exhaust is analyzed.

O-RING SEAL: Molded synthetic rubber seal designed with a circular cross-section. This type of seal is used primarily in static applications.

OBD II (ON-BOARD DIAGNOSTICS, SECOND GENERATION): Refers to the federal law mandating tighter control of 1996 and newer vehicle emissions, active monitoring of related devices, and standardization of terminology, data link connectors, and other technician concerns.

OCTANE RATING: A number, indicating the quality of gasoline based on its ability to resist knock. The higher the number, the better the quality. Higher compression engines require higher octane gas.

OEM: Original Equipment Manufactured. OEM equipment is that furnished standard by the manufacturer.

OFFSET: The distance between the vertical center of the wheel and the mounting surface at the lugs. Offset is positive if the center is outside the lug circle; negative offset puts the center line inside the lug circle.

OHM'S LAW: A law of electricity that states the relationship between voltage, current, and resistance. Volts = amperes x ohms

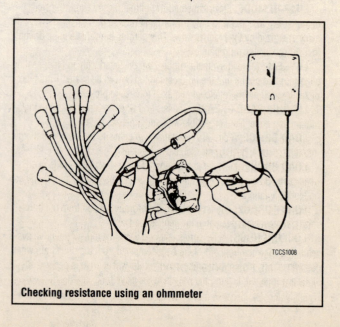

Checking resistance using an ohmmeter

OHM: The unit used to measure the resistance of conductor-to-electrical flow. One ohm is the amount of resistance that limits current flow to one ampere in a circuit with one volt of pressure.

OHMMETER: An instrument used for measuring the resistance, in ohms, in an electrical circuit.

ONE-WAY CLUTCH: A mechanical clutch of roller or sprag design that resists torque or transmits power in one direction only. It is used to either hold or drive a planetary member.

ONE-WAY ROLLER CLUTCH: A mechanical device that transmits or holds torque in one direction only.

OPENCIRCUIT: A break or lack of contact in an electrical circuit, either intentional (switch) or unintentional (bad connection or broken wire).

ORIFICE: Located in hydraulic oil circuits, it acts as a restriction. It slows down fluid flow to either create back pressure or delay pressure buildup downstream.

OSCILLOSCOPE: A piece of test equipment that shows electric impulses as a pattern on a screen. Engine performance can be analyzed by interpreting these patterns.

OUTPUT SHAFT: The shaft which transmits torque from a device, such as a transmission.

OUTPUT SPEED SENSOR (OSS): Identifies transmission/transaxle-output shaft speed for shift timing and may be used to calculate TCC slip; often functions as the VSS (vehicle speed sensor).

OVERDRIVE: (1.) A device attached to or incorporated in a transmission/transaxlethat allows the engine to turn less than one full revolution for every complete revolution of the wheels. The net effect is to reduce engine rpm, thereby using less fuel. A typical overdrive gear ratio would be .87:1, instead of the normal 1:1 in high gear. (2.) A gear assembly which produces more shaft revolutions than that transmitted to it.

OVERDRIVE PLANETARY GEARSET: A single planetary gearset designed to provide a direct drive and overdrive ratio. When coupled to a three-speed transmission/transaxleconfiguration, a four-speed/overdrive unit is present.

OVERHEAD CAMSHAFT (OHC): An engine configuration in which the camshaft is mounted on top of the cylinder head and operates the valve either directly or by means of rocker arms.

OVERHEAD VALVE (OHV): An engine configuration in which all of the valves are located in the cylinder head and the camshaft is located in the cylinder block. The camshaft operates the valves via lifters and pushrods.

OVERRUN CLUTCH: Another name for a one-way mechanical clutch. Applies to both roller and sprag designs.

OVERSTEER: The tendency of some vehicles, when steering into a turn, to over-respond or steer more than required, which could result in excessive slip of the rear wheels. Opposite of understeer.

OXIDATION STABILIZERS: Absorb and dissipate heat. Automatic transmission fluid has high resistance to varnish and sludge buildup that occurs from excessive heat that is generated primarily in the torque converter. Local temperatures as high as 6000F (3150C) can occur at the clutch plates during engagement, and this heat must be absorbed and dissipated. If the fluid cannot withstand the heat, it burns or oxidizes, resulting in an almost immediate destruction of friction materials, clogged filter screen and hydraulic passages, and sticky valves.

OXIDES OF NITROGEN: See nitrogen oxide (NOx).

OXYGEN SENSOR: Used with a feedback system to sense the presence of oxygen in the exhaust gas and signal the computer which can use the voltage signal to determine engine operating efficiency and adjust the air/fuel ratio.

PARALLEL CIRCUIT: (See circuit, parallel.)

PARTS WASHER: A basin or tub, usually with a built-in pump mechanism and hose used for circulating chemical solvent for the purpose of cleaning greasy, oily and dirty components.

PART-TIME FOUR WHEEL DRIVE: A system that is normally in the two wheel drive mode and only runs in four-wheel drive when the system is manually engaged because more traction is desired. Two or four wheel drive is normally selected by a lever to engage the front axle, but if locking hubs are used, these must also be manually engaged in the Lock position. Otherwise, the front axle will not drive the front wheels.

PASSIVE RESTRAINT: Safety systems such as air bags or automatic seat belts which operate with no action required on the part of the driver or passenger. Mandated by Federal regulations on all vehicles sold in the U.S. after 1990.

PAYLOAD: The weight the vehicle is capable of carrying in addition to its own weight. Payload includes weight of the driver, passengers and cargo, but not coolant, fuel, lubricant, spare tire, etc.

PCM: Powertrain control module.

PCV VALVE: A valve usually located in the rocker cover that vents crankcase vapors back into the engine to be reburned.

PERCOLATION: A condition in which the fuel actually "boils," due to excessive heat. Percolation prevents proper atomization of the fuel causing rough running.

PICK-UP COIL: The coil in which voltage is induced in an electronic ignition.

PINION GEAR: The smallest gear in a drive gear assembly. piston: A disc or cup that fits in a cylinder bore and is free to move. In hydraulics, it provides the means of converting hydraulic pressure into a usable force. Examples of piston applications are found in servo, clutch, and accumulator units.

PING: A metallic rattling sound produced by the engine during acceleration. It is usually due to incorrect ignition timing or a poor grade of gasoline.

PINION: The smaller of two gears. The rear axle pinion drives the ring gear which transmits motion to the axle shafts.

PISTON RING: An open-ended ring which fits into a groove on the outer diameter of the piston. Its chief function is to form a seal between the piston and cylinder wall. Most automotive pistons have three rings: two for compression sealing; one for oil sealing.

PITMAN ARM: A lever which transmits steering force from the steering gear to the steering linkage.

PLANET CARRIER: A basic member of a planetary gear assembly that carries the pinion gears.

PLANET PINIONS: Gears housed in a planet carrier that are in constant mesh with the sun gear and internal gear. Because they have their own independent rotating centers, the pinions are capable of rotating around the sun gear or the inside of the internal gear.

PLANETARY GEAR RATIO: The reduction or overdrive ratio developed by a planetary gearset.

PLANETARY GEARSET: In its simplest form, it is made up of a basic assembly group containing a sun gear, internal gear, and planet carrier. The

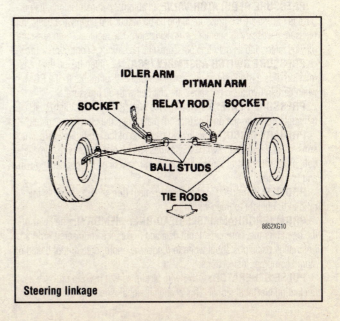

Steering linkage

IDLER ARM

PITMAN ARM

SOCKET

RELAY ROD

SOCKET

BALL STUDS

TIE RODS

8852XG10

gears are always in constant mesh and offer a wide range of gear ratio possibilities.

PLANETARY GEARSET (COMPOUND): Two planetary gearsets combined together.

PLANETARY GEARSET (SIMPLE): An assembly of gears in constant mesh consisting of a sun gear, several pinion gears mounted in a carrier, and a ring gear. It provides gear ratio and direction changes, in addition to a direct drive and a neutral.

PLY RATING: A. rating given a tire which indicates strength (but not necessarily actual plies). A two-ply/four-ply rating has only two plies, but the strength of a four-ply tire.

POLARITY: Indication (positive or negative) of the two poles of a battery.

PORT: An opening for fluid intake or exhaust.

POSITIVE SEALING: A sealing method that completely prevents leakage.

POTENTIAL: Electrical force measured in volts; sometimes used interchangeably with voltage.

POWER: The ability to do work per unit of time, as expressed in horsepower; one horsepower equals 33,000 ft.lbs. of work per minute, or 550 ft.lbs. of work per second.

POWER FLOW: The systematic flow or transmission of power through the gears, from the input shaft to the output shaft.

POWER-TO-WEIGHT RATIO: Ratio of horsepower to weight of car.

POWERTRAIN: See Drivetrain.

POWERTRAIN CONTROL MODULE (PCM): Current designation for the engine control module (ECM). In many cases, late model vehicle control units manage the engine as well as the transmission. In other settings, the PCM controls the engine and is interfaced with a TCM to control transmission functions.

Ppm: Parts per million; unit used to measure exhaust emissions.

PREIGNITION: Early ignition of fuel in the cylinder, sometimes due to glowing carbon deposits in the combustion chamber. Preignition can be damaging since combustion takes place prematurely.

PRELOAD: A predetermined load placed on a bearing during assembly or by adjustment.

PRESS FIT: The mating of two parts under pressure, due to the inner diameter of one being smaller than the outer diameter of the other, or vice versa; an interference fit.

PRESSURE: The amount of force exerted upon a surface area.

PRESSURE CONTROL SOLENOID (PCS): An output device that provides a boost oil pressure to the mainline regulator valve to control line pressure. Its operation is determined by the amount of current sent from the PCM.

PRESSURE GAUGE: An instrument used for measuring the fluid pressure in a hydraulic circuit.

PRESSURE REGULATOR VALVE: In automatic transmissions, its purpose is to regulate the pressure of the pump output and supply the basic fluid pressure necessary to operate the transmission. The regulated fluid pressure may be referred to as mainline pressure, line pressure, or control pressure.

PRESSURE SWITCH ASSEMBLY (PSA): Mounted inside the transmission, it is a grouping of oil pressure switches that inputs to the PCM when certain hydraulic passages are charged with oil pressure.

PRESSURE PLATE: A spring-loaded plate (part of the clutch) that transmits power to the driven (friction) plate when the clutch is engaged.

PRIMARY CIRCUIT: The low voltage side of the ignition system which consists of the ignition switch, ballast resistor or resistance wire, bypass, coil, electronic control unit and pick-up coil as well as the connecting wires and harnesses.

PROFILE: Term used for tire measurement (tire series), which is the ratio of tire height to tread width.

PROM (PROGRAMMABLE READ-ONLY MEMORY): The heart of the computer that compares input data and makes the engineered program or strategy decisions about when to trigger the appropriate output based on stored computer instructions.

PULSE GENERATOR: A two-wire pickup sensor used to produce a fluctuating electrical signal. This changing signal is read by the controller to determine the speed of the object and can be used to measure transmission/transaxle input speed, output speed, and vehicle speed.

PSI: Pounds per square inch; a measurement of pressure.

PULSE WIDTH DUTY CYCLE SOLENOID (PULSE WIDTH MODULATED SOLENOID): A computer-controlled solenoid that turns on and off at a variable rate producing a modulated oil pressure; often referred to as a pulse width modulated (PWM) solenoid. Employed in many electronic automatic transmissions and transaxles, these solenoids are used to manage shift control and converter clutch hydraulic circuits.

PUSHROD: A steel rod between the hydraulic valve lifter and the valve rocker arm in overhead valve (OHV) engines.

PUMP: A mechanical device designed to create fluid flow and pressure buildup in a hydraulic system.

QUARTER PANEL: General term used to refer to a rear fender. Quarter panel is the area from the rear door opening to the tail light area and from rear wheelwell to the base of the trunk and roof-line.

RACE: The surface on the inner or outer ring of a bearing on which the balls, needles or rollers move.

RACK AND PINION: A type of automotive steering system using a pinion gear attached to the end of the steering shaft. The pinion meshes with a long rack attached to the steering linkage.

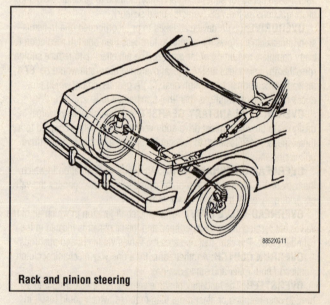

8852XG11

Rack and pinion steering

RADIAL TIRE: Tire design which uses body cords running at right angles to the center line of the tire. Two or more belts are used to give tread strength. Radials can be identified by their characteristic sidewall bulge.

RADIATOR: Part of the cooling system for a water-cooled engine, mounted in the front of the vehicle and connected to the engine with rubber hoses. Through the radiator, excess combustion heat is dissipated into the atmosphere through forced convection using a water and glycol based mixture that circulates through, and cools, the engine.

RANGE REFERENCE AND CLUTCH/BAND APPLY CHART: A guide that shows the application of clutches and bands for each gear, within the selector range positions. These charts are extremely useful for understanding how the unit operates and for diagnosing malfunctions.

RAVIGNEAUX GEARSET: A compound planetary gearset that features matched dual planetary pinions (sets of two) mounted in a single planet carrier. Two sun gears and one ring mesh with the carrier pinions.

REACTION MEMBER: The stationary planetary member, in a planetary gearset, that is grounded to the transmission/transaxlecase through the use of friction and wedging devices known as bands, disc clutches, and one-way clutches.

REACTION PRESSURE: The fluid pressure that moves a spool valve against an opposing force or forces; the area on which the opposing force acts. The opposing force can be a spring or a combination of spring force and auxiliary hydraulic force.

REACTOR, TORQUE CONVERTER: The reaction member of a fluid torque converter, more commonly called a stator. (See stator.)

REAR MAIN OIL SEAL: A synthetic or rope-type seal that prevents oil from leaking out of the engine past the rear main crankshaft bearing.

RECIRCULATING BALL: Type of steering system in which recirculating steel balls occupy the area between the nut and worm wheel, causing a reduction in friction.

RECTIFIER: A device (used primarily in alternators) that permits electrical current to flow in one direction only.

REDUCTION: (See gear reduction.) regulator valve: A valve that changes the pressure of the oil in a hydraulic circuit as the oil passes through the valve by bleeding off (or exhausting) some of the volume of oil supplied to the valve.

REFRIGERANT 12 (R-12) or 134 (R-134): The generic name of the refrigerant used in automotive air conditioning systems.

REGULATOR: A device which maintains the amperage and/or voltage levels of a circuit at predetermined values.

RELAY: A switch which automatically opens and/or closes a circuit.

RELAY VALVE: A valve that directs flow and pressure. Relay valves simply connect or disconnect interrelated passages without restricting the fluid flow or changing the pressure.

RELIEF VALVE: A spring-loaded, pressure-operated valve that limits oil pressure buildup in a hydraulic circuit to a predetermined maximum value.

RELUCTOR: A wheel that rotates inside the distributor and triggers the release of voltage in an electronic ignition.

RESERVOIR: The storage area for fluid in a hydraulic system; often called a sump.

RESIN: A liquid plastic used in body work.

RESIDUAL MAGNETISM: The magnetic strength stored in a material after a magnetizing field has been removed.

RESISTANCE: The opposition to the flow of current through a circuit or electrical device, and is measured in ohms. Resistance is equal to the voltage divided by the amperage.

RESISTOR SPARK PLUG: A spark plug using a resistor to shorten the spark duration. This suppresses radio interference and lengthens plug life.

RESISTOR: A device, usually made of wire, which offers a preset amount of resistance in an electrical circuit.

RESULTANT FORCE: The single effective directional thrust of the fluid force on the turbine produced by the vortex and rotary forces acting in different planes.

RETARD: Set the ignition timing so that spark occurs later (fewer degrees before TDC).

RHEOSTAT: A device for regulating a current by means of a variable resistance.

RING GEAR: The name given to a ring-shaped gear attached to a differential case, or affixed to a flywheel or as part of a planetary gear set.

ROADLOAD: grade.

ROCKER ARM: A lever which rotates around a shaft pushing down (opening) the valve with an end when the other end is pushed up by the pushrod. Spring pressure will later close the valve.

ROCKER PANEL: The body panel below the doors between the wheel opening.

ROLLER BEARING: A bearing made up of hardened inner and outer races between which hardened steel rollers move.

ROLLER CLUTCH: A type of one-way clutch design using rollers and springs mounted within an inner and outer cammed race assembly.

ROTARY FLOW: The path of the fluid trapped between the blades of the members as they revolve with the rotation of the torque converter cover (rotational inertia).

ROTOR: (1.) The disc-shaped part of a disc brake assembly, upon

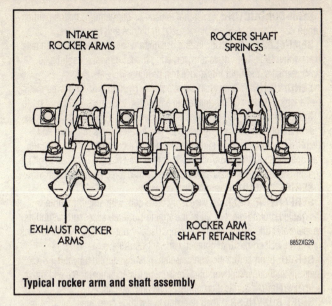

Typical rocker arm and shaft assembly

which the brake pads bear; also called, brake disc. (2.) The device mounted atop the distributor shaft, which passes current to the distributor cap tower contacts.

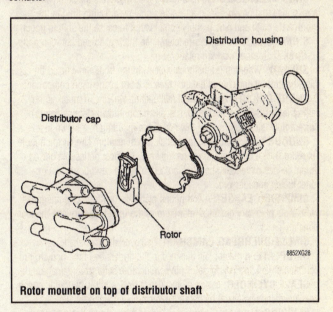

Rotor mounted on top of distributor shaft

ROTARY ENGINE: See Wankel engine.

RPM: Revolutions per minute (usually indicates engine speed).

RTV: A gasket making compound that cures as it is exposed to the atmosphere. It is used between surfaces that are not perfectly machined to one another, leaving a slight gap that the RTV fills and in which it hardens. The letters RTV represent room temperature vulcanizing.

RUN-ON: Condition when the engine continues to run, even when the key is turned off. See dieseling.

SEALED BEAM: A automotive headlight. The lens, reflector and filament from a single unit.

SEATBELT INTERLOCK: A system whereby the car cannot be started unless the seatbelt is buckled.

SECONDARY CIRCUIT: The high voltage side of the ignition system, usually above 20,000 volts. The secondary includes the ignition coil, coil wire, distributor cap and rotor, spark plug wires and spark plugs.

SELF-INDUCTION: The generation of voltage in a current-carrying wire by changing the amount of current flowing within that wire.

SEMI-CONDUCTOR: A material (silicon or germanium) that is neither a good conductor nor an insulator; used in diodes and transistors.

SEMI-FLOATING AXLE: In this design, a wheel is attached to the axle shaft, which takes both drive and cornering loads. Almost all solid axle passenger cars and light trucks use this design.

SENDING UNIT: A mechanical, electrical, hydraulic or electromagnetic device which transmits information to a gauge.

SENSOR: Any device designed to measure engine operating conditions or ambient pressures and temperatures. Usually electronic in nature and designed to send a voltage signal to an on-board computer, some sensors may operate as a simple on/off switch or they may provide a variable voltage signal (like a potentiometer) as conditions or measured parameters change.

SERIES CIRCUIT: (See circuit, series.)

SERPENTINE BELT: An accessory drive belt, with small multiple v-ribs, routed around most or all of the engine-powered accessories such as the alternator and power steering pump. Usually both the front and the back side of the belt comes into contact with various pulleys.

SERVO: In an automatic transmission, it is a piston in a cylinder assembly that converts hydraulic pressure into mechanical force and movement; used for the application of the bands and clutches.

SHIFT BUSYNESS: When referring to a torque converter clutch, it is the frequent apply and release of the clutch plate due to uncommon driving conditions.

SHIFT VALVE: Classified as a relay valve, it triggers the automatic shift in response to a governor and a throttle signal by directing fluid to the appropriate band and clutch apply combination to cause the shift to occur.

SHIM: Spacers of precise, predetermined thickness used between parts to establish a proper working relationship.

SHIMMY: Vibration (sometimes violent) in the front end caused by misaligned front end, out of balance tires or worn suspension components.

SHORT CIRCUIT: An electrical malfunction where current takes the path of least resistance to ground (usually through damaged insulation). Current flow is excessive from low resistance resulting in a blown fuse.

SHUDDER: Repeated jerking or stick-slip sensation, similar to chuggle but more severe and rapid in nature, that may be most noticeable during certain ranges of vehicle speed; also used to define condition after converter clutch engagement.

SIMPSON GEARSET: A compound planetary geartrain that integrates two simple planetary gearsets referred to as the front planetary and the rear planetary.

SINGLE OVERHEAD CAMSHAFT: See overhead camshaft.

SKIDPLATE: A metal plate attached to the underside of the body to protect the fuel tank, transfer case or other vulnerable parts from damage.

SLAVE CYLINDER: In automotive use, a device in the hydraulic clutch system which is activated by hydraulic force, disengaging the clutch.

SLIPPING: Noticeable increase in engine rpm without vehicle speed increase; usually occurs during or after initial clutch or band engagement.

SLUDGE: Thick, black deposits in engine formed from dirt, oil, water, etc. It is usually formed in engines when oil changes are neglected.

SNAP RING: A circular retaining clip used inside or outside a shaft or part to secure a shaft, such as a floating wrist pin.

SOFT: Slow, almost unnoticeable clutch apply with very little shift feel.

SOFTCODES: DTCs that have been set into the PCM memory but are not present at the time of testing; often referred to as history or intermittent codes.

SOHC: Single overhead camshaft.

SOLENOID: An electrically operated, magnetic switching device.

SPALLING: A wear pattern identified by metal chips flaking off the hardened surface. This condition is caused by foreign particles, overloading situations, and/or normal wear.

SPARK PLUG: A device screwed into the combustion chamber of a spark ignition engine. The basic construction is a conductive core inside of a ceramic insulator, mounted in an outer conductive base. An electrical charge from the spark plug wire travels along the conductive core and jumps a pre-set air gap to a grounding point or points at the end of the conductive base. The resultant spark ignites the fuel/air mixture in the combustion chamber.

SPECIFIC GRAVITY (BATTERY): The relative weight of liquid (battery electrolyte) as compared to the weight of an equal volume of water.

SPLINES: Ridges machined or cast onto the outer diameter of a shaft or inner diameter of a bore to enable parts to mate without rotation.

SPLIT TORQUE DRIVE: In a torque converter, it refers to parallel paths of torque transmission, one of which is mechanical and the other hydraulic.

SPONGY PEDAL: A soft or spongy feeling when the brake pedal is depressed. It is usually due to air in the brake lines.

SPOOLVALVE: A precision-machined, cylindrically shaped valve made up of lands and grooves. Depending on its position in the valve bore, various interconnecting hydraulic circuit passages are either opened or closed.

SPRAG CLUTCH: A type of one-way clutch design using cams or contoured-shaped sprags between inner and outer races. (See one-way clutch.)

SPRUNG WEIGHT: The weight of a car supported by the springs.

SQUARE-CUT SEAL: Molded synthetic rubber seal designed with a square- or rectangular-shaped cross-section. This type of seal is used for both dynamic and static applications.

SRS: Supplemental restraint system

STABILIZER (SWAY) BAR: A bar linking both sides of the suspension. It resists sway on turns by taking some of added load from one wheel and putting it on the other.

STAGE: The number of turbine sets separated by a stator. A turbine set may be made up of one or more turbine members. A three-element converter is classified as a single stage.

STALL: In fluid drive transmission/transaxle applications, stall refers to engine rpm with the transmission/transaxle engaged and the vehicle stationary; throttle valve can be in any position between closed and wide open.

STALL SPEED: In fluid drive transmission/transaxle applications, stall speed refers to the maximum engine rpm with the transmission/transaxle engaged and vehicle stationary, when the throttle valve is wide open. (See stall; stall test.)

STALL TEST: A procedure recommended by many manufacturers to help determine the integrity of an engine, the torque converter stator, and certain clutch and band combinations. With the shift lever in each of the forward and reverse positions and with the brakes firmly applied, the accelerator pedal is momentarily pressed to the wide open throttle (WOT) position. The engine rpm reading at full throttle can provide clues for diagnosing the condition of the items listed above.

STALL TORQUE: The maximum design or engineered torque ratio of a fluid torque converter, produced under stall speed conditions. (See stall speed.)

STARTER: A high-torque electric motor used for the purpose of starting the engine, typically through a high ratio geared drive connected to the flywheel ring gear.

STATIC: A sealing application in which the parts being sealed do not move in relation to each other.

STATOR (REACTOR): The reaction member of a fluid torque converter that changes the direction of the fluid as it leaves the turbine to enter the impeller vanes. During the torque multiplication phase, this action assists the impeller's rotary force and results in an increase in torque.

STEERING GEOMETRY: Combination of various angles of suspension components (caster, camber, toe-in); roughly equivalent to front end alignment.

STRAIGHT WEIGHT: Term designating motor oil as suitable for use within a narrow range of temperatures. Outside the narrow temperature range its flow characteristics will not adequately lubricate.

STROKE: The distance the piston travels from bottom dead center to top dead center.

SUBSTITUTION: Replacing one part suspected of a defect with a like part of known quality.

SUMP: The storage vessel or reservoir that provides a ready source of fluid to the pump. In an automatic transmission, the sump is the oil pan. All fluid eventually returns to the sump for recycling into the hydraulic system.

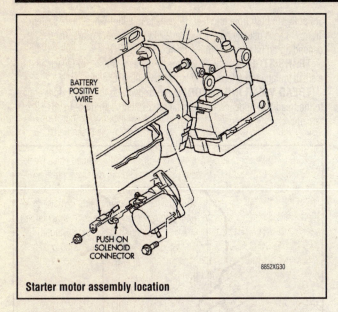

Starter motor assembly location

SUN GEAR: In a planetary gearset, it is the center gear that meshes with a cluster of planet pinions.

SUPERCHARGER: An air pump driven mechanically by the engine through belts, chains, shafts or gears from the crankshaft. Two general types of supercharger are the positive displacement and centrifugal type, which pump air in direct relationship to the speed of the engine.

SUPPLEMENTAL RESTRAINT SYSTEM: See air bag.

SURGE: Repeating engine-related feeling of acceleration and deceleration that is less intense than chuggle.

SWITCH: A device used to open, close, or redirect the current in an electrical circuit.

SYNCHROMESH: A manual transmission/transaxle that is equipped with devices (synchronizers) that match the gear speeds so that the transmission/transaxle can be downshifted without clashing gears.

SYNTHETIC OIL: Non-petroleum based oil.

TACHOMETER: A device used to measure the rotary speed of an engine, shaft, gear, etc., usually in rotations per minute.

TDC: Top dead center. The exact top of the piston's stroke.

TEFLON SEALING RINGS: Teflon is a soft, durable, plastic-like material that is resistant to heat and provides excellent sealing. These rings are designed with either scarf-cut joints or as one-piece rings. Teflon sealing rings have replaced many metal ring applications.

TERMINAL: A device attached to the end of a wire or cable to make an electrical connection.

TEST LIGHT, CIRCUIT-POWERED: Uses available circuit voltage to test circuit continuity.

TEST LIGHT, SELF-POWERED: Uses its own battery source to test circuit continuity.

THERMISTOR: A special resistor used to measure fluid temperature; it decreases its resistance with increases in temperature.

THERMOSTAT: A valve, located in the cooling system of an engine, which is closed when cold and opens gradually in response to engine heating, controlling the temperature of the coolant and rate of coolant flow.

THERMOSTATIC ELEMENT: A heat-sensitive, spring-type device that controls a drain port from the upper sump area to the lower sump. When the transaxle fluid reaches operating temperature, the port is closed and the upper sump fills, thus reducing the fluid level in the lower sump.

THROTTLE POSITION (TP) SENSOR: Reads the degree of throttle opening; its signal is used to analyze engine load conditions. The ECM/PCM decides to apply the TCC, or to disengage it for coast or load conditions that need a converter torque boost.

THROTTLE PRESSURE/MODULATOR PRESSURE: A hydraulic signal oil pressure relating to the amount of engine load, based on either the amount of throttle plate opening or engine vacuum.

THROTTLE VALVE: A regulating or balanced valve that is controlled mechanically by throttle linkage or engine vacuum. It sends a hydraulic signal to the shift valve body to control shift timing and shift quality. (See balanced valve; modulator valve.)

THROW-OUT BEARING: As the clutch pedal is depressed, the throwout bearing moves against the spring fingers of the pressure plate, forcing the pressure plate to disengage from the driven disc.

TIE ROD: A rod connecting the steering arms. Tie rods have threaded ends that are used to adjust toe-in.

TIE-UP: Condition where two opposing clutches are attempting to apply at same time, causing engine to labor with noticeable loss of engine rpm.

TIMING BELT: A square-toothed, reinforced rubber belt that is driven by the crankshaft and operates the camshaft.

TIMING CHAIN: A roller chain that is driven by the crankshaft and operates the camshaft.

TIRE ROTATION: Moving the tires from one position to another to make the tires wear evenly.

TOE-IN (OUT): A term comparing the extreme front and rear of the front tires. Closer together at the front is toe-in; farther apart at the front is toe-out.

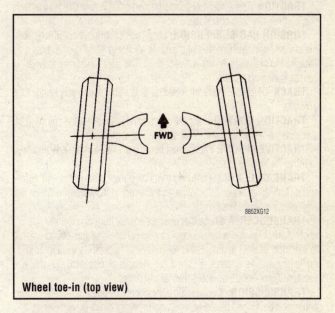

Wheel toe-in (top view)

TOP DEAD CENTER (TDC): The point at which the piston reaches the top of its travel on the compression stroke.

TORQUE: Measurement of turning or twisting force, expressed as foot-pounds or inch-pounds.

TORQUE CONVERTER: A turbine used to transmit power from a driving member to a driven member via hydraulic action, providing changes in drive ratio and torque. In automotive use, it links the driveplate at the rear of the engine to the automatic transmission.

TORQUE CONVERTER CLUTCH: The apply plate (lockup plate) assembly used for mechanical power flow through the converter.

TORQUE PHASE: Sometimes referred to as slip phase or stall phase, torque multiplication occurs when the turbine is turning at a slower speed than the impeller, and the stator is reactionary (stationary). This sequence generates a boost in output torque.

TORQUE RATING (STALL TORQUE): The maximum torque multiplication that occurs during stall conditions, with the engine at wide open throttle (WOT) and zero turbine speed.

TORQUE RATIO: An expression of the gear ratio factor on torque effect. A 3:1 gear ratio or 3:1 torque ratio increases the torque input by the ratio factor of 3. Input torque (100 ft. lbs.)x 3 = output torque (300 ft. lbs.)

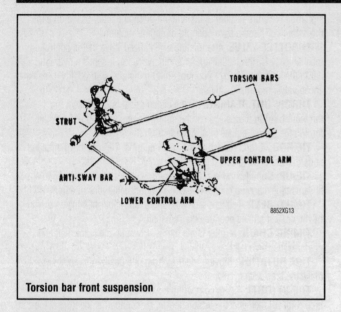

Torsion bar front suspension

TRACTION: The amount of usable tractive effort before the drive wheels slip on the road contact surface.

TORSION BAR SUSPENSION: Long rods of spring steel which take the place of springs. One end of the bar is anchored and the other arm (attached to the suspension) is free to twist. The bars' resistance to twisting causes springing action.

TRACK: Distance between the centers of the tires where they contact the ground.

TRACTION CONTROL: A control system that prevents the spinning of a vehicle's drive wheels when excess power is applied.

TRACTIVE EFFORT: The amount of force available to the drive wheels, to move the vehicle.

TRANSAXLE: A single housing containing the transmission and differential. Transaxles are usually found on front engine/front wheel drive or rear engine/rear wheel drive cars.

TRANSDUCER: A device that changes energy from one form to another. For example, a transducer in a microphone changes sound energy to electrical energy. In automotive air-conditioning controls used in automatic temperature systems, a transducer changes an electrical signal to a vacuum signal, which operates mechanical doors.

TRANSMISSION: A powertrain component designed to modify torque and speed developed by the engine; also provides direct drive, reverse, and neutral.

TRANSMISSION CONTROL MODULE (TCM): Manages transmission functions. These vary according to the manufacturer's product design but may include converter clutch operation, electronic shift scheduling, and mainline pressure.

TRANSMISSION FLUID TEMPERATURE (TFT)SENSOR: Originally called a transmission oil temperature (TOT) sensor, this input device to the ECM/PCM senses the fluid temperature and provides a resistance value. It operates on the thermistor principle.

TRANSMISSION INPUT SPEED (TIS) SENSOR: Measures turbine shaft (input shaft) rpm's and compares to engine rpm's to determine torque converter slip. When compared to the transmission output speed sensor or VSS, gear ratio and clutch engagement timing can be determined.

TRANSMISSION OIL TEMPERATURE (TOT) SENSOR: (See transmission fluid temperature (TFT) sensor.)

TRANSMISSION RANGE SELECTOR (TRS) SWITCH: Tells the module which gear shift position the driver has chosen. turbine: The output (driven) member of a fluid coupling or fluid torque converter. It is splined to the input (turbine) shaft of the transmission.

TRANSFER CASE: A gearbox driven from the transmission that deliv-

ers power to both front and rear driveshafts in a four-wheel drive system. Transfer cases usually have a high and low range set of gears, used depending on how much pulling power is needed.

TRANSISTOR: A semi-conductor component which can be actuated by a small voltage to perform an electrical switching function.

TREAD WEAR INDICATOR: Bars molded into the tire at right angles to the tread that appear as horizontal bars when 1/16th in. of tread remains.

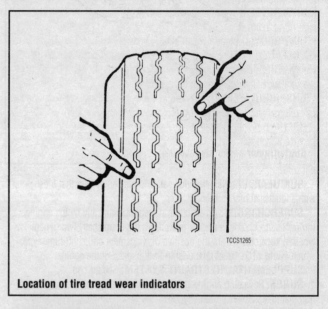

Location of tire tread wear indicators

TREAD WEAR PATTERN: The pattern of wear on tires which can be "read" to diagnose problems in the front suspension.

TUNE-UP: A regular maintenance function, usually associated with the replacement and adjustment of parts and components in the electrical and fuel systems of a vehicle for the purpose of attaining optimum performance.

TURBOCHARGER: An exhaust driven pump which compresses intake air and forces it into the combustion chambers at higher than atmospheric pressures. The increased air pressure allows more fuel to be burned and results in increased horsepower being produced.

TURBULENCE: The interference of molecules of a fluid (or vapor) with each other in a fluid flow.

TYPE F: Transmission fluid developed and used by Ford Motor Com-

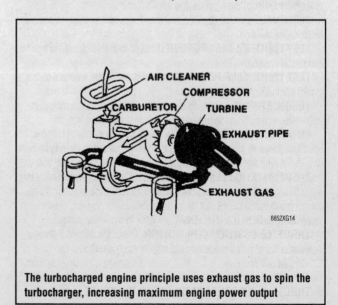

The turbocharged engine principle uses exhaust gas to spin the turbocharger, increasing maximum engine power output

pany up to 1982. This fluid type provides a high coefficient of friction.

TYPE 7176: The preferred choice of transmission fluid for Chrysler automatic transmissions and transaxles. Developed in 1986, it closely resembles Dexron and Mercon. Type 7176 is the recommended service fill fluid for all Chrysler products utilizing a lockup torque converter dating back to 1978.

U-JOINT (UNIVERSAL JOINT): A flexible coupling in the drive train that allows the driveshafts or axle shafts to operate at different angles and still transmit rotary power.

UNDERSTEER: The tendency of a car to continue straight ahead while negotiating a turn.

UNIT BODY: Design in which the car body acts as the frame.

UNLEADED FUEL: Fuel which contains no lead (a common gasoline additive). The presence of lead in fuel will destroy the functioning elements of a catalytic converter, making it useless.

UNSPRUNG WEIGHT: The weight of car components not supported by the springs (wheels, tires, brakes, rear axle, control arms, etc.).

UPSHIFT: A shift that results in a decrease in torque ratio and an increase in speed.

VACUUM: A negative pressure; any pressure less than atmospheric pressure.

VACUUM ADVANCE: A device which advances the ignition timing in response to increased engine vacuum.

VACUUM GAUGE: An instrument used for measuring the existing vacuum in a vacuum circuit or chamber. The unit of measure is inches (of mercury in a barometer).

VACUUM MODULATOR: Generates a hydraulic oil pressure in response to the amount of engine vacuum.

VALVES: Devices that can open or close fluid passages in a hydraulic system and are used for directing fluid flow and controlling pressure.

VALVE BODY ASSEMBLY: The main hydraulic control assembly of the transmission/transaxlethat contains numerous valves, check balls, and other components to control the distribution of pressurized oil throughout the transmission.

VALVE CLEARANCE: The measured gap between the end of the valve stem and the rocker arm, cam lobe or follower that activates the valve.

VALVE GUIDES: The guide through which the stem of the valve passes. The guide is designed to keep the valve In proper alignment.

VALVE LASH (clearance): The operating clearance in the valve train.

VALVE TRAIN: The system that operates intake and exhaust valves, consisting of camshaft, valves and springs, lifters, pushrods and rocker arms.

VAPOR LOCK: Boiling of the fuel in the fuel lines due to excess heat. This will interfere with the flow of fuel in the lines and can completely stop the flow. Vapor lock normally only occurs in hot weather.

VARIABLE DISPLACEMENT (VARIABLE CAPACITY) VANE PUMP: Slipper-type vanes, mounted in a revolving rotor and contained within the bore of a movable slide, capture and then force fluid to flow. Movement of the slide to various positions changes the size of the vane chambers and the amount of fluid flow. Note: GM refers to this pump design as variable displacement, and Ford terms it variable capacity.

VARIABLE FORCE SOLENOID (VFS): Commonly referred to as the electronic pressure control (EPC) solenoid, it replaces the cable/linkage style of TV system control and is integrated with a spool valve and spring assembly to control pressure. A variable computer-controlled current flow varies the internal force of the solenoid on the spool valve and resulting control pressure.

VARIABLE ORIFICE THERMAL VALVE: Temperature-sensitive hydraulic oil control device that adjusts the size of a circuit path opening. By altering the size of the opening, the oil flow rate is adapted for cold to hot oil viscosity changes.

VARNISH: Term applied to the residue formed when gasoline gets old and stale.

VCM: See Electronic Control Unit (ECU).

VEHICLE SPEED SENSOR (VSS): Provides an electrical signal to the computer module, measuring vehicle speed, and affects the torque converter clutch engagement and release.

VESPEL SEALING RINGS: Hard plastic material that produces excellent sealing in dynamic settings. These rings are found in late versions of the 4T60 and in all 4T60-E and 4T80-E transaxles.

VISCOSITY: The ability of a fluid to flow. The lower the viscosity rating, the easier the fluid will flow. 10 weight motor oil will flow much easier than 40 weight motor oil.

VISCOSITY INDEX IMPROVERS: Keeps the viscosity nearly constant with changes in temperature. This is especially important at low temperatures, when the oil needs to be thin to aid in shifting and for cold-weather starting. Yet it must not be so thin that at high temperatures it will cause excessive hydraulic leakage so that pumps are unable to maintain the proper pressures.

VISCOUS CLUTCH: A specially designed torque converter clutch apply plate that, through the use of a silicon fluid, clamps smoothly and absorbs torsional vibrations.

VOLT: Unit used to measure the force or pressure of electricity. It is defined as the pressure

VOLTAGE: The electrical pressure that causes current to flow. Voltage is measured in volts (V).

VOLTAGE, APPLIED: The actual voltage read at a given point in a circuit. It equals the available voltage of the power supply minus the losses in the circuit up to that point.

VOLTAGE DROP: The voltage lost or used in a circuit by normal loads such as a motor or lamp or by abnormal loads such as a poor (high-resistance) lead or terminal connection.

VOLTAGE REGULATOR: A device that controls the current output of the alternator or generator.

VOLTMETER: An instrument used for measuring electrical force in units called volts. Voltmeters are always connected parallel w ith the circuit being tested.

VORTEX FLOW: The crosswise or circulatory flow of oil between the blades of the members caused by the centrifugal pumping action of the impeller.

WANKEL ENGINE: An engine which uses no pistons. In place of pistons, triangular-shaped rotors revolve in specially shaped housings.

WATER PUMP: A belt driven component of the cooling system that mounts on the engine, circulating the coolant under pressure.

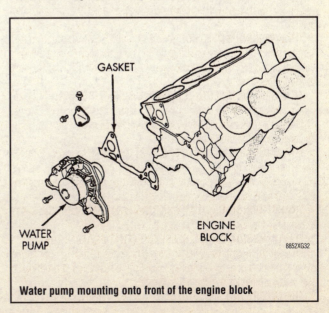

Water pump mounting onto front of the engine block

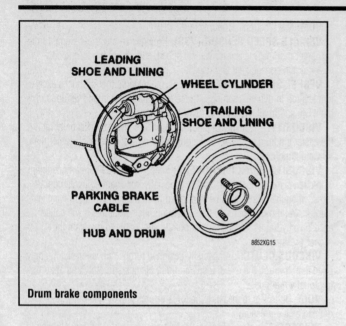

LEADING SHOE AND LINING

WHEEL CYLINDER

TRAILING SHOE AND LINING

PARKING BRAKE CABLE

HUB AND DRUM

8852XG15

Drum brake components

WATT: The unit for measuring electrical power. One watt is the product of one ampere and one volt (watts equals amps times volts). Wattage is the horsepower of electricity (746 watts equal one horsepower).

WHEEL ALIGNMENT: Inclusive term to describe the front end geometry (caster, camber, toe-in/out).

WHEEL CYLINDER: Found in the automotive drum brake assembly, it is a device, actuated by hydraulic pressure, which, through internal pistons, pushes the brake shoes outward against the drums.

WHEEL WEIGHT: Small weights attached to the wheel to balance the wheel and tire assembly. Out-of-balance tires quickly wear out and also give erratic handling when installed on the front.

WHEELBASE: Distance between the center of front wheels and the center of rear wheels.

WIDE OPEN THROTTLE (WOT): Full travel of accelerator pedal.

WORK: The force exerted to move a mass or object. Work involves motion; if a force is exerted and no motion takes place, no work is done. Work per unit of time is called power. Work = force x distance = ft.lbs. 33,000 ft.lbs. in one minute = 1 horsepower

ZERO-THROTTLE COAST DOWN: A full release of accelerator pedal while vehicle is in motion and in drive range.

MASTER INDEX